LONDON MATHEMATICAL SOCIETY LECTURE NOTE SERIES

Managing Editor: Professor J.W.S. Cassels, Department of Pure Mathematics and Mathematical Statistics, University of Cambridge, 16 Mill Lane, Cambridge CB2 1SB, England

The titles below are available from booksellers, or, in case of difficulty, from Cambridge University Press.

34 Representation theory of Lie groups, M.F. ATIYAH *et al*
36 Homological group theory, C.T.C. WALL (ed)
39 Affine sets and affine groups, D.G. NORTHCOTT
46 p-adic analysis: a short course on recent work, N. KOBLITZ
50 Commutator calculus and groups of homotopy classes, H.J. BA
59 Applicable differential geometry, M. CRAMPIN & F.A.E. PIR
66 Several complex variables and complex manifolds II, M.J. FIEI
69 Representation theory, I.M. GELFAND *et al*
74 Symmetric designs: an algebraic approach, E.S. LANDER
76 Spectral theory of linear differential operators and comparison algebras, H.O. CORDES
77 Isolated singular points on complete intersections, E.J.N. LOOIJENGA
79 Probability, statistics and analysis, J.F.C. KINGMAN & G.E.H. REUTER (eds)
83 Homogeneous structures on Riemannian manifolds, F. TRICERRI & L. VANHECKE
86 Topological topics, I.M. JAMES (ed)
87 Surveys in set theory, A.R.D. MATHIAS (ed)
88 FPF ring theory, C. FAITH & S. PAGE
89 An F-space sampler, N.J. KALTON, N.T. PECK & J.W. ROBERTS
90 Polytopes and symmetry, S.A. ROBERTSON
91 Classgroups of group rings, M.J. TAYLOR
92 Representation of rings over skew fields, A.H. SCHOFIELD
93 Aspects of topology, I.M. JAMES & E.H. KRONHEIMER (eds)
94 Representations of general linear groups, G.D. JAMES
95 Low-dimensional topology 1982, R.A. FENN (ed)
96 Diophantine equations over function fields, R.C. MASON
97 Varieties of constructive mathematics, D.S. BRIDGES & F. RICHMAN
98 Localization in Noetherian rings, A.V. JATEGAONKAR
99 Methods of differential geometry in algebraic topology, M. KAROUBI & C. LERUSTE
100 Stopping time techniques for analysts and probabilists, L. EGGHE
101 Groups and geometry, ROGER C. LYNDON
103 Surveys in combinatorics 1985, I. ANDERSON (ed)
104 Elliptic structures on 3-manifolds, C.B. THOMAS
105 A local spectral theory for closed operators, I. ERDELYI & WANG SHENGWANG
106 Syzygies, E.G. EVANS & P. GRIFFITH
107 Compactification of Siegel moduli schemes, C-L. CHAI
108 Some topics in graph theory, H.P. YAP
109 Diophantine analysis, J. LOXTON & A. VAN DER POORTEN (eds)
110 An introduction to surreal numbers, H. GONSHOR
113 Lectures on the asymptotic theory of ideals, D. REES
114 Lectures on Bochner-Riesz means, K.M. DAVIS & Y-C. CHANG
115 An introduction to independence for analysts, H.G. DALES & W.H. WOODIN
116 Representations of algebras, P.J. WEBB (ed)
117 Homotopy theory, E. REES & J.D.S. JONES (eds)
118 Skew linear groups, M. SHIRVANI & B. WEHRFRITZ
119 Triangulated categories in the representation theory of finite-dimensional algebras, D. HAPPEL
121 Proceedings of *Groups - St Andrews 1985*, E. ROBERTSON & C. CAMPBELL (eds)
122 Non-classical continuum mechanics, R.J. KNOPS & A.A. LACEY (eds)
124 Lie groupoids and Lie algebroids in differential geometry, K. MACKENZIE
125 Commutator theory for congruence modular varieties, R. FREESE & R. MCKENZIE
126 Van der Corput's method of exponential sums, S.W. GRAHAM & G. KOLESNIK
127 New directions in dynamical systems, T.J. BEDFORD & J.W. SWIFT (eds)
128 Descriptive set theory and the structure of sets of uniqueness, A.S. KECHRIS & A. LOUVEAU
129 The subgroup structure of the finite classical groups, P.B. KLEIDMAN & M.W.LIEBECK
130 Model theory and modules, M. PREST
131 Algebraic, extremal & metric combinatorics, M-M. DEZA, P. FRANKL & I.G. ROSENBERG (eds)
132 Whitehead groups of finite groups, ROBERT OLIVER
133 Linear algebraic monoids, MOHAN S. PUTCHA
134 Number theory and dynamical systems, M. DODSON & J. VICKERS (eds)
135 Operator algebras and applications, 1, D. EVANS & M. TAKESAKI (eds)
136 Operator algebras and applications, 2, D. EVANS & M. TAKESAKI (eds)
137 Analysis at Urbana, I, E. BERKSON, T. PECK, & J. UHL (eds)

138 Analysis at Urbana, II, E. BERKSON, T. PECK, & J. UHL (eds)
139 Advances in homotopy theory, S. SALAMON, B. STEER & W. SUTHERLAND (eds)
140 Geometric aspects of Banach spaces, E.M. PEINADOR & A. RODES (eds)
141 Surveys in combinatorics 1989, J. SIEMONS (ed)
142 The geometry of jet bundles, D.J. SAUNDERS
143 The ergodic theory of discrete groups, PETER J. NICHOLLS
144 Introduction to uniform spaces, I.M. JAMES
145 Homological questions in local algebra, JAN R. STROOKER
146 Cohen-Macaulay modules over Cohen-Macaulay rings, Y. YOSHINO
147 Continuous and discrete modules, S.H. MOHAMED & B.J. MÜLLER
148 Helices and vector bundles, A.N. RUDAKOV *et al*
149 Solitons, nonlinear evolution equations and inverse scattering, M. ABLOWITZ & P. CLARKSON
150 Geometry of low-dimensional manifolds 1, S. DONALDSON & C.B. THOMAS (eds)
151 Geometry of low-dimensional manifolds 2, S. DONALDSON & C.B. THOMAS (eds)
152 Oligomorphic permutation groups, P. CAMERON
153 L-functions and arithmetic, J. COATES & M.J. TAYLOR (eds)
154 Number theory and cryptography, J. LOXTON (ed)
155 Classification theories of polarized varieties, TAKAO FUJITA
156 Twistors in mathematics and physics, T.N. BAILEY & R.J. BASTON (eds)
157 Analytic pro-*p* groups, J.D. DIXON, M.P.F. DU SAUTOY, A. MANN & D. SEGAL
158 Geometry of Banach spaces, P.F.X. MÜLLER & W. SCHACHERMAYER (eds)
159 Groups St Andrews 1989 volume 1, C.M. CAMPBELL & E.F. ROBERTSON (eds)
160 Groups St Andrews 1989 volume 2, C.M. CAMPBELL & E.F. ROBERTSON (eds)
161 Lectures on block theory, BURKHARD KÜLSHAMMER
162 Harmonic analysis and representation theory, A. FIGA-TALAMANCA & C. NEBBIA
163 Topics in varieties of group representations, S.M. VOVSI
164 Quasi-symmetric designs, M.S. SHRIKANDE & S.S. SANE
165 Groups, combinatorics & geometry, M.W. LIEBECK & J. SAXL (eds)
166 Surveys in combinatorics, 1991, A.D. KEEDWELL (ed)
167 Stochastic analysis, M.T. BARLOW & N.H. BINGHAM (eds)
168 Representations of algebras, H. TACHIKAWA & S. BRENNER (eds)
169 Boolean function complexity, M.S. PATERSON (ed)
170 Manifolds with singularities and the Adams-Novikov spectral sequence, B. BOTVINNIK
171 Squares, A.R. RAJWADE
172 Algebraic varieties, GEORGE R. KEMPF
173 Discrete groups and geometry, W.J. HARVEY & C. MACLACHLAN (eds)
174 Lectures on mechanics, J.E. MARSDEN
175 Adams memorial symposium on algebraic topology 1, N. RAY & G. WALKER (eds)
176 Adams memorial symposium on algebraic topology 2, N. RAY & G. WALKER (eds)
177 Applications of categories in computer science, M. FOURMAN, P. JOHNSTONE, & A. PITTS (eds)
178 Lower K- and L-theory, A. RANICKI
179 Complex projective geometry, G. ELLINGSRUD *et al*
180 Lectures on ergodic theory and Pesin theory on compact manifolds, M. POLLICOTT
181 Geometric group theory I, G.A. NIBLO & M.A. ROLLER (eds)
182 Geometric group theory II, G.A. NIBLO & M.A. ROLLER (eds)
183 Shintani zeta functions, A. YUKIE
184 Arithmetical functions, W. SCHWARZ & J. SPILKER
185 Representations of solvable groups, O. MANZ & T.R. WOLF
186 Complexity: knots, colourings and counting, D.J.A. WELSH
187 Surveys in combinatorics, 1993, K. WALKER (ed)
189 Locally presentable and accessible categories, J. ADAMEK & J. ROSICKY
190 Polynomial invariants of finite groups, D.J. BENSON
191 Finite geometry and combinatorics, F. DE CLERCK *et al*
192 Symplectic geometry, D. SALAMON (ed)
193 Computer algebra and differential equations, E. TOURNIER (ed)
195 Arithmetic of blowup algebras, WOLMER VASCONCELOS
196 Microlocal analysis for differential operators, A. GRIGIS & J. SJÖSTRAND
197 Two-dimensional homotopy and combinatorial group theory, C. HOG-ANGELONI, W. METZLER & A.J. SIERADSKI (eds)
198 The algebraic characterization of geometric 4-manifolds, J.A. HILLMAN
199 Invariant potential theory in the unit ball of C^n, MANFRED STOLL
201 Singularities, JEAN-PAUL BRASSELET (ed)

London Mathematical Society Lecture Note Series. 201

Singularities

Lille 1991

Edited by

Jean-Paul Brasselet
CIRM, Luminy

CAMBRIDGE
UNIVERSITY PRESS

CAMBRIDGE UNIVERSITY PRESS
Cambridge, New York, Melbourne, Madrid, Cape Town,
Singapore, São Paulo, Delhi, Mexico City

Cambridge University Press
The Edinburgh Building, Cambridge CB2 8RU, UK

Published in the United States of America by Cambridge University Press, New York

www.cambridge.org
Information on this title: www.cambridge.org/9780521466318

First published 1994

A catalogue record for this publication is available from the British Library

Library of Congress Cataloguing in Publication Data

ISBN 978-0-521-46631-8 Paperback

Contents

INTRODUCTION

The international congress "Singularities in Geometry and Topology", organized by the "Unité de Recherche Associée au CNRS" Geometry, Analysis and Topology of Lille (France) was held from 3 to 8 June 1991. More than 230 researchers and students coming from 28 countries met to discuss themes concerning singularities in Geometry, Algebraic Topology and Algebraic Geometry. The Scientific Committee of the congress was composed of A. Aroca (Valladolid), H. Hamm (Münster), J.L. Koszul (Grenoble), B. Teissier (ENS, Paris), A. Verjovsky (ICTP, Trieste) and C.T.C. Wall (Liverpool). The congress was sponsored by the CNRS, the French Ministry of Education, the French Ministry of Research and Technology, the European Project "Singularities", the Region "Nord-Pas de Calais", the Lille and Villeneuve d'Ascq Municipalities and the local University Authorities : the "Université des Sciences et Techniques de Lille-Flandres-Artois", its Mathematics Department and the "Geometry-Analysis-Topology" Unit of the CNRS. There were many participants from Eastern European countries and for many of them it was the first opportunity to take part in a congress outside their country.

The Scientific Committe proposed 12 plenary conferences and 36 specialized conferences. In addition, a system of free conferences was organized in such a way that everyone willing to speak had the possibility of giving a short talk. These free conferences have been published in a special issue of the IRMA publications of Lille University*.

The papers of this volume concern 16 of the plenary and specialized conferences and deal with different points of view about singularities in geometry and topology.

---

* Publications IRMA, UFR de Mathématiques, USTL, 59655 Villeneuve d'Ascq Cedex, France

A.A. du Plessis and C.T.C. Wall give a survey on the geometry of topological stability for C^∞ maps. D.Mond and J. Montaldi extend a result of J.Damon about the deformations of mappings of complete intersections and bifurcations. As an application, they compute the discriminant Milnor number of a map. F. Delgado studies the topological type of the polar curve. He exhibits an arithmetical decomposition of the polar curve and obtains relations with the description of branches.

I. Scherback looks at the metamorphosis of wavefronts and caustics : if the initial wavefront is a surface with a boundary, new singularities arise. She gives a description and a classification of such singularities in terms of bifurcation sets of boundary singularities. The geometry of singularities is presented in the paper of J.W. Bruce, he describes applications to the differential geometry of surfaces in Euclidean and projective 3-space.

B. Malgrange gives a survey of the main known results about meromorphic connexions (in the regular and irregular cases) and local monodromy of algebraic equations, particularly results about the Riemann problem. The paper of B. Lichtin concerns asymptotics of lattice points and D-modules. The analytic theory of singularities and tools like the Newton polyhedron at infinity are used in the study of volumes of families of bounded semi-algebraic sets. The main result gives a proof of a conjecture of L.Ehrenpreis concerning hypoelliptic polynomials. L. Narvaez-Macarro describes the vanishing cycles and perverse sheaves relative to reducible curves.

In his paper on systems of micro-differential equations, O. Neto proves desingularization theorems for Lagrangian curves and for regular holonomic systems with support on a Lagrangian curve. This is achieved by showing that the blow-up of a regular holonomic system of microdifferential equations is regular holonomic and computing its support.

J.P. Henry and M. Merle give new results about equisingularity of the discriminant. Following ideas of T.Gaffney, they translate equisingularity conditions for isolated singularities of complete intersections in terms of integral dependance and multiplicities of modules. In this vocabulary, the conormal space is the geometric space associated to the jacobian module and the Buchsbaum-Rees multiplicity of the jacobian module plays an important role.

R. Thom gives a survey of canonical stratifications, especially stratifications of morphisms, for example, the existence of canonical stratification of the Tarski-Seidenberg projection, stratification of the space of jets, stratification of algebraic sets... In the case of spaces, M. Ferrarotti proves the existence of trivializations of a riemannian stratified space with differential of bounded norm. This result is applied to the local finiteness of volume.

G.M. Greuel and G. Pfister give a construction of moduli spaces for singular objects in projective theory. They study moduli spaces for curve singularities and for modules over the local ring of a fixed curve singularity. This provides a classification of irreducible plane curve singularities. M. Vaquié generalizes a result of Zariski about the computation of the irregularity of a cyclic covering of the complex projective plane ramified over a curve whose singularities are ordinary double points and cuspidal points. Using resolution of singularities, he computes the irregularity in the general case and proves vanishing theorems for the irregularity.

G. Barthel and A. Dimca study the topology of hypersurfaces which are projective homology manifolds. Their construction provides interesting examples of homology **P**n$'s$ with isolated singularities and allows one to classify such varieties. L. Kaup studies connectedness of a space ; he gives a general statement about Lefschetz theorems for singular varieties and the relation with topological completeness for complex spaces.

I would like to thank the Scientific Committee, all speakers, participants and administrative staff who made this congress a success, also the London Mathematical Society for publishing these proceedings.

Jean-Paul BRASSELET

Organizer of the Congress

Plenary Conferences

* J.W. BRUCE, Generic Geometry and Duality.

F. CANO, Polyèdre de Newton dans les équations différentielles.

T. GAFFNEY, Equisingularity of mappings : the future through the past.

M. GORESKY, Lefschetz fixed point formula on singular spaces.

D.T. LE, Topologie des singularités et entrelacs complexes.

R. MACPHERSON, Lefschetz numbers of Hecke correspondences.

* B. MALGRANGE, Connexions à singularités irrégulières.

C. SABBAH, Intégration dans les fibres d'un morphisme analytique ou algébrique.

B. TEISSIER, Classes caractéristiques.

* R. THOM, Quid des stratifications canoniques ?

* C.T.C. WALL, Stabilité topologique.

S. ZUCKER, $\mathcal{L}^2$-cohomology and singularities.

* Published in this volume.

Specialized Conferences

B. CENKL, Cohomologie d'intersection modérée.

D. CERVEAU, Cohomologie relative des formes différentielles.

J. DAMON, Singular Milnor fibers and higher multiplicities for non isolated Hypersurface Singularities.

* F. DELGADO, An arithmetical Decomposition for the polar Curve.

* A. DIMCA, Topology of hypersurfaces, results and open problems.

* M. FERRAROTTI, Trivializations of stratified Spaces with bounded Differential.

* G.M. GREUEL, On tame and wild Curve Singularities.

* L. KAUP, Theorems of Lefschetz type and topological q-completeness.

A. KHOVANSKII, Polyèdres de Newton.

* B. LICHTIN, Asymptotic of Lattice Points and D-modules.

F. LOESER, Déterminants et monodromies.

I. LUENGO, Strates de l'espace de paramètres de courbes.

C. MCCRORY, Secant Planes of Space Curves.

* M. MERLE, Equisingularité du discriminant.

* J. MONTALDI, Critical points of $\mathbf{C}^*$-invariant Functions.

* L. NARVAEZ-MACARRO, Cycles évanescents et faisceaux pervers.

* O. NETO, A duality and asymetry in symplectic Geometry and Systems of micro-differential Equations.

G. PFISTER, Local moduli problems and Singularities.

J.J. RISLER, Germes de courbes planes.

* I. SCHERBACK, Boundary of wave fronts and caustics and their metamorphosis.

A. SEBBAR, Equations différentielles monodromiques et variétés abéliennes.

R. SJAMAAR, Espaces de phase réduits et quantification géométrique.

* M. VAQUIE, Irrégularité des revêtements cycliques des surfaces

V. VERSHININ, Cobordism of Manifolds with Singularities.

* Published in this volume.

List of participants

Mohammed ABLY Lille France
Abdelhak ABOUQATEB Lille France
Abdallah AL AMRANI Strasbourg France
Klaus ALTMANN Kaiserslautern Germany
Christian ANGHEL Orsay France
Enrique ARTAL-BARTOLO Genève Swiss
Eric AURISSERGUES Lille France
Reynir AXELSSON Reyjavik Iceland
Daniel BARLET Nancy France
Raymond BARRE Valenciennes France
Gottfried BARTHEL Konstanz Germany
Belkacem BENDIFFALAH Marseille France
Michèle BENYOUNES Lille France
Isabel BERMEJO La Laguna Spain
Lev BIRBRAIR Jerusalem Israel
Havald BJAR Oslo Norway
Jean-Paul BRASSELET Marseille France
James William BRUCE Liverpool UK
Alexandre BRUNO Moscow CEI
Ragnar BUCHWEITZ Toronto Canada
Javolim BURES Prague Czechoslovakia
Bernard CALLENAERE Lille France
Antonio CAMPILLO Valladolid Spain
Felipe CANO Valladolid Spain
Josi CANO Valladolid Spain
Manuel CARNICIER Valladolid Spain
Pierrette CASSOU-NOGUES Talence France
Francisco CASTRO-JIMENEZ Sevilla Spain
Bohumil CENKL Boston USA
Dominique CERVEAU Rennes France
Marc CHARDIN Palaiseau France
Denis CHENIOT Marseille France
Sergey CHMUTOV Pereslavl-Zalessky CEI
Jan CHRISTOPHERSEN Oslo Norway
Gérard COEURE Lille France
Cyrille CRAPSKY Nancy France
James DAMON Chapel-Hill USA
Edouard DATRINIDADE Valenciennes France
Félix DELGADO DE LA MATA Valladolid Spain
Alexandru DIMCA Bonn Germany
Benaouda DJAMAI Lille France
Philippe DU BOIS Angers France
Andrew DU PLESSIS Aarhus Denmark

Nicolas DUPONT Lille France
Anne DUVAL Lille France
Mohammed EL AMRANI Angers France
Mohammed EL HAOUARI Lille France
Aziz EL KACIMI Valenciennes France
Abdelghani EL MAZOUNI Lille France
Fouad EL ZEIN quadis France
Yves FELIX Louvain-la-Neuve Belgium
Massimo FERRAROTTI Pisa Italy
Karl-Heinz FIESELER Konstanz Germany
Javier FINAT Lille France
Laurence FOURRIER Toulouse France
Louis FUNAR Orsay France
Terence GAFFNEY Boston USA
Françoise GEANDIER Nancy France
Peter GIBLIN Liverpool UK
Philippe GIMENEZ Grenoble France
Bouchra GMIRA Tetouan Marocco
François GOICHOT Valenciennes France
Mark GORESKY Boston USA
Victor V. GORYUNOV Moscow CEI
Jean-Michel GRANGER Angers France
Gert-Martin GREUEL Kaiserslautern Germany
Boguslaw HAJDUK Wrocław Poland
Helmut HAMM Münster Germany
Youssef HANTOUT Lille France
Alain HENAUT Bordeaux France
Jean-Pierre HENRY Palaiseau France
Michel HICKEL Talence France
Johannes HUEBSCHMANN Lille France
Marc HUTTNER Lille France
Tony IARROBINO Boston USA
Patrick IGLESIAS Lyon France
Schynichi IZUMIYA Sapporo Japan
Michel JAMBU Nantes France
Piotr JAWORSKI Warszawa Poland
Ahmed JEDDI Nancy France
Trygve JOHNSEN Tromsø Norway
Abdelhak KABILA Valenciennes France
Jean-Michel KANTOR quadis France
Yakov KARPISHPAN Baltimore USA
Ludger KAUP Konstanz Germany
Askold KHOVANSKII Moscow CEI
Martine KLUGHERTZ Toulouse France
Robert Jan KOOMAN Lille France

Vladimir KOSTOV Utrecht Netherlands
Jean-Louis KOSZUL Grenoble France
Tzee-Char KUO Epping Australia
Michał KWIECINSKI Marseille France
Rafael LABRERO Valladolid Spain
Serguei K. LANDO Pereslavl-Zalessky CEI
Olav Arnfinn LAUDAL Oslo Norway
Yves LAURENT Grenoble France
Dũng Tràng LE quadis France
Van Thanh LE Hanoï Vietnam
Danièle LEFEBVRE Lille France
André LEGRAND Toulouse France
Daniel LEHMANN Lille France
Josiane LEHMANN-LEJEUNE Lille France
François LESCURE Lille France
Noureddine LHALOUI Lille France
Ben LICHTIN Bordeaux France
François LOESER Palaiseau France
Stanisław LOJASIEWICZ Krakow Poland
Ignacio LUENGO Madrid Spain
Robert MACPHERSON Cambridge USA
Jon MAGNUSSON Reykjavik Iceland
Philippe MAISONOBE Nice France
Bernard MALGRANGE Grenoble France
Martin MARKL Prague Czechoslovakia
Pierre MARRY quadis France
Bernard MARTIN Berlin Germany
David MASSEY Boston USA
Jean-François MATTEI Toulouse France
Danielle MATTHYS Lille France
Kodjo MAWOUSSI quadis France
Mostafa MBEKTHTA Lille France
Clint MCCRORY Athens USA
Michel MERLE Nice France
Françoise MICHEL Angers France
Nicolae MIHALACHE Bucharest Roumania
Michel MINICONI Nice France
David MOND Coventry UK
James MONTALDI Nice France
Marcel MORALES Grenoble France
Piotr MORMUL Warsaw Poland
Witold MOZGAWA Lublin Poland
Gerd MULLER Mainz Germany
Luis NARVAEZ MACARRO Sevilla Spain
Vicente NAVARRO AZNAR Barcelone Spain

Jan NEKOVAR Prague Czechoslovakia
Orlando NETO Lisboa Portugal
Marcel NICOLAU Barcelone Spain
Takashi NISHIMURA Yokohama Japan
Carolina Ana NUMEZ-JIMENEZ Valladolid Spain
Juan NUNO BALLESTEROS Burjasot Spain
Philippe NUSS Strasbourg France
Armand NWATSOCK Lille France
Donal O'SHEA South Hadley USA
Karl OELJEKLAUS Lille France
Mutsuo OKA Tokyo Japan
Patrice ORRO Chambéry France
Victor PALAMODOV Moscow CEI
Stefan PAPADIMA Bucharest Roumania
William PARDON Durham USA
Emmanuel PAUL Toulouse France
Wieslaw PAWLICKI Krakow Poland
Rosa PERAIRE Barcelone Spain
Tomas PEREZ Valladolid Spain
Gerhard PFISTER Toronto Canada
Joan PORTI PIQUE Barcelone Spain
Marie-Thérèse POURPRIX Lille France
Ana REGUERA Valladolid Spain
Jean-Jacques RISLER quadis France
Gérard ROBERT quadis France
Claude ROCHE Dijon France
Marko ROCZEN Berlin Germany
Félice RONGA Genève Swiss
Ile RUNAR Oslo Norway
Claude SABBAH Palaiseau France
Jordi SALUDES Barcelone Spain
Martin SARALEGI Madrid Spain
Jacques SAULOY Toulouse France
Pierre SCHAPIRA quadis France
Daniel SCHAUB Angers France
Inna SCHERBACK Pereslavl-Zalessky CEI
Marie-Hélène SCHWARTZ Lille France
Jose SEADE Mexico Mexico
Ahmed SEBBAR Bordeaux France
A. SEVENSTER Amsterdam Netherlands
Dirk SIERSMA Utrecht Netherlands
Arvid SIQUELAND Oslo Norway
Reyer SJAMAAR Cambridge USA
Arne B. SLETSJOE Oslo Norway
Valeri SMYSHLAEV Leningrad CEI

Ahmajohn SOLEEV Samarkand CEI
Dag Einar SOMMERVOLL Oslo Norway
Kim SOV Lille France
Mark SPIVAKOVSKY Cambridge USA
Henri STERK Amsterdam Netherlands
Jan STEVENS Hamburg Germany
Aïda SURUBARU Lille France
Jacek SWIATKOWSKI Wrocław Poland
Aviva SZPIRGLAS Villetaneuse France
Daniel TANRE Lille France
Joseph TAPIA Toulouse France
Farid TARI Aarhus Denmark
Bernard TEISSIER quadis France
Tomohide TERASOMA Mannheim Germany
René THOM quadis France
Jean-Claude THOMAS Lille France
Mihai TIBAR Utrecht Netherlands
Pedro TSHINGI Lille France
Gijs TUYNMAN Lille France
Arkady VAINTRIOB Austin USA
Michel VAQUIE quadis France
Valerio VASSALLO Lille France
Vladimir VERSHININ Novosibirsk CEI
Wim VEYS Leuven Belgium
Nguyen VIET DUNG Hanoï Vietnam
C.T.C. WALL Liverpool UK
Georges G. WEILL New York USA
Klaus WIRTHMULLER Kaiserslautern Germany
Zdzistaw WOJTKOWIAK Nice France
Robert WOLAK Krakow Poland
David YAVIN Cambridge USA
Yosi YOMDIN Rehovot Israel
Xuan-Being YU Pekin China
Alexandru ZAHARIA Bucharest Roumania
Vladimir M. ZAKALYUKIN Moscou CEI
Abdellatif ZEGGAR Lille France
Jianyi ZHOU Marseille France
Steven ZUCKER Baltimore USA
Maria Angeles ZURRO Valladolid Spain

On Complex Projective Hypersurfaces which are Homology-Pn's

GOTTFRIED BARTHEL AND ALEXANDRU DIMCA

Konstanz University (GB)
Sydney University (AD)

Dedicated to Friedrich Hirzebruch

Abstract. We discuss hypersurfaces in $\mathbf{P}_{n+1}$ that are homology-$\mathbf{P}_n$'s, i.e., they have the integral homology of $\mathbf{P}_n$. The only cohomology ring-$\mathbf{P}_n$'s are the hyperplanes. Using singularity theoretic methods, we construct examples of homology-$\mathbf{P}_n$'s with isolated singularities in any dimension $n \geq 2$ and for any degree $d \geq 3$. In the odd-dimensional case, these are topological manifolds. Our methods yield interesting examples in singularity theory, e.g., isolated hypersurface singularity links that are topological spheres, but that are not associated to polynomials of the familiar "Pham-Brieskorn" type. Furthermore, we classify normal homology-$\mathbf{P}_2$'s in $\mathbf{P}_3$ with $\mathbf{C}^*$-action up to isomorphy and homology-$\mathbf{P}_3$'s in $\mathbf{P}_4$ with isolated singularities up to homeomorphy, and we construct examples of homology-$\mathbf{P}_n$'s with non-isolated singularities.

1: INTRODUCTION AND STATEMENT OF RESULTS

Considering the importance of the complex projective n-space $\mathbf{P}_n = \mathbf{P}_n(\mathbf{C})$ in algebraic geometry and topology, it is obvious that characterizing that space by algebro-geometric or topological properties always has been a matter of great interest. Therefore, it is quite natural to investigate spaces that share some of these properties. In this paper, we look for hypersurfaces in $\mathbf{P}_{n+1}$ (where $n \geq 2$) with normal or even isolated singularities which are *homology-$\mathbf{P}_n$'s*, i.e., which have the same integral homology as $\mathbf{P}_n$. As the integral homology $H_*(V, \mathbf{Z})$ of any n-dimensional projective variety V always contains a graded subgroup isomorphic to $H_*(\mathbf{P}_n, \mathbf{Z})$, homology-$\mathbf{P}_n$'s are characterized among such varieties by an obvious minimality property. By the universal coefficient formula, the property of being an homology-$\mathbf{P}_n$ is equivalent to having the integral cohomology *groups* of $\mathbf{P}_n$. The condition that the integral cohomology *ring* agrees with that of $\mathbf{P}_n$ turns out to be rather restrictive, as shows our first result:

Theorem 1. *(Cohomology Ring-$\mathbf{P}_n$'s are Actual $\mathbf{P}_n$'s.) Let V be a closed subvariety of dimension $\dim V = n \geq 2$ in some projective space $\mathbf{P}_N$ which can be described by a system of at most $N-2$ homogeneous polynomials. If the cohomology rings $H^\bullet(V, \mathbf{Z})$ and $H^\bullet(\mathbf{P}_n, \mathbf{Z})$ are isomorphic as abstract graded rings, then V is a linear subspace of $\mathbf{P}_N$.*

Actually, the condition on the cohomology ring structure can be slightly weakened; see the precise statement in section 1 below. Note that the condition on the number of defining equations is always satisfied for complete intersection varieties, and only for these in the surface case $n = 2$. The example of the Veronese surface V (i.e., the projective plane embedded by quadratic forms in $\mathbf{P}_5$) shows that this condition is sharp, as V just can be described by 4 quadratic polynomials. Moreover, the example of the smooth plane quadric curve (i.e., the projective line embedded by quadratic forms in $\mathbf{P}_2$) shows that the condition $\dim V = n \geq 2$ is sharp, too. (Of course, in the case of curves, the number of defining equations is at least $N - 1$).

Turning now to the study of homology-$\mathbf{P}_n$'s, our main results are as follows:

Theorem 2. *(Examples of Homology-$\mathbf{P}_n$'s with Isolated Singularities.) For any dimension $n \geq 2$, degree $d \geq 3$, and integer a with $1 \leq a < d - 1$, we consider the hypersurface $V := V^a_{n,d} : (f_{d,a} = 0)$ in $\mathbf{P}_{n+1}$ defined by*

$$f_{d,a}(x_0, x_1, \ldots, x_n, x_{n+1}) := x_0^a x_1^{d-a} + x_1 x_2^{d-1} + \ldots + x_{n-1} x_n^{d-1} + x_{n+1}^d .$$

This hypersurface has isolated singularities and satisfies
(i) $H_\bullet(V, \mathbf{Q}) \cong H_\bullet(\mathbf{P}_n, \mathbf{Q})$ for $(a, d) = 1$;
(ii) $H_\bullet(V, \mathbf{Z}) \cong H_\bullet(\mathbf{P}_n, \mathbf{Z})$ for $(a, d) = (a, d - 1) = 1$.

The proof uses singularity theoretic arguments; it is given in section 2. It provides examples for the following phenomena that may be of interest in singularity theory and topology:

(i) Examples of hypersurface singularities with one-dimensional singular locus and with monodromy operator equal to the identity (section 2, (ii), Lemma 2 and Remark). This contrasts with the situation for isolated hypersurface singularities, as described by A'Campo.

(ii) Examples of isolated complex hypersurface singularity links in all real dimensions $2n - 1 \geq 5$ which are integral homology spheres (and hence topological spheres), defined by positively weighted homogeneous polynomials which are not equivalent to polynomials of the familiar "Pham-Brieskorn" type $\sum_{j=0}^{n} x_j^{a_j}$ (see section 2, Corollary 1). This contrasts with the situation in dimension 2 (see section 3, Appendix).

(iii) Examples of projective hypersurfaces in all odd dimensions $n \geq 3$ with (one or two) isolated singularities and with minimal homology which are topological manifolds. (In fact, if an odd-dimensional projective hypersurface with isolated singularities has the integral homology of $\mathbf{P}_n$, then it is a topological manifold; see section 2, (v), Proposition. Such a variety even has the integral cohomology groups and the rational homotopy type of $\mathbf{P}_n$, but is not homotopy equivalent to $\mathbf{P}_n$, e.g., by Theorem 1.) Again, this contrasts with the situation in dimension 2: By a famous result of Mumford, a surface with normal singularities (e.g., a complete intersection surface with isolated singularities) never is a topological manifold.

The cases $n = 1$ or $d = 2$ not covered by the theorem are easy to deal with: The plane curve $V_{1,d}^a$ defined by $x_0^a x_1^{d-a} + x_2^d$ for $0 < a < d$ and $(a, d) = 1$ actually is homeomorphic to the projective line. For degree $d = 2$, it is well known that the only homology-$\mathbf{P}_n$'s among the quadric hypersurfaces with isolated singularities are the even-dimensional quadratic cones, and that, moreover, the odd-dimensional smooth quadrics Q_n are the only non-linear smooth hypersurfaces which are homology-$\mathbf{P}_n$'s.

After these examples, we turn to classification results for surfaces and threefolds. Note that the hypersurfaces $V_{n,d}^a$ admit a natural algebraic $\mathbf{C}^*$-action, as the affine equation at $\mathfrak{o}_0 := (1:0:0:\ldots:0)$ (and also at $\mathfrak{o}_1 := (0:1:0:\ldots:0)$) is weighted homogeneous. It turns out that in the class of normal surfaces in $\mathbf{P}_3$ with such a $\mathbf{C}^*$-action, our $V_{2,d}^a$ are the only homology-$\mathbf{P}_2$'s; moreover, they are pairwise non homeomorphic. In the case of threefolds, the topological type of an arbitrary homology-$\mathbf{P}_3$ in $\mathbf{P}_4$ with isolated singularities is completely determined by the degree, so such hypersurfaces with analytically different singularities may be homeomorphic. The precise statement is given in the following result (see section 3):

Theorem 3. *(Classification Results for Homology-$\mathbf{P}_2$'s and -$\mathbf{P}_3$'s.)*
($n = 2$) Let V be a homology-$\mathbf{P}_2$ in $\mathbf{P}_3$ of degree $d \geq 3$ with isolated singularities which admits an algebraic $\mathbf{C}^*$-action. Then V is equal to $V_{2,d}^a$ for a unique integer a satisfying $1 \leq a < d-1$ and $(a, d-1) = (a, d) = 1$. These surfaces are pairwise non homeomorphic.
($n = 3$) Let V, V' be homology-$\mathbf{P}_3$'s in $\mathbf{P}_4$ with isolated singularities. Then V and V' are homeomorphic if and only if they have the same degree.

Examples of homology-$\mathbf{P}_n$'s in dimensions $n \geq 3$ with singular locus of positive dimension can be obtained by more elementary methods than in the isolated singularity case. Such examples will be presented in section 4 (see Theorem 4).

We mention that Theorem 1 in the hypersurface case and some of the examples in Theorem 2 in the two-dimensional case (namely, the case $a = 1$, slightly disguised) have already appeared in [ChDi]. Moreover, the varieties $V_{n,d}^1$ play a key rôle in the work of Libgober on the connected sum decomposition of smooth hypersurfaces, and the fact that they have the integral homology of $\mathbf{P}_n$ is stated in [LiWo: § 2]. — Smooth projective manifolds with the rational cohomology ring of $\mathbf{P}_n$ have been investigated by several authors. For results, mainly for $n = 4$, and further references, the reader is referred to Wilson's article.

It is a pleasure for us to thank Ludger Kaup for his stimulating interest. In particular, section 1 was strongly influenced by him through discussions with one of us. Moreover, in sections 3 and 4, we closely follow ideas of earlier joint work of his and the first-named author. His remarks on a preliminary version of the present paper were most helpful. We would also like to thank Karl Fieseler who carefully read the text. Finally, we thank the referee for his valuable comments and his indications of some rather fine inexactitudes. His pertinent remarks encouraged us to keep the paper accessible to non-specialists. For them, the recent book [STH] might be useful as an introduction to and reference for many results and techniques used in the sequel.

Both authors enjoyed the hospitality of the "Max Planck -Institut für Mathematik" in Bonn, the second-named one twice during the time when this paper was written, and the first-named one at some earlier occasions. We both appreciated very much the stimulating and friendly atmosphere. It is our great pleasure to thank that institution, its members and staff, and in particular its director, Friedrich Hirzebruch. We dedicate this article to him.

A preliminary version appeared in the institute's preprint series as MPI 89–53.

NOTATIONS AND CONVENTIONS

Most of the varieties to be considered in the sequel are—in suitable affine coordinates—defined by *weighted homogeneous* (or quasihomogeneous) polynomials. Recall that by definition, such a polynomial $p(y_0, y_1, \ldots, y_m)$ satisfies an identity $p(t^{q_0}y_0, t^{q_1}y_1, \ldots, t^{q_m}y_m) = t^N \cdot p(y_0, y_1, \ldots, y_m)$ for a suitable vector $\mathfrak{q} = (q_0, q_1, \ldots, q_m)$ of integers q_j and an integer $N =: \mathfrak{q}\text{-deg}(p)$, the $\mathfrak{q}$-degree, so p is a homogeneous element of degree $\mathfrak{q}\text{-deg}(p) = N$ with respect to the grading of the polynomial algebra $\mathbf{C}[y_0, y_1, \ldots, y_m]$ given by $\mathfrak{q}\text{-deg}(y_j) = q_j$. We do not assume that the q_j's are necessarily positive; in fact, the "mixed" case occurs in the proof of Lemma 2 in section 2.

We adopt here the convention to call the q_j's the *weights*. They are just the weights of the $\mathbf{C}^*$-action on $\mathbf{C}^{m+1}$ given by

$$t_\bullet(y_0, y_1, \ldots, y_m) = (t^{q_0}y_0, t^{q_1}y_1, \ldots, t^{q_m}y_m)$$

that is associated to the grading. Dividing by the greatest common divisor of the weights if necessary, we may always assume that the weight vector $\mathfrak{q}$ is primitive, i.e., $\gcd(q_0, q_1, \ldots, q_m) = 1$ or, equivalently, that the action is effective. We sometimes call p a $\mathfrak{q}$-homogeneous polynomial. The pair $(\mathfrak{q}, \mathfrak{q}\text{-deg}(p))$ is called the *type* of p.

Concerning the notion of "weight", there are different conventions used in the literature, especially in the case of a strictly positive grading (i.e., all $q_j > 0$, corresponding to a "good" $\mathbf{C}^*$-action). In addition to conventions discussed in [TRCS: Ch. 7, § 1], we mention the one adopted in several papers (by Milnor, Orlik, and some others, including earlier papers of the first-named author) where the positive rational numbers $w_j := \mathfrak{q}\text{-deg}(f)/q_j$ are called weights. Instead, we call these w_j the *coweights* in the sequel. If we want to emphasize that we are in the case of a strictly positive grading, we sometimes call such a $\mathfrak{q}$-homogeneous polynomial *positively* weighted homogeneous.

References [..] to articles in journals, conference proceedings, etc., are given by the Names of author(s) (or abbreviations), using Small Capitals. References to monographs are given by Title Initials in Capitals [TIC]. The sign [-] refers to the last preceding reference.

1: PROJECTIVE VARIETIES WITH THE COHOMOLOGY RING OF $\mathbf{P}_N$.

In this section, we prove the result mentioned in the introduction above: A hypersurface with the same integral cohomology ring as $\mathbf{P}_n$ is a hyperplane, so it just *is* $\mathbf{P}_n$. The actual, slightly more general, statement is as follows.

Theorem 1. *Let V be a closed subvariety of dimension $\dim V = n \geq 2$ in some projective space $\mathbf{P}_N$ which (set-theoretically) can be described by a system of at most $N-2$ homogeneous polynomials. If the cohomology group $H^2(V,\mathbf{Z})$ is generated (up to torsion) by a class u such that u^n generates $H^{2n}(V,\mathbf{Z})$, then V is a linear subspace of $\mathbf{P}_N$.*

Proof. Denote with $j : V \hookrightarrow \mathbf{P}_N$ the inclusion mapping and with $\omega \in H^2(\mathbf{P}_N)$ the canonical generator. Then there is an integer $\alpha \neq 0$ (without loss of generality $\alpha > 0$) such that $j^*\omega = \alpha u$ holds in $H^2(V)$ (up to torsion), and hence $j^*\omega^n = \alpha^n u^n$ holds in $H^{2n}(V)$. By a well known property of the degree (see, e.g., [PAG: p. 171]), we have

$$\deg V = \langle \omega^n, j_*[V]\rangle = \langle j^*\omega^n, [V]\rangle = \alpha^n \langle u^n, [V]\rangle = \alpha^n$$

(where $\langle -, - \rangle$ denotes the usual pairing between homology and cohomology, sometimes called "Kronecker product"). In order to show that $\alpha = 1$, look at the following part of the exact cohomology sequence of the pair $(\mathbf{P}_N, V)$:

$$H^2(\mathbf{P}_N) \xrightarrow{j^*} H^2(V) \xrightarrow{\delta^*} H^3(\mathbf{P}_N, V) .$$

By Lefschetz duality, the last group is isomorphic to $H_{2N-3}(\mathbf{P}_N \setminus V)$. As V can be defined by at most $N-2$ equations ($f_j = 0$), the complex manifold $\mathbf{P}_N \setminus V$ is the union of at most $N-2$ affine open subsets ($f_j \neq 0$) and thus is *topologically* ($N-3$)*-complete* (see Fieseler and Kaup [$-_1$: § 1 and 2.3] for the definition of topological completeness and for the result). It follows from the theorem stated in the introduction of [$-_2$] that $H_{2N-3}(\mathbf{P}_N \setminus V)$ has no torsion (note that we may assume $N \geq 3$, so $2N-3 \geq N$ holds). Using the exactness of the sequence, we infer that $H^2(V)$ is free, generated by u, and that δ^* is the zero homomorphism as $\delta^* j^*\omega = \alpha\delta^* u$ vanishes. Hence $j^*\omega$

is a generator of $H^2(V)$, which yields $\alpha = 1$ and thus $\deg V = 1$. •

Remark. The first part of the previous argument proves the well-known fact that cohomology ring of V contains the truncated polynomial ring

$$\mathbf{Z}[j^*\omega]/(j^*\omega^{n+1}) \cong H^*(\mathbf{P}_n, \mathbf{Z})$$

as a graded subring. — The notion of a topologically q-complete space used in the second part is modeled after the topological properties of *analytically* q-complete spaces. In fact, by a theorem of. Hamm, an analytically q-complete complex space of dimension n has the homotopy type of a CW-complex of (topological) dimension at most $n + q$. The topological completeness has a much nicer behaviour and better permanence properties with respect to standard operations; in particular, it is a homeomorphy invariant.

2: PROJECTIVE HYPERSURFACES WITH ISOLATED SINGULARITIES WHICH ARE HOMOLOGY-$\mathbf{P}_N$'S.

In this section, we prove Theorem 2 as stated in the introduction. In order to show that those hypersurfaces $V^a_{n,d} : (f_{d,a} = 0)$ in $\mathbf{P}_{n+1}$ (with $n \geq 2$) which satisfy the conditions on d and a have the integral homology of $\mathbf{P}_n$, we proceed in several steps. First, in steps *(i)–(iii)*, we use duality and monodromy arguments to check that they are rational homology-$\mathbf{P}_n$'s. In *(iv)*, we state necessary and sufficient conditions (in terms of Milnor lattices) for a rational homology-$\mathbf{P}_n$ in $\mathbf{P}_{n+1}$ with isolated singularities to be an integral homology-$\mathbf{P}_n$. That turns out to have interesting consequences, and we briefly digress to discuss some of them in *(v)* (e.g., in odd dimensions, such a variety is a topological manifold; see the Proposition below). Finally, using results of Milnor, Orlik, and Randell on the monodromy of certain weighted homogeneous singularities, we show in *(vi)–(viii)* that our examples satisfy these conditions.

(i) We begin with a characterization of *rational* homology-$\mathbf{P}_n$'s in terms of the monodromy operator of the defining homogeneous polynomial f (for the moment, we do not assume $f = f_{d,a}$). Let $V : (f = 0)$ be a hypersurface of degree $d \geq 2$ in $\mathbf{P}_{n+1}$.

Lemma 1. *The following statements are equivalent:*
(α) $H_\bullet(V, \mathbf{Q}) \cong H_\bullet(\mathbf{P}_n, \mathbf{Q})$;
(β) Let $F \subset \mathbf{C}^{n+2}$ *be the Milnor fibre* $(f = 1)$ *of* f, *and let* $h_f^* : \tilde{H}^\bullet(F, \mathbf{Q}) \to \tilde{H}^\bullet(F, \mathbf{Q})$ *be the monodromy operator. Then all eigenvalues of* h_f^* *are different from* 1.

Proof (cf. [Oka$_2$: Thm.1, proof]). Dualizing the argument used in the proof of Theorem 1, we infer from the exact homology sequence of the pair $(\mathbf{P}_{n+1}, V)$ and from the relative (Alexander-Lefschetz type) duality theorem that statement (α) is equivalent to the vanishing of $\tilde{H}^\bullet(\mathbf{P}_{n+1} \setminus V, \mathbf{Q})$. The affine variety $\mathbf{P}_{n+1} \setminus V$ is the quotient of $\mathbf{C}^{n+2} \setminus (f = 0)$ under the standard action of $\mathbf{C}^*$ by multiplication. By homogeneity, the surjective mapping $f| : \mathbf{C}^{n+2} \setminus (f = 0) \to \mathbf{C}^*$ is equivariant with respect to the action $(\tau, t) \mapsto \tau^d \cdot t$ on the base. Hence, the stabilizer subgroup of the Milnor fibre F (and of any other fibre) is the group C_d of d-th roots of unity, so the orbit space F/C_d is canonically identified with $\mathbf{P}_{n+1} \setminus V$. The total space $E : (|f| = 1)$ of the Milnor fibration $f|_E : E \to S^1$ is invariant under the free S^1-action that is induced from the standard action of $\mathbf{C}^*$, and the action of the standard generator $\zeta := \exp(2\pi i/d)$ of C_d on $F \subset \mathbf{C}^{n+2}$ given by $(x_0, \ldots, x_{n+1}) \mapsto (\zeta x_0, \ldots, \zeta x_{n+1})$ is the natural geometric monodromy h_f of the Milnor fibration. Hence, the reduced cohomology of $\mathbf{P}_{n+1} \setminus V$ is isomorphic to $\tilde{H}^\bullet(F, \mathbf{Q})^{h_f^*}$, the fixed part under h_f^*, and the latter is $\ker(\mathrm{id} - h_f^*)$, the eigenspace of 1. •

(ii) To obtain examples of homogeneous polynomials $f(x_0, x_1, \ldots, x_n, x_{n+1})$ that satisfy condition (β) from above, we first search for a homogeneous polynomial $g(x_0, x_1, \ldots, x_n)$ with trivial monodromy. Adding to g the monomial x_{n+1}^d with $d := \deg g$ will yield f which has the required properties by a Thom-Sebastiani type argument due to Oka (see the proof of Lemma 3 below). So let us consider the homogeneous polynomial

$$g = g_{d,a}(x_0, x_1, \ldots, x_n) := x_0^a x_1^{d-a} + x_1 x_2^{d-1} + \ldots + x_{n-1} x_n^{d-1}$$

of degree $d \geq 3$ with $n \geq 2$ and $1 \leq a < d - 1$, and prove:

Lemma 2. *The monodromy operator h_g^* associated to g is the identity operator if a and d are coprime.*

Proof. We denote with G the Milnor fibre $(g = 1)$ of g in $\mathbf{C}^{n+1}$. We want to find a $\mathbf{C}^*$-action on $\mathbf{C}^{n+1}$ such that G is invariant and the natural geometric monodromy $h_g : G \to G$ is given by "multiplication" $x \mapsto \lambda \bullet x$ (with respect to that action) by some element $\lambda \in \mathbf{C}^*$. Since $\mathbf{C}^*$ is connected, this will imply that h_g is homotopy equivalent to the identity, thus proving the lemma.

As g is homogeneous of degree d, the geometric monodromy takes the same nice form $h_g(x_0, \ldots, x_n) = (\zeta x_0, \ldots, \zeta x_n)$ with $\zeta := \exp(2\pi i/d)$ as above. The $\mathbf{C}^*$-action will be of diagonal form $t\bullet(x_0, \ldots, x_n) = (t^{q_0} x_0, \ldots, t^{q_n} x_n)$ given by a vector $\mathfrak{q} = (q_0, \ldots, q_n)$ of integral weights $q_j = \mathfrak{q}\text{-}\deg(x_j)$. As G has to be invariant under that action, $\mathfrak{q}$ must be so chosen that g is q-homogeneous with $\mathfrak{q}\text{-}\deg(g) = 0$. Hence, we have the condition

$$aq_0 + (d - a)q_1 \;=\; q_1 + (d - 1)q_2 \;=\; \ldots \;=\; q_{n-1} + (d - 1)q_n \;= 0$$

which is clearly satisfied by taking $q_n = a$, $q_{n-1} = (1-d)a, \ldots, q_1 = (1-d)^{n-1}a$ and $q_0 = (1 - d)^{n-1}(a - d)$. As a and d are coprime by assumption, we can find an integer b with $ab \equiv 1 \bmod d$. Since all weights satisfy $q_i \equiv a \bmod d$, the element $\lambda := \zeta^b$ has the property that $\lambda^{q_i} = \zeta$, so $\lambda_\bullet(x_0, \ldots, x_n) = h_g(x_0, \ldots, x_n)$ as claimed at the beginning. •

Remark. Note that the affine hypersurface $(g_{d,a} = 0)$ in $\mathbf{C}^{n+1}$ has a one-dimensional singular locus, namely, the x_0-axis for $a = 1$, and the union of the x_0- and the x_1-axis for $a > 1$. — For *isolated* hypersurface singularities, the monodromy operator is the identity only in the case of an odd-dimensional A_1-singularity, as follows from the results of A'Campo [–: Thme. 2].

(iii) As indicated above, we now add the monomials x_{n+1}^d to the polynomials $g = g_{d,a}$, thus obtaining the homogeneous polynomials $f := f_{d,a}$ that define our hypersurfaces.

Lemma 3. *Denote with $f := f_{d,a}$ the polynomial*

$$f_{d,a}(x_0, x_1, \ldots, x_n, x_{n+1}) := g_{d,a}(x_0, x_1, \ldots, x_n) + x_{n+1}^d$$

and with $V := V_{n,d}^a$ the hypersurface in $\mathbf{P}_{n+1}$ defined by $(f = 0)$. Then V has isolated singularities. Moreover, V is a rational homology-$\mathbf{P}_n$, i.e., we have $H_\bullet(V_{n,d}^a, \mathbf{Q}) \cong H_\bullet(\mathbf{P}_n, \mathbf{Q})$, if a and d are coprime.

Proof. The singularities of V are easily determined: Denote with $\mathfrak{o}_i$ (for $i = 0, \ldots, n+1$) the origin of the standard affine coordinate system $(x_i = 1)$ on $\mathbf{P}_{n+1}$. The affine equation for V at $\mathfrak{o}_0$ is $f_0 = x_1^{d-a} + x_1 x_2^{d-1} + \ldots + x_{n-1} x_n^{d-1} + x_{n+1}^d$, so $\mathfrak{o}_0$ is always an isolated singular point. At $\mathfrak{o}_1$, we have the affine equation $f_1 = x_0^a + x_2^{d-1} + x_2 x_3^{d-1} + \ldots + x_{n-1} x_n^{d-1} + x_{n+1}^d$, so $\mathfrak{o}_1$ is a singular point if (and only if) $a > 1$. It is easy to see that there are no other singularities.

To show that $V_{n,d}^a$ is a rational homology-$\mathbf{P}_n$ if $\gcd(a, d) = 1$ holds, it suffices to verify that all eigenvalues of the monodromy operator h_f^* are different from 1 (Lemma 1, (β)). To that end, we can apply results of Oka and use Lemma 2, as f is the sum $g \oplus r$ of the parts $g(x_0, \ldots, x_n)$ and $r(x_{n+1}) := x_{n+1}^d$ (with distinct variables). By [$-_1$: Thm.1, Cor.2], the Milnor fibre F of f is homotopy equivalent to the join $G * R$ of the Milnor fibres of g and r, and the monodromy operator h_f^* on $\tilde{H}^{\bullet+1}(F, \mathbf{Q}) \cong (\tilde{H}^\bullet(G, \mathbf{Q}) \otimes \tilde{H}^\bullet(R, \mathbf{Q}))$ is induced from the join of the geometric monodromies. As R is zero-dimensional, that implies the equality $h_f^* = h_g^* \otimes h_r^*$ (which would follow from the Thom-Sebastiani theorem if g had isolated singularities). Since h_g^* is the identity on $\tilde{H}^\bullet(G, \mathbf{Q})$ by Lemma 2, whereas all eigenvalues of h_r^* on $\tilde{H}^\bullet(R, \mathbf{Q})$ are different from 1, we are done. •

Remark. As is well known (cf. the remark following Theorem 1), the rational cohomology ring of any n-dimensional projective variety V contains a graded subring isomorphic to $H^*(\mathbf{P}_n, \mathbf{Q})$. Hence, if V has the rational homology of $\mathbf{P}_n$, then it even has the same rational cohomology ring as $\mathbf{P}_n$, so in particular, rational Poincaré duality holds. Thus, if V has isolated singularities, it follows from L. Kaup's long exact Poincaré duality sequence (see the introduction in [$-_1$]) that V is a rational homology manifold, i.e., all the singularities of V have links that are rational homology spheres (see also [Di$_2$: Cor.(2.9)]). Moreover, a rational homology-$\mathbf{P}_n$ has the same rational homotopy type as $\mathbf{P}_n$, as the latter is determined by the rational cohomology ring (see Babenko [-: §2]).

(iv) In order to show that we can actually obtain *integral* homology-$\mathbf{P}_n$'s among these varieties $V^a_{n,d}$, we now consider an arbitrary hypersurface $V \subset \mathbf{P}_{n+1}$ with isolated singularities. It is a well known consequence of Lefschetz type theorems, duality theory, and universal coefficient formulae that $H_j(V, \mathbf{Z})$ and $H_j(\mathbf{P}_n, \mathbf{Z})$ are isomorphic except for the two middle dimensions $j = n$ and $n+1$ (see, e.g., [STH: 5.2.6, 5.2.11]). Hence, we only have to discuss these two exceptional cases.

To that end, we use results of [Di$_1$]. By [-: Thm. 2.1] (or [STH: 5.4.3]), there are isomorphisms

$$H_j(V, \mathbf{Z}) \cong H_j(\mathbf{P}_n, \mathbf{Z}) \oplus \begin{cases} \operatorname{coker} \varphi_V & \text{for } j = n \ , \\ \ker \varphi_V & \text{for } j = n+1 \ , \end{cases}$$

where $\varphi_V : \bigoplus_i L_i \to \overline{L}$ denotes a natural homomorphism of Milnor lattices associated to $V \subset \mathbf{P}_{n+1}$. (Recall that the Milnor lattice of an isolated n-dimensional hypersurface singularity is the reduced integral homology $\tilde{H}_n(F, \mathbf{Z})$ of the corresponding Milnor fibre F, endowed with the intersection form. The lattice—i.e., the form—is symmetric if the dimension n is even, and skew-symmetric if n is odd.) The source of φ_V is the (orthogonal) direct sum of the Milnor lattices L_i at the singular points of V. The target is the *reduced* Milnor lattice $\overline{L} := L/(\mathrm{Rad}\ L)$ associated to the singularity at the origin of the affine cone corresponding to a generic (and thus smooth) hyperplane section of V. Hence, by deformation, that lattice $\overline{L} = \overline{L}_{n,d}$ depends only on n and on the degree d of V. In particular, the middle homology $H_{n,d} := H_n(\tilde{V}, \mathbf{Z})$ of a *smooth* hypersurface $\tilde{V} = \tilde{V}_{n,d} \subset \mathbf{P}_{n+1}$ is isomorphic (as group) to $\overline{L}_{n,d} \oplus H_n(\mathbf{P}_n, \mathbf{Z})$. In fact, $\overline{L}_{n,d}$ can be identified with the sublattice $\ker(j_* : H_n(\tilde{V}_{n,d}) \to H_n(\mathbf{P}_{n+1})) \subset H_{n,d}$ of "vanishing cycles"; cf. [KuWo: 6]. As a consequence, if the dimension n is odd, then $\overline{L}_{n,d}$ and $H_{n,d}$ agree, so $\overline{L}$ is unimodular. On the other hand, if n is even, then $\overline{L}$ is the orthogonal complement $h^\perp$ to the corresponding iterated hyperplane section class h, so it has determinant $\pm d$ (see [Di$_1$: Cor. 1.4, 1.5, and Rem. 2.4]). As a side-remark, we mention that the lattice homomorphism φ_V also determines the cohomology ring structure of V (see [-: Prop. 6.1]).

To apply the result, we note that $H_{n+1}(V, \mathbf{Z})$ is always torsion free (see [–: Cor. 2.3]). Hence, if V is a rational homology-$\mathbf{P}_n$, then φ_V is a monomorphism—so all L_i are nondegenerate—and its cokernel is a finite torsion group whose order satisfies $|\operatorname{coker}\varphi_V|^2 = \pm(\prod_i \det L_i) / \det \overline{L}$. This yields the following criterion:

Lemma 4. *Let $V \subset \mathbf{P}_{n+1}$ be a rational homology-$\mathbf{P}_n$ of degree d with isolated singularities. The following conditions are equivalent:*
(α) V is an integral homology-$\mathbf{P}_n$;
(β) the cokernel of the lattice homomorphism φ_V is trivial;
(γ) $\prod_i \det L_i = \pm \det \overline{L} = \begin{cases} \pm d & \text{if } n \text{ is even,} \\ \pm 1 & \text{if } n \text{ is odd.} \end{cases}$ •

We can replace condition (γ) by an equivalent one, using the relation between the intersection form, the monodromy operator, and the "variation operator" (or Seifert form) of an isolated hypersurface singularity defined by a polynomial equation ($p = 0$) (see, e.g., Lamotke's paper [–: §6, Hauptsatz] or the book [SDM II: 2.5] for that relation): The determinant of the Milnor lattice satisfies $\det L_p = \pm\Delta_p$, where $\Delta_p := \Delta_p(1)$ is the value at $t = 1$ of the characteristic polynomial $\Delta_p(t) := \det(t \cdot I - h_p^*)$ of the monodromy operator. Hence, (γ) is equivalent to the condition

(γ') $\prod_i \Delta_i = \begin{cases} \pm d & \textit{if } n \textit{ is even,} \\ \pm 1 & \textit{if } n \textit{ is odd} \end{cases}$

(where Δ_i corresponds to L_i, of course).

(v) The previous lemma has interesting consequences, so we briefly digress to discuss some of them. By one of the classical results in Milnor's book [SPCH: Thm. 8.5], an isolated hypersurface singularity link $K := (p = 0) \cap S_\varepsilon^{2n+1}$ is an integral homology $(2n-1)$-sphere if (and only if) $\Delta_p = \pm 1$. In (complex) dimensions $n \geq 3$, it follows from the generalized Poincaré conjecture that such an integral homology $(2n-1)$-sphere actually is a topological sphere, so the corresponding singularity is a topological manifold point. Moreover, the link K always bounds a parallelizable manifold (see [–: 5.1, 6.1]). Hence, such a singularity link with $\Delta_p = \pm 1$ is h-cobordant to the standard sphere S^{2n-1} if bP_{2n} (the group of h-cobordism classes of $(2n-1)$-dimensional homology spheres that bound a parallelizable $2n$-manifold) is trivial, and that is known to be true for $n = 3, 7, 15$, and 31 (see [KeMi: p. 512] and [HiMa: 10.4, pp. 74/75] for the orders of bP_{2n} with $n \leq 10$, and [Bro: Cor. 2] and [BaJoMa] for $bP_{30} = \{S^{29}\}$ and $bP_{62} = \{S^{61}\}$.) In that case, the smooth structure on the punctured ε-neighbourhood of the singularity can be naturally extended to the singular point (use the h-cobordism to the standard sphere and the fact that a punctured ε-neighbourhood of an isolated singularity is diffeomorphic

to the punctured open cone over the link. Note that this does not contradict the remark in [SPCH: p. 13], as the latter refers to the embedding as complex subvariety). Putting together that information, we have the following result.

Proposition. *Let V be an n-dimensional projective hypersurface with isolated singularities which is a homology-$\mathbf{P}_n$. If n is odd, then*
(α) V is a topological manifold, and
(β) all singularity links are topological $(2n-1)$-spheres.
Moreover, if the group bP_{2n} is trivial (e.g., for $n = 3, 7, 15$, and 31), then V has a natural smooth structure. •

Note that for degrees $d \geq 2$, these topological manifolds are not homotopy equivalent to $\mathbf{P}_n$, though they have the integral homology and the rational homotopy type of $\mathbf{P}_n$. The difference lies in the cohomology ring structure, which is easily described as follows:

Supplement. *Let V be as above, of dimension $n = 2m - 1$ for some integer $m \geq 2$; denote with $j : V \hookrightarrow \mathbf{P}_{2m}$ the embedding and with d the degree. Then the integral cohomology ring $H^*(V, \mathbf{Z})$ is isomorphic to the truncated polynomial ring $\mathbf{Z}[u_1, u_m]/(u_1^m - du_m, u_m^2)$ generated by $u_1 := j^*(\omega) \in H^2(V, \mathbf{Z})$ and another class $u_m \in H^{2m}(V, \mathbf{Z})$.*

Proof. By an argument as used in the proof of theorem 1, the fact that the homology $H_{4m-k}(\mathbf{P}_{2m} \setminus V, \mathbf{Z})$ vanishes for $k < 2m$ implies that $j^* : H^k(\mathbf{P}_{2m}, \mathbf{Z}) \to H^k(V, \mathbf{Z})$ is an isomorphism for $k \leq 2m - 2 = n - 1$. Hence, the class $u_1 := j^*(\omega)$ generates the cohomology ring in these degrees. As V is a topological manifold, the Poincaré duality homomorphism $H^{2n-k}(V, \mathbf{Z}) \overset{\cap[V]}{\to} H_k(V, \mathbf{Z})$ is an isomorphism, so in particular, $u_1^l \cap [V]$ is a natural generator of $H_{2(n-l)}(V, \mathbf{Z})$ for $0 \leq l \leq m - 1$. Denote with $u_{n-l} \in H^{2(n-l)}(V, \mathbf{Z})$ the ("Kronecker") dual generator. For $l = m - 1$, we thus obtain the generator u_m of $H^{n+1}(V, \mathbf{Z})$. We obviously have $u_m^2 \in H^{2n+2}(V, \mathbf{Z}) = 0$, and the equalities

$$\langle u_1^{m-1-l} u_m, u_1^l \cap [V] \rangle = \langle u_m, u_1^{m-1} \cap [V] \rangle = 1 = \langle u_{n-l}, u_1^l \cap [V] \rangle$$

and

$$\langle u_1^{n-l}, u_1^l \cap [V] \rangle = \langle u_1^n, [V] \rangle = d$$

for $0 \leq l \leq m-1$ immediately imply $u_{n-l} = u_1^{m-1-l} u_m$ and $u_1^{n-l} = du_{n-l}$, thus proving our claim. •

In the terminology of Libgober and Wood [-: § 2], such a variety is called a "d-twisted" homology-$\mathbf{P}_n$.)

Note that in the case of even dimension n, the condition (γ') from above shows that a projective hypersurface V of degree $d \geq 2$ which is an integral homology-$\mathbf{P}_n$ can never be a topological manifold. Except for the lack of Poincaré duality, the cohomology ring structure is quite similar to the odd-dimensional case described above. Details will be leaft to the reader.

(vi) We now return to the problem of finding integral homology-$\mathbf{P}_n$'s among our varieties $V^a_{n,d}$. By condition (γ') of Lemma 4, we are left with the computation of the determinants $\Delta_i := \det(\mathrm{id} - h_i^*)$ corresponding to the two (possibly) singular points o_0 and o_1 (cf. Lemma 3, proof). To that end, we shall proceed as follows: We observe that both affine equations f_0 and f_1 can be decomposed in the form $f_i = f_i' \oplus f_i''$—i.e., we may rearrange the affine variables as $y = (y', y'')$ and write $f_i(y', y'') = f_i'(y') + f_i''(y'')$—with positively weighted homogeneous parts f_i' and f_i''. If their weighted degrees N_i' and N_i'' are relatively prime, we can use formula (1) below given by Milnor and Orlik to reduce the computation of Δ_i to determining Δ and another invariant κ for these parts. Now, in both cases, we may choose the decomposition such that f_i'' is of "Pham-Brieskorn" type, where κ and Δ are easily determined, whereas f_i' belongs to another special class of positively weighted homogeneous polynomials that has been investigated by Orlik and Randell. Their results provide the necessary information to complete the computation.

To discuss the formula of Milnor and Orlik, let $p(y_0, y_1, \ldots, y_m)$ be any positively weighted homogeneous polynomial with an isolated singularity at the origin. We consider the invariant $\kappa := \kappa_p := \dim\ker(I - h_p^*)$, i.e., the multiplicity of 1 as an eigenvalue of the monodromy operator. We obviously have $\kappa = 0$ if and only if $\Delta \neq 0$. — We remark that κ is the Betti number $b_{m-1}(K) = b_m(K)$ of the singularity link K; in particular, it vanishes if and only if K is a rational homology sphere or, equivalently, if the singularity is a rational homology manifold point (cf. [STH: 3.4.2/7]).

Now let us assume that we have a decomposition $p = p' \oplus p''$ into positively weighted homogeneous parts as above. Denote with κ', κ'' and with Δ', Δ'' the respective invariants. According to Milnor and Orlik, we have the following formula [-: §4, Lemma 3]:

$$(1) \qquad \kappa_p = \kappa' \cdot \kappa'' \quad \text{and} \quad \Delta_p = (\Delta')^{\kappa''} \cdot (\Delta'')^{\kappa'} \qquad \text{if} \quad (N', N'') = 1$$

(with $0^0 := 1$. Note that $\Delta = \Delta(1)$ and the invariant ρ occuring in [-: §4, Lemmas 2 and 3] agree if Δ is non-zero, the only case of interest for us.) — If $(\mathfrak{q}', N')$ and $(\mathfrak{q}'', N'')$ are the types of p' and p'', respectively, then the type $(\mathfrak{q}, N)$ of p is $\frac{1}{(N',N'')} \cdot (N''\mathfrak{q}' \oplus N'\mathfrak{q}'', N' \cdot N'')$.

(vii) To apply these formulae to the affine equations f_0 and f_1 in our case, we make use of their decomposition $f_i = f_i' \oplus f_i''$ into the parts

$$(2_0)\ f_0'(x_1, x_2, \ldots, x_n) := x_1^{d-a} + x_1 x_2^{d-1} + \ldots + x_{n-1} x_n^{d-1}, \quad f_0''(x_{n+1}) := x_{n+1}^d$$

and

$$(2_1) f_1'(x_2, \ldots, x_n) := x_2^{d-1} + x_2 x_3^{d-1} + \ldots + x_{n-1} x_n^{d-1}, \quad f_1''(x_0, x_{n+1}) := x_0^a + x_{n+1}^d$$

(with $n \geq 3$ for f_1; in the case $n = 2$, the polynomial $f_1(x_0, x_2, x_3) = x_0^a + x_2^{d-1} + x_3^d$ already is of "Pham-Brieskorn" type, and not just by accident; see

Theorem 3A below). The "main" parts f_0' and f_1' both belong to the class of (positively) weighted homogeneous polynomials

(3) $$p_{\mathfrak{a}}(y_0, y_1, \ldots, y_m) := y_0^{a_0} + y_0 y_1^{a_1} + \ldots + y_{m-1} y_m^{a_m}$$

with $m \geq 1$ and with exponents $\mathfrak{a} = (a_0, a_1, \ldots, a_m)$ satisfying $a_0 \geq 2$ and $a_j \geq 1$ for $1 \leq j \leq m$. That class has been investigated by Orlik and Randell in [–: 2]. Using the integers $r_k := \prod_{j=0}^{k} a_j$ and their alternating sum $\mu_k := \sum_{j=-1}^{k} (-1)^{k-j} r_j = r_k - \mu_{k-1}$ (for $-1 \leq k \leq m$, with $r_{-1} = \mu_{-1} = 1$ and $\mu_{-2} = 0$) defined in that paper [–: p. 203], the type $(\mathfrak{q}, N)$ of $p_{\mathfrak{a}}$ is easily expressed: We have $N = \tilde{N} / \gcd(\tilde{q}_0, \ldots, \tilde{q}_m)$ and $q_i = \tilde{q}_i / \gcd(\tilde{q}_0, \ldots, \tilde{q}_m)$, where

$$\tilde{N} = r_m = \prod_{j=0}^{m} a_j \quad \text{and} \quad \tilde{q}_k = \frac{\mu_{k-1} \cdot r_m}{r_k} = \mu_{k-1} \cdot \prod_{j=k+1}^{m} a_j \quad (\text{for } k = 0, \ldots, m) .$$

In the case of our polynomials f_i', the exponents $\mathfrak{a} = (a_0, a_1, \ldots, a_m)$ have the special form $(b, c, \ldots, c)$ with $b = d - a$, $c = d - 1$, $m = n - 1$ for f_0', and $b = c = d - 1$, $m = n - 2$ for f_1'. That yields the values

$$r_k = bc^k, \quad \mu_k = b \cdot \frac{c^{k+1} - (-1)^{k+1}}{c+1} - (-1)^k, \quad \tilde{q}_k = \mu_{k-1} \cdot c^{m-k}, \quad \text{and} \quad \tilde{N} = bc^m ;$$

in particular, the weighted degrees N_i of the f_i''s divide $\tilde{N}_0' = (d-a)(d-1)^{n-1}$ and $\tilde{N}_1' = (d-1)^{n-1}$.

To determine the invariants κ_p and Δ_p associated with a polynomial $p = p_{\mathfrak{a}}$, one can use the formula

$$\Delta_p(t) = \delta(t) = \prod_{j=-1}^{m} (t^{r_j} - 1)^{\varepsilon_j} \qquad \text{with} \quad \varepsilon_j := (-1)^{m-j}$$

for the characteristic polynomial of the monodromy operator (see [–: (2.12)]). For our purpose, is suffices to conclude that the multiplicity κ_p of 1 as eigenvalue depends only on the dimension m of the hypersurface $(p_{\mathfrak{a}} = 0)$, namely,

(4) $$\kappa_p = \begin{cases} 0 & \text{if } m \text{ is even,} \\ 1 & \text{if } m \text{ is odd.} \end{cases}$$

(viii) In order to apply formula (1) from above to the decompositions $f_i = f_i' \oplus f_i''$ of f_0 and f_1 in (2_i), we have to assume that the respective weighted degrees N_i' and N_i'' are coprime; moreover, we ¡need the values κ_i'' and Δ_i'' for the "remainders" $f_0'' = x_{n+1}^d$ (with $N_0'' = d$) and $f_1'' = x_0^a + x_{n+1}^d$ (with $N_1'' = ad/(a,d)$). Simple direct computation yields

(5_0) $$\kappa_0'' = 0 \quad \text{and} \quad \Delta_0'' = d$$

as well as

(5_1) $$\kappa_1'' = 0 \quad \text{and} \quad \Delta_1'' = 1 \quad \text{if} \quad \gcd(a, d) = 1 .$$

The greatest common divisors (N_i', N_i'') of the weighted degrees obviously divide (a, d) and $(a, d-1)$, respectively. Hence, assuming that the condition

$(a, d-1) = (a, d) = 1$ holds, it follows immediately from formula (1) that we have $\kappa_0 = \kappa_1 = 0$ (so both local equations f_i have nondegenerate Milnor lattices L_i) and $\Delta_i = (\Delta_i'')^{\kappa_i'}$. By (4) and ($5_0$), we obtain two different values for $\Delta_0 = \pm\det(L_0)$ according to the parity of n, namely

$$\Delta_0 = \begin{cases} d & \text{if } n \text{ is even,} \\ 1 & \text{if } n \text{ is odd.} \end{cases}$$

Using (5_1), we immediately get $\Delta_1 = 1$, so L_1 is always unimodular. It follows that condition (γ') of Lemma 4 in *(iv)* is satisfied, so $V = V_{n,d}^a$ has the integral homology of $\mathbf{P}_n$.

That completes the proof of Theorem 2 as stated in the introduction. • •

We mention explicitly the following consequence of the proof, as announced in the introduction.

Corollary 1. *For some integers $1 \leq m < n$, let $p(x_0, \ldots, x_n)$ be a positively weighted homogeneous polynomial with an isolated singularity at the origin. Assume that there is a splitting $p = p' \oplus p''$ into positively weighted homogeneous parts $p'(x_0, \ldots, x_m)$ and $p''(x_{m+1}, \ldots, x_n)$ with $p' = p_a$ is as in (vii),* (3) *such that their weighted degrees N' and N'' are relatively prime. Denote with κ', κ'', Δ' and Δ'' the respective invariants. Then the singularity link defined by p is a $(2n-1)$-dimensional homology sphere (and hence actually a topological sphere) if one of the following conditions hold:*
a) m is even, and either $\kappa'' = 0$ or $\Delta' = \pm 1$;
b) m is odd, and $\Delta'' = \pm 1$. •

From the Proposition in *(v)*, we have the following

Corollary 2. *If the dimension $n \geq 3$ is odd and if the hypersurface $V_{n,d}^a$: $(f_{d,a} = 0)$ of the theorem is an integral homology-$\mathbf{P}_n$, then it is a topological manifold. For n =* 3, 7, 15, *and* 31, *that manifold even admits a smooth structure.* •

A much more complete statement concerning smooth structures follows from the evaluation of the characteristic polynomial at $t = -1$:

Complement. *Under the same assumptions ($n \geq 3$ odd, $V_{n,d}^a$ a $\mathbf{Z}$-homology-$\mathbf{P}_n$), the value $\Delta_i(-1)$ of the characteristic polynomial of f_i for $i = 0$ or $i = 1$ is as follows:*

$$\Delta_0(-1) = \begin{cases} d & \textit{if } d \textit{ is odd,} \\ (d-a)(d-1)^{(n-1)/2} & \textit{if } d \textit{ is even;} \end{cases}$$

$$\Delta_1(-1) = \begin{cases} 1 & \textit{if } d \textit{ is odd,} \\ a & \textit{if } d \textit{ is even.} \end{cases}$$

Hence, for all degrees $d \equiv \pm 1$ mod 8, the variety $V_{n,d}^a$ admits a smooth struc-

ture, whereas for $d \equiv \pm 3 \bmod 8$ *and* $n \neq 2^k - 1$, *it doesn't.*

The conclusion follows from the fact that for an isolated complex hypersurface singularity of odd dimension n with $\Delta(1) = \pm 1$, the condition $\Delta(-1) \equiv \pm 1 \bmod 8$ implies that the singularity link is h-cobordant to the standard sphere S^{2n-1} (see [SPCH: Rem. 8.7]), whereas $\Delta(-1) \equiv \pm 3 \bmod 8$ implies that the link is the Kervaire sphere which is known to be exotic for $n \neq 2^k - 1$. We leave it to the reader to formulate the conclusions for even degrees.

The proof of the formula for $\Delta(-1)$ may be given elsewhere. — We remark that for the odd-dimensional varieties $V^1_{n,d}$ with only one singularity (i.e., $a = 1$), the value for $\Delta_0(-1)$ has been obtained by Libgober, together with $\Delta_0(1) = 1$ (see [$-_1$: Thm. 2, (iv)]). The fact that $V^1_{n,d}$ is a "d-twisted" homology- $\mathbf{P}_n$ is stated in [LiWo: § 2]. — We are grateful to Anatoly Libgober for pointing out to us the relation to his work.

3: CLASSIFICATION RESULTS FOR HOMOLOGY-$\mathbf{P}_2$'S AND -$\mathbf{P}_3$'S.

In this section, we discuss two classification results of different kind for homology- $\mathbf{P}_n$'s in dimensions $n = 2$ and 3. For the surface case, note that our examples $V^a_{n,d}$ of homology- $\mathbf{P}_n$'s with isolated singularities constructed in section 2 admit a natural algebraic $\mathbf{C}^*$-action, as their affine equation f_0 (and f_1 as well) is weighted homogeneous. Normal homology- $\mathbf{P}_2$'s in $\mathbf{P}_3$ that admit such an action can be classified up to isomorphy: It turns out that for degrees $d \geq 3$, the only such surfaces are our examples $V^a_{2,d}$, and these are pairwise non homeomorphic. The result in the threefold case is a classification of homology- $\mathbf{P}_3$'s in $\mathbf{P}_4$ with isolated singularities up to homeomorphy, and even diffeomorphy with respect to the natural smooth structure. The actual statement, slightly more general than in the introduction, is as follows:

Theorem 3. *(a: Surface Case) Let* V *be a normal surface of degree* d *in* $\mathbf{P}_3$ *which has the same* **Z***-homology groups as* $\mathbf{P}_2$ *and which admits an algebraic* $\mathbf{C}^*$*-action. Then* V *is one of the following surfaces:*
$(d = 1)$ *the projective plane* $\mathbf{P}_2$;
$(d = 2)$ *the quadratic cone* $(q = 0)$ *(where* $q(x_0, \dots, x_3)$ *is a quadratic form of rank 3);*
$(d \geq 3)$ *the surface* $V^a_{2,d}$ *for a unique positive integer* $a < d-1$ *relatively prime to* $d-1$ *and* d.
These surfaces are pairwise non homeomorphic.
(b: Threefold Case) Let V, V' *denote homology-* $\mathbf{P}_3$*'s in* $\mathbf{P}_4$ *with isolated singularities, endowed with their natural smooth structure. The following conditions are equivalent:*
(α) V *and* V' *are diffeomorphic;*
(β) V *and* V' *are homeomorphic;*

(γ) *V and V' are oriented homotopy equivalent;*
(δ) *V and V' have isomorphic cohomology ring structures;*
(ε) $\deg V = \deg V'$.

Proof (a: Surface Case). As the plane and the quadratic cone are the only surfaces of degree $d \leq 2$ in $\mathbf{P}_3$ satisfying the assumptions, we may restrict to the case $d \geq 3$. The $\mathbf{C}^*$-action on V is induced from an action on the ambient space, where it is diagonalizable. Hence, there is a system $(x_0 : x_1 : x_2 : x_3)$ of homogeneous coordinates and a triple $\mathfrak{q} := (q_1, q_2, q_3)$ of integral weights with $q_1 \geq q_2 \geq q_3 \geq 0$ and $\gcd(q_1, q_2, q_3) = 1$ such that the action is of the form $t{\bullet}x := (x_0 : t^{q_1}x_1 : t^{q_2}x_2 : t^{q_3}x_3)$ (e.g., see [Bar_3: 1.1]). The corresponding affine equation $f(1, x_1, x_2, x_3)$ defining $V \cap (x_0 = 1)$ is $\mathfrak{q}$-homogeneous of some $\mathfrak{q}$-degree N (recall our conventions from the beginning). It is easy to see (e.g., in [$-_3$: 1.4]) that, up to the only exception of the smooth quadric, every normal $\mathbf{C}^*$-surface in $\mathbf{P}_3$ has one or two *elliptic* fixed points, i.e., fixed points lying in the closure of every orbit passing through a suitable neighbourhood. By taking such a point as centre of the invariant affine chart $(x_0 = 1)$, we may assume $q_3 > 0$ (up to reversing the action, i.e., replacing t by t^{-1}). Then the affine equation f is positively weighted homogeneous of type $(\mathfrak{q}, N)$.

We note first that this affine equation f is not homogeneous: Otherwise, V would be the cone over the smooth plane curve $(x_0 = f = 0)$ of degree d and hence have the third Betti number $b_3(V) = (d-1)(d-2)$ strictly positive (as $d > 2$). The proof of the claim is an easy consequence of the following lemma and of Theorem 3A in the appendix to this section.

Lemma. *For a surface V of degree $d \geq 3$ as in Theorem 3 (a), there is a system of homogeneous coordinates $(w:x:y:z)$ with the following properties:*
a) The origin $\mathfrak{o} := (1:0:0:0)$ of the affine chart $(w = 1)$ is an elliptic fixed point of the action;
b) the second integral local homology $\mathcal{H}_{2,\mathfrak{o}}$ at $\mathfrak{o}$ is trivial;
c) the curve at infinity $V_\infty := V \cap (w = 0)$ is a projective line.

To continue the proof of the theorem, we note that the second local homology group $\mathcal{H}_{2,\mathfrak{o}}$ at $\mathfrak{o}$ is isomorphic to the first homology of the corresponding singularity link K, so the latter is an integral homology sphere. Hence, by Theorem 3A in the Appendix below, the affine equation in the chart $(w = 1)$ of the lemma is $x^a + y^b + z^c = 0$, where the exponents a, b, c are pairwise coprime. We now observe that the affine surface defined by a polynomial of "Pham-Brieskorn" type has a normal projective closure (i.e., isolated singularities at infinity) if and only if the two highest exponents differ by at most 1. Assuming without loss of generality that $a \leq b \leq c = d$ holds, we must thus have $b = d - 1$. It follows that V has the equation $w^{d-a}x^a + wy^{d-1} + z^d = f_{d,a}(x, w, y, z) = 0$.

The fact that these surfaces are pairwise non homeomorphic is easy to verify: First note that the cup product square of the canonical generator $j^*\omega$ of $H^2(V,\mathbf{Z})$ yields the degree d. Next, observe that for fixed $d \geq 3$, we can distinguish topologically between the surfaces $V^a_{2,d}$ for different values of a as follows: By the famous result of Mumford, normal surface singularities have non-trivial local fundamental groups and hence actually are topological singularities. Recall from step (viii) in section 2 the values of the invariant Δ_i at the points $\mathfrak{o}_i$, namely, $\Delta_0 = d$ and $\Delta_1 = 1$. These are the just the orders of the first homology groups of the singularity links or, equivalently, of the second local homology groups $\mathcal{H}_{2,\mathfrak{o}_i}$, so they are topological invariants of the singularities. It follows that $\mathfrak{o}_0$ is always a singular point; hence, if it is the only one, then $\mathfrak{o}_1$ is a regular point, so we have $a = 1$. Otherwise, the singularity at $\mathfrak{o}_1$, being of "Pham-Brieskorn" type, is a (positively) weighted homogeneous surface singularity with relatively prime integer coweights $(a, d-1, d)$ with $2 \leq a < d-1$. In this situation, the singularity link cannot be a lens space, and hence, by a result of Orlik, the fundamental group of the link determines the coweights (see [$-_1$: § 3]). •

Proof of the Lemma. As we have seen above, the affine equation in the coordinate system ($x_0 = 1$) is not homogeneous. We may thus apply the results of [Bar_1] (keeping the different notation in mind). We have $b_2(V) = b_3(V) + b_2(A)$, according to [-: (3.5.4)(i)], and hence $b_2(A) = 1$ for the curve $A := V \cap (x_0 = 0)$, so A is irreducible. The argument preceding [-: (3.5.5)] yields that A is not only homeomorphic to a projective line, but that, interchanging the roles of x_0 and x_1 if necessary, we may even assume that A actually is a projective line. (Note that the condition of [-: (2.3.1)] only concerns the affine singularities.) Then [-: (3.5.5)(i)] yields $H_2(V) = \mathbf{Z} \oplus \mathcal{H}_{2,\mathfrak{o}}$ and hence $\mathcal{H}_{2,\mathfrak{o}} = 0$ for the local homology at the affine origin, so we have proved our claim. •

Remark. It is easy to see that the surfaces V_d of [ChDi] and our $V^1_{2,d}$ are isomorphic. In fact, after renaming the coordinates for $\mathbf{P}_3$ so that V_d is defined by $w^{d-1}x + w^d + wy^{d-1} + z^d$, the linear transformation $(w:x:y:z) \mapsto (w:x-w:y:z)$ takes V_d into $V^1_{2,d}$. — The fact that the surfaces $V^1_{2,d}$ have the integral homology of the projective plane has been mentioned in [Bar_2: 2].)

Proof (b: Threefold Case). According to the results of Wall [-: Thm. 5] and Jupp [-: Thm. 1 and Cor.] (see also the paper by Žubr), the classifications of smooth simply connected closed 6-manifolds with torsion-free homology up to diffeomorphy and up to homeomorphy coincide, and they are given by the cohomology ring structure and the characteristic classes $w_2 \in H^2(V, \mathbf{Z}/(2))$ and $p_1 \in H^4(V,\mathbf{Z})$. By the supplement to the proposition in section 2, (v), the cohomology ring structure and the degree d uniquely determine each other. It thus suffices to show that the classes w_2 and p_1 only depend on the degree d. That follows by the standard computation of the characteristic classes of smooth hypersurfaces in $\mathbf{P}_n$ (see [TMAG: 22.1, p. 159]) with only a slight variation: Denote with S the singular locus of V and with $\mathring{V}$ and $\mathring{\mathbf{P}}_4$ the complement of S in V and in $\mathbf{P}_4$, respectively. By duality theory, the inclusion mappings $\mathring{X} \hookrightarrow X$ for $X = V$ and $X = \mathbf{P}_4$, respectively, induce isomorphisms $H^q(X, \mathbf{Z}) \to H^q(\mathring{X}, \mathbf{Z})$ in cohomology for $q \leq 4$. Hence, it suffices to investigate the induced characteristic classes $\mathring{w}_2$ and $\mathring{p}_1$ of $\mathring{V}$.

Now $\mathring{V}$ is a smooth complex hypersurface in $\mathring{\mathbf{P}}_4$. The classes $\mathring{w}_2$ and $\mathring{p}_1$ are thus determined by the Chern classes $\mathring{c}_1$ and $\mathring{c}_2$ of the tangent bundle $T\mathring{V}$ in the usual way, namely: $\mathring{w}_2$ is $\mathring{c}_1$ reduced mod 2, and $\mathring{p}_1$ equals $\mathring{c}_1^2 - 2\mathring{c}_2$. We consider the standard short exact sequence

$$0 \to T\mathring{V} \to T\mathring{\mathbf{P}}_4|_{\mathring{V}} \to N \to 0$$

of vector bundles on $\mathring{V}$, where N denotes the normal bundle. Now $T\mathring{\mathbf{P}}_4$ and N obviously are the restrictions of well-known bundles on $\mathbf{P}_4$, namely, the tangent bundle $T\mathbf{P}_4$ and the line bundle $\{V\}$ associated to the divisor V (with total Chern classes $(1+\omega)^5$ and $1 + d\omega$, respectively). Applying the Whitney sum formula and using the naturality properties of the total Chern class $c := 1 + c_1 + c_2 + \ldots$, we obtain

$$\begin{aligned} \mathring{c} = c(T\mathring{V}) &= \mathring{\jmath}^* \left(c(T\mathbf{P}_4) c\{V\}^{-1} \right) \\ &= \mathring{\jmath}^* \left((1+\omega)^5 (1+d\omega)^{-1} \right) \\ &= \mathring{\jmath}^* \left(1 + (5-d) \cdot \omega + (10 - 5d + d^2) \cdot \omega^2 + \ldots \right) \end{aligned}$$

(where $\mathring{\jmath}$ denotes the inclusion of $\mathring{V}$ in $\mathring{\mathbf{P}}_4 \subset \mathbf{P}_4$). That yields $\mathring{w}_2 \equiv 0$ or $\mathring{w}_2 \equiv \mathring{u}_1 := \mathring{\jmath}^*\omega$ mod 2, according to the parity of d, and $\mathring{p}_1 = (5 - d^2)\mathring{u}_1^2$, so $\mathring{w}_2$ and $\mathring{p}_1$ are determined by the degree, thus proving our claim. — We are grateful to Terry Wall for helpful discussions with one of us during a conference in Oberwolfach. •

3A. Appendix: Weighted Homogeneous Surface Singularity Links that are Homology Spheres.

In this appendix, we discuss the classification of isolated two-dimensional weighted homogeneous hypersurface singularities whose link is a homology sphere. That result was crucial for our proof of Theorem 2, (a).

Theorem 3A. *Let $p(x, y, z)$ be a positively weighted homogeneous polynomial with an isolated singularity at the origin. Assume that the link K of the singularity is an integral homology sphere. Then we have $p(x, y, z) = x^a + y^b + z^c$ in a suitable coordinate system, and the exponents a, b, c (which agree with the coweights) are pairwise coprime.*

Proof. For weighted homogeneous polynomials with integral coweights—as represented by the "Pham-Brieskorn" type—, the result is a special case of Brieskorn's characterization of homology spheres (see [-: 2, Satz 1]). (A thorough discussion of the corresponding surface singularity links is found in Milnor's article.) To exclude the various classes of weighted homogeneous polynomials with at least one fractional coweight, we give three arguments.

Maybe the simplest—but also the least illuminating—way is by checking that for polynomials in these classes, the group $H := H_1(K) \cong \mathcal{H}_{2,\circ}$ never vanishes. That group can be computed using a general formula given by Orlik (see [$-_2$: 2.6, 3.3, 3.4]), which is made explicit in our case as follows: Write the coweights w_j of $p(x_1, x_2, x_3)$ as reduced fractions u_j/v_j. Then the rank $\kappa := b_1(K) = b_{2,\circ}$ of H is

$$\kappa = \frac{w_1 w_2 w_3}{\operatorname{lcm}(u_1, u_2, u_3)} - \sum_{i \neq j} \frac{w_i w_j}{\operatorname{lcm}(u_i, u_j)} + \sum_k \frac{w_k}{u_k} - 1.$$

To compute the torsion subgroup T of H, introduce the numbers

$$\kappa_{ij} := \frac{w_i w_j}{\operatorname{lcm}(u_i, u_j)} - \frac{w_i}{u_i} - \frac{w_j}{u_j} + 1$$

and define integers c, c_i, and c_{ij} by the factorization of the denominators u_j as follows: Let $c := \gcd(u_1, u_2, u_3)$,

$$c_i := \frac{\gcd(u_j, u_k)}{c}, \quad \text{and} \quad c_{ij} := \frac{u_k}{c c_i c_j} \quad \text{for} \quad \{i, j, k\} = \{1, 2, 3\}$$

(so $u_i = c c_j c_k c_{jk}$). Finally, introduce the integers

$$t_l := \prod_{\kappa_{ij} \geq l} c_{ij} \quad \text{for} \quad 1 \leq l \leq m := \max\{\kappa_{ij}\}.$$

Then the torsion subgroup is the direct sum of cyclic groups

$$T \cong \mathbf{Z}/(ct_1) \oplus \mathbf{Z}/(t_2) \oplus \ldots \oplus \mathbf{Z}/(t_m).$$

Using the explicit formulae in terms of the exponents of typical monomials for the different classes of weighted homogeneous polynomials that are listed in [TSCC: pp. 285–286], we get the result.

The next approach is somewhat more conceptual; in fact, it shows the background of the formulae above. For every weighted homogeneous surface V in $\mathbf{C}^3$ with an isolated singularity, the link $K = V \cap S^5_\epsilon \cong (V \setminus 0)/\mathbf{R}_{>0}$ is a closed oriented three-dimensional manifold with a fixed point free S^1-action. As such, it has the structure of a Seifert fiber space (see, e.g., Orlik's Lecture Notes [SM]). It follows from Seifert's computation of the fundamental group [SM: 5.3] that if $H_1(K)$ vanishes, then necessarily, the genus g of the "decomposition surface" $K/S^1 \cong (V \setminus 0)/\mathbf{C}^*$ vanishes, the number of exceptional orbits is at least three (unless $K \cong S^3$, i.e., $V \cong \mathbf{C}^2$), and their orders are pairwise coprime (see [Sei: §12, Satz 12], where such homology spheres are called "Poincarésche Räume", or [OrWa_1: p. 280]). Now all the exceptional orbits lie in the intersection of V with the coordinate hyperplanes ($x_i = 0$). Orlik and Wagreich show in [$-_2$: 3.5] how the orders of such orbits and the numbers of orbits of a given order can be expressed explicitly in terms of the type $(\mathfrak{q}, N)$. From their results, it follows that for weighted homogeneous polynomials with fractional coweights, the necessary condition for K to be a homology sphere is never fulfilled.

The most satisfactory argument comes from a result of W. Neumann, and we are grateful to him for pointing this out to us. If the genus of the decomposition surface K/S^1 vanishes, then the homology group H is a finite abelian group. There is a corresponding finite unramified covering K' of K, the "universal abelian covering", having H as group of decktransformations. This can be extended to a covering $V' \to V$ of normal weighted homogeneous surfaces, ramified only at the fixed point. By Neumann's result [–: Thm.1], that universal abelian covering surface V' is always a complete intersection $V_{a_1,\dots,a_N} : (\sum_{i=1}^N \lambda_{ij} x_i^{a_i} = 0)_{j=1,\dots,N-2}$ defined by "Pham-Brieskorn" type polynomials, where the integers a_i are the orders and N is the number of the exceptional orbits of V. If K is a homology sphere, then, of course, the covering is trivial, i.e., $K = K'$ and $V = V'$. Hence, if in the case $V \subset \mathbf{C}^3$ under consideration, the link K is a homology sphere, then the defining polynomial is of "Pham-Brieskorn" type $x_1^{a_1} + x_2^{a_2} + x_3^{a_3} = 0$ and there are exactly three exceptional orbits, so the exponents a_i are pairwise coprime. — An explicit (and earlier) reference for the characterization of Seifert fibred homology spheres as complete intersections $V_{a_1,\dots,a_N}$ with coprime exponents a_i is Theorem 4.1 in the article by Neumann and Raymond.•

4: HOMOLOGY OF "ASYMPTOTICALLY LINEAR" HYPERSURFACES IN $\mathbf{P}_{N+1}$ WITH $\mathbf{C}^*$-ACTION AND EXAMPLES OF HOMOLOGY-$\mathbf{P}_N$'S WITH NON-ISOLATED SINGULARITIES.

In the two-dimensional case, the affine equation of our examples $V_{n,d}^a$ at $\mathfrak{o}_1$ is $x_0^a + x_2^{d-1} + x_3^d$, so the *leading form*, i.e., the homogeneous part of the highest degree d, is x_3^d, the d-th power of a coordinate function. Accordingly, the corresponding (reduced) hyperplane section $V \cap (x_1 = 0)$ "at infinity" is the linear subspace $(x_1 = x_3 = 0)$. By the natural good $\mathbf{C}^*$-action, the affine part $V \cap (x_1 = 1)$ is contractible. Using singular duality theory, this decomposition into homologically simple pieces allowed us to translate the homology-$\mathbf{P}_2$ condition in section 3 to the homology sphere condition for the singularity link at $\mathfrak{o}_1$.

This situation may be generalized to higher dimensions as follows. For $x = (x_1, \dots, x_{n+1})$ with $n \geq 2$, let $p(x)$ be a positively weighted homogeneous polynomial of degree $d \geq 2$ with an isolated singularity at the affine origin $\mathfrak{o} \in \mathbf{C}^{n+1}$, and assume that the leading form is $p_d = x_{n+1}^d$ (up to a non-zero scalar factor). Denote with U the normal affine weighted homogeneous hypersurface $(p = 0)$ in $\mathbf{C}^{n+1}$ and with V its projective closure $(\hat{p} = 0)$ in

$\mathbf{P}_{n+1}$, where $\hat{p}(x_0,\ldots,x_{n+1}) := x_0^d \cdot p(x/x_0)$ is the (usual) homogenization of p. The part at infinity $V_\infty := V \setminus U = V \cap (x_0 = 0)$ is the projective subvariety $(p_d = 0)$ of the hyperplane $(x_0 = 0)$ at infinity that is defined by the leading form, so topologically, it is the linear subspace $(x_0 = x_{n+1} = 0) \cong \mathbf{P}_{n-1}$. (Its points correspond to the asymptotic directions on U, hence the name "asymptotically linear".) In general, the singularities of V at infinity are non-isolated. If the "sub-leading form" p_{d-1} is not divisible by x_{n+1}, however, they have codimension at least two and are thus normal. The projective variety V is invariant under the $\mathbf{C}^*$-action on $\mathbf{P}_{n+1}$ given by $t_\bullet(x_0 : x_1 : \ldots : x_{n+1}) := (x_0 : t^{q_1}x_1 : \ldots : t^{q_{n+1}}x_{n+1})$ that extends the natural good $\mathbf{C}^*$-action on $\mathbf{C}^{n+1} \cong (x_0 = 1)$ corresponding to the weight vector $\mathfrak{q}$ of p. By that action, the affine part U gets contracted to the affine origin $\mathfrak{o} \in \mathbf{C}^{n+1}$ for $t \to 0$. We show that the (co-)homology of such a projective hypersurface V coincides with that of $\mathbf{P}_n$ except for the two middle dimensions, where the correction term is determined by the homology of the the affine singularity link $K := U \cap S_\varepsilon^{2n+1}$:

Theorem 4. *For V as above, the integral cohomology groups are*

$$H^k(V) \cong \begin{cases} H^k(\mathbf{P}_n) & \text{for } k \neq n,\ n+1; \\ H^k(\mathbf{P}_n) \oplus H_{2n-k}(K) & \text{for } k = n,\ n+1. \end{cases}$$

The cohomology ring structure can be described by means of the homomorphisms $j^k : H^k(\mathbf{P}_{n+1}) \to H^k(V)$ induced by inclusion:

- For $k \neq 2n+2$, these mappings are injective.
- For $k \neq n,\ n+1,\ 2n,\ 2n+2$, they are isomorphisms.
- For $k = n,\ n+1$, the image of j^k has the direct complement

$$\ker(i^k : H^k(V) \to H^k(V_\infty)) \cong H_{2n-k}(K),$$

and all cup products with positive-dimensional classes vanish on that complement.

- For $k = 2n$, the canonical generator ω^n of $H^{2n}(\mathbf{P}_{n+1})$ is mapped onto $d \cdot u_n$, where $u_n \in H^{2n}(V)$ is the canonical generator (dual to the fundamental class).

Hence, if the link K is a rational homology sphere, then V has the rational cohomology ring and hence also the rational homotopy type of $\mathbf{P}_n$. That holds in particular for (integral) homology-$\mathbf{P}_n$'s among these hypersurfaces, which are obtained in the obvious manner:

Corollary. *If the singularity link is an integral homology sphere, then the hypersurface V has the integral homology of $\mathbf{P}_n$. In that case, the integral cohomology ring is isomorphic to the truncated polynomial ring* $\mathbf{Z}[u_1, u_n]/(u_1^n - d \cdot u_n, u_1 u_n, u_n^2)$.

To give two simple examples, note that K is an integral homology sphere if f is regular at $\mathfrak{o}$, or has integral coweights $w_j \geq 2$ which are pairwise coprime. In the regular case, the affine variety U is isomorphic to $\mathbf{C}^n$, and V is thus a singular compactification of $\mathbf{C}^n$ with $\mathbf{P}_{n-1}$ as part at infinity. In the second case, the polynomial p is necessarily of the "Pham-Brieskorn" type $p(x) = \sum x_j^{w_j}$ (up to non-zero scalar factors). Of course, there are much more such examples, e.g., those obtained by polynomials satisfying the conditions of [Bri: 2, Satz 1], or by applying the results of Corollary 1 at the end of section 2.

Proof of Theorem 4. The result is an application of the "APL" (Alexander-Poincaré-Lefschetz) type duality theory for singular varieties, as developped by L. Kaup in his papers [$-_1$],[$-_2$], and of J. Milnor's classical results on the topology of hypersurface singularities in his book [SPCH]. Essentially, the proof follows the lines of [BaKp: 3.5]. The affine part $U = V \setminus V_\infty$ is either smooth, or it has an isolated singularity at the origin $\mathfrak{o}$ of $\mathbf{C}^{n+1}$. Hence, the pair (V, V_∞) is a "relative variety with isolated singularities", so relative (Lefschetz type) duality theory yields a long exact sequence

$$0 \to H^1(V,V_\infty) \to H_{2n-1}(U) \to \mathcal{H}_{2n-1,\mathfrak{o}} \to H^2(V,V_\infty) \to H_{2n-2}(U) \to \dots$$
$$\dots \to H_{2n-k+1}(U) \to \mathcal{H}_{2n-k+1,\mathfrak{o}} \to H^k(V,V_\infty) \to H_{2n-k}(U) \to \dots$$
$$\dots \to H_1(U) \to \mathcal{H}_{1,\mathfrak{o}} \to H^{2n}(V,V_\infty) \to H_0(U) \to \mathcal{H}_{0,\mathfrak{o}} \to 0$$

(see [Kp$_1$: Bsp. 2.1, p. 14]), where $\mathcal{H}_{l,\mathfrak{o}}$ is the l-th integral local homology at the affine origin $\mathfrak{o}$. As U is contractible, that long exact sequence yields $H^1(V,V_\infty) = 0$, and it breaks into isomorphisms $H^k(V,V_\infty) \cong \mathcal{H}_{2n-k+1,\mathfrak{o}}$ for $2 \le k \le 2n-1$. As the affine variety U is locally near $\mathfrak{o}$ (and even globally) isomorphic to the open real cone over the singularity link K, there are isomorphisms $\mathcal{H}_{l,\mathfrak{o}} \cong \tilde{H}_{l-1}(K)$ for all l. By [SPCH: Thm. 5.2, p. 45], an n-dimensional hypersurface singularity link is $(n-2)$-connected. This implies that the homology groups $\mathcal{H}_{l,\mathfrak{o}} \cong \tilde{H}_{l-1}(K)$ vanish for $l \ne n$, $n+1$, $2n$. It follows that $H^k(V,V_\infty)$, being isomorphic to $H_{2n-k}(K)$ for $2 \le k \le 2n-1$, vanishes for all k with $0 \le k \le 2n-1$ and $k \ne n, n+1$.

We now consider the long exact cohomology sequence

$$\dots \longrightarrow H^k(V,V_\infty) \xrightarrow{r^k} H^k(V) \xrightarrow{i^k} H^k(V_\infty) \xrightarrow{\delta^k} H^{k+1}(V,V_\infty) \longrightarrow \dots$$

of the pair (V, V_∞). The (reduced) part at infinity $V_\infty := V \cap (x_0 = 0)$ is the linear subspace $(x_0 = x_{n+1} = 0) \cong \mathbf{P}_{n-1}$. Hence, the composition $i^k j^k$: $H^k(\mathbf{P}_{n+1}) \to H^k(V) \to H^k(V_\infty)$ yields an isomorphism of free cyclic groups in all even dimensions $k < 2n$; in all other cases, $H^k(V_\infty)$ vanishes. In particular, the homomorphism $i^k : H^k(V) \to H^k(V_\infty)$ is always split surjective, so the exact cohomology sequence breaks into split short exact sequences

$$0 \to H^k(V,V_\infty) \xrightarrow{r^k} H^k(V) \xrightarrow{i^k} H^k(V_\infty) \to 0$$

for all k. The result describing the cohomology group structure and the first two statements concerning the homomorphisms j^k now follow immediately.

The fact that $j^{2n}\omega^n$ equals $d{\cdot}u_n$ has already been used several times.

To prove the remaining assertion, we look at the subgroups $\operatorname{im} r^k = \ker i^k$ of $H^k(V)$ for $k = n$ and $n+1$, which are the direct complement of $\operatorname{im} j^k$. To show that all cup products with positive-dimensional classes vanish on them, it suffices to check that the products with $j^2\omega$ are trivial, and that $\operatorname{im} r^n$ is self-orthogonal. To that end, we use again singular duality theory: By [Kp$_1$: Thm. 2.1, p. 10], the long exact sequences of (V, V_∞) in cohomology and of (V, U) in homology are joined to a "ladder", i.e., there is a commutative diagram

$$\begin{array}{ccccccc} \ldots \longrightarrow & H^k(V, V_\infty) & \longrightarrow & H^k(V) & \longrightarrow & H^k(V_\infty) & \longrightarrow \ldots \\ & \downarrow & & \downarrow & & \downarrow & \\ \ldots \longrightarrow & H_{2n-k}(U) & \longrightarrow & H_{2n-k}(V) & \longrightarrow & H_{2n-k}(V, U) & \longrightarrow \ldots \end{array}$$

where the vertical arrows are the singular duality homomorphisms. Hence, by the contractibility of U, the image of $H^k(V, V_\infty)$ lies in the kernel of the "absolute" Poincaré duality homomorphism $P_{2n-k} : H^k(V) \to H_{2n-k}(V)$. As in the smooth case, that homomorphism is the cap product with the fundamental homology class (see [BaKp: 2.5]). By the well known relation $(\alpha \cup \beta) \cap [V] = \alpha \cap (\beta \cap [V])$ between cup and cap product, it thus suffices to prove that P_{2n-k} is injective for $k \geq n+2$. Using the results obtained so far, that is an immediate consequence of the fact that $P_{2(n-l)}$ maps $j^{2l}\omega^l$ to $d \cdot i_{2(n-l)}(g_{n-l})$ for $1 \leq l \leq n$, where $g_m \in H_{2m}(V_\infty)$ $(= H_{2m}(\mathbf{P}_{n-1}) \cong H_{2m}(\mathbf{P}_{n+1}))$ for $0 \leq m \leq n-1$ denotes the canonical generator represented by $[\mathbf{P}_m]$. To see that result, put $m := n - l$ and note that one has $j_{2m}(j^{2l}\omega^l \cap [V]) = \omega^l \cap j_{2n}[V] = \omega^l \cap d \cdot [\mathbf{P}_n] = j_{2m} i_{2m}(d \cdot g_m)$. To conclude, we still have to check that the restriction of j_{2m} to the image of the Poincaré homomorphism P_{2m} is injective. That finally follows from the fact that $\operatorname{im} P_{2m}$ lies in the image of i_{2m}, which can be seen by considering the analogous duality "ladder" to the one above, but where the roles of U and V_∞ are interchanged. •

Remarks. (i) Most of the results on the (co-)homology of the varieties V are already contained in L. Kaup's examples [–$_2$: Kor. 3.6, 3.7, pp. 502/503] as special case: put $r = 0$ and interchange the roles of x_0 and x_{n+1}. Note that the variety ${}_0F_{n-1}$ occuring there is just $\mathbf{P}_{n-1}$.

(ii) In the general duality theory of [Kp$_1$], one has to be careful about supports. If no supports are explicitly noted, compact supports are understood in homology and closed supports in cohomology. As these two families agree on the compact varieties V and V_∞, there is no problem in our case.

References

[A'C] N. A'Campo: Le nombre de Lefschetz d'une monodromie. Indag. math. 76 (1973), 113–118

[Bab] I. K. Babenko: On Real Homotopy Properties of Complete Intersections. Izv. Akad. Nauk SSSR, ser. Matem. 43:5 (1979), 1004–1024; Transl.: Math. USSR Izvestija 15 (1980), 241–258

[BaJoMa] M. Barrat, J. Jones, M. Mahowald: Relations Amongst Toda Brackets and the Kervaire Invariant in Dimension 62. J. London Math. Soc. 30 (1984), 553–550

[BaKp] G. Barthel, L. Kaup: Homotopieklassifikation einfach zusammenhängender normaler kompakter komplexer Flächen. Math. Ann. 212 (1974), 113–144

[Bar_1] G. Barthel: Topologie normaler gewichtet homogener Flächen. In: Real and Complex Singularities, Oslo 1976 (Proc. Nordic Summer School in Math., P. Holm, ed.). Sijthoff & Noordhoff, Alphen a.d. Rijn 1977, pp. 99–126

[$-_2$] —: Complex Surfaces of Small Homotopy Type. Singularities (Arcata, 1981), Proc. Symp. Pure Math. **40**,1 (1983), 71–80

[$-_3$] —: Homeomorphy Classification of Normal Surfaces in $\mathbf{P}_3$ with $\mathbf{C}^*$-action. Singularities (Arcata, 1981), Proc. Symp. Pure Math. **40**,1 (1983), 81–103

[Bri] E. Brieskorn: Beispiele zur Differentialtopologie von Singularitäten. Invent. Math. 2 (1966), 1–14

[Bro] W. Browder: The Kervaire Invariant of Framed Manifolds and its Generalizations. Ann. of Math. 90 (1969), 157–186

[ChDi] A.D.R. Choudary, A. Dimca: Singular Complex Surfaces in $\mathbf{P}_3$ Having the Same $\mathbf{Z}$-Homology and $\mathbf{Q}$-Homotopy Type as $\mathbf{P}_2$. Bull. London Math. Soc. 22 (1990), 145–147

[Di_1] A. Dimca: On the Homology and Cohomology of Complete Intersections with Isolated Singularities. Compositio Math. 58 (1986), 321–339

[$-_2$] —: Betti Numbers of Hypersurfaces and Defects of Linear Systems. Duke Math. Journal 60 (1990), 285–298

[$FiKp_1$] K.-H. Fieseler, L. Kaup: Intersection Cohomology of q-Complete

Complex Spaces. In: Proc. Conf. Algebraic Geometry Berlin 1985 (H. Kurke, M. Roczen, ed.), Teubner-Texte zur Math. 92, Teubner, Leipzig 1986, pp. 83–105

[$-_2$] —, —: Vanishing Theorems for the Intersection Homology of Stein Spaces. Math. Zeitschr. 197 (1988), 153–176

[Ha] H. Hamm: Zum Homotopietyp q-vollständiger Räume. J. Reine u. Angew. Math. 364 (1986), 1–9

[HiMa] F. Hirzebruch, K.-H. Mayer: $O(n)$-Mannigfaltigkeiten, exotische Sphären und Singularitäten. Lecture Notes in Math. 57, Springer-Verlag, Berlin 1968

[Ju] P.E. Jupp: Classification of Certain 6-Manifolds. Proc. Camb. Phil. Soc. 73 (1973), 293–300

[Kp_1] L. Kaup: Poincaré-Dualität für Räume mit Normalisierung. Ann. Scuola Normale Sup. Pisa XXVI (26, 1972), 1–31

[$-_2$] —: Zur Homologie projektiv algebraischer Varietäten. Ann. Scuola Normale Sup. Pisa XXVI (26, 1972), 479–513

[KeMi] M. Kervaire, J. Milnor: Groups of Homotopy Spheres I. Ann. of Math. 77 (1963), 504–537

[KuWo] R. Kulkarni, J. Wood: Topology of Nonsingular Complex Hypersurfaces. Adv. Math. 35 (1980), 239–263

[La] K. Lamotke: Die Homologie isolierter Singularitäten. Math. Zeitschr. 143 (1975), 27–44

[Li_1] A. Libgober: A Geometrical Procedure for Killing the Middle Dimensional Homology Groups of Algebraic Hypersurfaces. Proc. Amer. Math. Soc. 63 (1977), 198–202

[$-_2$] —: On the Topology of Some Even-Dimensional Algebraic Hypersurfaces. Topology 18 (1979), 217–222

[LiWo] A. Libgober, J. Wood: Differentiable Structures on Complete Intersections–I. Topology 21 (1982), 469–482

[Mil] J. Milnor: On the 3-Dimensional Brieskorn Manifolds $M(p,q,r)$. In: Knots, Groups, and 3-Manifolds—Papers Dedicated to the Memory of R.H. Fox (L. Neuwith, ed.), Annals of Math. Studies 84, Princeton Univ. Press

1975, pp. 175–225

[MiOr] J. Milnor, P. Orlik: Isolated Singularities Defined by Weighted Homogenous Polynomials. Topology 9 (1970), 385–393

[Mu] D. Mumford: The Topology of Normal Singularities of an Algebraic Surface and a Criterion for Simplicity. Publ. Math. I.H.E.S. 9 (1961), 5–22

[NeRa] W. Neumann, F. Raymond: Seifert Manifolds, Plumbing, μ-Invariant and Orientation Reversing Maps. In: Algebraic and Geometric Topology (Proc. Santa Barbara 1977), Lecture Notes in Math. 664, Springer-Verlag, Berlin 1978, pp. 163–196

[Neu] W. Neumann: Abelian Covers of Quasihomogeneous Surface Singularities. Singularities (Arcata, 1981), Proc. Symp. Pure Math. **40**,2 (1983), pp. 233–243

[Or] P. Orlik: Weighted Homogeneous Polynomials and Fundamental Groups. Topology 9 (1970), 267–273

[$-_2$] —: On the Homology of Weighted Homogeneous Manifolds. In: Proc. Second Conf. on Compact Transformation Groups I, Lecture Notes in Math. 298, Springer-Verlag, Berlin 1972, pp. 260–269

[OrRa] P. Orlik, R. Randell: The Monodromy of Weighted Homogeneous Singularities. Invent. math. 39 (1977), 199–211

[$OrWa_1$] P. Orlik, Ph. Wagreich: Equivariant Resolution of Singularities with C^*-Action. In: Proc. Second Conf. on Compact Transformation Groups I, Lecture Notes in Math. 298, Springer-Verlag, Berlin 1972, pp. 270–290

[$-_2$] —, —: Algebraic Surfaces with k^*-Action. Acta math. 138 (1977), 43–81

[Oka_1] M. Oka: On the Homotopy Type of Hypersurfaces Defined by Weighted Homogeneous Polynomials. Topology 12 (1973), 19–32

[$-_2$] —: On the Cohomology Structure of Projective Varieties. Manifolds-Tokyo 1973, Univ. of Tokyo Press 1975, pp. 137–143

[Sei] H. Seifert: Topologie dreidimensionaler gefaserter Räume. Acta math. 60 (1933), 147–238

[Wa] C.T.C. Wall: Classification Problems in Differential Topology V. On Certain 6-Manifolds. Invent. Math. 1 (1966), 355–374

[Wi] P.M.H. Wilson: On Projective Manifolds with the Same Rational Cohomology as $\mathbf{P}_4$. In: Algebraic Varieties of Small Dimension, Turin 1985. Rend. Sem. Mat. Univers. Politec. Torino, fasc. spez. 1986/87, 15–23

[Žu] A.V. Žubr: Classification of Simply Connected Topological 6-Manifolds. In: Topology and Geometry — Rohlin Seminar (O.Ya. Viro, ed.). Lecture Notes in Math. 1346, Springer-Verlag, Berlin 1988

[PAG] Ph. Griffiths and J. Harris: Principles of Algebraic Geometry. Wiley, New York 1978

[SDM II] V.I. Arnold, S.M. Gusein-Zade, A.N. Varchenko: Singularities of Differentiable Maps, vol. II. Monographs in Math. vol. 83, Birkhäuser, Boston-Basel-Berlin 1988

[SM] P. Orlik: Seifert Manifolds. Lecture Notes in Math. 291, Springer-Verlag, Berlin 1972

[SPCH] J. Milnor: Singular Points of Complex Hypersurfaces. Annals of Math. Studies 61, Princeton Univ. Press 1968

[STH] A. Dimca: Singularities and Topology of Hypersurfaces. Universitext, Springer-Verlag, New York 1992

[TMAG] F. Hirzebruch: Topological Methods in Algebraic Geometry (Translation of "Neue topologische Methoden ..."). Grundlehren 131, Springer-Verlag, Berlin 1966

[TRCS] A. Dimca: Topics on Real and Complex Singularities. Advanced Lectures in Math., Vieweg, Braunschweig 1987

[TSCC] S. Kilambi, G. Barthel, L. Kaup: Sur la Topologie des Surfaces Complexes Compactes. Sém. Math. Sup. 80, Les Presses de l'Univ. de Montréal 1982

Addresses

Gottfried Barthel
Fakultät für Mathematik
Universität Konstanz
Postfach 5560<D 203>
D–78434 Konstanz

Alexandru Dimca
School of Mathematics
and Statistics
Sydney University
Sydney NSW 2006, Australia

E-mail
barthel@cantor.mathe.uni-konstanz.de
dimca_a@maths.su.oz.au

Generic Geometry and Duality

J.W.BRUCE

Liverpool University

1: INTRODUCTION

Singularity theory is not a theory in the usual (axiomatic) sense. Indeed it is precisely its width, its vague boundaries and its interaction with other branches of mathematics, and science, which makes it so attractive. Our subject then rather defies a neat definition. This nebulous nature can, however, lead to identity crises and I then find it useful to think of singularity theory as the direct descendant of the differential calculus. It has, for example, the same concerns with Taylor series, and one can view much current research as natural extensions of problems which our forefathers laboured with, and considered central. The calculus is the tool, par excellence (sadly my only concession to the French language) for studying physics, differential equations in general and the geometry of curves and surfaces. Consequently one might hope that singularity theory will have applications in these fields. Indeed one might judge the vigour of the subject by its success in these areas.

In this paper I will focus on applications to the differential geometry of surfaces in Euclidean and projective 3-space, for three reasons. First its familiarity: this is a classical area known to all professional mathematicians, with a high intuitive content. Secondly any originality here is a good measure of success, this is a well worn path to take. And finally because it illustrates in a quite striking way the benefits which accrue when one moves from studying linear and quadratic phenomena to those of higher order. (I should add that the subject does have some beautiful consequences for the other areas mentioned above, and we refer the interested reader to the books [8],[40]). My main aim is to introduce the subject of generic geometry to those with a solid background in singularity theory, although I have attempted to widen the article's accessibility by giving relevant definitions and references. To get a broader picture of the subject at a variety of levels and from a variety of viewpoints I recommend as further reading [6],[8],[10],[22],[46],[64],[74],[78].

The subject of generic geometry is now over 20 years old, and even given the rather severe restriction of dealing only with surfaces in 3-space there is no possibility of giving a complete survey. Instead I will give some basic background material and then concern myself largely with various duality theorems that have emerged recently. I shall try to give due credit where possible and apologise for any omission. At risk of further embarrassment through an oversight I should like to state my debt to the following who have all made important contributions to this area: Arnold, Banchoff, Callahan, Damon, Gaffney, Goryunov, Giblin, Gibson, Kergosien, Koenderink, Landis, Mather, McCrory, Mond, Montaldi, Platonova, Porteous, Romero-Fuster, Ruas, O.P.Shcherbak, I.G.Shcherbak, Thom, Wall, Zakalyukin. (The list, like our set of references, although running from A through Z, is doubtless incomplete.)

The author was Five College Visiting Professor of Geometry while this paper was written, and also held both a Fulbright and an SERC Award. I am grateful to these institutions for their financial support, and in particular to the Mathematics Department at Mount Holyoke for its kind hospitality.

2: CONTACT

In old fashioned texts on differential geometry much use is made of the notion of contact. Two plane curves X and Y have k-point contact at p if infinitesimally they have k points in common at p. More precisely:

Definition 2.1 Suppose without loss of generality that the point $p = 0$, that X is locally the zero set of a submersion $f : \mathbf{R}^2, 0 \to \mathbf{R}, 0$ and Y is parametrised, again locally, as the image of an immersion $i : \mathbf{R}, 0 \to \mathbf{R}^2, 0$. Then X and Y have k-point contact if and only if $f \circ i : \mathbf{R}, 0 \to \mathbf{R}, 0$ has a Taylor series with leading term cx^k for some (non-zero c).

Of course having made the definition one needs to check a couple of things. Namely that the definition is independent of the choices of f and i, and that we obtain the same answer if we interchange the roles of X and Y. This is not difficult to do.

If X is a hypersurface and Y a curve in say $\mathbf{R}^n$ then clearly the same definition can be made. But even dealing with a pair of surfaces in 3-space we run into problems. Perhaps the first thing to note is that in 2.1 we could, rather pretentiously, have asked that $f \circ i$ have an A_{k-1} singularity, rather than the straightforward condition on the Taylor series, to define k-point contact. Once this observation is made the way forward is clear. So singularity theory's first gift to differential geometry is the correct generalisation of the notion of k-point contact. The key notion here is that of contact equivalence, which is due to Mather,

Definition 2.2 (i) Let $f_i : \mathbf{R}^n, 0 \to \mathbf{R}^p, 0$, $i = 1, 2$, be two smooth germs. We say that f_1 and f_2 are contact or ($\mathcal{K}$-) equivalent if the following holds. There should exist a germ of a diffeomorphism $H : \mathbf{R}^{n+p}, 0 \to \mathbf{R}^{n+p}, 0$ of the form $H(x, y) = (\Phi(x), \Psi(x, y))$ with $\Psi(x, 0) = 0$ and $\Psi(x, f_1(x)) = f_2(\Phi(x))$.
(ii) Let X and Y be submanifolds of $\mathbf{R}^n$ through $p = 0 \in \mathbf{R}^n$. If X is defined

locally by a submersion $F : \mathbf{R}^n, 0 \to \mathbf{R}^{n-r}, 0$, and Y locally by an immersion $\mathbf{R}^s, 0 \to \mathbf{R}^n, 0$ then the *contact* of X and Y at p is the contact ($\mathcal{K}$-) class of the composite $f \circ i : \mathbf{R}^s, 0 \to \mathbf{R}^{n-r}, 0$.

Further details concerning contact equivalence can be found in [77] which I will take as my basic reference for the singularity theory. The definition appears to be rather complicated, but the key consequence to bear in mind is that any two germs which are contact-equivalent clearly determine diffeomorphic varieties. It is important to note however that one gets more information both in the real case and when the dimension of the source is at most the dimension of the target.

Just as for curves there are a couple of things to check.

Proposition 2.3(Montaldi [55]) This contact class does not depend on the choice of f and i. It also yields the same answer, up to *suspension*, if the roles of X and Y are interchanged.

Intuitively the basic invariant is the (complete intersection) singularity $(X \cap Y, p) \subset (X, p)$ or equally $(X \cap Y, p) \subset (Y, p)$. If X and Y are of different dimensions, say $dim(X) = r > dim(Y) = s$ then by suspension above we mean that if we extend the inclusion of $X \cap Y$ in Y trivially to one in $Y \times \mathbf{R}^{r-s}$ the resulting variety is diffeomorphic to that obtained from $(X \cap Y, p) \subset (X, p)$. See [55] for details.

With our new generalisation of contact we can now describe an important idea, first formulated by Gaffney.

Basic Principle 2.4 Given a surface in X in $\mathbf{R}^3$ we can measure how flat or how round that surface is by considering the contact with archetypical flat or round objects.

So for example to measure how flat a surface is we should measure its contact with planes. The basic idea is that the more degenerate the singularity of contact the more plane-like i.e. flatter the surface at the given point. Other aspects of flatness would be measured by considering contact with lines, while roundness is measured by contact with spheres or circles. These archetypical objects we can refer to as *model* submanifolds.

Example 2.5 It is interesting to see what the singularity theory approach tells us in the case of plane and space curves. Here classical differential geometers did have the tools (the standard notion of contact) for the job. For functions of a single variable the $\mathcal{K}$-classes of (non-flat) germs have as representatives $f : \mathbf{R}, 0 \to \mathbf{R}, 0$, $f(x) = \pm x^{k+1}$ labelled A_k singularities. So the notion of having A_k contact is equivalent to the classical notion of

$(k+1)$-point contact. The key new idea is that contact should be measured by a singularity type, and not by an integer.

For plane curves we can consider contact with lines. A line having two point contact with the curve at a point (an A_1 singularity) is simply a tangent line, while 3-point contact (an A_2) corresponds to an inflexional tangent. Note that "most" curves do not have any higher order contact lines, in the old terminology there are no undulations on a generic curve. For contact with circles A_1 singularities correspond to the 2-parameter family of tangent circles. We have A_2 contact if the circle is the circle of curvature, and A_3 contact if the point in question is, in addition, a vertex. (Recall that a vertex is an extremem of curvature, for example the points of intersection of an ellipse with its axes.) Again this a complete list of phenomenon for the generic case. We shall have more to say on what we mean by a generic curve or surface later. It is interesting to note that Apollonius' work on conics essentially centred around the family of distance (squared) functions in the plane. This accounts for what at first sight appears to be a rather strange result of his. Namely that the normal to a conic is perpendicular to the corresponding tangent! For Apollonius a normal is a line joining a point of the plane to an extreme point on the conic of the corresponding distance function. The vertices are singled out here as those points giving rise to singularities on the envelope of normals to the curve; see figure 1.

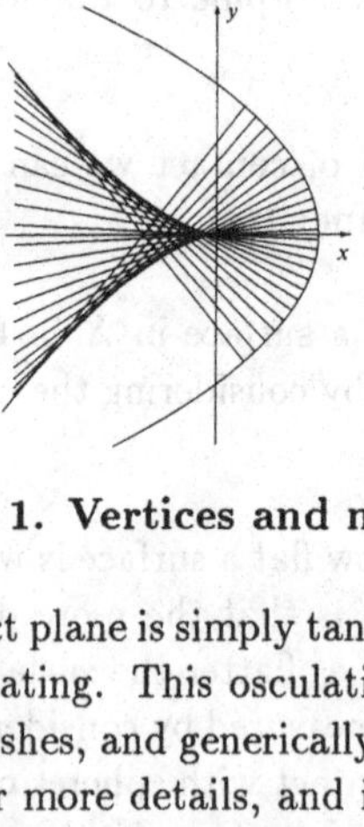

Figure 1. Vertices and normal

For space curves an A_1 contact plane is simply tangent, while an A_2 singularity emerges if the plane is osculating. This osculating plane has A_3 contact at points where the torsion vanishes, and generically this is the most degenerate singularity one expects. (For more details, and in particular for an analysis of the contact between spheres and space curves see [22].)

So in the case of curves the singularity theory approach ties in very well with the classical view, as we would expect. This is also true initially when we move to surfaces. What perhaps is rather astonishing however, is that here the modern approach quickly turns up phenomena which describe important geometric properties of the generic surfaces of which differential geometers of today seem totally ignorant.

Examples 2.6

(1) Contact with Planes

When measuring contact of a surface X with planes clearly we have a singularity precisely when the plane is tangent. One then expects to encounter contact of type A_1 generally, A_2 along a curve of points and A_3 at isolated points for a generic surface. (Roughly speaking there are a 3-parameter family of planes, and the subscript measures the number of conditions one needs to impose to obtain a singularity of this type.) Of course the conditions for contact corresponding to the two types of real A_1's, or Morse singularities have long been recognised, these determine whether a point is elliptic or hyperbolic. Compare also the classical indicatrix of Dupin. Consequently the parabolic points are those whose tangent planes have degenerate contact, in other words they are the flat points on the surface. Again this was classically well-known, although some modern geometers seem to be unaware that the tangent plane at a generic parabolic point meets the surface in a cusp (figure 2).

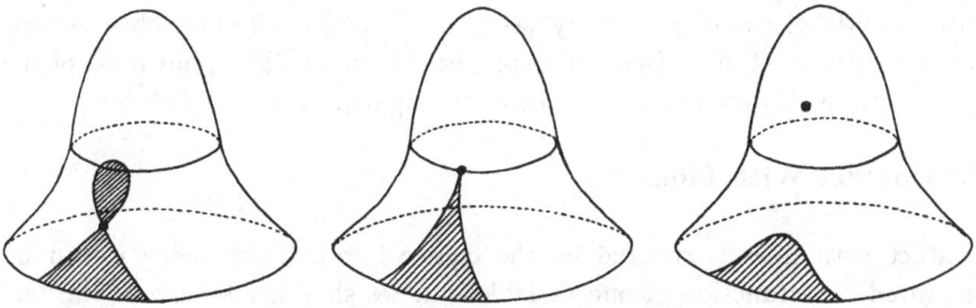

Figure 2. Tangency singularities

The A_3 singularities occur at even flatter points of the surface, here then the plane meets the surface in curve given in local co-ordinates by $x^2 \pm y^4 = 0$. (Note that in one case the tangent plane and surface meet (locally)in only a point; the notion of $\mathcal{K}$-equivalence distinguishes this contact from that at a standard elliptic point.) These special parabolic points have been labelled cusps of Gauss (or swallowtail points) in the literature. The first label should not mislead you however, since these points, and their basic properties, seem to have gone unnoticed until the mid 1970's. In retrospect it may seem rather surprising that they had failed to appear by the mid 1870's. For references here see [47],[10],[12].

(2) Contact With Spheres

Generally spheres tangent to a surface have contact of type A_1. At a general point (the non-umbilics as it turns out) there will be two spheres which have A_2 contact (or worse) with the surface, these are the spheres of curvature with centres the given centres of curvature. These centres of curvature come with corresponding orthogonal principal directions down on the surface. All this is classical. There is then a curve of points on the surface where one of these spheres of curvature has A_3 contact with surface. These have been labelled by Porteous the ridge points of the surface, and they are of great interest geometrically. Again it seems rather astonishing that they are unknown to differential geometers. At special ridge points the contact will be of type A_4. As before the more degenerate the singularity the more "round" the surface, in one direction at least. We have corank 2 contact with the surface precisely at umbilics, and generically one of the two real types of D_4 singularity occurs. Here the intersection of the surface and unique sphere of curvature is locally diffeomorphic to three (concurrent) distinct lines, two of which may be complex. Of course umbilics are old chapeau to geometers, since they are quadratic phenomena. But the fact that there are two generic types appears to have emerged from the singularity theory viewpoint, again because the distinction depends on higher order phenomena. (There is a further, and not unrelated, classification according to the configuration of the lines of curvature in a neighbourhood of the umbilics. This was well known by traditional geometers, and goes back at least as far as work of Darboux, although it was only recently discussed rigorously, see [43] and [21].) The first observations concerning contact of surfaces with spheres are due to Thom, but most of the basic work in this area is due to Porteous [63],[64].

(3) Contact With Lines

Contact with lines is covered by the classical techniques, since it can be measured by a function of one variable, but we shall need some of the terminology later on. One expects contact of type A_1, A_2, A_3, A_4 for a generic surface. Singular contact corresponds to tangent lines, which are asymptotic if the contact is of type A_2 or worse (which we write as $A_{\geq 2}$). Contact of type $A_{\geq 3}$ occurs at the so-called flecnodal points, and A_4 at special flecnodal points known as biflecnodes, though it might be better to call them undunodal. This terminology is explained as follows. At a hyperbolic point of a surface the tangent plane meets that surface in a node (a node is an A_1 with a pair of real branches). The tangents to these branches are the asymptotic lines, and these generally have 3 point contact with the surface at the tangency point, 2 from one branch and 1 from the other. We have A_3 or 4-point contact if one of the branches of the node has an inflexion, hence flecnodal. If it has an undulation we have A_4 contact, hence undunodal. (The disadvantage of the term biflecnode is that it is natural to interpret this as a node both of whose branches have inflexions). See [5],[10],[46],[47],[58],[59],[60],[72].

Remarks 2.7 (a) How does one account for these enormous gaps, especially those concerning (2), in the traditional approach? It really is simply that classical differential geometers only have linear tools at their disposal, so they

can only study linear (and hence quadratic) objects. The modern theory of singularities allows us to study non-linear or higher order phenomena, and discover some important and central facts concerning the geometry of curves and surfaces in Euclidean and projective 3-space. Why else, for example, could the determination of the local structure of the tangent developable of a space curve at a zero of torsion have been delayed by 200 years? (See [30].)
(b) It is worthwhile stating that although the fundamental theorems of differential geometry assure us that any geometric condition we emerge with can be interpreted in terms of the first and second fundamental forms, it is largely pointless to seek such descriptions. One is simply attempting to describe higher order phenomena in linear and quadratic terms. Why describe colours in words when you have a canvas and full palette at your disposal?

3: VERSALITY AND DISCRIMINANTS

At some stage we need to justify the repeated use of the word generic above. What we have in mind is pretty much the dictionary definition i.e. related or characteristic of a whole group or class. The first thing to note is that what is or is not generic depends on the type of geometry one is considering. So the unit circle in the plane is very generic if one is considering contact with lines, since each tangent line has A_1 contact. On the other hand it could not be less generic when one is studying contact with circles! So it is more accurate to describe generic properties of surfaces, and then a generic surface is one satisfying the given property. The precise definition hinges on transversality results, one of which I will describe in section 8. However a consequence is that a residual and often open and dense set of surfaces satisfy any "generic" condition. This is true for example of the space of proper embeddings of surfaces in 3-space which have contact of type A_k, $1 \leq k \leq 3$ with their tangent planes.

We now take a slightly different viewpoint. So far we have considered the contact between individual tangent planes and the surface. Given such a contact singularity we note that by moving the tangent plane about we obtain an "unfolding" of the singularity. By an unfolding of a germ $f : \mathbf{R}^n, 0 \to \mathbf{R}^p, 0$ we simply mean a family of mappings $F : \mathbf{R}^n \times \mathbf{R}^s, 0 \to \mathbf{R}^p, 0$ with $F(x, 0) = f(x)$. The s variables used to vary the original germ are referred to as unfolding parameters. An unfolding F is versal if every unfolding G of f is contained in F. In other words for each parameter u the mapping $G_u = G(-, u)$ is $\mathcal{K}$-equivalent to some $F_{\alpha(u)} = F(-, \alpha(u))$ for some smooth α, and this happens in a smoothly parametrised way. Of course the existence of a versal unfolding is far from clear. Roughly speaking a given singularity has a versal unfolding if and only if the corresponding complexified variety has an isolated singularity. (For details we refer the reader to [77]; in the terminology there we are actually discussing $\mathcal{K}_e$-versal unfoldings.)

General principles tell us that we expect most unfoldings to *versally unfold* their constituent singularities, at least when those singularities have no moduli. For example it turns out that most (in fact an open and dense set of) surfaces have tangency singularities of type A_k, $1 \leq k \leq 3$, and they are versally unfolded by the contact of the surface with the 3-parameter family of all planes. Given an unfolding of a singularity the fundamental geometric invariant is the discriminant, in other words the set of parameter values corresponding to singular fibres. So with reference to the unfolding F we are considering the set

$$\{u \in \mathbf{R}^s : F(x,u) = \partial F/\partial x_1 = \ldots = \partial F/\partial x_n = 0 \text{ for some } x\}.$$

If the unfolding is versal the diffeomorphism type of the discriminant depends only on the singularity and the dimension of the unfolding space. As an elementary example we can consider the unfolding of an A_k singularity $f : \mathbf{R},0 \to \mathbf{R},0$ given by $f(x) = x^{k+1}$. A versal unfolding here is given by $F : \mathbf{R} \times \mathbf{R}^{k+1},0 \to \mathbf{R},0$, where $F(x,u) = x^{k+1} + u_1 x^k + \ldots + u_{k+1}$. The discriminant in this case can be thought of as the set of polynomials of degree $k+1$ with repeated roots. (Actually using the standard Tschirnhausen transformation we can get rid of the x^k term, and obtain a versal unfolding with the minimal number of unfolding parameters.)

What do these discriminants tell us about the geometry of the surfaces for the families of contact discussed above?

Examples 3.1

(1) Planes

The discriminant here is a venerable object indeed, it is the dual of the surface; for a plane meets the surface in a singulartity if and only if it is tangent at the corresponding point. For a generic surface the contact singularities we discussed above will be $\mathcal{K}$-versally unfolded by the family of planes. Since we have local models for the discriminants of versal unfoldings of low-dimensional singularities we can describe the *local* appearance of the dual of a generic surface. For unfoldings of 3 dimensions the discriminant of an A_1 singularity is a piece of smooth surface. For an A_2 we obtain the product of a cusp with a line and for an A_3 the swallowtail pictured in figure 3. Hence the terminology swallowtail point. This already allows us to deduce some useful information about the A_3 point on the surface. For the swallowtail point lies in the closure of the line of self-intersections of the discriminant. However a self intersection point of the dual corresponds to a bitangent plane, so this tells us that there are planes which are bitangent to the surface at points arbitrarily close to these A_3 or cusps of Gauss. Indeed for a generic surface this property characterises these points. (For this and many other characterisations see [10], [47].) Of course there are various types of "generic" self-intersections of the discriminant possible. They are also pictured in figure 3; we leave their geometric interpretation to the reader.

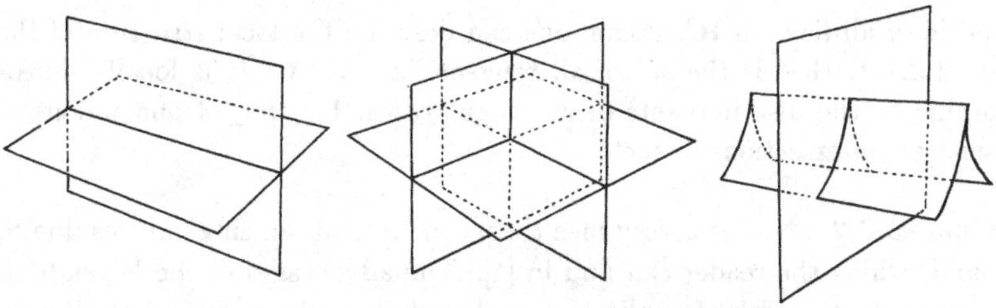

Figure 3a. Self-intersections of discriminants

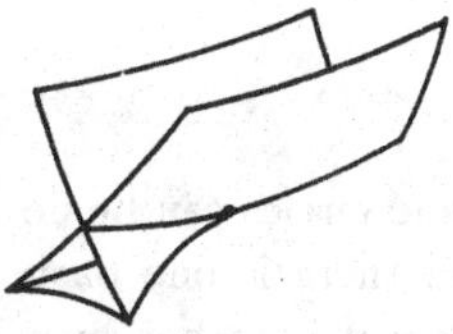

Figure 3b. Swallowtail

(2) Spheres

The discriminant for contact with spheres lives in a 4-dimensional space. It is of some interest, but it is easier to visualise the discriminant of a related family. So instead of considering the family of all spheres we shall consider those of a fixed radius, say $t > 0$. We can parametrise these spheres by their centres, that is by $\mathbf{R}^3$. It is clear that the discriminant here consists of the distance t parallels of X, that is the locus of points obtained by moving a distance t out along the normals, in both directions. If we view X as an initial wavefront, in a homogeneous medium which propogates at unit speed, these are just the time t subsequent wavefronts. The same local models apply here as in (1) above. We deduce that a generic wavefront can have cuspidal edges and swallowtails, together with the various self-intersections pictured in figure 3. Note that these results do not describe the local structure of all wavefronts corresponding to an initial generic surface X. To do this one needs to consider all values of t simultaneously, in other words the projection from the full discriminant of all tangent spheres to the line given by assigning to each sphere its radius. This important problem, of describing wavefront evolution from generic initial fronts was solved by Arnold [2] and Zakalyukin [85].

(3) Lines

Here the discriminant again lives in a 4-dimensional space, namely the Klein quadric of all lines in $\mathbf{R}^3$. Again one can describe the local structure of the discriminant, that is the set of all tangent lines to X. It is locally diffeomorphic to the discriminants of A_k singularities, $1 \leq k \leq 4$ and unions of transverse intersections of such.

Remarks 3.2 There is a very nice theory of Legendrian singularities due to Arnold, which the reader can find in [4]. The advantages of the Legendrian formalism is its width of application, although if one is concerned at all with verifying transversality conditions it is generally true that each different example requires a different transversality result. Very roughly speaking one can think of Legendrian singularities as those connected with discriminants. Concrete examples include wavefronts, duals and envelopes (see [14]).

(4) Circles

Contact between a surface and circles can be described with the naive notion of contact given in 2.1, and there is one basic classical result here due to Meusnier, 3.3(i) below. Again it is rather surprising that there seems to be nothing else in the classical literature beyond this, although the tools for further exploration were available. The question of contact between a surface and circles was raised by Porteous in [64], and the following theorem is given there. A formal proof together with a number of fascinating further results appear in a paper of Montaldi published in 1986. First some notation. Let X be the surface in $\mathbf{R}^3$, p a point of X, and u a tangent vector at p. The Meusnier sphere M_u is the sphere through p centred at the centre of curvature of the curve obtained by slicing X by the plane containing the normal and u.

Theorem 3.3 (Montaldi-Porteous) (i) A circle tangent at p to u (a u-circle) has $A_{\geq 2}$ contact at p if and only if it lies in M_u.
(ii) If u is not principal there is a unique u-circle with contact of type $A_{\geq 3}$. If p is not an umbilic, and u is principal then
(a) If u is not associated to a ridge there is no u-circle of type $A_{\geq 3}$
(b) If u is associated to a ridge then every u-circle is of type $A_{\geq 3}$.

For a complete statement of the results concerning contact with circles and their proofs see [57]. Of course since differential geometers did not have the notion of ridge one cannot expect the result to have appeared in this form in days of old. On the other hand (iii) gives a characterisation of the missing notion of ridge point which would have been accessible since it uses only the notion of contact between a curve and surface.

4: MAPPINGS

It is often useful to bundle our model submanifolds, for example our planes or our lines, into certain families. We then consider the simultaneous contact between the surface and members of the given family, for example all planes parallel to a given one. This involves a different set of ideas.

Example 4.1
(i) The function $h_0 : \mathbf{R}^3 \to R^2$, given by $h_0(x) = x_1$ when restricted to X measures the contact between X and all planes of the form x_1 =constant.
(ii) The function $d_0 : \mathbf{R}^3 \to \mathbf{R}$ given by $d_0(x) = x_1^2 + x_2^2 + x_3^2 = ||x||^2$ when restricted to X measures the contact between X and all spheres centred at the origin.
(iii) The mapping $\pi_0 : \mathbf{R}^3 \to \mathbf{R}^2$ given by $\pi_0(x) = (x_1, x_2)$ when restricted to X measures the contact between X and all lines parallel to the x_3-axis.

Since we now need to keep track of all the fibres of our mappings we need to change our equivalence relation to $\mathcal{A}$-equivalence. Here two germs are regarded as equivalent if and only if a change of co-ordinates in the source and target takes the one to the other. More precisely:

Definition 4.2 Let $f_i : \mathbf{R}^n, 0 \to \mathbf{R}^p, 0$, $i = 1, 2$, be two smooth germs. We say that f_1 and f_2 are right-left or ($\mathcal{A}$-) equivalent if the following holds. There should exist germs of a diffeomorphisms $\Phi : \mathbf{R}^n, 0 \to \mathbf{R}^n, 0$ and $\Psi : \mathbf{R}^p, 0 \to \mathbf{R}^p, 0$ such that $\Psi \circ f_1 \circ \Phi(x) = f_2(x)$.

This is in many ways a more natural equivalence relation to work with than $\mathcal{K}$-equivalence. The extra detailed information one obtains is paid for by a corresponding theory, which is generally considerably more complicated than that associated with $\mathcal{K}$-equivalence. Again we refer to [77] for details. Relatively recent advances in our understanding of classification and determinacy techniques, and topological equivalence have made applications possible which would otherwise have been hopeless. See [26], [27],[31]-[34],[36]-[39],[62] and [77].

Of course there is nothing sacrosanct about the choices of h_0, d_0, π_0 above. If we allow any choice of plane in (i), any centre for our spheres in (ii), any choice of line in (iii) we actually generate families

(i) $h : X \times S^2 \to \mathbf{R}$, $h(x, a) = x \cdot a$, the usual dot product. We refer to this as the family of height functions.

(ii) $d : X \times \mathbf{R}^3 \to R$, $d(x, u) = ||x - u||^2$. This is the family of distance squared functions.

(iii) In the last case we need to work a little harder. We let B denote the tangent bundle to S^2 i.e. $B = \{(a, b) \in \mathbf{R}^3 \times S^2 : a \cdot b = 0\}$ and define a map $\pi : X \times S^2 \to B$ by $\pi(x, b) = (x - x \cdot b, b)$. The map $\pi_b(x) = \pi(x, b)$ has target the tangent space $T_b S^2$, it is simply projection in the direction b to the plane orthogonal to b. We shall rather sloppily replace the range of π

by $\mathbf{R}^2$, a sin we shall repeat in other contexts below. The excuse is that our concerns are almost exclusively local, and locally we can write π in this form by composing its first component with projection to a suitable fixed plane. Of course the result is certainly $\mathcal{A}$–equivalent to the original. Indeed the $\mathcal{A}$-type of any linear projection is determined by its kernel.

We shall see that this bundling of fibres allows us to obtain information on the Gauss map of X in case (i), the focal and symmetry sets in case (ii), the outlines of X and its so-called viewsphere in case (iii). We will explain the terminology below, but before we proceed I want to augment this list with two further families.

(iv) We define $c_0 : \mathbf{R}^3 \to \mathbf{R}^2$ by $c(x) = (x_1^2 + x_2^2, x_3)$. When restricted to X this measures contact between X and all circles centred on, and orthogonal to the line L_0 given by $x_1 = x_2 = 0$. If K denotes the set of all oriented lines in $\mathbf{R}^3$, then each line L determines such a map, so we have a family $c : X \times K \to \mathbf{R}^2$ with c_L measuring the contact between X and circles centred on, and orthogonal to, L.

(v) We define $f_0 : \mathbf{R}^3 \to \mathbf{R}^3$ by $f(x) = (x_1^2, x_2, x_3)$. When restricted to X this measures contact between X and all pairs of points "centred on" the plane $x_1 = 0$. If P denotes the set of all oriented planes in $\mathbf{R}^3$, then each plane determines such a map, and we have a family $f : X \times P \to \mathbf{R}^3$ with f_π measuring the contact between X and points centred on the plane π. We can expect the singularities of the family f_π to give us some information on the local and semi-local reflectional symmetry of X in π.

Remarks 4.3 In a sense which I will try to make more precise later on, simultaneous contact does not play any real role in (i) and (ii). It is rather the fact that the one value of the parameter, for example a choice of normal in (i) or centre in (ii), can correspond to more than one singular contact which brings new benefits.

5: $\mathcal{A}$-CLASSIFICATIONS

Just as before we can now list singularities of low codimension, this time up to $\mathcal{A}$ -equivalence, and interpret them for the various families above. As mentioned above $\mathcal{A}$-classifications are considerably harder to produce. The style of the classifications vary according to the dimensions of the source and target. We consider first target dimension 1.

5.1 Functions

The first remark is that when dealing with functions there is initially no difference between the $\mathcal{A}$ and $\mathcal{K}$-classifications. Indeed perfectly adequate classifications are possible using $\mathcal{R}$-equivalence, that is changes of co-ordinates in the source alone, which is technically much simpler than $\mathcal{A}$-equivalence to deal with. As a consequence when looking at the individual mappings there is nothing new to report beyond what we learnt above. The advantages, for examples (i) and (ii), of working with the finer equivalence relation will show up in the next section where we consider unfoldings.

5.2 Mappings from the plane to the plane

For mappings from the plane to the plane there have been extensive classifications done [38], [62],[65]. Here are some of the germs which arise in families with few parameters.

(a) Stable maps: for mappings from a surface to a plane the stable mappings were classified by Whitney in one of the great seminal papers of the subject [79]. A stable map has only fold and cusp singularities, in local co-ordinates $(x_1, x_2), (x_1, x_2^2), (x_1, x_2^3 + x_1x_2)$ respectively.

(b) Codimension 1 singularities: here we have

$$(x_1, x_2^3 - x_1^2x_2), (x_1, x_2^3 + x_1^2x_2), (x_1, x_2^4 + x_1x_2)$$

labelled beaks, lips and swallowtail respectively. Note that up to $\mathcal{K}$-equivalence the first two and the cusp are indistinguishable. Indeed one can think of beaks and lips as non-transverse cusps. They form part of an infinite family of maps all $\mathcal{K}$-equivalent to (x_1, x_2^3), namely those equivalent to $(x_1, x_2^3 \pm x_1^kx_2)$ which we label 4_k. Although these germs are all $\mathcal{K}$-equivalent they are distinguished, for example, by the fact that their critical sets have (plane curve) singularities of type A_{k-1}. The finer classification is capturing the failure of transversality of the jet-extension to the singular subset of the space of 1-jets. Here is an example of a situation where "simultaneous" contact is giving information. The swallowtail on the other hand is reflecting higher order contact, although again there are an infinite family of $\mathcal{A}$-types amongst germs $\mathcal{K}$-equivalent to (x_1, x_2^3), of which the swallowtail is merely the simplest.

(c) It is worthwhile noting that multi-germs of mappings are important when studying the differential geometry of surfaces in 3-space. The only stable multi-germs here occur as pairs of transverse folds, that is fold points having the same critical value with images of their singular sets transverse. Of course in higher codimension these fold curves may be tangent. Indeed we shall refer to k-point contact folds, by which we mean that their critical sets have A_{k-1} tangency. It turns out, as seems reasonable, that this determines the bi-germ up to $\mathcal{A}$-equivalence.

(iii) Applications to projections to planes

These singularities have direct geometrical interpretations when applied to projections. So generally we see a cusp on the surface at a point where one of the asymptotic directions is parallel to the viewing direction. A beaks/lips or worse corresponds to asymptotic directions at parabolic points. Similarly swallowtail singularities correspond to asymptotic directions at flecnodes. The various multi-germs also have their interpretations. For example tangent folds correspond to secants joining pairs of points sharing a tangent plane. Note that there are no corank 2 singularities.

(iv) Applications to circle mappings

Here some of the singularities have an immediate interpretation. Swallowtails generally simply correspond to 4-point contact circles, triple fold points to tritangent circles. We shall deal with the more interesting singularities, like beaks/lips or tangent folds in Sections 6 and 8.

5.3 Mappings from a surface to 3-space

Again there are lists of singularities of mappings $\mathbf{R}^2, 0 \to \mathbf{R}^3, 0$ see for example [15],[52], [62]. The second of these references is the most useful for our needs.

(a) Stable maps: The stable mappings were again identified by Whitney [80]. Such a map is an immersion at most points, with isolated crosscaps, a cross-cap having local normal form (x_1, x_2^2, x_1x_2). The image can self-intersect transversely and have isolated triple points. Of course these forms were well known to nineteenth century geometers.

(b) Higher codimension singularities. We shall only be interested in germs which are $\mathcal{K}$-equivalent to $(x_1, x_2^2, 0)$, for these are exactly the singularities which arise from the folding maps. There is a very nice connection here with singularities of functions on manifolds with boundary which is discussed below. The classification of Mond uses this connection, and includes the following simple singularities:

$B_k^\pm$ with normal form $(x_1, x_2^2, x_1^2x_2 \pm x_2^{2k+1})$, $k \geq 1$.
$S_k^\pm$ with normal form $(x_1, x_2^2, x_2^3 \pm x_1^{k+1}x_2)$, $k \geq 2$.
$C_k^\pm$ with normal form $(x_1, x_2^2, x_1x_2^3 \pm x_1^kx_2)$, $k \geq 3$.

(c) Again there are a whole multitude of multi-germs to consider. Because of the form of the folding map we shall only be interested in bi-germs, indeed in bi-germs of immersions. Mond showed that the type of such a bi-germ is determined by the contact of their images, that is the $\mathcal{K}$-type of their intersection.

(v) Applications to folding maps

Again the various degenerate singularities have nice geometric interpretations.

(i) The map f will have a degenerate singularity, that is a B_1 or worse at a point $p \in X$ if and only if the folding takes place across a plane containing the normal and principal direction at p. In other words since the singularities of the folding map are measuring the reflective symmetry of the surface it is most (reflectionally) symmetric at a point p across one of these *principal planes.* We shall delay interpreting the more degenerate singularities until Sections 6 and 8. We do however note that there is a natural surface singularity corresponding to these foldings. For reflecting X in the plane π we obtain a second smooth surface X_π, and if π contains the normal at p then X and X_π are tangent at p. Their tangency is measured as in Section 2. Indeed the resulting singularity is, in a natural way, Z_2-equivariant, and is directly related to the singularity emerging in the classification of Mond. See [29].

(ii) Tangent images occur when we fold in a plane which corresponds to a non-local reflectional symmetry of X. In other words if there is a pair of points p, q on X which are interchanged along with their tangent plane by such a reflection.

Remark 5.4 We should remark that the approach described in the papers cited above is not the only line one can take to obtain classification results. Arnold and his school have, for example, made many interesting classifications of projections of surfaces using rather different techniques. See the paper [5], which was the first to be published giving a complete list of the singularities which arise from orthogonal projections of generic surfaces in $\mathbf{R}^3$. (The equivalence relation used looks a little finer than $\mathcal{A}$-equivalence, although they turn out to coincide [42].) Here the basic intuitive idea is to consider the set of lines tangent to the surface X, which, as we stated above is locally described by discriminants of A_k-singularities in $\mathbf{R}^4$. A given direction of projection corresponds to a 2-parameter family of such lines, and the outline can be thought of as the corresponding plane section of this discriminant. More degenerate views correspond to more degenerate sections. The classification uses similar techniques to those employed for studying wavefront evolution. See for example [2],[5],[41],[42[,[59],[60].

6:VERSALITY AND BIFURCATION SETS

For $\mathcal{K}$-equivalence the key invariant for the family was the discriminant. We saw how for our families of model submanifolds this subset of the deformation space has interesting geometric interpretations. Moreover generic surfaces generally give rise to versal deformations, and we then have local universal models for these discriminants. When we consider mappings, rather than varieties, and work with $\mathcal{A}$- rather than $\mathcal{K}$-equivalence there is a similar, if more complicated, theory of deformations or unfoldings. The relevant subset of the unfolding space (for example the sphere S^2 when considering height functions) is the *bifurcation set.*

Definition 6.1 Let $G : X \times U \to Y$ be a family of mappings, $G_u : X \to Y$.

The set $\{u : G_u \text{ is not stable}\}$ is the *bifurcation set* of G, and is denoted Bif(G).

Again detailed definitions would take us too far afield. A map G_u is stable if any sufficiently small perturbation is indistinguishable from it. More accurately, is equal to it after composing with diffeomorphisms in the source and target. Basically then Bif(G) is the set of parameter values where something new is happening. Of course in low dimensions stable mappings are dense, so Bif(G) is generally a proper subset of U which, if G has some geometric significance, carries useful information about X. One can usually describe a finite number of elementary ways in which a map can fail to be stable and Bif(G) will have a component for each such "catastrophe". Again we promise to shed further light on this terminology below. Since Mather's characterisation of stable mappings is essentially local in nature the bifurcation sets are local or at least semilocal in the same sense. We next seek to interpret the bifurcation sets for the families (i)-(v) above.

6.2 Functions

For functions, that is maps with target the real line, stable means Morse, and a function can fail to be Morse in two ways, namely
(a) via a degenerate singularity or
(b) through two critical points having the same critical value.

In an unfolding the smooth parts of these sets are the stratum of functions giving A_2 points, and the set of parameter values corresponding to two A_1 singularities at the same level, which we label $2A_1$.

(i) Contact with planes

Here (a) degenerate contact corresponds to critical values of the Gauss map, while (b) corresponds to normals to bitangent planes.

(ii) Contact with spheres

This time (a) corresponds to centres of spheres of curvature of X, in other words the focal set of X. If we think of R^3 as a homogeneous medium and X as a wavefront then this is simply the caustic of X . On the other hand (b) corresponds to centres of bitangent spheres, the so-called symmetry set, which is of some interest in pattern recognition, see [25],[46]. The focal set is a well-studied object, and has a particularly interesting structure at the centres of curvature corresponding to umbilics. At such points the two sheets of the focal set come together, and the contact of the surface with the corresponding sphere is one of the two real types of D_4. The local form of the bifurcation sets are illustrated in figure 4.

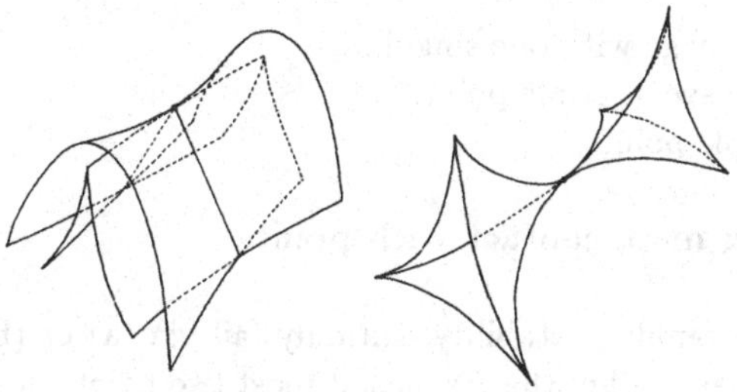

Figure 4. Focal sets of umbilics

6.3 Mappings $X \to \mathbf{R}^2$

Given our list of stable phenomena a little experimentation shows that stability can fail in five elementary ways.
(a) Through contact of type $A_{\geq 3}$. The corresponding subset of the parameter space is referred to as the *swallowtail set.*
(b) Through A_2 contact which fails to be a cusp, this yields the *beaks/lips set.*
(c) Via a pair of fold points on X whose critical sets have tangent images, the *tangent fold set.*
(d) Via a triple of fold points with coincident images, the *triple fold set.*
(e) Via a cusp and fold point with coincident images. Lacking imagination we label this the *cusp+fold set.*
(f) Via a corank 2 singularity.

The more numerate will remark that this is a list of 6 not 5 catastrophes. In fact corank 2 singularities are not "elementary" in the sense that they are codimension 2 and not 1. They often correspond to interesting geometric phenomena however, and are worthwhile adding to our list.

(iii) Contact with lines

The set of critical values here is simply the outline of the object when viewed from the given direction. The bifurcation set on the other hand consists of the set of viewing directions where this ouline changes. The sphere together with this subset is referred to as the *viewsphere*, and carries a great deal of information about the surface (in fact usually too much to handle comfortably).

6.4 Mappings $X \to \mathbf{R}^3$

There are 5 ways in which stability can fail.
(a) Through singularities worse than a cross-cap.
(b) Via tangent images.

(c) Via a double image with one singular.
(d) Via a non-transverse triple point.
(e) Via a quadruple point.

(iv) The folding map, contact with points

Note that for the family f stability can only fail via (a) or (b) above since the map f_0, and hence all of the f's, has at most two points in any fibre. The bifurcation set for the family of folding maps turn out to be rather interesting.

Proposition 6.5 [29] That part of the bifurcation set corresponding to degenerate singularities is the closure of the set of tangent planes to the focal set of X, while the subset corresponding to tangent images is the closure of the set of tangent planes to the symmetry set.

This follows directly from the interpretations of the singularities of the folding map given in Section 5.

(v) Contact with circles

Here we return to the case of mappings from a surface to the plane. A certain subset of the bifurcation set for our family of circle maps has a similar interpretation.

Proposition 6.6 [20] (i) For the family of circle maps the beaks/lips stratum corresponds to lines that are tangent to the focal set of X.
(ii) The tangent fold stratum consists of lines which are tangent to the symmetry set.
(iii) The corank 2 stratum corresponds to the normals of X.

As we remarked above the other strata in the bifurcation set have straightforward, and less interesting, interpretations.

Remarks 6.7

(a) This raises the possibility of studying the differential geometry of very singular objects, which focal and symmetry sets clearly are. It seems to be true that the differential geometry of these (and many other singular sets) is remarkably like that of a smooth surface. See [17], [20], [29]. The interesting geometric phenomena tend to be associated to the points on the surface corresponding to *both* the focal and the symmetry sets, that is ridge points. (This fact follows by considering the bifurcation set of an A_3 singularity, which is made up of a cuspidal edge of A_2's and a cuspidal tangent plane of $2A_1$ points.)

(b) It appears that the folding maps f, which are dual to the distance squared functions d, actually have some advantages over the latter when it comes to revealing interesting geometry of X. In particular they yield a natural blow-up of the geometry at an umbilic [29]. For at an umbilic the surface is locally symmetric across every plane containing the normal, in other words the bifurcation set contains a corresponding line of planes of symmetry. The point is that this is, in its own right, an interesting way to study the geometry of the surface.

(c) We have not yet taken advantage of the local models for the bifurcation sets which singularity theory provides. We shall give one application of this type in section 7.

7: DUALITY

One of the great advances in geometry was the early realisation that there is much profit in studying spaces of objects other than points. So when studying a surface X in projective 3-space, we have seen that it is useful to consider the set of tangent planes to X in the dual space. This dual reveals some interesting geometry concerning X. We start this section by proving a result concerning the contact of surfaces and their duals, which again seems surprisingly recent. This result exhibits a useful idea in its simplest (and its original) context. It is the idea, rather than the proposition itself, which is the key to proving most of the duality results below.

Proposition 7.1 [13] Let X, Y be smooth surfaces in P^3 tangent at p with the contact of finite $\mathcal{K}$-type. Suppose further that that p is not a parabolic point on either X or Y. Then the duals X^* and Y^* are tangent at the dual point p^* and the contact is, after complexifying, the same as that between X and Y at p.

Proof After a certain amount of fiddling one can assume that the contact between X^* and Y^* is also $\mathcal{K}$-finite and suppose that X, Y are analytic near p. (If you feel unhappy at this sleight of hand just place these in the set of hypotheses.) Next note that both X^* and Y^* are smooth at the corresponding dual point p^*. We now embed X in an analytic family X_t with $X_t \cap Y$ a $\mathcal{K}$-versal unfolding of the singularity $X \cap Y \subset Y$. We similarly embed Y^* in a smooth family Y_s^* with $X^* \cap Y_s^*$ yielding a versal unfolding of $X^* \cap Y^*$. Here t and s belong to some smooth deformation spaces T and S respectively.

We define Y_s to be $(Y_s^*)^*$, this gives a deformation of Y, and we consider $X_t \cap Y_s$ as a versal unfolding of $X \cap Y$ and $X_t^* \cap Y_s^*$ as a versal unfolding of $X^* \cap Y^*$. In other words we have versal unfoldings of the original contact singularity and its dual with the same parameter space $S \times T$. We now consider the discriminant in the two cases. The crucial observation is that X_t and Y_s are tangent at q if and only if X_t^* and Y_s^* are tangent at the dual point q^*. As a consequence the discriminants of these two unfoldings coincide. It follows from a beautiful theorem of Wirthmuller [81] that the two singularities are $\mathcal{K}$-equivalent.

Remarks 7.2 (a) It is worthwhile isolating the ingredients in this proof. Basically all we needed above was (i) the dual of a dual is the original object and (ii) the fact that two surfaces are tangent if and only if their duals are. Note that these key geometric facts are really very easy to check, although the conclusion would be rather difficult to verify directly.
(b) This result, which clearly generalises to the case of arbitrary hypersurfaces, first arose in connection with work on geometric optics, by Giblin, Gibson and the author. A bare-hands proof for the case of simple singularities was given in [24], while the proof above appeared in [13]. See [16] for further applications of this idea .
(c) Even in the case of curves this result appears to have been unknown until its (straightforward) proof appeared in [23]. The curve case was rediscovered in [72].

We now state some duality results which are established by this type of argument. For the family of circle maps we have

Proposition 7.3 [20] (1) The circle map C_L with axis L has a 4_k singularity at $p \in X$, $k \geq 2$, if and only if the circle with this axis through p has tangent a principal direction, and L is tangent to the focal set at the corresponding centre of curvature and has k-point contact i.e. an A_{k-1} singularity.
(2) The circle map C_L with axis L has a k-point tangent fold singularity corresponding to the points p, $q \in X$, if and only if there is a sphere bitangent at p and q with centre a say, and L is tangent to the symmetry set at a and has k-point contact i.e. an A_{k-1} singularity.

So we can obtain information about contact between the (very singular) focal set and lines by considering the family of circle maps on the surface X.

This Proposition allows us to extend Montaldi's results in various ways.

Addition to Montaldi's Theorem. 7.4 Let v be a principal direction at a non-ridge. The Meusnier sphere is made up of 3-point contact circles with beaks/lips singularities or worse (i.e. type $4_{\geq 2}$). There will be 0, 1 or 2 singularities of type $4_{\geq 3}$ according as the focal set is elliptic/parabolic/hyperbolic at the corresponding focal point.

Our investigation of duality results started with joint work with my student Wilkinson in [29] on the family of folding maps f. An example of what one can prove here is

Proposition 7.5 The folding map f corresponding to a plane π has an S_k singularity at $p \in X$ if and only if π is tangent to the focal set of X and has contact of type A_k.

The models supplied by bifurcation sets of $\mathcal{A}$-versal unfoldings also have some interesting consequences. We have just seen that an S_3 singularity of a folding at a point $p \in X$ corresponds to a Gauss cusp or swallowtail on the focal set. In other words if X is very symmetric across the corresponding principal plane then the focal set is very flat at the corresponding point. Now for a generic surface the S_3 singularity will be versally unfolded, and its bifurcation set is the same as the discriminant of the A_3, a swallowtail. The inner pocket of the swallowtail corresponds to polynomials with 4 distinct roots, which in this case has an interpretation as the set of planes containing normals at 4 points of the surface. Indeed S_3 points p are characterised as having planes which contain 4 such points inside any neighbourhood of p. The line of self-intersection gives an alternative characterisation, namely given any neighbourhood U of an S_3 point there are planes which contain pairs of points inside U, their normals and one of their principal directions. Such planes are bitangent to the focal set. See [29].

Remarks 7.6

(a) Again there are analogous results at the bi-germ level relating the singularities of f with the contact between the corresponding plane and the symmetry set of X.

(b) In a sense Proposition 7.5 gives a method of studying the dual of the singular focal set by considering an auxilliary family on X itself. One might consider the reverse question: is there any way of studying the family of distance-squared functions on the dual of X via an auxilliary family on X. Of course the first objection is that the dual does not live in a Euclidean space. There are however Euclidean versions of the dual. For example fixing a point in 3-space as the origin we can consider the locus formed by reflecting this point in the tangent planes to X. This is the so-called orthotomic and it can be regarded as an affine dual of X. It usually has a rather degenerate singularity at the origin, corresponding to tangent planes of X passing through there; see [22]. Of course it is easy to see why one can think of this as a dual. For in turn the tangent planes can be recovered from the orthotomic as the perpendicular bisectors of the segments joining points of the orthotomic and the origin.

In the case when X is compact and convex and the origin lies in its interior the focal set of the orthotomic of X is the caustic by first reflection formed by regarding X as the mirror, and the base point as a light source. In other words the orthotomic can be taken as the corresponding wavefront. This was exactly the situation studied in [23], [24] by Giblin, Gibson and the author. Moreover 7.1 was motivated precisely by the problem of how could one study this caustic by considering functions on the mirror X itself rather than its orthotomic. In fact one does this by considering the contact between X and quadrics of revolution with one focus at the light source. The reason is simply that the orthotomic of the quadric will be a sphere. The contact of the sphere with the orthotomic is the same as that between the quadric and the mirror.

With the next result we shall return to duals. First a little terminology. Given a line l in projective space P there is a pencil of planes containing this line and hence a "dual" line l^* in the dual space P^*.

Theorem 7.7 [72], [18]. Let X be a smooth compact surface in R^3 with the property that the complement of its parabolic set is dense in X. Let $K(X)$ denote the set of lines tangent to X and let $K^*(X)$ denote the set of lines in the dual space tangent to the dual of X. Then
(1) the dual map $K \to K^*$ is a diffeomorphism taking $K(X)$ to $K^*(X)$.
(2) Moreover if p is not a parabolic point of X and the contact between X and l at p is finite, i.e. of type A_r for some finite r, then the contact between X^* and the dual line l^* at p^* is of the same type.
As a consequence we can deduce that:
(3) If l is an asymptotic line to X at the non-parabolic point p then the dual line l^* is asymptotic to X^* at p^*. In particular p is elliptic (resp. hyperbolic) if and only if p^* is elliptic (resp. hyperbolic).
(4) The non-parabolic part of the flecnodal curve of X corresponds, under the duality map, to the flecnodal curve of X^* (or rather the part of the flecnodal curve on the smooth part of X^*).
(5) The special hyperbolic points on the flecnodal curves corresponding to fourth order contact of one of the branches of the intersection of the surface and tangent plane with its tangent line are also preserved.

Remarks 7.8 Since giving this talk I have discovered that a result of this type was first obtained by Shcherbak [72] in the important case when X is a generic surface; in particular he obtained results (3)-(5) above. He gives no proofs. The proof of the general theorem given in [18], which uses a variant of the duality technique above, naturally avoids a case by case enumeration.

In the same paper Shcherbak also proved some beautiful results relating the bifurcation set of the family of central projections of a generic projective surface X and its dual. The Euclidean version of this result relating the bifurcation set of the family of height functions h and a subset of the bifurcation set of the family π is due to Romero-Fuster and the author in [28]. A more general version of the Euclidean result is the following

Proposition 7.9 [19] (i) Let $a \in S^2$ be a smooth point of the A_2 stratum

in Bif(h) being the normal at the parabolic point p of X. The tangent great circle to the stratum at a determines (a pair of) dual point(s) $\pm a^*$of S^2, the poles corresponding to the given equatorial circle. Suppose further that the contact of the tangent line with the A_2 stratum at a is finite (i.e. of type A_k for some k) and the singularity of the dual projection(s) is $\mathcal{A}$-finitely-determined. Then the tangent line has k-point contact with the A_2 stratum at a if and only if the the dual points $\pm a^*$ determine a projection of X at p of type C_k.
(ii) Let a be a smooth point of the $2A_1$ stratum in $Bif(h)$ coresponding to the points p, q of X, and let $\pm a^*$ be the corresponding dual points. Suppose further that the contact of the tangent line with the $2A_1$ stratum is finite (i.e. of type A_k for some k) and the bi-germ singularity at p, q of the dual projection is $\mathcal{A}$-finitely-determined. Then the tangent line has k-point contact with the $2A_1$-stratum if and only if the dual point a^* determines a projection of X which is a k-point contact fold.

Remarks 7.10 There are corresponding results for central projections, and in the Euclidean case some dual (dual) results too. See [19].

8: GENERICITY

A basic ingredient in all investigations of generic geometry are transversality results. One of the most general and powerful results, which deals admirably with all current applications to submanifolds of Euclidean and projective space, is the following result of Montaldi.

Theorem 8.1 [56] Let X, Y,Z, U be smooth manifolds, and let $\mathcal{G}$ be one of the standard groups arising in singularity theory: $\mathcal{R}, \mathcal{R}^+, \mathcal{L}, \mathcal{C}, \mathcal{A}, \mathcal{K}$; see for example [77]. If $F : Y \times U \to Z$ is smooth then given a mapping $i : X \to Y$ we can define a composite $F_i : X \times U \to Z$ by $F_i(x, u) = F(i(x), u)$.
(i) Suppose that $F : Y \times U \to Z$ is a locally $\mathcal{G}$-versal mapping and let S be a $\mathcal{G}^{(k)}$-invariant submanifold of $J^k(X, Z)$. Then for a residual set of immersions $X \to Y$ the k-jet-extension $j_1^k F_i : X \times U \to J^k(X, Z)$ is transverse to S.
(ii) Suppose that $F : Y \times U \to Z$ is a $\mathcal{G}$-versal mapping and let S be a $\mathcal{G}$-invariant submanifold of the multi-jet space ${}_rJ^k(X, Z)$. Then for a residual set of embeddings $X \to Y$ the multi-jet extension ${}_rj_1^k F_i : X^{(r)} \times U \to_r J^k(X, Z)$ is transverse to S.

Moreover, in both parts, the set of "good" immersions/embeddings is open and dense if in addition U is compact, and S is closed.

This theorem had some important but less general precursors, notably in [48] and [78]. See also [15] for an elementary calculation driven approach. Ideally Theorem 8.1 is applied as follows. First one makes a classification of the relevant map-germs up to a certain $\mathcal{A}_e$-codimension, namely the dimension of U. One then stratifies the corresponding (multi-)jet-space by the resulting hopefully finite number of orbits and their complement. This complement one shows can be decomposed into a finite number of manifolds of codimension greater than dim(X)+dim(U). One then applies the Theorem to this stratification. Then a generic immersion/embedding gives rise to a composite exhibiting only those singularities which turned up in the classification. Transversality to the given orbits also ensures that these singularities are versally unfolded. The words "ideally" and "hopefully" is an indirect reference to the problem of moduli. See Remarks 8.3 (d) (e) below. Let us see how the theorem works out for the families we have been considering.

Examples 8.2

The following maps are all actually stable

$$(x_1, x_2, x_3) \mapsto x_1; x_1^2 + x_2^2 + x_3^2; (x_1, x_2); (x_1^2 + x_2^2, x_3); (x_1^2, x_2, x_3).$$

The transversality results we require for the families h, d, π, c, f now all follow from the theorem above.

Remarks 8.3

(a) One could restate Montaldi's Theorem roughly as saying that a generic embedding $X \hookrightarrow Y$ yields a generic family $F| : X \times U \to Z$.The intuitive content is simply that F imposes no hidden structure, no hidden symmetry on this resulting family. In other words although the original mapping F may have, indeed will have, some geometric significance the restriction is just a general family of smooth mappings $X \times U \to Y$from the viewpoint of singularity theory .

(b) It is not difficult to cook up mappings F which do impose some hidden structure. A rather trivial example would be a family of constant mappings. Less trivial is the map $(x_1, x_2, x_3) \mapsto x_1^2 + x_2^2$ which could be used to describe the contact of X with cylinders centred on the x_3-axis. We now consider the family of such functions parametrised by the set of lines in R^3. This family is certainly not $\mathcal{A}$-versal. Moreover it is not difficult to check that the lines tangent to X give rise to $A_{\geq 3}$ singularities, so that this stratum has codimension 1 in the space of all lines, whereas in a generic family they would form a subset of codimension 2.

(c) We actually need a slight extension of this result to prove our duality results in [18], [19], [20]. Roughly what is true is that any finitely determined germ which arises from a composition as in 8.1, whether generic or not, is versally unfolded if as well as the U unfolding variables we also allow deformations by some finite dimensional family of embeddings. The details appear in [19], but it requires only trivial changes in Montaldi's proof to obtain the result.

(d) Clearly there is no reason why the classification should not turn up uncountably many distinct orbits. We then have to collect these orbits into strata, and consider not the codimension of the individual orbits, but that of the associated stratum. One also no longer expects the jet-extension mapping to be transverse to the individual orbits making up the stratum, so one loses the local models for the discriminants and bifurcation sets. This is bad news indeed, for we have seen that these sets often hold the key to the geometry. Thankfully it is often true that in low codimension the singularities which arise are simple. Nevertheless these problems do arise in even quite simple situations. Some sterling work by Damon over the last decade on topological equivalence [34] provides the major source of aid when moduli do crop up.

(e) In many ways when moduli occur one should consider not smooth but topological stability. The reader is referred to the forthcoming book by du Plessis and Wall on this subject. The good news is that for many dimension ranges they provide the sort of explicit information required to do make concrete applications which one needs for generic geometry. However although one can define the topological analogue of the smooth bifurcation set one should not feel too optimistic about our understanding in this area. As far as I am aware a complete understanding of the bifurcation sets of functions of type $\tilde{E}_k$, the simplest non-simple singularities, is still not in sight. The calculations here may (just) be do-able, but (famous last words?) the outlook in general is pretty hopeless.

References

[1] V.I.Arnold, Singularities of smooth mappings, Uspekhi Math. Nauk 23:1 (1968), 3-44.

[2] V.I.Arnold, Wave front evolution and equivariant Morse lemma, Commun. Pure and Appl. Math. 29 (1976), 557-582.

[3] V.I.Arnold, Critical points of functions on a manifold with boundary, the simple Lie groups B_k, C_k, F_4 and singularities of evolutes, Russian Math. Surveys 33:5 (1978), 99-116.

[4] V.I.Arnold, Mathematical Methods of Classical Mechanics, - Graduate Texts in Mathematics 60, Springer-Verlag, New York-Heidelberg-Berlin, 1978.

[5] V.I. Arnold, Indices of singular points of 1-forms on a manifold with boundary, convolution of invariants of reflection groups, and singularities of projections of smooth surfaces, Russian Math. Surveys 34:2 (1979), 1-42.

[6] V.I.Arnold, Singularities of systems of rays, Russian Math. Surveys 38

(1983), 87-176.

[7] V.I.Arnold, Singularities of ray systems. In: Proceedings of the International Congress of Mathematicians, August 16-24, 1983, Warszawa vol. 1. North Holland 1984, 27-49.

[8] V.I.Arnold, Catastrophe Theory, Springer-Verlag, New York-Heidelberg-Berlin, 1986.

[9] V.I.Arnold, S.M.Gussein-Zade and A.N.Varchenko, Singularities of Differentiable Maps, Volume I, Birkhauser, Monographs in Mathematics, Vol. 82, Birkhauser, Boston, 1985.

[10] T.Banchoff, T.J.Gaffney and C.McCrory, Cusps of Gauss Mappings, Pitman, New York, 1982.

[11] D.Bleeker and L.C.Wilson, Stability of Gauss maps, Illinois J. Math. 22 (1978), 279-289.

[12] J.W.Bruce, The duals of generic hypersurfaces, Math. Scand. 49 (1981), 36-69.

[13] J.W.Bruce, On contact of hypersurfaces, Bull. London Math. Soc. 13 (1981), 51-54.

[14] J.W.Bruce, On singularities, envelopes and elementary differential geometry, Math. Proc. Camb. Phil. Soc. 89 (1981), 43-48.

[15] J.W.Bruce, Generic reflections and projections, Math. Scand. 54 (1984), 262-278.

[16] J.W.Bruce, Envelopes duality and contact structures, Proceedings of Symposia in Pure Mathematics, Volume 40, Part1,195-202, American Mathematical Society,1983, Providence.

[17] J.W.Bruce, Geometry of singular sets, Math. Proc. Camb. Phil. Soc.106 (1989), 495-509.

[18] J.W.Bruce, Lines, surfaces and duality, Preprint,1991.

[19] J.W.Bruce, Generic geometry, transversality and projections, Preprint 1991.

[20] J.W.Bruce, Lines, circles, focal and symmetry sets, Preprint 1991.

[21] J.W.Bruce and D.L.Fidal, On binary differential equations and umbilics, Proc. Royal Soc. Edin. 111A (1989), 147-168.

[22] J.W.Bruce and P.J.Giblin, Curves and Singularities, Cambridge University Press,1984.

[23] J.W.Bruce P.J.Giblin and C.G.Gibson, On caustics of plane curves, American Math. Monthly, 88 (1981), 651-667.

[24] J.W.Bruce and P.J.Giblin, Caustics by reflexion, Topology 21 (1982), 179-199.

[25] J.W.Bruce, P.J.Giblin and C.G.Gibson, Symmetry sets, Proc. Royal Soc. Edin. 104A (1986), 179-204.

[26] J.W.Bruce and A.A.du Plessis, Complete transversals, Preprint, 1991.

[27] J.W.Bruce, A.A.du Plessis and C.T.C.Wall, Determinacy and unipotency, Invent. math. 88 (1987), 521-554.

[28] J.W.Bruce and M.C.Romero-Fuster, Duality and projections of curves and surfaces in 3-space, Quarterly Journal of Mathematics, to appear.

[29] J.W.Bruce and T.C.Wilkinson, Folding maps and focal sets, To appear in Proceedings of Warwick Symposium on Singularities, Springer Lecture Notes in Math., vol 1462, p. 63-72, Springer-Verlag, Berlin and New York, 1991.

[30] J.Cleave, On the tangent developable of a space curve, Math. Proc. Camb. Phil. Soc., 88 (1980), 403-407.

[31] J.N.Damon, Finite determinacy and topological triviality I, Invent. Math. 62 (1980), 249-324.

[32] J.N.Damon, Finite determinacy and topological triviality II, Sufficient conditions and topological stability. Compositio Math. 47 (1982), 101-132.

[33] J.N.Damon, The unfolding and determinacy theorems for subgroups of $\mathcal{A}$ and $\mathcal{K}$, Mem. Amer. Math. Soc. 50 (1984).

[34] J.N.Damon, Topological triviality and versality for subgroups of $\mathcal{A}$ and $\mathcal{K}$, Mem. Amer. Math. Soc. 75, 389 (1988).

[35] J.M.S.David, Projection-generic curves, J.London Math. Soc. 27 (1983), 552-562.

[36] T.J.Gaffney, On the order of determination of a finitely determined germ, Invent. Math. 37 (1976), 83-92.

[37] T.J.Gaffney, A note on the order of determination of a finitely determined germ, Invent. Math. 52 (1979), 127-135.

[38] T.J.Gaffney, The structure of TA(f), classification and an application to differential geometry, Proceedings of Symposia in Pure Mathematics, Volume 40, Part1,409-428, American Mathematical Society,1983, Providence.

[39] T.J.Gaffney and A.A.du Plessis, More on the determinacy of smooth map-germs, Invent. math. 66 (1982),137-163.

[40] M.Golubitsky and D.Schaeffer, Singularities and Groups in Bifurcation Theory,Vol.1. Springer-Verlag, New York-Berlin-Heidelberg-Tokyo, 1985.

[41] V.V.Goryunov, Singularities of projections of complete intersections. IN: Contemporary Problems in Mathematics (in Russian), vol. 22 (Itogi Nauki i Tekh., Sovrem. Probl. Mat.), VINITI, Moscow, 1983, 130-166. (In English, J.Sov.Math. 27 (1984), 2785-2811.)

[42] V.V.Goryunov, Projections of generic surfaces with boundary. In: The theory of singularities and its applications. Edited by V.I.Arnold. Advances in Soviet Mathematics. AMS. Providence RI, 1990.

[43] C.Guttierrez and J.Sotomayer, Structurally stable configurations of lines of principal curvature, Asterisque 98-99 (1982).

[44] Y.L.Kergosien et R.Thom, Sur les points paraboliques des surfaces, C. R. Acad. Sci.,Paris, Set. A 290 (1980), 705-710.

[45] Y.L.Kergosien, La famille des projections orthogonales d'une surface et ses singularities, C.R.Acad.Sci.,Paris, Set. A 292 (1981), 929-932.

[46] J.J.Koenderink, Solid Shape, MIT Press, Cambridge, Massachusetts, 1990.

[47] E.E.Landis, Tangential singularities, Funct. Anal. Appl. 15 (1981), 103-114.

[48] E.J.N.Looijenga, Structural stability of smooth families of C^∞-functions, Doctoral thesis, Universiteit van Amsterdam, 1974.

[49] C.McCrory, Profiles of surfaces, Preprint University of Warwick, 1980.

[50] C.McCrory, Generic curves and surfaces in 3-space, Preprint University of Warwick, 1980.

[51] J.N.Mather, Generic projections, Ann. of Math. 98 (1973), 226-245.

[52] D.M.Q.Mond, On the classification of germs of maps from R^2 to R^3, Proc. London Math. Soc. 50 (1985), 333-369.

[53] D.M.Q.Mond, Singularities of the exponential map of the tangent bundle associated with an immersion, Proc. London Math. Soc. (3), 53 (1986), 357-384.

[54] D.M.Q.Mond, On the tangent developable of a space-curve, Math. Proc. Camb. Phil. Soc., 91 (1982), 351-355.

[55] J.A.Montaldi, On contact between submanifolds, Michigan Math. J. 33 (1986), 195-199.

[56] J.A.Montaldi, On generic composites of maps, Bull. London Math. Soc. 23 (1991), 81-85.

[57] J.A.Montaldi, Surfaces in 3-space and their contact with circles, J. Differential Geometry 23 (1986), 109-126.

[58] O.A.Platonova, Singularities of the mutual disposition of a surface and a line, Russian Math. Surveys 36:1 (1981), 248-249.

[59] O.A.Platonova, Singularities of projection of smooth surfaces, Russian Math. Surveys 39:1 (1984), 177-178.

[60] O.A.Platonova, Projections of smooth surfaces, Tr. Semin. Im. I.G.Petrovskogo 10 (1984), 135-149.

[61] O.A.Platonova, Singularities in the problem of the quickest by-passing of an obstacle, Functional Anal. Appl. 15 (1981), 147-148.

[62] A.A.du Plessis, On the determinacy of smooth map-germs, Invent. Math. 58 (1980), 107-160.

[63] I.R.Porteous, The normal singularities of a submanifold, Jour. Diff. Geom. 5 (1971), 543-564.

[64] I.R.Porteous, The normal singularities of surfaces in $\mathbf{R}^3$, Proceedings of Symposia in Pure Mathematics, Volume 40, Part2, 379-394, American Mathematical Society,1983, Providence.

[65] J.H.Rieger, Families of maps from the plane to the plane, J. London Math. Soc. (2) 36 (1987), 351-369.

[66] M.C.Romero-Fuster, Sphere stratifications and the Gauss map, Proc. Royal Soc. Edinburgh, 95A (1983), 115-136.

[67] M.C.Romero-Fuster, Convexly generic curves in $\mathbf{R}^3$, Geometriae Dedicata 28 (1988), 7-29.

[68] V.D.Sedykh, Singularities of the convex hull of a curve in $\mathbf{R}^3$, Functional Anal. Appl. 11:1 (1977), 72-51.

[69] V.D.Sedykh, Singularities of convex hulls, Sib. Math. J. 24:3 (1983), 158-175.

[70] I.G.Shcherbak, Duality of boundary singularities, Russian Math. Surveys 39:2 (1984), 195-196.

[71] I.G.Shcherbak, Focal set of a surface with boundary, and caustics of groups generated by reflections B_k, C_k and F_4, Functional Anal. Appl. 18 (1984), 84-85.

[72] O.P.Shcherbak, Projectively dual space curves and Legendre singularities, Trudy Tbiliss. Univ. 232-233 (1982), 280-336.

[73] O.P.Shcherbak, Singularities of evolvents in the neighbourhood of a point of inflection of a curve and the group H_3 generated by reflections, Functional Anal. Appl.17 (1983), 301-302.

[74] O.P.Shcherbak, Wavefronts and reflection groups, Russian Math. Surveys 43:3 (1988), 149-194.

[75] R.Thom, Stabilité Structurelle et Morphogénèse. Reading, Massachusetts, Benjamin/Addison-Wesley (1973).

[77] C.T.C.Wall, Finite determinacy of smooth map germs, Bull. Lond. Math. Soc. 13 (1981), 481-539.

[78] C.T.C.Wall, Geometric properties of generic differentiable manifolds, Geometry and Topology, Rio de Janeiro 1976, SLNM 597, 707-774, Springer-Verlag, New York.

[79] H.Whitney, On singularities of mappings of Euclidean spaces I,Mappings from the plane to the plane to the plane, Ann. of Math. 62 (1955), 374-410.

[80] H.Whitney, The singularities of smooth n-manifolds in (2n-1)-space, Ann. of Math. 45 (1944), 247-293.

[81] K.Wirthmüller, Singularities determined by their discriminant, math. Ann. 252 (1980), 231-245.

[82] V.M.Zakalyukin, On Lagrangian and Legendrian singularities, Functional Anal. Appl. 10:1 (1976), 23-31.

[83] V.M.Zakalyukin, Reconstructions of wavefronts depending on one parameter, Functional Anal. Appl. 10:2 (1976), 139-140.

[84] V.M.Zakalyukin, Singularities of convex hulls of smooth manifolds, Functional Anal. Appl. 11 (1977), 225-227.

[85] V.M.Zakalyukin, Reconstructions of fronts and caustics depending on a parameter and versality of mappings. IN: Contemporary Problems in Mathematics (in Russian), vol. 22 (Itogi Nauki i Tekh., Sovrem. Probl. Mat.), VINITI, Moscow, 1983, 56-93. (In English, J.Sov. Math. 27 (1984), 2713-2735.)

The Department of Pure Mathematics,

The Department of Pure Mathematics,
The University,
PO BOX 147,
Liverpool, L69 3BX.
U.K.

AN ARITHMETICAL FACTORIZATION FOR THE CRITICAL POINT SET OF SOME MAP GERMS FROM $\mathbb{C}^2$ TO $\mathbb{C}^2$

FÉLIX DELGADO DE LA MATA

Universidad de Valladolid

0.- INTRODUCTION

Let $\varphi, \psi \in \mathbb{C}\{X,Y\}$ be two reduced analytic power series over the complex field $\mathbb{C}$. When we consider the germ of the analytic map given by

$$\phi = (\varphi, \psi) : (\mathbb{C}^2, 0) \to (\mathbb{C}^2, 0)$$

the critical point set of ϕ, $C(\phi)$, is the plane curve given by the vanishing of the jacobian determinant of ϕ: $j(\phi) = j(\varphi, \psi) = \varphi_x \psi_y - \varphi_y \psi_x$. The aim of this paper is to give some information about the topological type of the curve $C(\phi)$ which depends solely on the topological type of φ and ψ.

Note that if one takes ψ as a regular series, that is, if ψ defines a smooth plane curve, the curve $C(\phi)$ is the *polar curve* of φ with respect to ψ and, even if ψ is transversal to φ, the topological type of $C(\phi)$ depends on the analytic type of φ and not only on its topological type.

The techniques used here are of an arithmetical nature and very close to those used in **[15]**, [8], [7]. The semigroup of values, S, of the curve defined by $f = \varphi\psi = 0$ provides such techniques, in this terms we compute (in (2.1)) the intersection multiplicities of $j(\varphi, \psi)$ with the branches of $f = \varphi\psi = 0$, $\tau = \underline{v}(j(\varphi, \psi)) \in S$ and express it as a sum of some 'principal elements' of S (essentially the values of the maximal contact completed with some elements of S related with the intersection multiplicities between the branches of f). This decomposition is also related to the description of the branches of the polar curves given in **[11]**,**[13]**, [7] (see also [15] and [8] for the irreducible case) as we show in (2.4). Finally, the arithmetical properties of S give us – in some particular cases – the way in which the arithmetical decomposition of τ can be translated to a right factorization of $j(\varphi, \psi)$ in $\mathbb{C}\{X,Y\}$ (see section 3) with similar properties to the case of polars. However, one must note that the results are only partial and for some particular cases of φ and ψ. As in the case of polars, the arithmetical method is not sufficient for treating the general case successfully.

1.- Notations and preliminaries

This section is devoted to introduce the different sets of numerical invariants classically related to plane curve singularities. Also, we recall some properties of these invariants which will be used throughout the paper. Essentially all the results contained in this section are more o less known, so adequate references are given instead the proofs.

Because the techniques to be used in the paper are arithmetical we have preferred to introduce the numerical invariants in the most arithmetical way. However, to move into the maze of numbers attached to singular curves without any intuitive reference turns out to be very hard. To repair partially this lake of intuition frequently we give interpretations in more geometric contexts of the main invariants and of its different features and properties. In particular the resolution graph introduced in (1.2) provides a good instrument to understand how the arithmetics is used and to have a geometrical interpretation of the main results.

We work with formal series instead of the analytic ones. Of course all the results are true in the analytic context. In the same way, exception made of the computation of $\underline{v}(j(\varphi,\psi))$ in the section 2, the results are also valid for any algebraically closed coefficient field of arbitrary characteristic instead of the complex field.

1.1.- Numerical invariants for reduced curve singularities

(1.1.1) Let $f = \prod_1^d f_i \in \mathbb{C}[[X,Y]]$ be such that f_i is irreducible for any $i \in I := \{1,\ldots,d\}$ and $f_i \neq f_j$ if $i \neq j$. We will denote by C the algebroid plane curve defined by f, C is reduced with d irreducible components ("branches" for short): $C_1,\ldots,C_d$; the branch C_i being defined by the irreducible series f_i. Sometimes we will use terms defined for the algebroid curves referred to the equations defining them, for example the $f_i's$ are the 'branches' of f.

Let v_i be the normalized valuation corresponding to the branch f_i and denote by $\mathcal{O} = \mathbb{C}[[X,Y]]/(f)$ the local ring of C. The semigroup of values of f (or C) is the subsemigroup, S, of $\mathbb{N}^d$ given by

$$S := \{\underline{v}(h) := (v_1(h),\ldots,v_d(h)) \mid h \in \mathcal{O},\ h \text{ non-zero divisor }\}.$$

We will denote by δ the conductor of S, that is, the minimum element δ of S such that $\delta + \mathbb{N}^d \subseteq S$. The coordinates of δ are $\delta_i = c_i + \sum_{i\neq j}[f_i,f_j]$, $(1 \leqslant i \leqslant d)$, c_i being the conductor for the semigroup $S_i \subseteq \mathbb{N}$ corresponding to the branch f_i and $[f_i,f_j]$ the intersection multiplicity between the branches f_i and f_j ([**6**]). Also, make $\tau = \delta - (1,\ldots,1)$. We want to recall that τ belongs to S ([**6**]). When we need to show the curve precise we use $\mathcal{O}_f$, v_f, $S(f)$, $\delta(f)$ and $\tau(f)$ instead of $\mathcal{O}$, $\underline{v}$, S, δ and τ.

(1.1.2) Numerical invariants for an irreducible curve singularity. With notations as above, suppose $d = 1$, i.e. f is irreducible. The maximal contact values of C, $\bar\beta_0, \dots, \bar\beta_g$ (see [**17**], [**2**]) can be defined as the minimal set of generators of $S \subseteq \mathbf{N}$ in the following way:

$$\bar\beta_0 := \min(S - \{0\}) = m(f)$$

where $m(f)$ stands for the multiplicity of f, and for $i \geqslant 1$,

$$\begin{aligned}\bar\beta_i &= \min\{\gamma \in S \mid \gamma \notin \sum_{j=0}^{i-1} \mathbf{Z}_+\bar\beta_j\} \\ &= \min\{\gamma \in S \mid \gcd(\bar\beta_0, \dots, \bar\beta_{i-1}, \gamma) < \gcd(\bar\beta_0, \dots, \bar\beta_{i-1})\}.\end{aligned}$$

Assuming x is transversal to f, we denote by $n = m(f)$ the multiplicity of f and let $y = \sum_{i \geqslant 0} a_i x^{i/n}$ be a Puiseux expansion for f. The *characteristic exponents* of f, $\beta_0, \dots, \beta_{g'}$, are the ordered sequence of integers defined by: $\beta_0 = n$ and, recursively, β_i is the minimum integer k such that $a_k \neq 0$ and $\gcd(\beta_0, \dots, \beta_{i-1}, k) < \gcd(\beta_0, \dots, \beta_{i-1})$.

It is well known (see [**17**], [**2**]) that, in this context, $g = g'$ (g is called the genus of f), $e_i = \gcd(\beta_0, \dots, \beta_i) = \gcd(\bar\beta_0, \dots, \bar\beta_i)$,

$$\bar\beta_0 = \beta_0 \ , \ \bar\beta_1 = \beta_1 \quad \text{and}$$
$$\bar\beta_{i+1} = \frac{e_{i-1}}{e_i}\bar\beta_i + \beta_{i+1} - \beta_i \qquad 1 \leqslant i < g.$$

These equalities provide the equivalence between the two different sets of exponents. The topological type or equisingularity type of C is characterized by the set of characteristic exponents or by the set of maximal contact values ([**17**]), so, for two branches C and D, the sentence C and D are *equisingular* or C and D have the *same equisingularity type* means that they have the same set of exponents. When the curves involved have more than one branch the equisingularity type is characterized by the number of branches, the characteristic exponents (or the maximal contact values) of each branch and the intersection multiplicities between pairs of branches.

Let us assume $1 \leqslant i < g$. If $\varphi \in \mathbf{C}[[X, Y]]$ satisfies that $[f, \varphi] = \bar\beta_i$ we will say that φ has *maximal contact of genus* $i - 1$ with f. When $i = g$ it is convenient to say that f is the only curve having maximal contact of genus g with f and $\bar\beta_{g+1} = \infty$.

For $i = 1, \dots, g$ we define

$$N_i = \frac{e_{i-1}}{e_i} \quad , \quad n_i = N_i - 1 \quad , \quad \delta_i = \frac{\beta_0}{e_{i-1}} \quad , \quad \gamma_i = \frac{\bar\beta_i}{\delta_i} ;$$

and to simplify the notations in section 3 we also put $n_0 = -1$.

The following properties of the maximal contact values will be used frequently throughout the paper, the proofs can be found in [2], [18] and [19].

i) $N_i\bar{\beta}_i \in \langle\bar{\beta}_0, \ldots, \bar{\beta}_{i-1}\rangle := \sum_0^{i-1} \mathbb{Z}_+\bar{\beta}_j$ for $i \geqslant 1$.

ii) $N_i\bar{\beta}_i < \bar{\beta}_{i+1}$ for $i \geqslant 1$. As a consequence $\gamma_1 < \gamma_2 < \ldots < \gamma_g$.

iii) If $\alpha \in S$, then α can be written in a unique way as $\alpha = \sum_{i=0}^{g} \alpha_i\bar{\beta}_i$ with $\alpha_0 \geqslant 0$ and $0 \leqslant \alpha_i < N_i$ for $1 \leqslant i \leqslant g$.

iv) If c denotes the conductor of S, then $c - 1 = \sum_1^g n_i\bar{\beta}_i - \bar{\beta}_0$. Note that, in particular, $\sum_0^r n_i\bar{\beta}_i \notin S$ for $r \leqslant g$.

When we need to show the precise curve we use $\bar{\beta}_0(f), \ldots, \bar{\beta}_g(f), e_0(f), \ldots$ to denote the above numbers for the curve defined by f. Also, if $f = \prod_1^d f_i$ is a reduced curve (as in (1.1.1)) then we use $\bar{\beta}_0^i, \ldots, \bar{\beta}_{g_i}^i, e_0^i, \ldots, e_{g_i}^i, \ldots$ instead of $\bar{\beta}_0(f_i), \ldots, \bar{\beta}_g(f_i), e_0(f_i), \ldots$.

(1.1.3) The contact pair.

In [6,(3.3)] we have defined the contact pair, $(f|f')$, for two irreducible branches f and f' in terms of the Hamburger-Noether expansions of f and f'. This pair of integers, $(f|f') = (q, c)$, can also be characterized in the following way: Denote by $\bar{\beta}_0', \ldots, \bar{\beta}_{g'}', e_0', \ldots, e_{g'}'$ the numbers defined in (1.1.2) for the curve f' and let t be the minimum integer such that

$$[f, f'] \leqslant \min\left\{e_t'\bar{\beta}_{t+1}, e_t\bar{\beta}_{t+1}'\right\} = p(t).$$

(Note that this integer always exists, setting $\bar{\beta}_{g+1} = \bar{\beta}_{g'+1} = \infty$ if necessary.) Denote by ℓ_t (resp. ℓ_t') the integer part of $(\bar{\beta}_{t+1} - N_t\bar{\beta}_t)/e_t$ (resp. of $(\bar{\beta}_{t+1}' - N_t'\bar{\beta}_t')/e_t'$). Then

- If $[f, f'] < p(t)$ there exists an integer c, $0 < c \leqslant \min\{\ell_t, \ell_t'\}$ such that

$$[f, f'] = e_{t-1}'\bar{\beta}_t + ce_te_t' = e_{t-1}\bar{\beta}_t' + ce_te_t'.$$

  In this case $q = t$ and $(f|h) = (t, c)$.

- If $[f, f'] = p(t)$ and $e_t'\bar{\beta}_{t+1} \neq e_t\bar{\beta}_{t+1}'$ then $(q, c) = (t, \min\{\ell_t+1, \ell_t'+1\})$. Remember that, if $[f, f'] = e_t'\bar{\beta}_{t+1} < e_t\bar{\beta}_{t+1}'$ then $\ell_t \leqslant \ell_t'$ and if $\ell_q < \ell_q'$ then $[f, f'] = e_t'\bar{\beta}_{t+1} < e_t\bar{\beta}_{t+1}'$ ([6]).
- Finally, if $[f, f'] = p(t)$ and $e_t'\bar{\beta}_{t+1} = e_t\bar{\beta}_{t+1}'$ then $(q, c) = (t + 1, 0)$.

Note that, if we set

$$\rho = max\left\{n \mid \bar{\beta}_0\bar{\beta}_i' = \bar{\beta}_0'\bar{\beta}_i \text{ for } i = 1, \ldots, n\right\}$$

the possible intersection multiplicities between two branches having the same equisingularity types as f and f' are the elements in the set

$$\{\xi(q,c) \mid 0 \leqslant q \leqslant \rho;\ 0 \leqslant c \leqslant min\{\ell_q, \ell'_q\}\} \cup \{\xi(\rho, \min\{\ell_\rho + 1, \ell'_\rho + 1\})\}$$

where $\xi(q,c) = e'_{q-1}\bar{\beta}_q + ce_q e'_q = e_{q-1}\bar{\beta}'_q + ce_q e'_q$ and $\xi(\rho, \min\{\ell_\rho + 1, \ell'_\rho + 1\}) = \min\{e'_\rho \bar{\beta}_{\rho+1},\ e_\rho \bar{\beta}'_{\rho+1}\}$. There is a 1-1 correspondence between the above set and the set of possible contact pairs:

$$\{(q,c) \mid 0 \leqslant q \leqslant \rho;\ 0 \leqslant c \leqslant \min\{\ell_q, \ell'_q\}\} \cup \{(\rho, \min\{\ell_\rho + 1, \ell'_\rho + 1\})\}.$$

Moreover, one has (taking the lexicographic order in $\mathbf{N}^2$)

$$\xi(q,c) < \xi(q',c') \iff (q,c) < (q',c').$$

(1.1.4) Let $f = \prod_1^d f_i$ be as in (1.1.1) defining the reduced algebroid curve C. The *contact pair* $(f_1|f_2|\dots|f_d)$, of C is the pair of integers (see [**6**])

$$(f_1|f_2|\dots|f_d) = \min\{(f_i|f_j) \mid i \neq j\ ,\ i,j \in I\}$$

where the minimum is relative to the lexicographic order in $\mathbf{N}^2$.

As above, if

$$\rho = \max\{n \mid \bar{\beta}_0^i \bar{\beta}_k^j = \bar{\beta}_k^i \bar{\beta}_0^j\ , \forall i,j \in I, i \neq j \quad \forall k \leqslant n\}\ ,$$

then

$$e_k^i e_0^j = e_0^i e_k^j \quad , \quad N_k^i = N_k^j \quad , \quad \delta_k^i = \delta_k^j \quad , \quad \forall i,j \in I, \forall k \leqslant \rho \quad ;$$

and the possible contact pairs for f can be characterized in similar terms to the case of two branches.

(1.1.5) Remark. The contact pair of two branches f, f' can be seen as a way to measure their intersection multiplicity $[f, f']$. In fact, if one knows the equisingularity type of f and f', $(f|f')$ is equivalent to $[f, f']$. In a more geometric context, we recall that the infinitely near points of a branch C are the closed points of the successive quadratic transforms of C (see [**17**] or [**2**]). The intersection multiplicity between C and C' can be computed by the Noether formula using the multiplicity of the branches at the infinitely near points common to both branches, so, if one knows the sequences of multiplicities of C and C' at the infinitely near points (these sequences are equivalent to the equisingularity types) the knowledge of the number of points common to C and C' suffices to compute the intersection multiplicity. In fact only the so called *free* infinitely near singular points are needed (see the above references).

In the same terms, now for a reduced curve C with d branches, the contact pair can be also viewed as a way to indicate the number of infinitely near singular points which are common to the d branches, if one considers the set $B(C)$ of free infinitely near singular points common to the d branches of C then

$$\#B(C) = \sum_{n<q} (\ell_n + 1) + c$$

where $\ell_n = \ell_n^i$ for any $i \in I$ (note that the numbers ℓ_n^i do not depend on i for $n < \rho$). Similar computations can be made for the set of infinitely near (free and satellite) singular points common to the d branches.

(1.1.6) Maximal contact values (several branches). Let $f = \prod_1^d f_i$ be reduced, take the notations of (1.1.1) and (1.1.2). For each $i \in I$ and $n \leqslant g_i$ let $\mathcal{B}_n^i$ be the set of curves having maximal contact of genus n with f_i, we will say that $\varphi \in \mathbf{C}[[X, Y]]$ has *maximal contact of genus n* with f if (see [**14**]):

(a) The set $J_\varphi = \{i \in I \mid \varphi \in \mathcal{B}_n^i\}$ is non empty.

(b) J_φ is maximal for the inclusion ordering, that is, there exists no branch ϕ such that $J_\varphi \subset J_\phi$.

By definition the *set of values of the maximal contact of genus n* is the set $\{\underline{v}(\varphi)\}$ where φ has maximal contact of genus n with f.

Another feature of the maximal contact values is as follows: for each $n \in \mathbf{N}$ we set $\mathcal{A}_n = \{\mathcal{B}_n^i \mid i \in I\}$ with the inclusion ordering (note that, if $i, j \in I$ then the only possibilities are $\mathcal{B}_n^i \cap \mathcal{B}_n^j = \emptyset$ or one of them is contained in the other). Then φ has maximal contact of genus n with f if and only if $\varphi \in \mathcal{B}_n^i$ for some minimal set $\mathcal{B}_n^i$.

Let $\mathcal{M}_n \subseteq \mathcal{P}(I)$ be the set defined by

$$\mathcal{M}_n = \{K \subseteq I \mid \forall i, j \in K \ \mathcal{B}_n^i = \mathcal{B}_n^j \text{ and } \mathcal{B}_n^i \text{ is minimal in } \mathcal{A}_n\},$$

and for $K \in \mathcal{M}_n$, $i \in K$ we set $\mathcal{B}_n^K = \mathcal{B}_n^i$. With these notations it is obvious that φ has maximal contact of genus n with f if and only if $\varphi \in \mathcal{B}_n^K$ for some $K \in \mathcal{M}_n$; moreover one can check (see [**6 (3.7)**]) that for such a φ:

$$v_j(\varphi) = \begin{cases} \bar{\beta}_{n+1}^j & \text{if } j \in K \\ [f_i, f_j]/e_n^i & \text{if } j \notin K \text{ and } i \in K \end{cases}$$

In particular, the maximal contact value $\underline{v}(\varphi)$ does not depend on the particular element $\varphi \in \mathcal{B}_n^K$ chosen. We use B_0 to denote the element $(\bar{\beta}_0^1, \ldots, \bar{\beta}_0^d)$ and for the sake of completeness we include B_0 as a maximal contact value. With this notations the set of maximal contact values for f is the set

$$\overline{\mathcal{B}} = \{\underline{v}(\varphi) \mid \varphi \in \mathcal{B}_n^K,\ K \in \mathcal{M}_n,\ n \in \mathbf{N}\} \cup \{B_0\}.$$

Note that, if $K \in \mathcal{M}_n$ and $\varphi \in \mathcal{B}_n^K$ then it is obvious that $K \subseteq J_\varphi$; i.e. φ has maximal contact of genus n with the branches corresponding to the subset K of I. However, the equality between both sets is not true in general, for example: if $f_1 = (Y^2 - X^3)$, $f_2 = Y$ and $n = 0$ then $\mathcal{B}_0^2 = \{Y\} \subset \mathcal{B}_0^1$. As a consequence, $K = \{2\}$ is the only element in $\mathcal{M}_0$, but $J_Y = I = \{1, 2\}$, i.e. $\underline{v}(Y) = (\bar\beta_1^1, \bar\beta_1^2) = (3, \infty)$.

(1.1.7) Remark. It is not difficult to prove that the knowledge of the set of maximal contact values is equivalent to that of the equisingularity type of f. Also, with the above definition, the branches f_i, $(i \in I)$ of f appear as curves with maximal contact of genus g_i with f and obviously $\underline{v}(f_i) \notin S$. Usually, we take out these elements and the term **maximal contact values** stands for the elements properly defined, that is

$$\mathcal{B} = \left(\{\underline{v}(\varphi) \mid \varphi \in \mathcal{B}_n^J\ ,\ J \in \mathcal{M}_n,\ n \in \mathbb{N}\} \cap S\right) \cup \{B_0\}.$$

However, one must note that this set does not allow us, in general, to recover the equisingularity type of f, the curve at the end of the above paragraph provides an example. For this curve $\mathcal{B} = \{(2, 1)\}$ which, obviously, does not determine the equisingularity type of f.

Let (q, c) be the contact pair of f and $n < q$. Using (1.1.4) one can show that $\mathcal{M}_n = \{I\}$ (see [**6**]), so the maximal contact values of genus lower than q are the elements $B_0, \ldots, B_q$, where $B_i = (\bar\beta_i^1, \ldots, \bar\beta_i^d)$ if $0 \leqslant i \leqslant q$. For the index q we have that the condition $\mathcal{B}_q^i \subseteq \mathcal{B}_q^j$ is equivalent to say that $(f_i|f_j) \geqslant (q, \ell_q^j + 1)$. So, the necessary and sufficient condition to have $\mathcal{M}_q = \{I\}$ is that $c = \ell_q^i + 1$ for any $i \in I$.

(1.1.8) Contact coefficients. Let f be irreducible and $\varphi \in \mathbb{C}[[X, Y]]$ such that f does not divide φ, the *contact coefficient* of f with φ (in the sense of Hironaka [**10**]) is the rational number

$$h(f, \varphi) = \frac{[f, \varphi]}{m(\varphi)}.$$

In the same way, if $f = \prod_1^d f_i$ is reduced we define the *multi-contact coefficient* as

$$H(f, \varphi) = (h(f_1, \varphi), \ldots, h(f_d, \varphi)) \in \mathbb{Q}^d.$$

The contact coefficient provides a way to measure the intersection multiplicity between f and φ in such a way that the role of f stands out. Obviously $h(-,-)$ is not symmetrical in f and φ and usually one would think of f as fixed and φ as variable. ([**13**]). Also, as we show in the next section the contact coefficient admits a natural interpretation in terms of the resolution graph of f.

Now, let f, φ and ψ be irreducible, the relations between the possible contact coefficients for these branches will be useful in the rest of the paper. Firstly, it is known that in the set $\{h(f, \varphi)/m(f), h(f, \psi)/m(f), h(\varphi, \psi)/m(\varphi)\}$

at least two elements are equal, being the other greater or equal than the repeated one (see [21] and [7]). As a consequence

(i) $h(f,\varphi) > h(f,\psi) \implies h(\psi,\varphi) = h(\psi,f)$ and $h(\varphi,f) > h(\varphi,\psi)$.

(ii) $h(f,\varphi) = h(f,\psi) \implies h(\psi,\varphi) \geqslant h(\psi,f)$ and $h(\varphi,\psi) \geqslant h(\varphi,f)$.

Using (1.1.3) and with the notations of (1.1.2), an easy computation shows that if $(f|\varphi) < (q, \ell_q(f)+1)$ then $h(f,\varphi) < \gamma_{q+1}(f)$ and if $(f|\varphi) > (q+1,0)$ then $h(f,\varphi) > \gamma_{q+1}(f)$. Thus, if $h(f,\varphi) = \gamma_{q+1}(f)$ then the contact pair of f and φ, $(f|\varphi)$, must be $(q, \ell_q(f)+1)$ or $(f|\varphi) = (q+1,0)$. Now, by (1.1.3) again, we have

(iii) $h(f,\varphi) = \gamma_n(f) \implies [f,\varphi] = e_{n-1}(\varphi)\bar{\beta}_n(f)$.

(iv) $h(f,\varphi) < \gamma_n(f) \implies e_{n-1}(f)$ divide $[f,\varphi]$.

As the above properties suggest there are some relations between the contact pair and the contact coefficient, the following properties can proved with a little work with both invariants.

(v) $(f|\varphi) > (f|\psi) \implies h(f,\varphi) \geqslant h(f,\psi)$.

(vi) $h(f,\varphi) > h(f,\psi) \implies (f|\varphi) \geqslant (f|\psi)$.

Moreover if $h(f,\varphi) > h(f,\psi)$ and $(f|\varphi) = (f|\psi) = (q,c)$ then it must be that $c = \min\{\ell_q(f)+1, \ell_q(\varphi)+1, \ell_q(\psi)+1\})$ and $[f,\psi] = e_q(f)\bar{\beta}_{q+1}(\psi)$, $[\varphi,\psi] = e_q(\varphi)\bar{\beta}_{q+1}(\psi)$.

In similar way if $h(f,\varphi) = h(f,\psi)$ and $(f|\varphi) > (f|\psi)$ then $h(f,\varphi) = h(f,\psi) = \gamma_n(f)$, $h(\varphi,\psi) = \gamma_n(\varphi)$ and $(f|\psi) = (n-1, \ell_q(\psi)+1)$, $(f|\varphi) = (n,0)$.

1.2.- The resolution graph

Let C be the reduced algebroid plane curve given by $f = \prod_1^d f_i \in \mathbb{C}[[X,Y]]$. We shall assume f to be reduced with d irreducible factors. Let $\pi: X \to \mathrm{Spec}(\mathbb{C}[[X,Y]])$ be the canonical minimal resolution of C, $E = \pi^{-1}(0)$ the exceptional divisor of π, $\overline{C}$ the total transform of C and $\widetilde{C} = \overline{\pi^{-1}(C - \{0\})}$ the strict transform of C. Recall that $\overline{C}$ is a curve in X with only normal crossings and $\widetilde{C}$ has d connected components $\widetilde{C}_1, \ldots, \widetilde{C}_d$ corresponding to the d branches $C_1, \ldots, C_d$ of C.

Associated with π one can construct the *resolution graph* (also called the *dual graph*), $A(f)$, as the dual figure of E. More precisely:

(i) Each irreducible component of E is represented by a point in $A(f)$.

(ii) Two points in $A(f)$ are joined if and only if their corresponding components in E meet.

(iii) For each irreducible component $\widetilde{C}_i$ of $\widetilde{C}$ a little arrow is drawn joined to the point corresponding to the only component of E meeting $\widetilde{C}_i$.

Denote by $E(P)$ the component of E corresponding to the point P of $A(f)$ and let $w(P)$ be the number of blow-ups needed to build the divisor $E(P)$. The knowledge of the weighted tree $(A(f), w)$ is equivalent to that of the equisingularity type of C (see [1], [17]). In particular the resolution graphs $A(f_i)$ $(1 \leqslant i \leqslant d)$ for the branches of f can be computed from $A(f)$ as a certain subset as is described in [1]. Frequently $A(f)$ is also presented as an oriented graph starting in the only point P_0 with weight 1.

Some of the usual terminology for the resolution graph is the following: Let $P \in A(f)$, $v(P)$ denotes the *valence* of P, that is, if $w(P) > 1$ then $v(P)$ is the number of points or arrows joined with P in $A(f)$ and if $P = P_0$ is the point with $w(P_0) = 1$ then $v(P_0)$ is the above number plus one. Notice that the exception made for P_0 is not standard, we prefer to use this definition instead to made exceptions in the sequel, note that with this definition always $v(P_0) \geqslant 2$. The *ordinary* points of $A(f)$ are the points $P \in A(f)$ such that $v(P) = 2$ and the *special* (non ordinary) points are divided in *extremal points or end points* (if $v(P) = 1$) and *rupture points or star points* (if $v(P) \geqslant 3$). An *arc* is a geodesic in $A(f)$ joining two special points and with no more special points. In particular, a *dead arc* in $A(f)$ is an arc L such that one of its extremes is an end point, denoted by $P(L)$, (necessarily the other extreme must be a star point because the dual graph is connected). The set of dead arcs of $A(f)$ will be denoted by $\mathcal{D}(f)$ or simply $\mathcal{D}$ if confusion is not possible. The *proper rupture (or star) points* are the rupture points not belonging to any dead arc and the rupture points P belonging to a dead arc such that $v(P) \geqslant 4$.

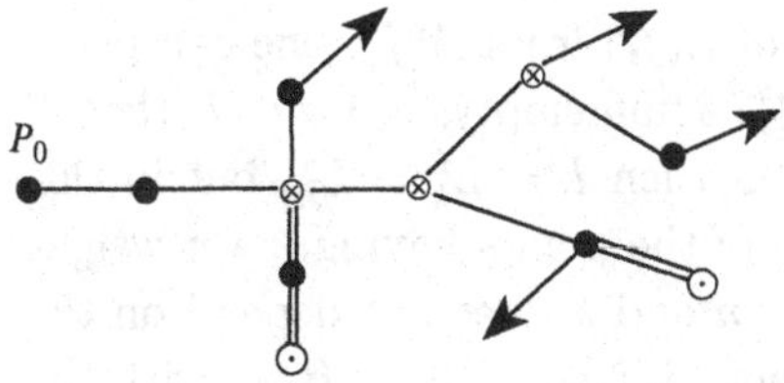

Example of a dual graph.
Symbols:
$\otimes$ = proper star points
$\odot$ = end points
$\bullet$ = other points
$\|$ = line for the dead arcs

Figure 1

Let $P \in A(f)$, a *curvette* at P is a smooth curve germ in X, Θ_P, transversal to $E(P)$ in a regular point of the exceptional divisor E. If $\varphi \in \mathbb{C}[[X,Y]]$ we denote by $\widetilde{\varphi}$ the strict transform of φ by π. We will say that φ *meets* (or goes through) a subset A of $A(f)$ if $\widetilde{\varphi} \cap E(P)$ is a regular point of E for some $P \in A$. Moreover, if φ meets $P \in A(f)$ and $\widetilde{\varphi}$ is smooth we say that φ becomes a curvette at P.

(1.2.1) Dead arcs and maximal contact values. Let f_i $(1 \leqslant i \leqslant d)$ be one of the branches of f. It is known that $A(f_i)$ has g_i dead arcs, so we can say $\mathcal{D}(f_i) = \{L^i_1, \ldots, L^i_{g_i}\}$ where we assume the dead arcs to be ordered by the weight, that is, $w(P(L^i_1)) < w(P(L^i_2)) < \ldots < w(P(L^i_{g_i}))$. With this notations, $[f_i, \pi(\Theta_{P(L^i_n)})] = \bar{\beta}^i_n$, or equivalently, $\varphi \in \mathbb{C}[[X,Y]]$ has maximal contact of genus $n-1$ with f_i if and only if φ becomes a curvette at the end point $P(L_n^i)$ of the dead arc L_n^i. On the other hand, $\bar{\beta}_0^i$ is the intersection multiplicity of f_i with a transversal curve and so $[f_i, \varphi] = \bar{\beta}_0^i$ if and only if φ becomes a curvette at the point P_0.

Taking into account the realization of the maximal contact values for the unibranch case and the definition of that values for the general case in (1.1.6) we have the following feature for the reduced curve defined by f:

Let $L \in \mathcal{D}$ be a dead arc and set $P(L)$ the end point of L. There exists $i \in I$ such that $P(L) \in A(f_i)$, then $\mathcal{B}_n^i$, for some n, is the set of curves which become curvettes at $P(L)$. If we have $\mathcal{B}_n^j \subset \mathcal{B}_n^i$ for some j, then all the branches of $\mathcal{B}_n^j$ go through $P(L)$ but only some curvettes among the curvettes at $P(L)$ are curves with maximal contact with f_j. As a consequence, f_j go through $P(L)$ in the resolution graph $A(f_i)$ and hence, $P(L)$ cannot be an end point. Thus the index j with the condition $\mathcal{B}_n^j \subset \mathcal{B}_n^i$ does not exist and $\mathcal{B}_n^i$ is minimal in $\mathcal{A}_n$. In other words we have proved that the maximal contact values for the reduced curve given by f can be realized by curvettes at the end points of $A(f)$ or at P_0, that is, are the elements in the set

$$\mathcal{B} = \{B_L = \underline{v}(\pi(\Theta_{P(L)})) \in S \mid L \in \mathcal{D}\} \cup \{B_0 = \underline{v}(\pi(\Theta_{P_0}))\}.$$

If $L \in \mathcal{D}$ then, using the determination of $A(f_i)$ from $A(f)$, one can prove that the set $J = J(L) = \{i \in I \mid L \in \mathcal{D}(f_i)\}$ is non empty, so if $i \in J$, then L must be $L = L_k^i$ for some $k \leqslant g_i$. If $i, j \in J$ then $L = L_k^i = L_n^j$, but in this case the sub graphs of f_i and f_j consisting of the points having lower weight than one in L agree. So, in particular $k = n$ and k does not depend on the index $i \in J$. We use L_k^J (resp. B_k^J) to denote L (resp. $B_L = B_{L_k^J}$) with the above definitions for J and k.

Using the Noether formula for the intersection multiplicity and (1.1.3) it can be shown that if $i, j \in J$ then $[f_i, f_j] \geqslant e_{k-1}^j \bar{\beta}_k^i$ and $(f_i|f_j) \geqslant (k, 0)$. That is, the first integer of the contact pair is exactly the number of dead arcs common to both branches. In particular, by (1.1.4), the natural numbers $N_k^i, n_k^i, \delta_k^i, \gamma_k^i$ defined in (1.1.2) do not depend on $i \in J$ and we can denote them by $N_L = N_k^J, n_L = n_k^J, \delta_L = \delta_k^J, \gamma_L = \gamma_k^J$. When $J = I$ we put simply $B_k, \delta_k, N_k, \ldots$ and if $J = \{i\}$ we use $B_k^i, \delta_k^i, \ldots$ instead $B_k^{\{i\}}, \delta_k^{\{i\}}, \ldots$.

If J is a subset of I and we denote by $\mathcal{D}^J$ the set of dead arcs L of $A(f)$ for which $J = J(L)$. It is obvious that the set of dead arcs $\mathcal{D}$ is the disjoint union of the subsets $\mathcal{D}^J$ when $J \in \mathcal{P}(I)$. The above comments give that $\mathcal{D}^I$ has q elements, (q, c) being the contact pair of f, $\mathcal{D}^I = \{L_1, \ldots, L_q\}$ where $L_n = L_n^i$ for any $i \in I$ and $n \leqslant q$.

Let $P(L)$ be the end point of $L = L_n^J \in \mathcal{D}$, it is obvious that the set $K = \{i \in I \mid P(L) \in A(f_i)\}$ contains the set J. In fact with the notations of

(1.1.6), K belongs to $\mathcal{M}_n$, as consequence

$$pr_k(B_n^J) = \begin{cases} \bar\beta_n^k & \text{if} \quad k \in K \\ \dfrac{[f_i, f_k]}{e_{n-1}^i} & \text{if} \quad k \notin K \quad \text{and} \quad i \in K. \end{cases}$$

In particular $pr_k(B_n^J) = \bar\beta_n^k$ for every $k \in J$. Also, note that it is possible to have $pr_k(B_n^J) = \bar\beta_n^k$ for some $k \notin K$. For example, if $f_1 = Y^2 - X^3$ and $f_2 = Y^2 - X^5$.

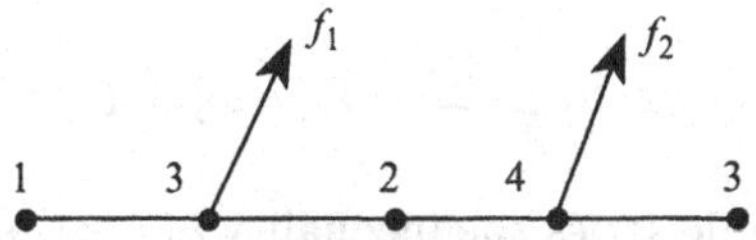

Figure 2

We have one dead arc: L_1^2 (corresponding to the end point with weight 3, this point belongs only to the resolution graph of f_2). However $B_1^2 = (\bar\beta_1^1, \bar\beta_1^2) = (3, 5)$ because every curve with maximal contact of genus 0 with f_2 also has maximal contact with f_1, note that $[f_1, f_2] = e_0^2 \bar\beta_1^1 = 2 \cdot 3$.

(1.2.2) Dual graph and contact coefficients.

Let $P \in A(f)$ and Θ_P be a curvette at P, we can consider the irreducible curve given by $\pi(\Theta_P)$ and define the *multi-coefficient of insertion* of P in f as

$$q(P) = H(f, \pi(\Theta_P)).$$

The following results are contained in [**13**] or can be easily proved with the definitions

a) $q(P)$ is constant when P varies in a dead arc L of $A(f)$, so we can call it $q(L)$

b) If φ is irreducible then φ meets a dead arc L if and only if $H(f, \varphi) = q(L)$.

The above properties, together with (1.2.1) permit us the explicit computation of $q(L)$ using curvettes at the end point of L, i.e. curves with maximal contact.

Firstly, given $L = L_n^J \in \mathcal{D}$, then $q(L) = B_L / m(\varphi)$ where φ becomes a curvette at $P(L)$. The properties in (1.1.4), (1.1.5) and (1.1.8) give us $m(\varphi) = \delta_n^i$, where $i \in J$. Now, using the same notations of (1.2.1) if $k \in K$ then $pr_k(q(L_n^J)) = \gamma_n^k = \bar\beta_n^k / \delta_n^k$ and if $k \notin K$ then $pr_k(q(L_n^J)) = [f_i, f_k] / \delta_n^i$ where $i \in J$.

Moreover, given $k \in K - J$ and $i \in J$, the end point of L, $P(L)$ is also the end point of $L_n^k \in \mathcal{D}(f_k)$, but $L_n^k \notin \mathcal{D}$ implies that f_i meets L_n^k in $A(f_k)$, hence $h(f_k, f_i) = \gamma_n^k$ and $[f_k, f_i] = e_{n-1}^i \bar\beta_n^k$. As a consequence, in the expressions of B_n^J and $q(L_n^J)$ the references to the subset K can be omitted, i.e. we can write

$$pr_k(B_n^J) = \begin{cases} \bar{\beta}_n^k & \text{if} \quad k \in J \\ \dfrac{[f_i, f_k]}{e_{n-1}^i} & \text{if} \quad k \notin J \end{cases} \qquad pr_k(q(L_n^J)) = \begin{cases} \gamma_n^k = \bar{\beta}_n^k / \delta_n^k & k \in J \\ \dfrac{[f_i, f_k]}{\delta_n^i} & \text{if} \quad k \notin J \end{cases}$$

where i is any index of J. We will denote $\Gamma_L = q(L) = B_L/\delta_L$ and, also, we frequently use Γ_n^J to denote $q(L_n^J)$.

With these notations property (b) above can be stated as follows: Let $\varphi \in \mathbb{C}[[X, Y]]$ be irreducible, we have that:

$$\varphi \text{ meets } L \iff H(f, \varphi) = \Gamma_L.$$

If f, φ, ψ are irreducible series the inequality $h(f, \varphi) > h(f, \psi)$ can be easily interpreted in $A(f\varphi\psi)$, in fact, it indicates that the branch ψ comes apart from f before φ, that is the separation point of f and ψ (by definition the last common point of the two geodesics joining the arrows corresponding to f and ψ with the starting point P_0 (see [13])) is closer to P_0 than the separation point of f and φ.

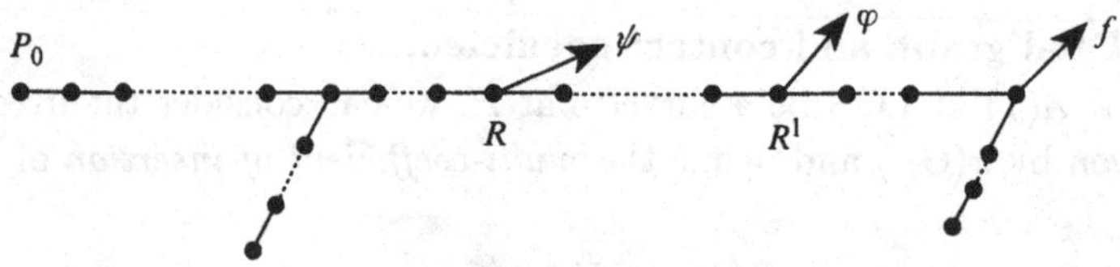

Figure 3

2.- The value of $j(\varphi, \psi)$ and its arithmetical expression in S.

Let $\varphi, \psi \in \mathbb{C}[[X, Y]]$ be as in the Introduction and suppose that $f := \varphi \cdot \psi$ is reduced. This section is mainly devoted to the computation of $\underline{v}(j(\varphi, \psi)) \in S(f)$ (2.1) and to proving its expression (see (2.3)) as a sum of more elementary elements of $S(f)$, the principal values. The set of principal values consists of the maximal contact values of f completed with some elements (defined in (2.2.2)) which have a completely natural interpretation in the light of the dual graph (2.2.6).

2.1.- The value of $j(\varphi, \psi)$

Let $\varphi = \prod_1^s \varphi_i \in \mathbb{C}[[X, Y]]$ be a reduced series. Denote by Z (resp. Z_i) the algebroid curve defined by φ (resp. φ_i) and let $\psi \in \mathbb{C}[[X, Y]]$ be such that φ_i does not divide ψ for any $i = 1, \ldots, s$. Let ω_ψ be the differential 1-form $d\psi / j(\varphi, \psi)$ restricted to Z, that is

$$\omega_\psi = \chi^* \left(\frac{d\psi}{j(\varphi, \psi)} \right),$$

χ being the natural morphism $\chi : Z = \mathrm{Spec}\, \mathcal{O}_\varphi \to \mathrm{Spec}\, \mathbb{C}[[X, Y]]$.

(2.1.1) Proposition. *The differential form ω_ψ does not depend on the element ψ. Moreover, one has that $v_\varphi(j(\varphi, \psi)) = \tau(\varphi) + v_\varphi(\psi)$.*

Proof: Let x be a transversal parameter for the curve Z, we shall prove that $\omega_\psi = \omega_x$. This is equivalent to proving that the differential form $\omega = j(\varphi, x)d\psi - j(\varphi, \psi)dx$ vanishes on Z, but

$$\omega = -\varphi_y(\psi_x dx + \psi_y dy) - (\varphi_x \psi_y - \varphi_y \psi_x)dx = -\psi_y(\varphi_x dx + \varphi_y dy) = -\psi_y d\varphi$$

which is identically zero on Z.

Let t be a uniformizing parameter for the branch Z_i, if one denotes the multiplicity of the branch Z_i by n, one can assume that $x = t^n$ and the valuation v_i associated to the branch φ_i is defined by $v_i = ord_t$. By the above result

$$v_i \left(\frac{d\psi}{j(\varphi, \psi)} \right) = v_i \left(\frac{dx}{j(\varphi, x)} \right).$$

Computing the different values involved, we find that $v_i(dx) = n - 1$, $v_i(d\psi) = ord_t(\psi'(t)) = (\varphi_i, \psi) - 1$. On the other hand

$$ord_t(j(\varphi, x)) = ord_t \left(\sum_{j=1}^{s} \frac{\varphi}{\varphi_j} j(\varphi_j, x) \right) = ord_t(\varphi/\varphi_i) + ord_t(j(\varphi_i, x)) =$$

$$= \sum_{j \neq i} [\varphi_i, \varphi_j] + c_i - 1 + n,$$

where c_i denotes the conductor of the semigroup $S(\varphi_i)$. In other words $v_\varphi(dx) = B_0(\varphi) - 1$, $v_\varphi(j(\varphi, x)) = \tau(\varphi) + B_0(\varphi)$ and $v_\varphi(d\psi) = v_\varphi(\psi) - 1$ where $B_0(\varphi)$ is used to denote the element B_0 defined in (1.1.6) for the curve φ.

So, for $j(\varphi, \psi)$ one has

$$v_\varphi(j(\varphi,\psi)) = \tau(\varphi) + v_\varphi(\psi).$$

(2.1.2) Assume that, under the above conditions, ψ is also reduced. Let f denote $\varphi \cdot \psi$ and take the notations of (1.1.1) for the reduced curve f. In particular, if $h \in \mathbb{C}[[X, Y]]$, $\underline{v}(h) = v_f(h) = (v_\varphi(h), v_\psi(h))$ and S is the semigroup of values of f. Then, an easy consequence of the Proposition is:

(2.1.3) Corollary. $\underline{v}(j(\varphi, \psi)) = \tau(f) \in S$.

2.2.- Principal elements of a reduced curve singularity

Let f be a reduced series. We use the same notations as in section 1. Let $(q, c) = (f_1 | \ldots | f_d)$ be the contact pair of the curve C, we will denote by $R_{(q,c)}$, or simply R if confusion is not possible, the equivalence relation on the set $I = \{1, \ldots d\}$ defined by

$$iR_{(q,c)}j \iff \begin{cases} (f_i|f_j) > (q,c) & \text{if} \quad c \leqslant \ell_q^i \\ (f_i|f_j) > (q+1,0) & \text{if} \quad c > \ell_q^i \end{cases}$$

Obviously the equivalence relation $R_{(q,c)}$ has more than one class.

(2.2.1) Remark. Taking into account the comments made in (1.1.5), this relation can be also defined in terms of the number of free infinitely near singular points common to the involved branches, namely, $iR_{(q,c)}j$ if and only if the branches f_i and f_j have in common strictly more free infinitely near points than all the branches of C, that is, if and only if $\#B(f) < \#B(f_i f_j)$.

Note that the above relation can be defined for any pair of integers (r, s). $R_{(r,s)}$ has exactly one equivalence class if $(r, s) < (q, c)$ and d classes if $r > g_i$ for every $i = 1, \ldots, d$. These equivalence relations are useful for inductively describing different features of the reduced singular curves. Notice that $R_{(q,c)}$ is similar to the equivalence relation given by Zariski in [**18**, 2-3] to describe the saturated rings, however $R_{(q,c)}$ is, in general, finer than the Zariski relation.

(2.2.2) Principal elements. Let (q, c) be the contact pair of f and assume that the classes for $R_{(q,c)}$ are $J_1, \ldots, J_t, I_{t+1}, \ldots, I_s$ in such a way that

$$c = \ell_q^i + 1 \quad \text{for} \quad i \in J := J_1 \cup \ldots \cup J_t$$
$$c \leqslant \ell_q^i \quad \text{for} \quad i \in I_{t+1} \cup \ldots \cup I_s$$

The rational number $\bar{\beta}^i_{q+1}/e^i_q$ is constant when i belongs to one of the classes J_n (see (1.1.4)), so one can assume that the minimum of the set $\{\bar{\beta}^i_{q+1}/e^i_q \mid i \in J\}$ is reached by the elements in the class J_1 and the maximum by the elements in J_t.

We shall define the element $D_I \in S$ as $D_I = \underline{v}(\varphi_I)$ for a certain series $\varphi_I \in \mathbb{C}[[X,Y]]$ defined as follows:

Suppose that $J \neq \emptyset$. Then φ_I is any irreducible algebroid curve of genus $q+1$ satisfying

$$\begin{aligned}(\varphi_I|f_i) &= (q+1,0) \quad \text{if} \quad i \in J_1\\ (\varphi_I|f_i) &\leqslant (q+1,0) \quad \text{if} \quad i \notin J_1.\end{aligned}$$

For φ_I, using (1.1.3), one can check that

$$pr_i(D_I) = v_i(\varphi_I) = \begin{cases} N^i_{q+1}\bar{\beta}^i_{q+1} & \text{if} \quad i \in J_1 \\ [f_i, f_j]/e^j_{q+1} & \text{if} \quad i \notin J_1 \text{ and } j \in J_1. \end{cases}$$

If $J = \emptyset$; i.e. if $c \leqslant \ell^i_q$ for any $i \in I$, then we choose φ_I from among the irreducible algebroid curves of genus q such that $(\varphi_I|f_i) = (q,c)$ for every $i \in I$. In this case, if $i \in I$

$$pr_i(D_I) = v_i(\varphi_I) = N^i_q\bar{\beta}^i_q + ce^i_q = [f_i, f_j]/e^j_q$$

where $j \in I$ is any index non-related with i.

Now, consider the algebroid curves $f_{J_1} = \prod_{i\in J_1} f_i$ and $f_{I-J_1} = \prod_{i\in I-J_1} f_i$ and repeat the above construction for these curves. That is, pick elements $\varphi_{J_1}, \varphi_{I-J_1}$ and define $D_{J_1} = \underline{v}(\varphi_{J_1}), D_{I-J_1} = \underline{v}(\varphi_{I-J_1})$. The procedure stops when we reach curves with only one branch. It can clearly be seen that by this method we have finally defined $d-1$ elements of S: $D^1 = D_I, D^2, \ldots, D^{d-1}$.

(2.2.3) Definition. The set of *principal values of* S is the set consisting of the maximal contact values and the $d-1$ elements $D^1, \ldots, D^{d-1}$.

(2.2.4) Remark. Taking into account the computation for the coordinates of the principal values it is not difficult to prove that the knowledge of the set of principal elements is equivalent to that of the equisingularity type of f. Also, we want to note that the above elements $D^1, \ldots, D^{d-1}$ are not necessarily different, for example, if for the contact pair (q,c) one has that $c \leqslant \ell^i_q$ for every $i \in I$ and the equivalence classes are $I_1, \ldots, I_s$ with $s \geqslant 3$ then

$$\underline{v}(\varphi_I) = \underline{v}(\varphi_{I-J_1}) = \ldots = \underline{v}(\varphi_{I_{s-1}\cup I_s}) = D_I$$

and the element $D_I \in S$ appears $s-1$ times in the set $\{D^1, \ldots, D^{d-1}\}$.

(2.2.5) Principal elements and dual graph. The elements $D^1, \ldots, D^{d-1}$ defined above admit an interpretation in the resolution graph of f:

Given $i, j \in I$ let Γ_i (resp. Γ_j) be the geodesic in $A(f)$ joining the only point P_0 with $w(P_0) = 1$ with the arrow corresponding to f_i (resp. f_j). The point $R \in \Gamma_i \cap \Gamma_j$ with maximal weight in $\Gamma_i \cap \Gamma_j$ is called the *separation point* of f_i and f_j.

Let us suppose that $[f_i, f_j] = e_q^j \bar{\beta}_{q+1}^i \leqslant e_q^i \bar{\beta}_{q+1}^j$, then $h(f_i, f_j) = \gamma_{q+1}^i$ and by (1.2.2) f_j goes through the (q+1)-dead arc L_{q+1}^i of $A(f_i)$. In consequence the separation point R is the only rupture point of $A(f_i)$ belonging to the dead arc L_{q+1}^i. Notice that the point R is a proper star point in $A(f_i f_j)$ and, as a result, also in $A(f)$.

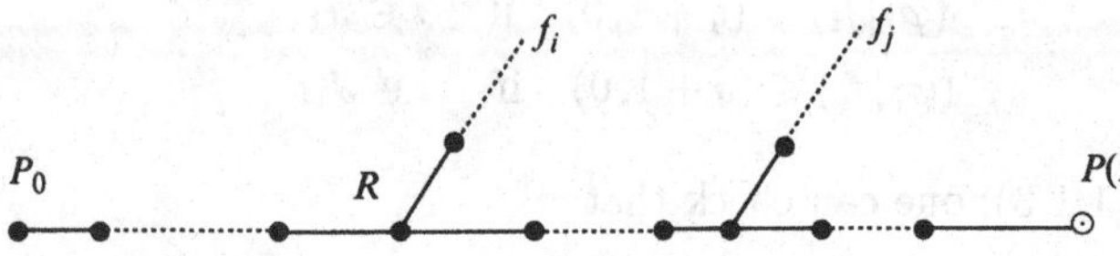

Figure 4

Let $\varphi = \pi(\Theta_{R)})$ be the image of a curvette at the separation point R, then φ has genus $q+1$, $(\varphi|f_i) = (q+1, 0)$ and $h(f_i, \varphi) = h(f_i, f_j) = \gamma_{q+1}^i$. So, using (1.2.1)(iii) and (1.1.4) one finds

$$v_i(\varphi) = e_q(\varphi)\bar{\beta}_{q+1}^i = N_{q+1}^i \bar{\beta}_{q+1}^i.$$

In a similar way one can compute that $v_j(\varphi) = [f_i, f_j]/e_{q+1}^i$.

On the other hand, if $[f_i, f_j] = e_{q-1}^i \bar{\beta}_q^j + c e_q^i e_q^j$ with $c > 0$ then the separation point R is an ordinary point on $A(f_i)$ (also in $A(f_j)$) out of the dead arcs. However R is a proper star point in $A(f)$. If φ becomes a curvette at R then, using $h(f_i, \varphi) = h(f_i, f_j)$ as above, one has that $v_i(\varphi) = [f_i, f_j]/e_q^j$, $v_j(\varphi) = [f_i, f_j]/e_q^i$.

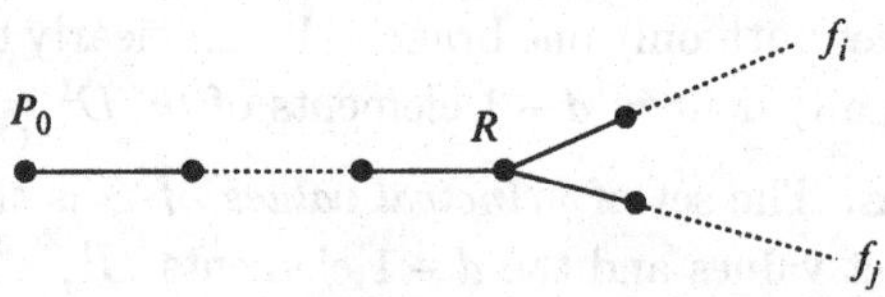

Figure 5

The above computations show that, if in (2.2.2) we choose $i \in J_1$ and $j \notin J_1$ (in the case $J = \emptyset$ simply take i, j non-related) then φ_I becomes a curvette at the separation point of f_i and f_j. Moreover this point, R, is exactly the proper star point of $A(f)$ nearest to the point P_0. We can follow this interpretation by induction on the number of branches, now taking f_{J_1} and f_{I-J_1}, as in the construction of the elements $D^1, \ldots, D^{d-1}$.

Also, the equivalence classes of R permits to understand the combinatorics of the resolution graph inductively. With the notations of (2.2.2), assume also the classes $J_1, \ldots, J_t$ ordered in such a way that $\bar{\beta}_{q+1}^{i_1}/e_q^{i_1} \leqslant \bar{\beta}_{q+1}^{i_2}/e_q^{i_2} \leqslant \ldots \leqslant \bar{\beta}_{q+1}^{i_t}/e_q^{i_t}$, where $i_k \in J_k$. We have that $h(f_{i_k}, f_{i_n}) = \gamma_{q+1}^{i_k}$ for $i_k \in J_k$, $i_n \in J_n$ and $k < n$. Moreover there exist branches having maximal contact of genus q

with all the branches of J simultaneously. On the other hand all the branches f_i with $i \in I_{t+1} \cup \ldots \cup I_s$ go through the end point $P(L^j_{q+1})$ of $A(f_j)$ where $j \in J$.

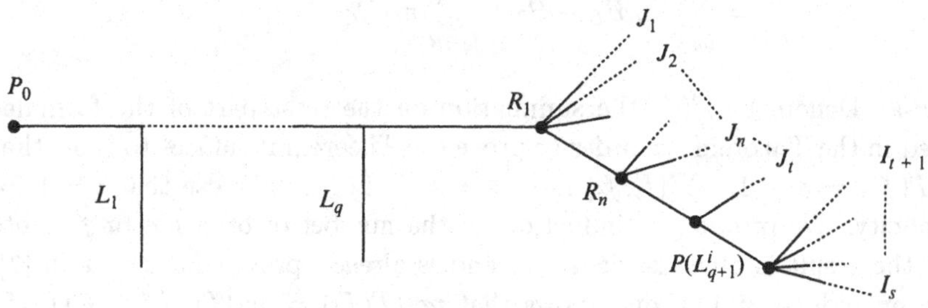

Figure 6

The behaviour of the dual graph at the levels greater than q can be understood only with that one of the curves $f_{J_i}, \ldots, f_{I_s}$, consisting on the branches of each equivalence class. In the above Figure, the sub graph of $A(f_{J_n})$ consisting of the points with weight greater than $w(R_n)$ is joined to R_n. The corresponding ones for f_{I_n} are all of them joined to $P(L^j_{q+1})$. In particular note that if $t < s$ then a curve has maximal contact of genus q with f if and only if has maximal contact of genus q with one of the curves f_{I_n}, $n = t+1, \ldots, s$. If $t = s$ then the curves with maximal contact of genus q with f_{J_s} are the curves with the same contact with f.

(2.2.6) Proposition. *The set $\{D^1, \ldots, D^{d-1}\}$ is the set of values reached by curvettes at the proper star points of $A(f)$. For each proper star point Q the corresponding $D_Q = \underline{v}(\pi(\Theta_Q))$ appears n_Q times, where $n_Q = v(Q) - 3$ if Q belongs to a dead arc and $n_Q = v(Q) - 2$ otherwise. As a consequence $\sum_{Q \in \mathcal{R}} n_Q = d - 1$, where $\mathcal{R}$ denotes the set of proper star points.*

The set of principal values is the set of values reached by curvettes at 'proper special' points, that is, at proper star points or at end points of $A(f)$.

PROOF: After the comments above the proof follows as an easy exercise with the combinatorics of the dual graph.

2.3.- Arithmetical decomposition of τ

(2.3.1) Let f be reduced and take the notations of section 1 and (2.2). Let $L \in \mathcal{D}$ be a dead arc of $A(f)$ and $i \in I$ such that $L = L^i_k$ (as in (1.2) L^i_k is the k-th dead arc in the dual graph of the branch f_i of f). Recall that in these conditions we put $n_L = N_L - 1 := N^i_k - 1 = n^i_k$. The purpose of this paragraph is to prove the following:

(2.3.2) Theorem. *With the above notations*

$$\tau = \sum_{L\in\mathcal{D}} n_L B_L - B_0 + D^1 + D^2 + \ldots + D^{d-1}$$
$$= \sum_{L\in\mathcal{D}} n_L B_L - B_0 + \sum_{Q\in\mathcal{R}} n_Q D_Q.$$

PROOF: Denote by $T(f)$ the summation on the right part of the formulae stated in the Theorem. In order to prove the Theorem it suffices to show that $pr_i(T(f)) = c_i - 1 + \sum[f_i, f_j]$ for $1 \leqslant i \leqslant d$. Let us suppose that $i = 1$ for simplicity. We proceed by induction on the number of branches of f. Note that the result is obvious for $d = 1$ and is already proved for $d = 2$ in [7]. Take an index $i \neq 1$, if one knows that $pr_1(T(f)) = pr_1(T(f/f_i)) + [f_1, f_i]$ then using the inductive hypothesis for f/f_i:

$$pr_1(T(f)) = pr_1(\tau(f/f_i)) + [f_1, f_i] = pr_1(\tau(f))$$

and the result is proved.

As a consequence one must prove that, if $i, j \in I$, $i \neq j$, then $pr_j(T(f)) = pr_j(T(f/f_i)) + [f_j, f_i]$. Let us suppose $i = d$ for the sake of simplicity and denote $\widetilde{f}$ the reduced curve $\widetilde{f} = f/f_d = f_1 \ldots f_{d-1}$. Assume, without loss of generality, that $h(f_d, f_{d-1}) \geqslant h(f_d, f_j)$ for every $j \in \{1, \ldots, d-1\}$, and let $(r, s) = (f_d|f_{d-1})$. Note that the equivalence relation $R_{(r,s)}$ has a class given by the single element d.

In order to compare $T(f)$ and $T(\widetilde{f})$ we use as a guide the comparison between the resolution graphs $A(f)$ and $A(\widetilde{f})$. Let Q denote the separation point of f_d and f_{d-1} and $D = D_Q$ the contribution of Q in $T(f)$ (obviously D does not appear in $T(\widetilde{f})$). The selection made for f_{d-1} permits one to consider only $A(f_{d-1}f_d)$ in order to understand the differences between $A(f)$ and $A(\widetilde{f})$. We distinguish several cases following the different possibilities for (r, s). The following picture shows the situation in cases A) and B) below.

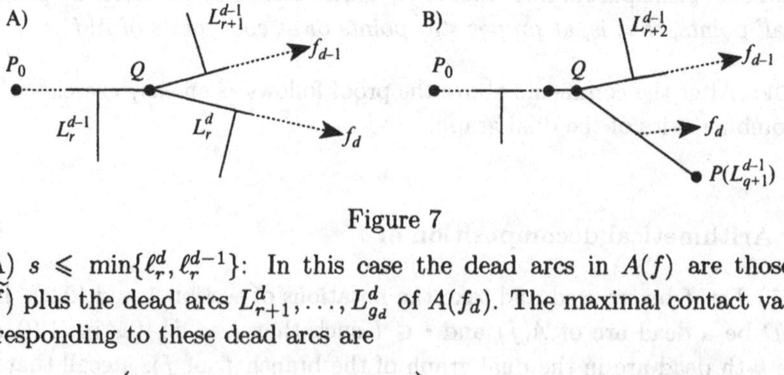

Figure 7

A) $s \leqslant \min\{\ell^d_r, \ell^{d-1}_r\}$: In this case the dead arcs in $A(f)$ are those of $A(\widetilde{f})$ plus the dead arcs $L^d_{r+1}, \ldots, L^d_{g_d}$ of $A(f_d)$. The maximal contact values corresponding to these dead arcs are

$$B^d_i = \left(\frac{[f_1, f_d]}{e^d_{i-1}}, \ldots, \frac{[f_{d-1}, f_d]}{e^d_{i-1}}, \bar{\beta}^d_i\right) \qquad \text{for} \quad i = r+1, \ldots, g_d.$$

The element D above is

$$D = \left(\frac{[f_1, f_d]}{e_r^d}, \ldots, \frac{[f_{d-1}, f_d]}{e_r^d}, \frac{[f_{d-1}, f_d]}{e_r^{d-1}} \right),$$

and so, for $i = 1, \ldots, d-1$:

$$pr_i(T(f)) = pr_i(T(\widetilde{f})) + \frac{[f_i, f_d]}{e_r^d} + \sum_{k=s+1}^{g_d} (N_i^d - 1)\frac{[f_i, f_d]}{e_{i-1}^d} = pr_i(T(\widetilde{f})) + [f_i, f_d].$$

B) $s = \min\{\ell_r^d + 1, \ell_r^{d-1} + 1\}$ and $[f_{d-1}, f_d] = e_r^d \bar{\beta}_{r+1}^{d-1} < e_r^{d-1} \bar{\beta}_{r+1}^d$. In this case the element of type D added to $T(\widetilde{f})$ is (see (2.2.5))

$$D = \left(\frac{[f_{d-1}, f_1]}{e_{r+1}^{d-1}}, \ldots, N_{r+1}^{d-1} \bar{\beta}_{r+1}^{d-1}, \frac{[f_{d-1}, f_d]}{e_{r+1}^{d-1}} \right).$$

Moreover, f_d goes through the dead arc L_{r+1}^{d-1} of $A(f_{d-1})$ and the separation point Q is the only star point of L_{r+1}^{d-1} (2.2.5). Notice that L_{r+1}^{d-1} is also a dead arc of $A(\widetilde{f})$, but every curve with maximal contact of genus r with f_d also has maximal contact of genus r with f_{d-1} i.e. L_{r+1}^{d-1} is not a dead arc of $A(f)$ but is replaced by L_{r+1}^d. Consequently, in $T(f)$ with respect to $T(\widetilde{f})$, one must add the elements

$$B_i^d = \left(\frac{[f_1, f_d]}{e_{i-1}^d}, \ldots, \frac{[f_{d-1}, f_d]}{e_{i-1}^d}, \bar{\beta}_i^d \right) \qquad \text{for} \quad i = r+1, \ldots, g_d$$

corresponding to the dead arcs $L_{r+1}^d, \ldots, L_{g_d}^d$ of $A(f_d)$ and subtract the element B_{r+1}^{d-1} of $S(\widetilde{f})$ given by

$$pr_i(B_{r+1}^{d-1}) = \begin{cases} [f_i, f_{d-1}]/e_r^{d-1} & \text{if} \quad (f_{d-1}|f_i) \leqslant (r, s) \\ \bar{\beta}_{r+1}^i & \text{if} \quad (f_{d-1}|f_i) > (r, s) \end{cases}$$

corresponding to L_{r+1}^{d-1}.

Using the above considerations and the facts

$$\begin{aligned} [f_i, f_d]/e_r^d = [f_i, f_{d-1}]/e_r^{d-1} \qquad & \text{if} \quad (f_i|f_{d-1}) \leqslant (r, s) \\ [f_i, f_d]/e_r^d = \bar{\beta}_{r+1}^i \qquad & \text{if} \quad (f_i|f_{d-1}) > (r, s) \end{aligned}$$

one finds that, for $i = 1, \ldots, d-1$: $pr_i(T(f)) = pr_i(T(\widetilde{f})) + [f_i, f_d]$.

C) $s = \min\{\ell_r^d + 1, \ell_r^{d-1} + 1\}$ and $(f_{d-1}, f_d) = e_r^{d-1} \bar{\beta}_{r+1}^d < e_r^d \bar{\beta}_{r+1}^{d-1}$. This case is similar to case B), but the role of both branches is interchanged. As a result no summand of $T(\widetilde{f})$ must be deleted and we only have to add the elements B_i^d, where $r + 2 \leqslant i \leqslant g_d$.

(2.3.3) Remark. The above decomposition provides interesting properties for the explicit computation of the semigroup of values. In [**6**] it is proved that S can be explicitly computed with the knowledge of some elements, (called absolute maximals) which are sums of the maximal contact values. In the same way, S can be computed if one knows the symmetrical ones of the absolute maximals (called relative maximals), where the symmetrical of α is $\tau - \alpha$.

In a simple formula the Theorem gives all the relative and absolute maximals, τ is in fact the greatest relative maximal of S, the element $D_I = D^1$ constructed here is the element σ given in [**6**, (3.20)] and one can prove (although not easily) that the element $D^2 + \ldots + D^{d-1}$ is the lowest relative maximal of S.

2.4.- The case of polars.

Let φ be reduced and $h \in \mathbb{C}[[X,Y]]$ regular and transversal to φ. Consider the map $\phi = (\varphi, h)$ as in the introduction, the critical point set of ϕ, $C(\phi)$, defined by $j(\varphi, h) = 0$ is the *polar curve*, P_h, of φ with respect to h. Take the same notations as above for $f = \varphi \cdot h$.

Note that the element D^1, the first principal element constructed in (2.2.2), satisfies that $D^1 = B_0$, because h is transversal to φ, and also $\mathcal{D}(f) = \mathcal{D}(\varphi)$. Thus, in the decomposition of $\tau(f)$ we can omit the element D^1 and the summand corresponding to $-B_0$, that is

$$\tau(f) = \sum n_L B_L + D^2 + \ldots + D^{d-1}$$

where the summation is extended to the dead arcs of $A(f)$ (or if one prefers to the dead arcs of $A(\varphi)$). The elements $D^2, \ldots, D^{d-1}$ are (except for the last coordinate, φ has $d-1$ branches!) the same elements appearing for φ instead of f and also note that the d-th coordinate of $\tau(f)$ is $[h, j(\varphi, h)] = m(j(\varphi, h)) = m(\varphi) - 1$.

(2.4.1) Assume that φ is irreducible, then the above decomposition has the explicit form

$$\tau(f) = \sum_{i=1}^{g} (N_i - 1)(\bar{\beta}_i, \delta_i),$$

where $\bar{\beta}_1, \ldots, \bar{\beta}_g, \delta_1, \ldots, \delta_g, N_1, \ldots, N_g$ are the sequences defined in (1.1.1) for the curve φ. This decomposition is used by Merle in [**15**] to give a factorization for $j(\varphi, h)$, namely, $j(\varphi, h) = A_1 \ldots A_g$ such that $\underline{v}(A_i) = (N_i - 1)(\bar{\beta}_i, \delta_i)$ and, if A_{ij} is an irreducible component of A_i, then $H(f, A_{ij}) = \Gamma_i = (\gamma_i, 1)$. In particular the polar quotients of φ, $h(\varphi, \psi)$ when ψ is an irreducible factor of $j(\varphi, h)$ ([**16**]), are exactly $\gamma_1, \ldots, \gamma_g$.

(2.4.2) Now, let φ be reduced, then in [**13**] it is proved, using topological arguments, that for the strict transform, $\widetilde{P}_h$, of P_h by π one has:

(a) $\widetilde{P}_h$ meets each dead arc of $A(\varphi)$.
(b) $\widetilde{P}_h$ meets each rupture point of $A(\varphi)$ not belonging to a dead arc.
(c) $\widetilde{C} \cap \widetilde{P}_h = \emptyset$ and $\widetilde{P}_h$ does not meet other different components.

These facts together with the realizations of the principal elements in the dual graph also permit the factorization of $j(\varphi, h)$ in packages corresponding to the elements in the decomposition of τ, in fact the above result has a translation in numerical terms:

For any ψ irreducible component of $j(\varphi, h)$ one of the two following possibilities is true

(a') $H(f, \psi) = \Gamma_L = q(L)$ for some $L \in \mathcal{D}(f)$.
(b') $H(f, \psi) = q(P)$ for some proper rupture point $P \in A(f)$.

So, we can construct the different factors using the equality of the contact coefficient to build them. The above computations provide the set of polar multi-quotients, in fact this set is: $\{q(P) \mid P \text{ is a rupture point of } A(f)\}$.

These facts show that the decomposition of $j(\varphi, h)$ is essentially described by the arithmetical decomposition of τ.

(2.4.3) Remark. When we take h regular but not necessarily transversal to φ one can prove a similar decomposition for τ in terms of a set of *principal values with respect to h* (see [7] for the case in which φ has two branches). The factorization for the special polar curve P_h, given in [8] for the irreducible case and in [7] for the case of two branches, is based on the arithmetical decomposition of τ and, as above, this one describes the factorization of P_h.

(2.4.4) Remark. If we take φ as a "general" element in the set of curves having a prefixed equisingularity type Casas in [3] and [4] proves that the behaviour of the generic polar of φ can be completely described in terms of the fixed equisingularity type, that is, one can describe the exact points of $A(\varphi)$ through which the polar branches go, the number of them and what its strict transforms are like. In particular the equisingularity type of $\varphi j(\varphi, h)$ can be determined. Note that this point of view is rather different from the one in the results above in which, roughly speaking, for any φ (non general) the "minimal" behaviour of the polars is described, although when one considers general curves φ and general transversal curves h one reaches the most concrete behaviour.

3.- Factorization Theorems

In this section we shall give partial results concerning the behaviour of the critical point set of two reduced curves in some particular cases. The methods used to prove these results are arithmetical and very close to those used in [5], [8], [7] for the problem of the behaviour of the polar curves.

The only Theorems for an arbitrary number of branches are (3.1.4) ('equi-separated curves') and (3.2.4) (curves with only one rupture point). In (3.3) is treated the case of curves with three branches giving the results that can be proved by this method for the different situations of the curve. The proofs always follow the same strategy, so after (3.1.4) the accent in the proofs is placed mainly in the differences with the first case (3.1.4). In particular in the case of three branches similar results require different proofs. In the same way results in this line can be proved in more particular cases, but, at my knowledge, the techniques are different in each case and do not permit the proof of more general results.

3.1.- Equiseparated case

(3.1.1) We shall say that f is *diagonal* if it satisfies the following conditions:

(1) The set of maximal contact values for f_i is independent of i $(1 \leqslant i \leqslant d)$.

(2) $[f_i, f_j] = \xi = e_{g-1}\bar{\beta}_g + c$ for any $i, j \in I$.

The condition (1) is equivalent to saying that the f_i's are equisingular. We denote by $\bar{\beta}_0, \ldots, \bar{\beta}_g$ the sequence of maximal contact values for any $i \in I$. Note that with the notations of [7] *diagonal* means diagonal with respect to a transversal curve. The maximal contact values of f are $B_i = (\bar{\beta}_i, \ldots, \bar{\beta}_i)$ $(0 \leqslant i \leqslant g)$; there exists only one proper star point Q in $A(f)$ and the corresponding principal element $D = D_Q$ is given by $D = (\xi, \ldots, \xi)$. Note that $B_0, \ldots, B_g, D$ are placed in the positive part of the diagonal of $\mathbb{N}^d$ and this is the reason for the name *diagonal*.

(3.1.2) Proposition. *Assume that f is diagonal and let $A \in \mathbb{C}[[X, Y]]$ be such that*

$$\underline{v}(A) = \tau(f) = \sum_1^g (N_i - 1)B_i + (d-1)D - B_0$$

Then for each irreducible component φ of A one has that $(f_i|\varphi) \leqslant (g, c)$ for every $i \in I$ and as a consequence $h(\varphi, f_i) = h(\varphi, f_j)$ for any $i, j \in I$.

Proof: Suppose that there exists an irreducible component φ of A such that $(\varphi|f_1) > (g, c) = (f_1|f_j)$. Then, in particular, one has $(\varphi|f_i) = (f_1|f_i) = (q, c)$ and, using (1.1.4) and the fact $e_g = 1$, $v_i(\varphi) = e_{g-1}(\varphi)\bar{\beta}_g + ce_g(\varphi) = e_g(\varphi)\xi$ for $i \neq 1$. On the other hand, in a similar way, $v_1(\varphi) > e_g(\varphi)\xi$. Taking into account that $v_i(A) = c_i - 1 + (d-1)\xi$ (see (1.1.2) iv)), it must be that $e_g(\varphi) < d - 1$ (note that $c_i - 1 \notin S_i$, c_i being the conductor of S_i)

Now we will proceed by induction on the number of branches d. If $d = 2$, we have a contradiction because the above condition for φ is $e_g(\varphi) \leqslant 0$, but always $e_g(\varphi) \geqslant 1$.

Assume $d > 2$. In this case take $\widetilde{A} = A/\varphi$ and put $e = d - e_g(\varphi) > 1$. We find that $v_j(\widetilde{A}) = \sum_1^g (N_i - 1)\bar{\beta}_i - \bar{\beta}_0 + (e-1)\xi$ for $j \neq 1$. In particular, if $\widetilde{f} = f_2 \dots f_{e+1}$ we have that

$$v_{\widetilde{f}}(\widetilde{A}) = (v_2(\widetilde{A}), \dots, v_{e+1}(\widetilde{A})) = \tau(\widetilde{f}).$$

By the induction hypothesis applied to $\widetilde{f}$, each irreducible component ϕ of $\widetilde{A}$ satisfies $(\phi|f_i) \leqslant (g, c)$ for $i \in \{2, \dots, e+1\}$, then it must be that $(\phi|f_1) \geqslant (\phi|f_i)$ and so: $v_1(\phi) \geqslant v_i(\phi)$. Consequently, $v_1(\widetilde{A}) \geqslant v_2(\widetilde{A}) = \sum(N_i - 1)\bar{\beta}_i - \bar{\beta}_0 + (e-1)\xi$. But this fact gives a contradiction because

$$v_1(A) = v_1(\varphi\widetilde{A}) > e_g(\varphi)\xi + v_1(\widetilde{A}) = v_1(A)$$

Thus, the proof is finished.

(3.1.3) Equiseparated curves. Let $f = \prod_{i=1}^d f_i$, as in (1.1.1), define a reduced algebroid curve. We will say that f is *equiseparated* if it satisfies that

$$(f_1|f_2|\dots|f_d) = (f_i|f_j) = (q, c) \text{ with } c \leqslant \ell_q^i \quad \text{for any} \quad i, j \in I.$$

The above definition has a natural interpretation in terms of the dual graph $A(f)$ of f: $A(f)$ has q dead arcs in common for all the branches of f, using the order given by weight we call them $L_1, \dots, L_q$. There is only one proper star point Q such that its valence is $v(Q) = d+1$ if Q does not belong to any dead arc or $d+2$ if Q belongs to the dead arc L_q (this situation is characterized by $c = 0$ in the definition). Joined to the point Q we have d trees, each one corresponding to the dual graph of each branch not common to the others. The rest of the dead arcs can be named $L_{q+1}^1, \dots, L_{g_1}^1, \dots, L_{q+1}^d, \dots, L_{g_d}^d$ in such a way that all are different and $L_i^j \subseteq A(f_j)$ is the i-th dead arc of f_j.

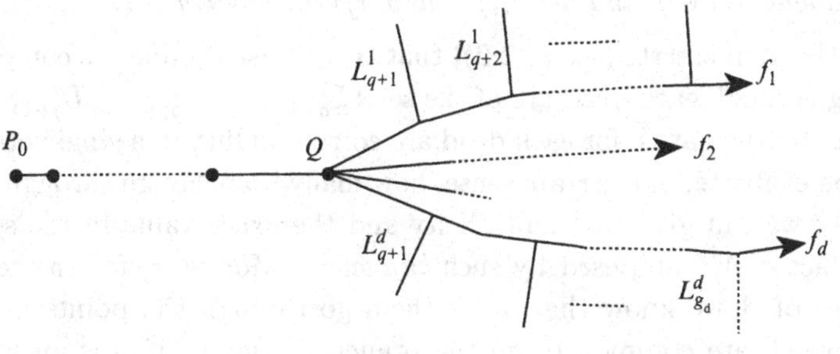

Figure 8

Let $D = D_Q$ be the only principal element attached to a proper star point. Let us denote by $\xi_{ij} = [f_i, f_j]$ the intersection multiplicity between the branches f_i and f_j, from (1.1.3) $\xi_{ij} = e^i_{q-1}\bar{\beta}^j_q + ce^i_q e^j_q = e^j_{q-1}\bar{\beta}^i_q + ce^i_q e^j_q$ and so the i-th coordinate of D (see (2.2.2) and (2.2.5)) is

$$pr_i(D) = \xi_{ij}/e^j_q = N^i_q \bar{\beta}^i_q + ce^i_q$$

where $j \neq i$.

The maximal contact values for f can be explicitly given by

$$B_{L_i} = B_i = (\bar{\beta}^1_i, \ldots, \bar{\beta}^d_i) \qquad \text{for} \qquad i = 0, \ldots, q$$

$$B_{L^j_i} = B^j_i = \left(\frac{\xi_{1j}}{e^j_{i-1}}, \ldots, \frac{\xi_{j-1,j}}{e^j_{i-1}}, \bar{\beta}^j_i, \frac{\xi_{j+1,j}}{e^j_{i-1}}, \ldots, \frac{\xi_{dj}}{e^j_{i-1}} \right)$$
$$\text{for} \quad j \in I, \quad i = q+1, \ldots, g_j$$

Taking into account that $N^i_k = N_k(f_i)$ does not depend on i for $k \leqslant q$ (1.1.4) and putting $N_k = N^i_k$ if $k \leqslant q$, the decomposition given in (2.3) is, in this case,

$$\tau = \sum_{i=1}^{q}(N_i - 1)B_i + \sum_{j=1}^{d}\left(\sum_{i=q+1}^{g_j} (N^j_i - 1)B^j_i \right) + (d-1)D - B_0$$

In this paragraph we will prove the following result:

(3.1.4) Theorem. *Let $A \in \mathbb{C}[[X, Y]]$ be such that $\underline{v}(A) = \tau$. Then A can be factorized as*

$$A = A' A^1_{q+1} \cdots A^1_{g_1} A^2_{q+1} \cdots A^d_{q+1} \cdots A^d_{g_d}$$

in such a way that

(1) $\underline{v}(A^j_i) = (N^j_i - 1)B^j_i$ and $m(A^j_i) = (N^j_i - 1)\delta^j_i$, $(1 \leqslant j \leqslant d, q+1 \leqslant i \leqslant g_j)$.

(2) $H(f, A^j_{ik}) = \Gamma^j_i$ for any irreducible component A^j_{ik} of A^j_i.

(3) For any irreducible component ϕ of A' one has that $(\phi|f_i) \leqslant (q, c)$ is independent of $i \in I$ and $h(\phi, f_i) = h(\phi, f_j)$ for any $i, j \in I$.

The theorem asserts (see (1.2.2)) that there exist irreducible components of A going through each dead arc of the set $\{L^1_{q+1}, \cdots, L^1_{g_1}, \ldots, L^d_{q+1}, \cdots, L^d_{g_d}\}$ (see Figure 8), that is, for each dead arc corresponding to a single branch. We can even evaluate, in a certain sense, how many branches go through L^j_i, more precisely we can give the multiplicity and the exact value in the semigroup of the factor A^j_i composed by such branches. Moreover, for the rest of the branches of A we know that all of them go through the points in the dual graph which are common to all the branches, that is, if φ is an irreducible component of A' and φ goes through $P \in A(f)$ then $w(P) \leqslant w(Q)$.

(3.1.5) We will prove this Theorem by induction on the number of branches d of f. For the case $d = 2$ the proof can be carried out in the same way as the general case, thus we will take the two cases together, advising at each step when we assume $d = 2$ and when we use the inductive hypothesis.

First of all, let r be the minimum integer $r \geqslant q$ such that there exist factors of A: $A^1_{r+1}, \dots, A^1_{g_1}$ with the requirements in the Theorem. We must prove that $r = q$, so assuming that $r > q$ we must be able to construct the factor A^1_r. If $r > q$, then for the series $G = A/(A^1_{r+1} \dots A^1_{g_1})$ we have:

$$\underline{v}(G) = \sum_{i=1}^{q}(N_i - 1)B_i + \sum_{j=2}^{d}\left(\sum_{i=q+1}^{g_j}(N_i^j - 1)B_i^j\right) + (d-1)D - B_0 + \sum_{q+1}^{r}(N_i^1 - 1)B_i^1 \,;$$

or, computing each coordinate:

$$(*)\qquad \begin{aligned} v_1(G) &= \sum_{1}^{r}(N_i^1 - 1)\bar\beta_i^1 + \sum_{j\neq 1}\xi_{1j} - \bar\beta_0^1 \\ v_j(G) &= \sum_{i=1}^{g_j}(N_i^j - 1)\bar\beta_i^j + \sum_{k\neq 1,j}\xi_{jk} + \xi_{1j}/e_r^1 - \bar\beta_0^j \qquad \text{if } \; j \neq 1. \end{aligned}$$

unless otherwise stated we will assume that this conditions hold. Before continuing the proof we will prove the following:

(3.1.6) Lemma. *Let G be as in (3.1.5). Suppose that we have even (a) $d = 2$ or (b) $d > 2$ and that the Theorem (3.1.4) is true for less than d branches.*

Then, there does not exist a factor of G, Φ, such that

(1) $h(f_1, \phi) \geqslant \gamma_r^1$ for every irreducible factor ϕ of Φ.

(2) $v_1(\Phi) \geqslant N_r^1 \bar\beta_r^1$.

In particular an irreducible factor of G for which the inequality (1) is strict does not exist.

PROOF: Assume that there exists Φ with the requirements in the Lemma. If ϕ is an irreducible component of Φ then $(f_1|\phi) \geqslant (r-1, \ell_{r-1}^1 + 1)$ and by (1.1.8)(iii): $v_1(\phi) \geqslant m(\phi)\gamma_r^1 = e_{r-1}(\phi)\bar\beta_r^1$. The fact $h(f_1,\phi) \geqslant \gamma_r^1 > h(f_1, f_i)$ implies that $h(f_i, f_1) = h(f_i, \phi)$ and so $v_i(\phi) = e_{r-1}(\phi)\xi_{1i}/e_{r-1}^1$. Let $e = \sum_{\phi|\Phi} e_{r-1}(\phi)$, if we prove that $e < N_r^1$ then in particular $e_{r-1}(\phi) < N_r^1$ for any irreducible factor ϕ of Φ. The condition $h(f_1,\phi) > \gamma_r^1$ implies $(f_1|\phi) \geqslant (r+1, 0)$ and so, using (1.1.4), $e_{r-1}(\phi) \geqslant N_r^1$. Thus, as a consequence, all the irreducible factors of Φ satisfies the equality in (1) and $v_1(\Phi) = e\bar\beta_r^1 < N_r^1\bar\beta_r^1$ as we want to prove.

Hence we must prove that $e < N_r^1$, for it suppose that $e = N_r^1 + a$ with $a \geqslant 0$. Let φ be an irreducible curve having maximal contact of genus $r-1$ with f_1, then $\underline{v}(\varphi) = B_r^1$ and, in particular, $v_i(\varphi) = \xi_{1i}/e_{r-1}^1$ for $i \neq 1$. For the series $\widetilde{G} = \varphi^a G/\Phi$ one finds that, if $i \neq 1$ then

$$v_i(\widetilde{G}) = a\frac{\xi_{1i}}{e_{r-1}^1} + v_i(G) - (N_r^1 + a)\frac{\xi_{1i}}{e_{r-1}^1} = v_i(G) - \frac{\xi_{1i}}{e_r^1}.$$

If $d = 2$ this equality is a contradiction because $v_2(G) - \xi_{12}/e_r^1 = c_2 - 1 \notin S_2$. So, the Lemma is proved in case (a).

Now, assume $d > 2$ and that the Theorem (3.1.4) is true for less than d branches. For $\widetilde{G}$ one has $(v_2(\widetilde{G}), \ldots, v_d(\widetilde{G})) = \tau(f_2 \ldots f_d)$, thus Theorem (3.1.4) applied to f/f_1 implies that we can factorize $\widetilde{G}$ as $\widetilde{G} = \Psi A_{q+1}^2 \ldots A_{g_2}^2 \ldots A_{q+1}^d \ldots A_{g_d}^d$ satisfying the conditions in the Theorem. In particular, for the irreducible components ψ of Ψ one has $(\psi|f_i) \leqslant (q, c)$ if $i \neq 1$, so $h(\psi, f_1) \geqslant h(\psi, f_i)$, $v_1(\psi) \geqslant v_i(\psi)e_0^1/e_0^i$ and then $v_1(\Psi) \geqslant (e_0^1/e_0^i)v_i(\Psi)$ for any index $i \neq 1$. Fix the index $i = 2$ for the sake of simplicity, then:

$$v_1(\Psi) \geqslant \frac{e_0^1}{e_0^2} v_2(\Psi) = \sum_1^q (N_i^1 - 1)\bar{\beta}_i^1 - \bar{\beta}_0^1 + \sum_{j \neq 1,2} \frac{\xi_{1j}}{e_q^j},$$

because the condition $(f_1|f_2) = (q, c)$ implies (see (1.1.3) and (1.1.4)) $(e_0^1/e_0^2)\bar{\beta}_k^2 = \bar{\beta}_k^1$ for $k \leqslant q$ and $(e_0^1/e_0^2)(\xi_{2j}/e_q^j) = \xi_{1j}/e_q^j$. On the other hand, it is straightforward to see that $v_1(A_i^j) = (N_i^j - 1)\xi_{1j}/e_{i-1}^j$ and then

$$v_1(\widetilde{G}/\Psi) = \sum_{j=2}^d \left(\sum_{q+1}^{g_j} (N_i^j - 1)\frac{\xi_{1j}}{e_{i-1}^j} \right) = \sum_{j=2}^d \left(\xi_{1j} - \frac{\xi_{1j}}{e_q^j} \right).$$

As a result, for the value of $\widetilde{G}$, one finds:

$$v_1(\widetilde{G}) = v_1(\Psi) + \sum_{j=2}^d \left(\xi_{1j} - \frac{\xi_{1j}}{e_q^j} \right) \geqslant \sum_1^q (N_i^1 - 1)\bar{\beta}_i^1 + \sum_{j=2}^d \xi_{1j} - \frac{\xi_{12}}{e_q^2} - \bar{\beta}_0^1.$$

Now, taking into account that $\xi_{12}/e_q^2 < \bar{\beta}_{q+1}^1$ and applying (1.1.2)(ii):

$$v_1(\Phi) - v_1(\varphi^a) \geqslant N_r^1 \bar{\beta}_r^1 > \ldots > \sum_{q+1}^r (N_i^1 - 1)\bar{\beta}_i^1 + \frac{\xi_{12}}{e_q^2}.$$

Finally, as a consequence of the last two inequalities:

$$v_1(G) = v_1(\Phi\widetilde{G}/\varphi^a) > \sum_1^r (N_i^1 - 1)\bar{\beta}_i^1 + \sum_2^d \xi_{1j} - \bar{\beta}_0^1 = v_1(G)$$

is a contradiction. Thus, the lemma is proved.

(3.1.7) Proof of the Theorem. Using the same notations as in (3.1.5), the strategy to prove the existence of the factor A_r^1 is as follows: first of all note that if φ is an irreducible factor of G then $h(f_1, \varphi) = \gamma_r^1 \iff H(f, \varphi) = \Gamma_r^1$. Thus the first step is to prove that there exist irreducible factors of G such that $h(f_1, \varphi) = \gamma_r^1$ and to collect all of them into the factor Φ. After that if one proves that $\underline{v}(\Phi) = (N_r^1 - 1)B_r^1$ the package $A_r^1 = \Phi$ is constructed.

Note that e^1_{r-1} divides $v_1(G) - (N^1_r - 1)\bar{\beta}^1_r$ but e^1_{r-1} does not divide $(N^1_r - 1)\bar{\beta}^1_r$, consequently e^1_{r-1} does not divide $v_1(G)$ and there exists irreducible factors φ of G such that $e^1_{r-1} \nmid v_1(\varphi)$. On the other hand, if $h(f_1, \varphi) < \gamma^1_r$ then $(f_1|\varphi) < (r-1, \ell^1_{r-1}+1)$ and so (by (1.1.3)) $[f_1, \varphi] = e_{t-1}(\varphi)\bar{\beta}^1_t + ae_t(\varphi)e^1_t$ for some $t < r$ or $[f_1, \varphi] = e^1_{r-1}\bar{\beta}_r(\varphi)$. In any case e^1_{r-1} divides $[f_1, \varphi]$.

As a consequence there exists φ such that $h(f_1, \varphi) \geqslant \gamma^1_r$. In fact, according to the above lemma, it must be that $h(f_1, \varphi) = \gamma^1_r$ and we can define Φ as the product of such irreducible components φ. Putting $e = \sum_{\varphi|\Phi} e_{r-1}(\varphi)$ one can check that $\underline{v}(\Phi) = eB^1_r$. The above argument concerning the existence of φ such that $h(f_1, \varphi) = \gamma^1_r$ implies that $e \geqslant (N^1_r - 1)$ and by Lemma (3.1.6) $e = (N^1_r - 1)$. The statement about the multiplicity of Φ is trivial, because $m(\phi) = e_{r-1}(\phi)\delta^1_r$ and then $m(\Phi) = (N^1_r - 1)\delta^1_r$. Thus we have constructed the package $A^1_r = \Phi$ with the conditions required in the Theorem.

Obviously the same proof can be carried out for the branches $f_2, \ldots, f_d$ and then we have proved the assertion about the packages $A^1_{q+1}, \ldots, A^d_{g_d}$.

It only remains to prove the statement (3) in the Theorem. To do so, for $1 \leqslant i \leqslant d$ let $\widetilde{f}_i$ be an irreducible curve such that $\widetilde{f}_i$ has genus q and $v_i(\widetilde{f}_i) = \bar{\beta}^i_{q+1}$. One can check that the maximal contact values for $\widetilde{f}_i$ are $\bar{\beta}^i_0/e^i_q, \ldots, \bar{\beta}^i_q/e^i_q$ and that the reduced curve $\widetilde{f} = \prod \widetilde{f}_i$ is diagonal (note that the k-th maximal contact value $\widetilde{\beta_k} = \bar{\beta}^j_k/e^j_q$ of $\widetilde{f}_j$ is independent of j). Moreover, if we call $\widetilde{v}_i$ the valuation given by $\widetilde{f}_i$ then

$$v_{\widetilde{f}}(A') = (\widetilde{v}_1(A'), \ldots, \widetilde{v}_d(A')) = \sum_1^q (N_i - 1)\widetilde{B}_i + (d-1)\widetilde{D} - \widetilde{B}_0 = \tau(\widetilde{f})$$

where $\widetilde{B}_i = (\widetilde{\beta}_i, \ldots, \widetilde{\beta}_i)$ $(0 \leqslant i \leqslant q)$ and $\widetilde{D} = (N_q\widetilde{\beta_q} + c, \ldots, N_q\widetilde{\beta_q} + c)$ are the principal values for $\widetilde{f}$. By the Proposition (3.1.2) for the diagonal curves we know that $(\varphi|\widetilde{f}_i) \leqslant (q, c)$ for any $i \in I$ and any irreducible component φ of A', as a consequence one has the statement (3) in the Theorem, because, by the definition of $\widetilde{f}_i$, $(\varphi|\widetilde{f}_i) = (\varphi|f_i)$.

(3.1.8) Remark. Note that the behaviour of the components in A' depends solely on one of the branches f_i (statement (3) in the Theorem). The arithmetical methods does not provide any information (in general) for these branches. For example, if we put

$$v_1(A') = \sum_1^q (N_i - 1)\bar{\beta}^1_i + (d-1)(N^1_q\bar{\beta}^1_q + ce^1_q) - \bar{\beta}^1_0 = \sum_0^q a_i\bar{\beta}^1_i$$

then the last equality, i.e. a decomposition in terms of $\bar{\beta}^1_0, \ldots, \bar{\beta}^1_q$ for $v_1(A')$, is not unique. Each decomposition gives us a possible factorization for A'. Moreover, no arithmetical reason exists for which the components of A' should go through the dead arcs in $A(f_1)$, so, the possible decompositions in terms of $\bar{\beta}_0^1, \ldots, \bar{\beta}_q^1$ do not suffice. It is possible, for example, to have $l \leqslant d-2$ irreducible factors ψ such that $\underline{v}(\psi) = D = (N_q^1\bar{\beta}_q^1 + ce_q^1, \ldots, N_q^d\bar{\beta}_q^d + ce_q^d)$ and a factorization for the rest in terms of the maximal contact values as above (branches going through the dead arcs). As a result it is obvious that, if we want to know the behaviour of the branches of A', we must use more data than the values of A'.

The non-existence of a factorization Theorem for the factor A' is the main reason why it is not possible to prove a Theorem like (3.1.4) for the general case of d branches. However, for the case of 2 and 3 branches the above result can be extended to other cases (not equiseparated), so we shall give these results in order to support the kind of general results that can be expected. In fact partial results for special positions of the branches can be given, but frequently the method for proving the one similar to the Lemma (3.1.6) (the fundamental result to prove the Theorem) is specific for each case and, as a consequence, are not very interesting for a general result.

3.2.- Factorization in the two branches case

Let $f = f_1 f_2$ be a reduced curve with two branches. We will use the same notations as in sections 1 and 2 for f. The set of dead arcs $\mathcal{D}$ of $A(f)$ can be decomposed (using the notations of (1.2.1)) as the disjoint union $\mathcal{D} = \mathcal{D}^I \cup \mathcal{D}^1 \cup \mathcal{D}^2$. Let Q be the only proper star point in $A(f)$ and $D = D_Q$ the principal element attached to Q.

(3.2.1) The decomposition of τ given in (2.3) can be stated in this case as:

$$\tau = \sum_{L \in \mathcal{D}^I} n_L B_L + \sum_{L \in \mathcal{D}^1} n_L B_L + \sum_{L \in \mathcal{D}^2} n_L B_L + D - B_0.$$

(3.2.2) Theorem. *Let $A \in \mathbb{C}[[X,Y]]$ be such that $\underline{v}(A) = \tau$. Then A can be factorized as*

$$A = A' \prod_{L \in \mathcal{D}^1} A_L^1 \prod_{L \in \mathcal{D}^2} A_L^2$$

in such a way that

(1) $\underline{v}(A_L^j) = n_L B_L$ and $m(A_L^j) = n_L \delta_L$, ($j = 1,2$ and $L \in \mathcal{D}^j$).

(2) $H(f, A_{Lk}^j) = \Gamma_L$ for any A_{Lk}^j irreducible component of A_L^j.

(3) For any ϕ irreducible component of A' one has that $(\phi|f_i) \leqslant (q,c)$ is independent of $i = 1,2$ and $h(\phi, f_1) = h(\phi, f_2)$.

PROOF: Let $\xi = [f_1, f_2]$ be the intersection multiplicity between the branches f_1 and f_2 and $(q,c) = (f_1|f_2)$ the contact pair. Note that if $c \leqslant \ell_q^i$ for $i = 1,2$ then f is equiseparated and the Theorem is (3.1.4), so we can assume that $\xi = e_q^2\bar{\beta}_{q+1}^1 < e_q^1\bar{\beta}_{q+1}^2$. The decomposition for τ is then:

$$\tau = \sum_1^q n_i B_i + \sum_{q+2}^{g_1} n_i^1 B_i^1 + \sum_{q+1}^{g_2} n_i^2 B_i^2 + D - B_0$$

where $D=(N^1_{q+1}\bar\beta^1_{q+1},\xi/e^1_{q+1})$.

The proof for the factors $A^1_{q+2},\dots,A^1_{g_1},A^2_{q+2},\dots,A^2_{g_2}$ follows in, essentially, the same way as in (3.1.4). Thus, assume that the above factors exist and let G be the series $G=A/(A^1_{q+2}\cdots A^2_{q+2}\cdots A^2_{g_2})$. For G we have, putting $n^1_0=n^2_0=-1$,

$$v_1(G)=\sum_0^{q+1} n^1_i\bar\beta^1_i+\xi/e^2_{q+1}$$

$$v_2(G)=\sum_0^{q+1} n^2_i\bar\beta^2_i+\xi/e^1_{q+1}.$$

As in (1.1.7) e^2_q does not divides $v_2(G)$, so there exist irreducible components φ of G such that $h(f_2,\varphi)\geqslant\gamma^2_{q+1}$, denote by Φ the product of such branches. Because $\gamma^2_{q+1}>h(f_2,f_1)$, as in (3.1.6) we reach that, if $e=\sum_{\varphi|\Phi}e_q(\varphi)$ then $e\geqslant N^2_{q+1}-1=n^2_{q+1}$,

$$v_1(\Phi)=e\xi/e^2_q=e\bar\beta^1_{q+1}\quad\text{and}\quad v_2(\Phi)\geqslant e\bar\beta^2_{q+1}.$$

Suppose that $e\geqslant N^2_{q+1}$, multiplying by irreducible curves with maximal contact of genus q with f_2 as in (3.1.6) we can assume that $e=N^2_{q+1}$. In this case $v_1(\Phi)=\xi/e^2_{q+1}$ and $v_1(G/\Phi)=\sum_0^{q+1}n^1_i\bar\beta^1_i\notin S(f_1)$, so we find a contradiction. That is $e=N^2_{q+1}-1$ and Φ is just the factor A^2_{q+1} wanted.

Now we shall prove the statement (3) for the branches of A'. Note that if ϕ is a branch of A' then $h(f_2,\phi)<\gamma^2_{q+1}$ by the construction A^2_{q+1} and if $h(f_1,\phi)<\gamma^1_{q+1}=h(f_1,f_2)$ then $h(\phi,f_1)=h(\phi,f_2)$.

Let us suppose that $h(f_1,\phi)\geqslant\gamma^1_{q+1}$ for some branch φ of A'. We distinguish the following situations:

$$(a)\quad h(f_1,\phi)=\gamma^1_{q+1}\implies v_1(\phi)=e_q(\phi)\bar\beta^1_{q+1}$$
$$(b)\quad h(f_1,\phi)>\gamma^1_{q+1}\implies v_2(\phi)=e_{q+1}(\phi)\xi/e^1_{q+1}.$$

Computing $v_1(A')$ one finds $v_1(A')=\sum_0^q n^1_i\bar\beta^1_i+N^1_{q+1}\bar\beta^1_{q+1}$, as consequence the case (a) cannot occur, because $v_1(A')-n\bar\beta^1_{q+1}$ does not belong to S_1 for any n with $n\geqslant 1$ (see (1.1.2)(iv)). Now, in case (b), computing $v_2(A')=\sum_0^q n^1_i\bar\beta^2_i+\xi/e^1_{q+1}$ we find a contradiction, because $v_2(A')-v_2(\phi)\notin S_2$.

This concludes the proof of the Theorem.

(3.2.3) Remark. Let f be reduced with d branches, we use the same notations as in section 1, in particular (q,c) is the contact pair of f. We know that if f is equiseparated then $\#\mathcal{R}=1$, i.e. in $A(f)$ only one proper star point exists. The converse is not true, in fact the condition $\#\mathcal{R}=1$ is equivalent to

$$h(f_i,f_j)=h(f_i,f_k)\quad\text{for any}\quad i,j,k\in\{1,\dots,d\}.$$

Let Q be the only proper star point in $A(f)$, because each point of $A(f)$ is joined to at most two points with lower weight and $n_Q = d-1$, we have the following possibilities for Q:

a) Q is joined with $d-1$ points with greater weight than Q.

b) Q is joined with $d-2$ points with greater weight than Q and with two points with lower weight.

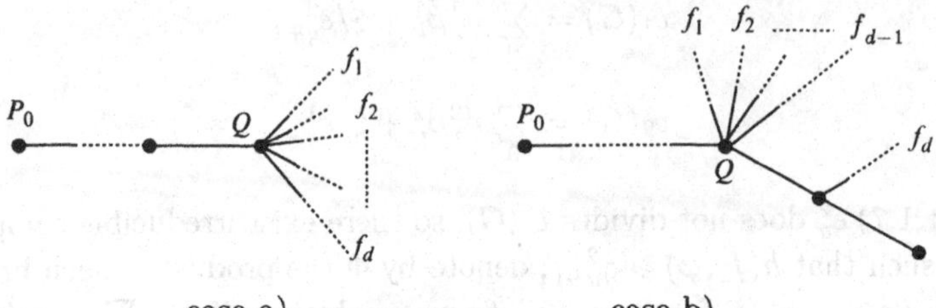

Figure 9

Looking to (2.2.5) and to the proof for the two branches case above it is not difficult to show that if case a) occurs then f is equiseparated and case b) can be also characterized by the existence of a branch f_i such that f/f_i is equiseparated and $h(f_j, f_i) = \gamma^j_n$, $h(f_i, f_j) < \gamma^i_n$ for any $j \neq i$ and some integer n (see Figure 9). For the sake of simplicity let us assume that $i = d$. In terms of the contact pair, $(f_k|f_j) = (n, 0) = (q+1, 0)$ for any $k, j \in \{1, \ldots, d-1\}$ and $(f_d, f_j) = (n-1, \ell^d_{n-1}+1) = (q, c)$.

The set of dead arcs is in this case the disjoint union of $\mathcal{D}^I = \{L_1, \ldots, L_q\}$, $\mathcal{D}^i = \{L^i_{q+2}, \ldots, L^i_{g_i}\}$ if $i \neq d$ and $\mathcal{D}^d = \{L^d_{q+1}, \ldots, L^d_{g_d}\}$ and the point $D = D_Q$ is $D = (N^1_{q+1}\bar{\beta}^1_{q+1}, \ldots, N^{d-1}_{q+1}\bar{\beta}^{d-1}_{q+1}, [f_1, f_d]/e^1_{q+1})$. The decomposition of τ is:

$$\tau = \sum_{i=1}^{q} n_i B_i + \sum_{j=1}^{d}\left(\sum_{i=q+2}^{g_j} n^j_i B^j_i\right) + n^d_{q+1}B^d_{q+1} + (d-1)D - B_0.$$

The proof of (3.1.4) can be adapted to this case without problems, the steps in which differences appear can be performed as in the proof of Theorem (3.2.2). So, we omit the proof of the following:

(3.2.4) Theorem. *Assume that f is reduced with only one proper star point. With the same notations as above, let $A \in \mathbb{C}[[X,Y]]$ be such that $\underline{v}(A) = \tau$. Then A can be factorized as*

$$A = A' A^d_{q+1} \cdots A^d_{g_1} \prod_{j=1}^{d-1} \prod_{i=q+2}^{g_j} A^j_i$$

in such a way that

(1) $\underline{v}(A^j_i) = (N^j_i - 1)B^j_i$ and $m(A^j_i) = (N^j_i - 1)\delta^j_i$, for $j = d$, $i = q+1$ and also for $1 \leqslant j \leqslant d$, $q+2 \leqslant i \leqslant g_j$.

(2) $H(f, A^j_{ik}) = \Gamma^j_i$ for any irreducible component A^j_{ik} of A^j_i.

(3) For any irreducible component ϕ of A' one has that $(\phi|f_i) \leqslant (q,c)$ is independent of $i \in I$ and $h(\phi, f_i) = h(\phi, f_j)$ for any $i, j \in I$.

3.3.- Factorization in the three branches case

Let $f = f_1 f_2 f_3$ be a reduced curve with three branches. We keep the same notations as in sections 1 and 2 for f. After (3.1) and (3.2.4) we can assume that f is not equiseparated and with more than one proper star point. So, in the sequel we suppose, without loss of generality, $h(f_3, f_2) > h(f_3, f_1)$, or equivalently (see (1.1.8)(i)) $h(f_2, f_3) > h(f_2, f_1)$. Denote by $\xi_{ij} = [f_i, f_j]$ the intersection multiplicity between f_i and f_j, let $(q,c) = (f_1|f_2|f_3) = (f_1|f_3)$ be the contact pair of f and $(r,s) = (f_2|f_3) \geqslant (q,c)$ the one of $f_2 f_3$.

Thus, the situation in the resolution graph, $A(f)$, is as follows: The set of dead arcs $\mathcal{D}$ can be decomposed (using the notations of (1.2.1)) as the disjoint union $\mathcal{D} = \mathcal{D}^I \cup \mathcal{D}^{\{2,3\}} \cup \mathcal{D}^1 \cup \mathcal{D}^2 \cup \mathcal{D}^3$. Moreover, in $A(f)$ exists two proper star points Q and Q'; we assume $w(Q) < w(Q')$ and let $D = D_Q$ and $D' = D_{Q'}$ be the principal values attached to Q and Q'.

The decomposition of τ given in (2.3) can be stated in this case as:

$$\tau = \sum_{L \in \mathcal{D}^I} n_L B_L + \sum_{L \in \mathcal{D}^{\{23\}}} n_L B_L + \sum_{j=1}^{3} \sum_{L \in \mathcal{D}^j} n_L B_L + D + D' - B_0 .$$

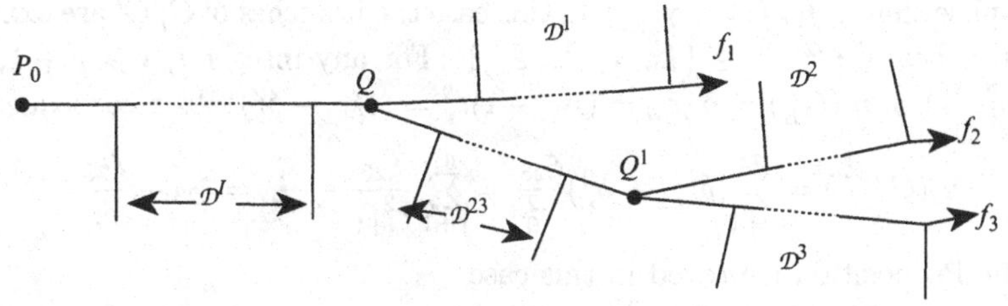

Figure 10

(3.3.1) Proposition. *Under the above conditions, put $\widetilde{f} = f_1 f_2$ and let $G \in \mathbb{C}[[X,Y]]$ be such that $v_{\widetilde{f}}(G) = \tau(\widetilde{f})$. Then*

$$v_3(G) \geqslant \sum_{0}^{r} n_i^3 \overline{\beta}_i^3 + \xi_{13} + \xi_{23} - \frac{\xi_{23}}{e_r^2},$$

and the equality occurs if and only if $h(f_2, f_3) < \gamma_{r+1}^2$.

PROOF: Theorem (3.2.2) applied to G and $\widetilde{f}$ gives to us a factorization

$$G = G' \prod_{L \in \mathcal{D}^1} G_L^1 \prod_{L \in \mathcal{D}^2} G_L^2 ,$$

with the requirements stated in (3.2.2) for the factors of G. Let φ be an irreducible factor of G, we will distinguish three possibilities for φ:

(a) If $h(f_2,\varphi) < h(f_2,f_3)$ then, by (1.1.8), $h(\varphi,f_2) = h(\varphi,f_3)$ and so $v_3(\varphi) = e_0^3 v_2(\varphi)/e_0^2$. Note that all the branches of $G'\prod_{L\in\mathcal{D}^1} G_L^1$ and all of G_L^2 for $L \in \{L_{q+1}^2,\ldots,L_r^2\}$ are in this situation, i.e. satisfies $h(f_2,\varphi) < h(f_2,f_3)$.

(b) If $h(f_2,\varphi) > h(f_2,f_3)$ then, $h(f_3,f_2) = h(f_3,\varphi)$ and so $v_3(\varphi) = e_0(\varphi)\xi_{23}/e_0^2$. Because $h(f_2,f_3) \leqslant \gamma_{r+1}^2$, at least all the branches of G_L^2, for $L \in \{L_{r+2}^2,\ldots,L_{g_2}^2\}$ satisfies condition (b).

(c) If $h(f_2,\varphi) = h(f_2,f_3)$ then $h(f_2,\varphi) > h(f_2,f_1)$. Hence, by (1.1.8), $h(\varphi,f_2) > h(\varphi,f_1)$ and looking to Theorem (3.2.2) φ must be an irreducible component of G_L^2, for some $L \in \mathcal{D}^2$. The requirements for f_3 implies in this case that $L = L_{r+1}^2$ and $h(f_2,\varphi) = h(f_2,f_3) = \gamma_{r+1}^2$.

The computations made in (c) implies that, if $h(f_2,f_3) < \gamma_{r+1}^2$ then condition (c) is satisfied for no branches in G. In particular condition (b) is also satisfied for the branches through L_{r+1}^2.

Denote by G_1 the product of the branches of G satisfying condition (a) above. A straightforward computation shows that $v_2(G_1) = \sum_0^r n_i^2\bar\beta_i^2 + \xi_{12}$. Now, by (1.1.4), $e_0^3\bar\beta_i^2 = e_0^2\bar\beta_i^3$ $(1 \leqslant i \leqslant r)$ and by (1.1.8) $e_0^2\xi_{13} = e_0^3\xi_{12}$. Finally, we reach, for G_1:

$$v_3(G_1) = \sum_0^r n_i^3\bar\beta_i^3 + \xi_{13}.$$

Suppose that $h(f_2,f_3) < \gamma_{r+1}^2$. In this case the branches of G/G' are exactly the branches of G_L^2, $L \in \{L_{r+1}^2,\ldots,L_{g_2}^2\}$. For any integer i, $i \geqslant r+1$, let $L = L_i^2$. Then $m(G_L^2) = n_L\delta_L = (N_i^2-1)\delta_i^2 = (\delta_{i+1}^2 - \delta_i^2)$. As a consequence,

$$v_3(G/G') = \sum_{r+1}^{g_2}(\delta_{i+1}^2-\delta_i^2)\frac{\xi_{23}}{e_0^2} = \sum_{r+1}^{g_2}\frac{\xi_{23}}{e_{i+1}^2} - \frac{\xi_{23}}{e_i^2} = \xi_{23} - \frac{\xi_{23}}{e_r^2}$$

and the Proposition is proved in this case.

Now, assume $h(f_2,f_3) = \gamma_{r+1}^2$. The above computation can be realized for the irreducible components of G_L^2 when $L \in \{L_{r+2}^2,\ldots,L_{g_2}^2\}$, so we have, putting $G_{L_{r+1}^2}^2 = G_{r+1}^2$,

$$v_3(G/(G'G_{r+1}^2)) = \xi_{23} - \frac{\xi_{23}}{e_{r+1}^2};$$

and it only remains to compute $v_3(G_{r+1}^2)$.

In the rest of the proof φ stands for a branch of G_{r+1}^2. By (3.2.2) we recall that $v_2(\varphi) = e_r(\varphi)\bar\beta_{r+1}^2 < e_r^2\bar\beta_{r+1}(\varphi)$ and $\sum e_r(\varphi) = (N_{r+1}^2 - 1) = n_{r+1}^2$, where the summation is for φ belonging to the set of branches of G_{r+1}^2. The condition $h(f_2,f_3) = \gamma_{r+1}^2$ implies that $\xi_{23} = e_r^3\beta_{r+1}^2 < e_r^2\bar\beta_{r+1}^3$. The branches of G_{r+1}^2 can be distributed in two sets:

$$A = \{\varphi \mid h(f_3,\varphi) \geqslant \gamma_{r+1}^3\}$$
$$B = \{\varphi \mid h(f_3,\varphi) < \gamma_{r+1}^3\}.$$

We have $v_3(\varphi) \geqslant e_r(\varphi)\bar{\beta}^3_{r+1}$ for the branches $\varphi \in A$, while $v_3(\varphi) = e^3_r\bar{\beta}_{r+1}(\varphi) < e_r(\varphi)\beta^3_{r+1}$ if $\varphi \in B$. Then, for G^2_{r+1} we find:

$$v_3(G^2_{r+1}) = \left(\sum_{\varphi\in A} e_r(\varphi)\right)\bar{\beta}^3_{r+1} + \sum_{\varphi\in B} e^3_r\bar{\beta}_{r+1}(\varphi) >$$
$$> \left(\sum_{\varphi\in A} e_r(\varphi)\right)\frac{e^3_r}{e^2_r}\bar{\beta}^2_{r+1} + \sum_{\varphi\in B} e^3_r\frac{e_r(\varphi)\bar{\beta}^2_{r+1}}{e^2_r} =$$
$$= \left(\sum_{\varphi\in A\cup B} e_r(\varphi)\right)\frac{e^3_r\bar{\beta}^2_{r+1}}{e^2_r} = (N^2_{r+1}-1)\frac{\xi_{23}}{e^2_r} = \frac{\xi_{23}}{e^2_{r+1}} - \frac{\xi_{23}}{e^2_r}.$$

Joining this inequality with the computations made above for the rest of the branches we finish the proof.

(3.3.2) Remark. Looking to the hypothesis in the above Proposition it is straightforward that the roles of f_2 and f_3 can be interchanged. So, the similar Proposition for $\tilde{f} = f_1 f_3$, $v_{\tilde{f}}(G) = \tau(f_1 f_3)$ and so on is also true. Now we will start with the Factorization Theorem for the case of three branches:

(3.3.3) Theorem. *Assume that $\underline{v}(A) = \tau$ for some $A \in C[[X,Y]]$, then A can be factorized as*

$$A = A' \prod_{L\in\mathcal{D}^2} A^2_L \prod_{L\in\mathcal{D}^3} A^3_L$$

in such a way that

(1) $\underline{v}(A^j_L) = n_L B_L$ *and* $m(A^j_L) = n_L\delta_L$, $(j = 2,3$ *and* $L \in \mathcal{D}^j)$.

(2) $H(f, A^j_{Lk}) = \Gamma_L$ *for any* A^j_{Lk} *irreducible component of* A^j_L *and* $L \in \mathcal{D}^j$

(3) *For any ϕ irreducible component of A' one has that* $(\phi|f_i) \leqslant (r,s)$ *(for $i = 2,3$) and* $h(\phi, f_2) = h(\phi, f_3)$.

Proof: The proof of (1) and (2) is the same for the indexes 2 and 3, so for the sake of simplicity let us only prove these statements for the dead arcs of $\mathcal{D}^3$. Following the notations of (1.2) we can put $\mathcal{D}^3 = \{L^3_{n+1}, \ldots, L^3_{g_3}\}$, where, by (2.2.5), $n = r+1$ if $h(f_3, f_2) < \gamma^3_{r+1}$ and $n = r+2$ if $h(f_3, f_2) = \gamma^3_{r+1}$.

Let $t \geqslant n$ be the minimum integer such that there exist factors $A^3_i = A^3_{L^3_i}$ $(t+1 \leqslant i \leqslant g_3)$ of A satisfying the requirements in the Theorem. We must prove that $t = n$, so assume $t > n$ and the proof consists in the construction of A^3_t. For the series $G = A/(A^3_{t+1}\ldots A^3_{g_3})$ we have $\underline{v}(G) = \tau(f) - \sum_{t+1}^{g_3} n^3_i B^3_i$, in particular,

$$(v_1(G), v_2(G)) = \tau(f_1 f_2) + (\xi_{13}/e^3_t, \xi_{23}/e^3_t)$$
$$v_3(G) = \sum_0^t n^3_i\bar{\beta}^3_i + \xi_{13} + \xi_{23}.$$

Before to continue with the proof we need to prove the analogous to the Lemma (3.1.6) for this case:

(3.3.4) Lemma. *Under the above conditions. Then, there does not exist a factor of G, Φ, such that*

(1) $h(f_3, \phi) \geqslant \gamma_t^3$ for every irreducible factor ϕ of Φ.

(2) $v_3(\Phi) \geqslant N_t^3 \bar{\beta}_t^3$.

In particular, an irreducible factor of G for which the inequality (1) is strict does not exist.

PROOF: Suppose that there exists Φ, a factor of G, as in the conditions of the Lemma. First of all, note that, if ϕ is an irreducible factor of Φ then $v_3(\phi) \geqslant e_{t-1}\phi\bar{\beta}_t^3$. Hence, because $h(f_3, \phi) \geqslant \gamma_t^3 > h(f_3, f_2) > h(f_3, f_1)$, we have: $v_1(\phi) = e_{t-1}(\phi)\xi_{13}/e_{t-1}^3$ and $v_2(\phi) = e_{t-1}(\phi)\xi_{23}/e_{t-1}^3$. As in (3.1.6) it suffices to show that $e = \sum_{\phi|\Phi} e_{t-1}(\phi)$ is lower than N_t^3.

Suppose that $e \geqslant N_t^3$, multiplying by irreducible series φ with maximal contact of genus $t-1$ with f_3 if it is necessary we can assume that $e = N_t^3$. Putting $\widetilde{f} = f_1 f_2$ we have $v_{\widetilde{f}}(\Phi) = (\xi_{13}/e_t^3, \xi_{23}/e_t^3)$ and so,

$$v_{\widetilde{f}}(G/\Phi) = \tau(\widetilde{f}).$$

Proposition (3.3.1) applied to G/Φ gives us $v_3(G/\Phi) \geqslant \sum_0^r n_i^3 \bar{\beta}_i^3 + \xi_{13} + \xi_{23} - \xi_{23}/e_r^2$. On the other hand $v_3(\Phi) \geqslant N_t^3 \bar{\beta}_t^3 \geqslant \sum_{r+1}^t n_i^3 \bar{\beta}_i^3 + \bar{\beta}_{r+1}^3$ and so, for G, joining both inequalities:

$$v_3(G) \geqslant \sum_0^t n_i^3 \bar{\beta}_i^3 + \xi_{13} + \xi_{23} + \bar{\beta}_{r+1}^3 - \xi_{23}/e_r^2 \geqslant v_3(G).$$

Now, note that the only possibility to have an equality is $t = r + 1$ (see (1.1.2)) and $h(f_3, f_2) < \gamma_{r+1}^3$ (by (3.3.1)); but in this case $\bar{\beta}_{r+1}^3 > \xi_{23}/e_r^2$. As a consequence, in any case we reach a contradiction and the Lemma is proved.

Returning to the proof of the Theorem, note that, looking to the expression for $v_3(G)$, e_{t-1} divide ξ_{13} and ξ_{23}, so, as in (3.1.7) there exists irreducible components φ of G such that $h(f_3, \varphi) \geqslant \gamma_t^3$. By the above Lemma these branches satisfy $h(f_3, \varphi) = \gamma_t^3$ and if $\Phi = \prod_{h(f_3,\varphi)=\gamma_t^3} \varphi$ then Φ satisfies the requirements for A_t^2 as we want to prove.

It only remains to prove the statement (3). To do this, assume that there exists a branch ϕ of A' such that $h(\phi, f_3) > h(\phi, f_2)$. Using (1.1.8) we can prove that $v_1(\phi) = \xi_{13}/e_r^3$, $v_2(\phi) = \xi_{23}/e_r^3$ and $v_3(\phi) > e_0^3 v_2(\phi)/e_0^2 = \xi_{23}/e_r^2$. As a consequence

$$v_1(A'/\phi) = \sum_0^{g_1} n_i^1 \bar{\beta}_i^1 + \xi_{12}/e_r^2$$

$$v_2(A'/\phi) = \sum_0^r n_i^2 \bar{\beta}_i^2 + \xi_{12}.$$

Let $\widetilde{f}_2$ be an irreducible curve of genus r with maximal contact of genus r with f_2, i.e., $[f_2, \widetilde{f}_2] = \bar{\beta}^2_{r+1}$. Note that we have $h(f_3, \widetilde{f}_2) > h(f_3, f_1)$ and $[f_3, \widetilde{f}_2] = \xi_{23}/e^2_r$. Using the formulae of section 1 and denoting by $\widetilde{v}_2$ the valuation of $\widetilde{f}_2$ we reach that $(v_1(A'/\phi), \widetilde{v}_2(A'/\phi)) = \tau(f_1\widetilde{f}_2)$. Now, applying Proposition (3.3.1) to $f_1, \widetilde{f}_2, f_3$ we have $v_3(A'/\phi) \geqslant \sum_0^r n^3_i \bar{\beta}^3_i + \xi_{13}$. Hence, for A':

$$v_3(A') \geqslant \sum_0^r n^3_i \bar{\beta}^3_i + \xi_{13} + v_3(\phi) > \sum_0^r n^3_i \bar{\beta}^3_i + \xi_{13} + \xi_{23}/e^2_r = v_3(A')$$

and we have the wanted contradiction.

This finish the proof of the Theorem.

(3.3.5) In the above Theorem there is no result for the dead arcs of $\mathcal{D}^1$ or $\mathcal{D}^{\{23\}}$. However, when $\mathcal{D}^{\{23\}} = \emptyset$ the Theorem can be extended for $\mathcal{D}^1$, proving the existence of the factors A^1_L with the usual requirements.

(3.3.6) Theorem. *With the same conditions and notations as in (3.3), assume that $\mathcal{D}^{\{23\}} = \emptyset$. Let $A \in \mathbb{C}[[X, Y]]$ be such that $\underline{v}(A) = \tau$. Then A can be factorized as*

$$A = A' \prod_{j=1}^{3} \prod_{L \in \mathcal{D}^j} A^j_L$$

in such a way that

(1) $\underline{v}(A^j_L) = n_L B_L$ and $m(A^j_L) \doteq n_L \delta_L$ for $j = 1, 2, 3$ and $L \in \mathcal{D}^j$.

(2) $H(f, A^j_{Lk}) = \Gamma_L$ for any A^j_{Lk} irreducible component of A^j_L and $L \in \mathcal{D}^j$.

(3) For any ϕ irreducible component of A' one has that $(\phi|f_i) \leqslant (q+1, 0)$ and $h(\phi, f_2) = h(\phi, f_3)$.

Proof: Note that the condition $(f_1|f_2|f_3) = (q, c)$ implies that $\mathcal{D}^I = \{L_1, \ldots, L_q\}$ where $L_i = L^j_i$ for any $j \in I$, $i \leqslant q$. Thus, the condition $\mathcal{D}^{\{23\}} = \emptyset$ implies in particular $(f_2|f_3) = (q, s)$ and so, $h(f_2, f_3) < \gamma^2_{q+1}$ or $h(f_3, f_2) < \gamma^3_{q+1}$. We assume $h(f_3, f_2) < \gamma^3_{q+1}$ and as a consequence we have $\mathcal{D}^3 = \{L^3_{q+1}, \ldots, L^3_{g_3}\}$ and $\mathcal{D}^2 = \{L^2_{n+1}, \ldots, L^2_{g_2}\}$ where $n = q$ if $h(f_2, f_3) < \gamma^2_{q+1}$ and $n = q+1$ if $h(f_2, f_3) = \gamma^2_{q+1}$. We put $L^1_{m+1}, \ldots, L^1_{g_1}$ to denote the dead arcs of $\mathcal{D}^1$, note that $m = q$ if $h(f_1, f_2) < \gamma^1_{q+1}$ and $m = q+1$ if $h(f_1, f_2) = \gamma^1_{q+1}$.

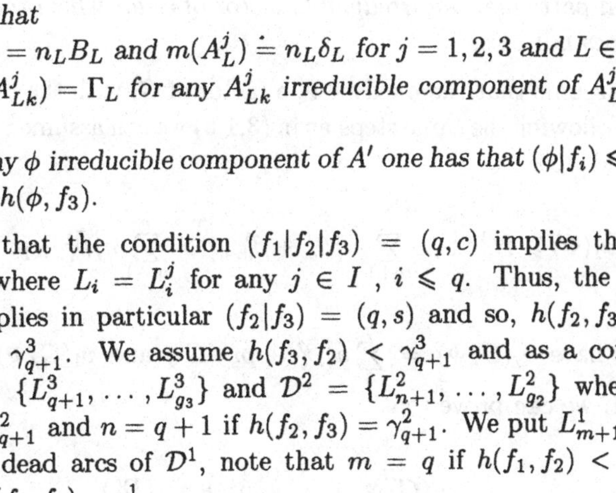

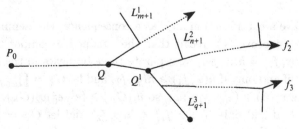

Figure 11

The existence of the factors $A_i^2 = A_{L_i^2}^2$ $(n+1 \leqslant i \leqslant g_2)$ and $A_j^3 = A_{L_j^3}^3$ $(q+1 \leqslant j \leqslant g_3)$ is proved in (3.3.3). A straightforward computation shows that $v_1(A_k^j) = \xi_{1j}/e_k^j - \xi_{1j}/e_{k-1}^j$ and so,

$$v_1(\prod_{L\in\mathcal{D}^2} A_L^2 \prod_{L\in\mathcal{D}^3} A_L^3) = \xi_{13} - \frac{\xi_{13}}{e_q^3} + \xi_{12} - \frac{\xi_{12}}{e_n^2}.$$

After to show the existence of the factors A_i^1, $i \geqslant m+1$, the statement (3) can be proved in the same way as in the above Theorems. So, we only shows the existence of $A_{m+1}^1, \ldots, A_{g_1}^1$. To do this, suppose, as usual, that we have $A_{t+1}^1, \ldots, A_{g_1}^1$ with the requirements in the Theorem and assume that $t > m$. We want to build the factor A_t^1. Let $G = A/(A_{t+1}^1 \cdots A_{g_1}^1 A_{n+1}^2 \cdots A_{g_3}^3)$. As in the proof of the above Theorems there exist branches φ of G such that $v_1(\varphi) \geqslant \gamma_r^1$ and we only need to show the analogous to Lemma (3.1.6) or (3.3.4) in order to finish the proof:

(3.3.7) Lemma. *Under the above conditions. Then, there does not exist a factor of G, Φ, such that*

(1) $h(f_1, \phi) \geqslant \gamma_t^1$ for every irreducible factor ϕ of Φ.

(2) $v_1(\Phi) \geqslant N_t^1 \bar\beta_t^1$.

In particular, an irreducible factor of G for which the inequality (1) is strict does not exist.

PROOF: Assume that there exists Φ, a factor of G, with the conditions of the Lemma. Following the same steps as in (3.1.6) we can assume $v_j(\Phi) = pr_j(D)$, $j = 2, 3$ and

$$v_1(\Phi) \geqslant N_r^1 \bar\beta_r^1 \geqslant \sum_{m+1}^{r} n_i^1 \bar\beta_i^1 + \bar\beta_{m+1}^1 > \sum_{m+1}^{r} n_i^1 \bar\beta_i^1 + pr_1(D).$$

Hence we have $v_j(G/\Phi) = \sum_0^q n_i^j \bar\beta_i^j + pr_j(D')$ and $v_1(G/\Phi) < \sum_0^q n_i^1 \bar\beta_i^1 + pr_1(D_1')$. If we can prove that

$$(*) \qquad v_1(G/\Phi) \geqslant \sum_0^q n_i^1 \bar\beta_i^1 + pr_1(D_1')$$

then we have a contradiction and, as a consequence, the Lemma is proved.

In order to prove $(*)$, recall that the irreducible components φ of G/Φ satisfies $h(\varphi, f_2) = h(\varphi, f_3)$, we separate these branches in the following way:

i) $\varphi \in F$ if and only if $h(\varphi, f_1) \geqslant h(\varphi, f_2)$ and let $G_F = \prod_{\varphi\in F}\varphi$. Note that in this case $v_1(\varphi) \geqslant (e_0^1/e_0^2) v_2(\varphi)$, so $v_1(G_F) \geqslant (e_0^1/e_0^2) v_2(G_F)$.

ii) $\varphi \in B$ if and only if $h(\varphi, f_1) < h(\varphi, f_2)$ and let $G_B = \prod_{\varphi\in B}\varphi$. The condition $\varphi \in B$ means that "f_2 come apart of φ after f_1", in this case we finds that $h(f_1,\varphi)=h(f_1,f_2)$, $h(f_2,\varphi)<h(f_2,f_3)$ and in particular one can check that

$$v_1(\varphi)=e_q(\varphi)\xi_{12}/e_q^2=e_q(\varphi)\xi_{13}/e_q^2 \qquad v_2(\varphi)<e_q(\varphi)\xi_{23}/e_q^3.$$

Denote $e=\sum_{\varphi\in B} e_q(\varphi)$, then $v_1(G_B)=e\xi_{13}/e_q^3$ and $v_2(G_B)<e\xi_{23}/e_q^3$.

Taking into account the value $v_2(G/\Phi)$ one has

$$v_2(G_F)=v_2(G)-v_2(G_B)>\sum_0^q n_i^2\bar\beta_i^2+pr_2(D')-e\xi_{23}/e_q^3$$

and then

$$v_1(G/\Phi)\geqslant e_0^1/e_0^2 v_2(G_F)+v_1(G_B)>\sum_0^q n_i^1\bar\beta_i^1+\frac{e_0^1}{e_0^2}(pr_2(D')-e\frac{\xi_{23}}{e_q^3})+e\frac{\xi_{13}}{e_q^3}.$$

If we can prove that

$$(**)\qquad T:=\frac{e_0^1}{e_0^2}\left(pr_2(D')-e\frac{\xi_{23}}{e_q^3}\right)+e\frac{\xi_{13}}{e_q^3}\geqslant pr_1(D')$$

then we find the inequality $(*)$ and the proof is finished. To do this we distinguish two cases:

a) Suppose $h(f_2,f_3)<\gamma_{q+1}^2$. Here $D'=(\xi_{12}/e_q^2=\xi_{13}/e_q^3,\xi_{23}3e_q^3,\xi_{23}/e_q^2)$. If $e\geqslant 2$ then

$$v_1(G/\Phi)\geqslant v_1(G_B)\geqslant 2\xi_{23}/e_q^3>\xi_{23}/e_q^3+\sum_0^q n_i^1\bar\beta_i^1$$

is the inequality $(*)$. So, assume $e\leqslant 1$. Then, using $h(f_2,f_1)<h(f_2,f_3)$:

$$T=\frac{e_0^1\xi_{23}}{e_0^2e_q^3}(1-e)+e\frac{\xi_{13}}{e_q^3}>T=\frac{e_q^1\xi_{12}}{e_q^2e_q^1}(1-e)+e\frac{\xi_{13}}{e_q^3}=\frac{\xi_{12}}{e_q^2}=pr_1(D').$$

b) Suppose $h(f_2,f_3)=\gamma_{q+1}^2$. Here $D'=(\xi_{12}/e_{q+1}^2,N_{q+1}^2\bar\beta_{q+1}^2,\xi_{23}/e_{q+1}^2)$. As above, if $e>N_{q+1}^3$ then

$$v_1(G/\Phi)\geqslant v_1(G_B)\geqslant\xi_{13}/e_{q+1}^3+\xi_{13}/e_q^3>pr_1(D')+\sum_0^q n_i^1\bar\beta_i^1$$

proves $(*)$. So, assume $e\leqslant N_{q+1}^2$. Then, using $h(f_2,f_1)<h(f_2,f_3)$:

$$T=\frac{e_q^1}{e_q^2}\bar\beta_{q+1}^2(N_{q+1}^2-e)+e\frac{\xi_{13}}{e_q^3}\geqslant\frac{\xi_{12}}{e_q^2}(N_{q+1}^2-e)+e\frac{\xi_{13}}{e_q^3}\geqslant=N_{q+1}^2\frac{\xi_{12}}{e_q^2}=pr_1(D').$$

Thus $(**)$ is proved and the proof for the Lemma is finished.

(3.3.8) Remark. Note that if we take two branches f_1 and f_2 and let h be a regular element of $\mathbb{C}[[X, Y]]$ then the polar curve of $f = f_1 f_2$ with respect to h, $P(f)$, verifies that $\underline{v}(P(f)) = \tau(fh)$ and the above decomposition Theorems provide the corresponding ones for $P(f)$ (see [**7** ,(3.6)]), but in this case the information given by $[P(f), h]$ (3rd coordinate of τ) suffices to show what happens to the components of A' (see [**7 (3.11) (3.12)**]).

(3.3.9) Remark. Let φ, ψ, ϕ be as in the Introduction and $f := \varphi\psi$. The method used in the paper does not made any difference between the branches of φ and ψ, so the results are valid for any combination of the branches of f and not only for the specific one given by φ and ψ. We think that techniques which stands out the functions φ and ψ separately can give more successfully results.

Although the arithmetical methods do not provide a way to explain (in general cases) the behaviour of $C(\phi)$ with respect to the topological type of $\varphi\psi$ the above results suggest that some regularity can occur, at least for the components of $C(\phi)$ with a high contact coefficient with respect to $\varphi\psi$. For example the following questions are natural: Do similar results exist for all the dead arcs corresponding to only one branch of $\varphi\psi$? and if this question has an affirmative answer, What happens to the components of the factor A' similar to the one in the above results?

(3.3.10) Remark. Given φ, ψ as at the beginning one can consider the pencil of germs $\Gamma = \{\lambda\varphi + \mu\psi \mid (\lambda : \mu) \in \mathbf{P}^1_{\mathbf{C}}\}$ generated by φ and ψ. If we take another system of generators of Γ, that is a pair of elements of Γ, φ', ψ' such that $\Gamma = \{\lambda\varphi' + \mu\psi'\}$, one can check that $j(\varphi, \psi)$ and $j(\varphi', \psi')$ differ only in the product by a constant, so the critical point set of the pair (φ, ψ) depends on the pencil and not on the chosen basis of it.

Moreover, consider the differential form $w = \varphi d\psi - \psi d\varphi$ associated to φ/ψ or to the pencil Γ, then $dw = 2j(\varphi, \psi)dx \wedge dy$ and then one can recover $j(\varphi, \psi)$ from w. These facts suggest that the information given for $j(\varphi, \psi)$ depends on weaker data than φ and ψ (the pencil Γ and the differential form w) so it is natural to question about the relations between the resolution of the singularities of the pencil Γ or of the differential form w and the behaviour of $j(\varphi, \psi)$. On the other hand $j(\varphi, \psi)$ also provides information about some pathologies of Γ and w ([**9**])

REFERENCES

[1] E. BRIESKORN - K. KNÖRRER *Plane algebraic curves.* Birkäuser Verlag, Basel 1986.

[2] CAMPILLO, A. *Algebroid curves in positive characteristic.* Lecture Notes in Math. 813, Springer Verlag, 1983.

[3] CASAS,E. *On the singularities of polar curves* . Manuscripta Mat. 43, 167-190 (1983).

[4] CASAS,E. *Infinitely near imposed singularities and singularities of polar curves* . Math. Ann. 287, 429-454, (1990).

[5] CASAS,E. *Base points of polar curves* . Ann. de l' Institut Fourier, 41 (1), 1-10, (1991)

[6] DELGADO,F. *The semigroup of values of a curve singularity with several branches.* Manuscripta Mat. 59, 347-374, (1987).

[7] DELGADO, F. *A factorization Theorem for the polar of a curve with two branches.* Preprint, Universidad Complutense, 1989/90.

[8] EPHRAIM,R. *Special polars and curves with one place at infinity* . Proc. of Symposia in Pure Math. 40 (I), 353-361, (1983).

[9] GARCIA DE LA FUENTE, J. *Geometria de los sistemas lineales de series de potencias en dos variables.* Thesis, Valladolid University 1989.

[10] HIRONAKA,H. *Introduction to the Theory of infinitely near singular points* . Memorias del Instituto Jorge Juan, CSIC, Madrid, 1974.

[11] KUO,T.-C. - LU, Y.-C. *On analytic function germs of two complex variables* . Topology 16, 299-310, (1977).

[12] LÊ-D. T. - MICHEL,F. - WEBER,C. *Courbes polaires et topologie des courbes planes* . Ann. scient. Ec. Norm. Sup. 4e série, t. 24, 141-169 (1991)

[13] LÊ-D. T. - MICHEL,F. - WEBER,C. *Sur le comportement des polaires associés aux germes de courbes planes* . Compositio Math. 72, 87-113 (1989).

[14] LEJEUNE, M *Sur l'equivalence des courbes algebroides planes. Coefficients de Newton.* Thesis, Univ. Paris VII (1973).

[15] MERLE,M. *Invariants polaires des courbes planes* . Inventiones Math. 41, 103-111, (1977).

[16] TEISSIER,B. *Varietés polaires I. Invariants polaires des singularités d'hypersurfaces* . Invent. Math. 40, 267-292, (1977).

[17] ZARISKI,O. *Studies in Equisingularity I,II y III* . (I): Am. J. Math. 87, 507-535, (1965). (II): Ibd. 87, 972-1006, (1965). (III): Ibd. 90, 961-1023, (1968).

[18] ZARISKI,O, *General theory of saturation and saturated local rings II.*. Am. J. Math. 93, 872-964 (1971)

[19] ANGERMULLER, G. *Die wertehalbgrupp einer ebenen irreduciblen algebroiden kurve.*. Math. Zeitchrift, 153, 267–282 (1977)

[20] HERZOG, J. *Generators and relations of abelian semigroups and semigroup rings.*. manuscripta math. 3, 175–193 (1970)

[21] PLOSKI, A. *Remarque sur la multiplicité d'intersection des branches planes.*.Bull. of the Polish Acad. of Sci. 33, 601–605 (1985).

Current address: Félix Delgado de la Mata. Dto. de ALgebra, Geometría y Topología. Universidad de Valladolid. E-47005 VALLADOLID. Spain

Trivializations of stratified spaces with bounded differential

MASSIMO FERRAROTTI

University of Pisa

In memory of Mario Raimondo

Abstract We study the existence of trivializations of a riemannian stratified space whose differential has bounded norm giving sufficient conditions. We also prove some consequences of this property, the most meaningful of which is local finiteness of volume.

INTRODUCTION

This paper is the development of some considerations already contained in $[F_3]$ about trivializations of stratified spaces. It is well known from Thom's isotopy lemmas that a *b*-regular (in the sense of Whitney) subset W of a smooth manifold is locally trivial in two senses : in one sense W is locally homeomorphic to the product of a smooth manifold with a stratified set, in the other the homeomorphism is with a mapping cylinder. Moreover, these homeomorphims are diffeomorphims on the strata. The abstract counterpart of these sets are the abstract or Thom-Mather stratifications (which we call stratified spaces). If we consider on a stratified space W a collection of riemannian metrics $g = \{g_X\}$, one for any stratum (in the embedded case g should be the restriction of a riemannian metric on the ambient manifold), it makes sense to ask if there exist trivializations such that the norm of the differential is bounded (D-bounded trivializations). We shall see that, in the mapping cylinder case, this has as a consequence that our space has locally finite volume. We point out that a *b*-regular stratified set has not in general locally finite volume, as shown in $[F_1]$, and that the positive results about finite volume given in $[F_1]$ and [B] are based on the boundedness of the derivatives of a trivialization. We remark that integration theories on stratified sets ([D],$[F_2]$) always need finite volume hypothesis and this could be one interest in looking for this kind of conditions.

We consider trivializations of mapping cylinder type. In detail, in section I we give some basic definitions and we show an extimate for the norm of the differential of a controlled flow which involves, as expected, second order derivatives (compare with [Ms], [W]). In section II D-bounded trivializations are introduced and consequences of their existence are proved : in particular

Lavoro svolto nell'ambito del GNSAGA del CNR con contibuto MURST fondi 40%.

we show that a space which has D-bounded trivializations at any point has locally finite volume. Section III contains sufficent conditions to have D-bounded trivializations : since our trivializations are obtained by integration of vector fields, we use section I result and we find that some of the conditions are related to the curvatures of the strata. Other necessary conditions are given too. At the end, in section IV we examine the case of an embedded stratified set with regularity conditions and we describe some applications.

I. STRATIFIED SPACES

By *stratified space* we mean an abstract (or Thom-Mather) stratification ([M], [T] ; see [V] for basic definitions and properties). If W is a stratified space, we shall denote with $\mathcal{S}_W$ the stratification (family of strata) of W and with $\mathcal{I}_W = \{(V_X, \phi_X, r_X); X \in \mathcal{S}_W\}$ the control data system. We recall some basic definitions (by "smooth" we mean of class C^r, $2 \leq r \leq \infty$).

a) W is a paracompact, locally compact, Hausdorff topological space with countable basis and $\mathcal{S}_W$ is locally finite partition of W into finite dimensional, locally closed smooth manifolds without boundary s.t. if $X, Y \in \mathcal{S}_W$, $X \cap \bar{Y} \neq \emptyset \Leftrightarrow X \subseteq \bar{Y}$. We write $X < Y$ for $X \subseteq \bar{Y}$ and we define $\operatorname{depth} X = \sup\{k; \exists \text{ a chain } X = X_1 < \cdots < X_k\}$, $\operatorname{depth} W = \sup\{\operatorname{depth} X\}$.

b) Let W, W' be topological spaces endowed with stratifications $\mathcal{S}_W$, $\mathcal{S}_{W'}$ and let $U \subset W$ be open ; $f : U \to W'$ is said to be a *stratified map* if $\forall X \in \mathcal{S}_W \quad \exists X_f \in \mathcal{S}_{W'}$ s.t. $f(X) \subseteq X_f$ and $f|_X$ is smooth.

c) Let $X \in \mathcal{S}_W$ and let $X \times [0, \infty)$ be stratified by $\mathcal{S} = \{X \times \{0\}, X \times \mathbf{R}^+\}$; then :

i) $\phi_X = (\pi_X, \rho_X) : V_X \to X \times [0, +\infty)$ is a continuous stratified map from a neighbourhood (nbd.) V_X of X in W with $\pi_X|_X = Id$ and $\rho_X^{-1}(0) = X$. If $X, Y \in \mathcal{S}_W$, $X \neq Y$, then $V_X \cap V_Y \neq \emptyset \Leftrightarrow X < Y$ or $Y < X$.

ii) $r_X : X \to \mathbf{R}^+$ is a smooth function (or $r_X \equiv +\infty$) s.t. the set

$$B_X = \{\rho_X(x) < r_X \circ \pi_X(x)\}$$

is a nbd. of X in V_X

iii) $\phi_{X|B_X}$ is a submersion on any stratum.

iv) If $Y > X, \pi_X \circ \pi_Y = \pi_X$ and $\rho_X \circ \pi_Y = \rho_X$ where the compositions are defined.

(if $\operatorname{depth} X = 0$, $V_X = X$, $\phi_X \equiv (Id, 0)$ and $r_X \equiv 0$).

d) A stratified subspace H of W is a locally closed subset such that the families $\mathcal{S}_H = \{X \cap H; X \in \mathcal{S}_W\}$ and $\mathcal{J}_H = \{(V_X \cap H, \phi_X|_{V_X \cap H}, r_X|_{X \cap H}\}$ make H a stratified space. Let $X \in \mathcal{S}_W$, $A \subseteq X$ a submanifold and $t \in (0, r_X)$; then $E(A,t) = \phi_X^{-1}(A \times \{t\})$ and $B(A,t) = \phi_X^{-1}(A \times [0,t))$ are stratified subspaces of W. In particular, for $x \in B_X$, we define $F_{x,t} = B(\{\pi_X(x)\}, t)$, and $K_x = E(\{\pi_X(x)\}, \rho_X(x))$ (the fibers of π_X and ϕ_X through x). If $y = \pi_X(x)$, we also write $F_{x,t} = F_t^y$.

e) A stratified C^0 map $f : U \subseteq W \to W'$ is said to be *controlled* if $\forall X \in \mathcal{S}_W$ we have (1) $\pi_{X_f} \circ f = f \circ \pi_X$ and (2) $\rho_{X_f} \circ f = \rho_X$. If (1) holds only we say f is *weakly controlled.* An isomorphism $f : W \to W'$ will be a stratified homeomorphism f such that $f|_X$ is a diffeomorphism for any $X \in \mathcal{S}_W$ and f, f^{-1} are controlled.

We call $\{\mathcal{S}_W, \mathcal{J}_W\}$ the *stratified structure* of W. We assume that $\dim W = \sup\{\dim X\} < \infty$, the strata are connected and $\phi_{X|B_X}$ is proper. Moreover, if $0 < r_X < \infty$, we substitute ρ_X with $\rho_X / r_X \circ \pi_X$ so that $r_X \equiv 1$ in these cases (see [GW]).

If $x \in W$, $X \in \mathcal{S}_W$ is s.t; $x \in X$ and f is a stratified map on W, we shall conventionally write $\dim(W, x)$, $T_x W$ and $d_x f$ instead of $\dim X$, $T_x X$ and $d_x(f|_X)$.

A product of a manifold with a stratified space admits a canonical stratified structure. If E is a stratified space, A a manifold, and $f : E \to A$ a weakly controlled submersion onto, we shall denote with $M(f)$ the (open) mapping cylinder of f, that is

$$M(f) = (E \times [0, +\infty) \amalg A) / \cong$$

where "$(x,t) \cong (x',t')$" $\Leftrightarrow$ "$(x,t) = (x',t')$" or "$t = t' = 0$" and $f(x) = f(x')$".

$M(f)$ is a stratified space in a natural way : if p is the canonical projection from $(E \times [0, +\infty)) \amalg A$ onto $M(f)$, we set $\mathcal{S}_{M(f)} = \{L' = p(L \times [0 + \infty)); L \in \mathcal{S}_E\} \cup \{A\}$, with $\mathcal{J}_{M(f)}$ given by

i) $V_{L'} = p(V_L \times (0, +\infty))$, $r_{L'} = r_L$, $\phi_{L'}([x,t]) = ([\pi_L(x), t], \rho_L(x))$.

ii) $V_A = M$, $r_A \equiv +\infty$, $\phi_A([X,t]) = (f(x), t)$.

($[x,t]$ is the equivalence class of (x,t)).

If $\varepsilon \in \mathbf{R}^+$, we denote $B(A, \varepsilon)$ with $M(f, \varepsilon)$; it is the (open) ε-mapping cylinder of f. If $A =$ one point, $M(f)$ is the cone $C(E)$ on E and $M(f, \varepsilon)$ is the ε-cone $C^\varepsilon(E)$ on E.

If $m = [z,t] \in M(f)$, we describe the elements of $T_m M(f) \simeq T_z E \oplus \mathbf{R}$ as follows. To $v \in T_z E$ we may associate a vector field $v(t)$ along the curve $t \to [z,t]$ defined as $v(t) = (v, 0) \in T_m M(f)$. Then for any $v \in T_m M(f)$, there are unique $\hat{v} \in T_z E$ and $\lambda \in \mathbf{R}$ s.t. $v = \hat{v}(t) + \lambda \frac{d}{dt}$.

A *stratified riemannian metric* g on W is a collection $\{g_X; X \in \mathcal{S}_W\}$, each g_X being a riemannian metric on X. A couple (W, g) where W is a stratified space and g is a stratified riemannian metric on W will be said a *riemannian stratified space* (r.s.s. ; compare with [M_0]). By convention we set $g_x = g_{X,x}$ if $x \in X$.

When we speak of "norm", "covariant derivative", "curvature"etc. on (W, g) we mean the ones induced on each stratum X by g_X and by the Levi-Civita connection of g_X.

Let $\Delta = \{\delta \in C^\infty(\mathbf{R}^+); \delta > 0, \delta(t)$ bounded as $t \to 0\}$ be the set of *weights*. We consider on Δ the ordering "$\delta \leq \delta' \Leftrightarrow \delta/\delta' \in \Delta$". We say that δ, δ' are *equivalent* ($\delta \cong \delta'$) iff $\delta \leq \delta'$ and $\delta' \leq \delta$.

We want now to build up "natural" stratified metrics on mapping cylinders parametrized by Δ. These metrics will generalize cone-like metrics of [C] and will coincide with them in the case of isolated singularities after taking $\delta(t) = t^{2a}, a \leq 0$.

1. Definition. Let (E, g) a r.s.s. , (A, g_A) a riemannian manifold and $f : E \to A$ a weakly controlled submersion. If $z \in E$, we set $\mathcal{K}_z = \ker(d_z f)$ and we denote with P_z and N_z the orthogonal projections of $T_z E$ on $\mathcal{K}_z$ and $\mathcal{K}_z^\perp$ with respect to g_z.

Given $\delta \in \Delta$, the *mapping-cylinder metric* (mc-metric) g^δ on $M(f)$ of weight δ is defined as follows :

i) If $m = [z, t] \in M(f)\backslash A$ and $v = \hat{v}(t) + \lambda \frac{d}{dt} \in T_m M(f)$, then

$$g^\delta_m(v, v) = \delta(t) g_z(P_z \hat{v}, P_z \hat{v}) + g_z(N_z \hat{v}, N_z \hat{v}) + \lambda^2 , \tag{1}$$

ii) $g^\delta_A = g_A$.

2. Definition. Let $f : (W, g) \to (W', g')$ be a stratified map between r.s.s.. If $X \in \mathcal{S}_W$ and $x \in W$, we set $\|df\|(x) = \sup\{\|dfv\|_{g'}; v \in T_x X, \|v\|_g = 1\}$. We say that f is with *bounded differential* (shortly "*f is D-bounded*") on a subset K of W if $\exists C_f > 0$ s.t. $\forall x \in K \quad \|d_x f\|(x) \leq C_f$ (if $K = W$, we simply say "f is D-bounded").

It is immediate that a composition of D-bounded maps is D-bounded.

3. Example. Let M be the mapping cylinder of the map $f : (E, g) \to (A, g_A)$. If $\delta, \delta' \in \Delta$, the map $Id : (M, g^\delta) \to (M, g^{\delta'})$ is D-bounded iff $\delta' \leq \delta$ (note that $\delta \cong 1 \Rightarrow$ Id-bounded C^1- bounded $\forall \delta'$).

A stratified arc in W will be a continuous map $\gamma : I = [0, 1] \to W$ such that $\forall X \in \mathcal{S}_W \; \gamma|_{\gamma^{-1}(X)}$ as a smooth arc in X. In (W, g) a stratified arc will have a length, eventually $= +\infty$.

A first metric property of D-bounded maps concerns geodesic diameter.

4. Definition. (see [Bu]). A subset of $K \subseteq (W, g)$ is said to have *finite* *geodesic diameter in the strong sense* (briefly f.g.d.) if $\exists C > 0$ s.t. any $x, y \in K$ are joined by a stratified arc $\gamma \subseteq K$ of length $< C$.

Then it is immediate from definitions that

5. Proposition. If $K \subseteq W$ and $f : (W, g) \to (W', g')$ is D-bounded on K, then "K has f.g.d $\Rightarrow f(K)$ has f.g.d.".

In the next sections we shall study the D-boundedness for the trivializations given by Thom's isotopy lemmas. The most important tool to build up local trivializations of stratified spaces are controlled vector fields. If W is a stratified space and $U \subseteq W$ is open, we recall the following definitions : a *stratified vector field* on U is a family $\xi = \{\xi_X; X \in \mathcal{S}_W\}$, where ξ_W is a smooth field on $X \cap U$; as before we let $\xi_{X,x} = \xi_x$ if $x \in X$. If we have $d\pi_X \xi_Y = \xi_X$ for any couple $X < Y$, we say that ξ is *weakly controlled* ; if $d\rho_X \xi_Y = 0$ holds too, ξ is *controlled.* We denote with $\mathcal{X}(U), \mathcal{X}_w(U), \mathcal{X}_c(U)$ respectively the stratified, the weakly controlled and the controlled vector fields on U. From the theory of integration of controlled vector fields ([GW],[V]) we have :

6. Lemma. Let W be a stratified space and $f : W \to \mathbf{R}$ be a weakly controlled submersion. If $\alpha < t < \beta$ are in Imf, the sets $B = f^{-1}((\alpha, \beta))$ and $E_t = f^{-1}(t)$ are statified subspaces of W. Let $\xi \in \mathcal{X}_c(B)$ such that $\xi(f) \equiv -1$; then

i) If $U_\xi = \{(x, u) \in W \times \mathbf{R} ; f(x) - \beta < u < f(x) - \alpha\}$, the flow $\Xi : U_\xi \to W$ of ξ is a controlled continuous map on U_ξ and $f(\Xi(x, u)) = -u + f(x)$.

ii) If $\varepsilon \in (\alpha, \beta)$ and $E = E_\varepsilon$, the map $\Xi_t = \Xi|_{E \times \{\varepsilon - t\}} : E \to E_t$ is an isomorphism for any $t \in (\alpha, \beta)$.

To pursue our aim we need to estimate the norm of the differential of the flow of ξ. Let $x \in W$, $\dim(W, x) = p$ and $f(x) = t \in (\alpha, \beta)$; we define on $T_x E_t$ the quadratic form $q_x^\xi(v) = g_x(D_v \xi, v)$. Then q_x^ξ is induced by the symmetric form

$$Q_x^\xi(v, w)_x = \frac{1}{2}(g_x(D_v \xi, w)_X + g_x(D_w \xi, v)), \quad v, w \in T_x E_t$$

Let $\lambda_1(x) \leq \ldots \leq \lambda_{p-1}(x)$ be the eigenvalues of Q_x^ξ ; we set

$$M_\xi(t) = \sup_{y \in E_t} \{\lambda_{p-1}(y)\}, \quad m_\xi(t) = \inf_{y \in E_t} \{\lambda_1(y)\}$$

Evidently $M_\xi(t), m_\xi(t) \in \mathbf{R} \cup (\pm\infty)$. Then we have defined two functions M_ξ, m_ξ from $(0, \varepsilon)$ to the extended line. We shall assume from now on that the integral of M_ξ and m_ξ exists on (t, ε) for any $t \in (0, \varepsilon)$, eventually being equal to $\pm\infty$.

7. Lemma. Let (W, g) be a r.s.s., $f, \alpha, \beta, \varepsilon$ as in Lemma 1.6. Then, for any

$z \in E$, $v \in T_z E$ and $t \in (\alpha, \beta)$, we have

$$\text{(2)} \qquad \exp(\int_t^\varepsilon m_\xi(s)ds)\|v\| \leq \|d_z \Xi_t v\| \leq \exp(\int_t^\varepsilon M_\xi(s)ds)\|v\|$$

Proof : Let $z \in E$ and $v \in T_z E$, $v \neq 0$; if $z \in X$, $X \in \mathcal{S}_W$, for $t \in (\alpha, \beta)$, $d_z \Xi_t v$ is a non-vanishing vector field along the curve $\gamma(t) = \Xi_t(z)$ tangent to the hypersurfaces $E_t \cap X$. If V is a nbd. of z in $E \cap X$ and $v(u)$ is a C^∞ vector field on V s.t. $v(z) = v$, then $\theta_{\Xi_t(u)} = d_u \Xi_t v(u)$ extends C^∞ the previous field in a nbd. of Imγ maintaining the same properties. We have (setting $\theta_s = d_z \Xi_s v$.)

$$\text{(3)} \qquad \frac{D}{ds}(\theta_s)_{\gamma(t)} = -(D_\xi \theta)_{\gamma(t)} = -(D_\theta \xi)_{\gamma(t)} - [\xi, \theta]_{\gamma(t)}$$

Now $[\xi, \theta] = L_\xi \theta = 0$ (compute the Lie derivative) and we get from (3)

$$\text{(4)} \qquad \frac{d}{ds}\|\theta_s\||_{s=t} = -g(D_\theta \xi, \theta/\|\theta\|)_{\gamma(t)}$$

for any t in (α, β). After setting $\eta = \theta/\|\theta\|$, we may write

$$\text{(5)} \qquad dlog\|\theta_s\|_{s=t} = -g(D_\eta \xi, \eta)_{\gamma(t)} = -q^\xi(\eta)_{\gamma(t)}$$

for any t in (α, ε). Finally, by integration and the fact that extremal values on the unit sphere of a quadratic form are given by the eigenvalues of the associated symmetric form, we obtain

$$\text{(6)} \qquad \|d_z \Xi_t v\| = \|\theta_t\| = \exp(\int_t^\varepsilon q^\xi(\eta)_{\gamma(s)}ds)\|v\| \quad \text{and}$$

$$\text{(7)} \quad \exp(\int_t^\varepsilon m_\xi(s)ds)\|v\| \leq \exp(\int_t^\varepsilon q^\xi(\eta)_{\gamma(s)}ds)\|v\| \leq \exp(\int_t^\varepsilon M_\xi(s)ds)\|v\|$$

for any t in (α, β). So (2) is proved.

It is difficult in general to estimate the differential of a vector field flow. In [W] a bound for the norm of a flow is given under the hypothesis that ξ is a unit vector field and $D_\xi \xi = 0$: the fields we use here do not in general fulfil these conditions.

II D-BOUNDED TRIVIALIZATIONS

We study mapping cylinder trivializations with bounded differential ; in particular we prove that their existence implies that volume is locally finite.

1. Definition. Let W be a stratified space, $X \in \mathcal{S}_W$ with depth$X > 0$, $A \subset X$ open and $\varepsilon \in (0, r_X)$. If $B(A, \varepsilon) = B$, $\xi \in \mathcal{X}_w(B)$ is said to be radial (compare with [S]) at X if :

i) $\xi \in \mathcal{X}_c(B \setminus A)$ and $\xi_A = 0$.

ii) $d\rho_X \xi \equiv -1$ on $B \setminus A$.

Radial fields at X on B are just the controlled ϕ_X-lifts on $B \setminus A$ of the field $(0, -\frac{d}{dt})$ on $A \times \mathbf{R}$; this ensures their existence (see [GW]).

Let W be a stratified space, let $X \in \mathcal{S}_W$ with depth$X > 0$ and let $A \subseteq X$ be an open connected subset. For $\varepsilon' \in (0, r_X]$ we set $B = B(A, \varepsilon')$; let $\xi \in \mathcal{X}(B)$ be radial at X. If Ξ denotes the flow of ξ on B and $\varepsilon \in (0, \varepsilon')$, then the expression $\Phi([x,t]) = \Xi(x, \varepsilon - t)$ defines an isomorphism $\Phi : M(\pi_{X|E(A,\varepsilon)}, \varepsilon') \to B$ such that $\Phi_X = Id_X$ and $\Phi|_{E(A,\varepsilon)} = Id_{E(A,\varepsilon)}$; we denote $M(\pi\ |_{E(A,\varepsilon)}, \varepsilon')$ with M. Then

2. Definition. We say that the map $\Phi : M \to B$ just defined is the mapping cylinder trivialization (mct) of W at X induced by ξ and ε. Now, let (W, g) be a r.s.s.. If $X \in \mathcal{S}_W$ and if Φ is a mct of W at X induced by ξ and ε, we have the map

$$\pi_X|_{E(A,\varepsilon)} : (E(A,\varepsilon), g|_{E(A,\varepsilon)}) \to (A, g_X|_A) ;$$

then we shall always consider on $\Phi : M(\pi_{X|E(A,\varepsilon)}, \varepsilon')$ the mc-metrics $g^\delta = (g|_{E(A,\varepsilon)})^\delta$ (Observe that $\mathcal{K}_z = \mathrm{Ker}(d_z\phi_X) = T_z K_z$).

3. Definition. A mct Φ of (W, g) is D-bounded of weight δ if $\Phi : (M, g^\delta) \to (W, g)$ is D-bounded.

4. Remark. By Example I-3, if Φ is D-bounded of weight δ, then Φ is D-bounded of any weight $\delta' \geq \delta$.

5. Proposition. Let Φ be a D-bounded mct of (W, g) of weight δ induced by ξ and ε. Then, for any $\varepsilon_\circ \in (0, \varepsilon')$ such that $\int_\varepsilon^{\varepsilon_\circ} M_\xi(s)ds < +\infty$, the mct $\Phi_\circ$ induced by ξ and $\varepsilon_\circ$ is D-bounded of weight δ.

Proof : Let $\Phi_\circ : M_\circ \to B$; then it is enough to show that $\psi = \Phi^{-1} \circ \Phi_\circ : M_\circ \to M$ is D-bounded with respect to g^δ. Now, from our definition we get $\psi([z,t]) = [\Xi(z, \varepsilon_\circ - \varepsilon), t]$ and our thesis follows from lemma I-7 (here $f = \rho$, $\alpha = 0$, $\beta = \varepsilon'$, $\varepsilon = \varepsilon_\circ$).

6. Corollary. Let $X \in \mathcal{S}_W$ with $depthX = 1$. If there is a D-bounded mct of W at X of weight δ which is induced by a radial field $\xi \in \mathcal{X}_w(B(A, \varepsilon'))$, A relatively compact, then for any $\varepsilon \in (0, \varepsilon')$ the mct induced by ξ and ε is D-bounded of weight δ.

Proof : If $depthX = 1$ and A relatively compact, $E(A, t)$ are relatively compact submanifolds and Prop. II-5 hypothesis if fulfilled for any $\varepsilon, \varepsilon_\circ$.

7. Definition. (W, g) is said to be a D-bounded stratified space if any $x \in W$ admits a nbd. which is the image of a D-bounded mct.

In force of Prop. I-5, a D-bounded compact stratified space has f.g.d.. An easily checked property of D-bounded stratified spaces is the following :

8. Lemma. If (W, g) is D-bounded and $H \subseteq W$ is a stratified subspace, then $(H, g|_H)$ is D-bounded.

We want to show now that D-bounded stratified spaces have metric properties stronger than f.g.d.. If $\dim W = p$ and W^p is the union of p-dimensional strata of W, for any σ-compact subset Q of W we set $\mathrm{Vol}_g(Q) = \mathrm{Vol}_g(Q \cap W^p)$ (compare with [F_1]). (W, g) will have *locally finite volume* (l.f.v.) if $\mathrm{Vol}_g(Q) < \infty$ for any $Q \subseteq W$ relatively compact. Then

9. Theorem. (W, g) D-bounded $\Rightarrow$ (W, g) has l.f.v..

Proof : It is enough to show that each point of W has a nbd. B such that $\mathrm{Vol}_g(B) < \infty$. Let $x \in \overline{W^p}$ (otherwise there is nothing to prove). Now x belongs to a stratum $X \in \mathcal{S}_W$ with $depth X = k$. If $k = 0$, our thesis is trivial. Then let $k > 0$ and let us consider a maximal chain of strata $X = X_0 < \ldots < X_k$; we set $dim X_1 = q_1$ $(q_k = p)$.

By hypothesis there are a relatively compact connected nbd. A_0 of x in X, $0 < \varepsilon_0 < \varepsilon_0' < r_X$, a radial field ξ_0 on $B = B(A_0, \varepsilon_0')$ and $\delta_0 \in \Delta$ s.t. the mct Φ_0 induced by ξ_0, ε_0 is D-bounded of weight δ_0.

Now $E_0 = E(A_0, \varepsilon_0)$ is a stratified subspace of W ; by an iterated use of Lemma II-8, we get $E_1, \ldots, E_{k-1}$ stratified subspaces of W s.t., for $0 \leq i \leq k-1$,

1) $\dim E_1 = p - i - 1$ and E_{k-1} is a submanifold of X_k ;

2) There are $\varepsilon_1 \in (0, r_{X_1})$ and $A_i \subseteq E_{i-1} \cap X_i$ relatively compact connected open subsets such that $E_i = E(A_i, \varepsilon_i) \subseteq E_{i-1}$;

3) There are $\varepsilon_i' \in (\varepsilon_i, r_{X_i})$, radial fields $\xi_i \in \mathcal{X}(B(A_i, \varepsilon_i'))$ and $\delta_i \in \Delta$ s.t. the mct's Φ_i of E_i generated by ξ_i, ε_i is D-bounded of weight δ_i.

We may assume by Remark II-4 that $\delta_i(\varepsilon_i) = 1$ in 3). Let now (U, h) be a relatively compact connected coordinate chart of E_{k-1} and let $h(U) = \Omega \subseteq \mathbf{R}^{p-k}$. We define recursively the maps $\Theta_i : \Omega \times (0, \varepsilon_{k-1}') \times \cdots \times (0, \varepsilon_i') \to X_k$ as :

$\Theta_k = h^{-1} : \Omega \to X_k$ and $\Theta_i(u, t_{k-1}, \ldots, t_i) = \Phi_i([\Theta_{i+1}(u, t_{k-1}, \ldots, t_{i+1}), t_i])$ for $i < k$ and $u \in \Omega$.

If we set $\Gamma = \Omega \times (0, \varepsilon'_{k-1}) \times \cdots \times (0, \varepsilon'_0) \subseteq \mathbf{R}^p$, the map $\Theta = \Theta_0 : \Gamma \to X_k$ is a local parametrization of X_k which is D-bounded with respect to the euclidean metric on $\mathbf{R}^p$ and g_{X_k}. Now

$$\mathrm{Vol}_g(\Theta(\Gamma)) = \int_\Gamma (\det\{\Theta^* g_{X_k}(\partial_a, \partial_b\})^{1/2} \leq \mathrm{Const}.\varepsilon'_0 \ldots \varepsilon'_{k-1}.\mathrm{Vol}(\Omega) < \infty,$$

where ∂_ℓ, $1 \leq \ell \leq p$, are the coordinate fields on $\mathbf{R}^p$.

In force of our compactness assumptions, we may cover $B \cap W^p$ with a finite number of open sets of the kind $\Theta(\Gamma)$; this ends our proof.

We observe that the opposite implication does not hold : spaces in $[F_1]$ with l.f.v. but not v-regular give counter-examples.

III EXISTENCE THEOREMS

At this point a natural question is to look for sufficient conditions to have D-bounded mct. Some of the conditions we find concern covariant derivatives of radial vector fields ; in some cases these conditions are expressed in terms of the principal curvatures of certain hypersurfaces.

Let (W, g) be a r.s.s., $\delta \in \Delta$ and let $X \in \mathcal{S}_W$ with $depth X > 0$. Given $A \subseteq X$ open, connected and relatively compact and ε, ε' with $0 < \varepsilon < \varepsilon' < r_X$, we set $\phi = \phi_X = (\pi_X, \rho_X) = (\pi, \rho)$, $B = B(A, \varepsilon')$, $E = E(A, \varepsilon)$, $F_x = F^{\pi(x)} = F_{x,\varepsilon'}$. If $x \in B$, we let $T_x K_x = \ker(d_x \phi) = \mathcal{K}_x$.

If $\xi \in \mathcal{X}(B)$ is radial at X and if Ξ denotes the flow of ξ, we set $\mathcal{H}_x = d_{z,t}\Xi(\mathcal{K}_z^\perp)$ for $x = \Xi(z, t)$. From the above considerations, we have $T_x E(A, t) = \mathcal{K}_x \oplus \mathcal{H}_x$; P_x and N_x will denote the projections on the first and second factor respectively. If no danger of confusion occurs, we write $g^\delta = g'$.

For any $x \in A$, $\xi^x = \xi|_{F^x} \in \mathcal{X}_w(F^x)$ and is radial ; then, up to take $f = \rho_X$, $\alpha = 0$, $\beta = \varepsilon'$, we may set $M_x(t) = M_{\xi^x}(t)$ and $m_x(t) = m_{\xi^x}(t)$.

We introduce now two "natural assumptions", in the sense that they hold if depth$X = 1$ and in the embedded case for any depth (see sec. IV) :

NA1. The map $\pi|_E : (E, g) \to (A, g_X)$ is D-bounded (with constant C_π),

NA2. $\exists C'_\pi > 0$ s.t. for any $z \in E$ and any $v \in \mathcal{K}^\perp$ the inequality $\|d_z \pi v\| \geq C'_\pi \|v\|$ holds.

With all these positions we have the following

1. Theorem. Let's assume that the following properties are verified :

(P_1) $\exists C_1 > 0$ s.t. $0 < \|\xi\| \leq C_1$ on $B \setminus A$,

(P_2) $\exists C_2$ s.t. $\forall x \in A$, $\forall t \in (0,\varepsilon')\ \int_t^\varepsilon M_x(s)ds < C_2$,

(P_3) $\exists C_3 > 0$ s.t. $\forall x \in B \setminus A$ the inequality $\|d_x\pi v\| \geq C_3\|v\|_g$ holds for any $v \in \mathcal{H}_x$.

Then the mct Φ of W at X induced by ξ, ε is D-bounded of any $\delta \in \Delta$ s.t.

$$(*) \quad \exists C > 0 \quad s.t. \quad \forall x \in A, \forall t \in (0,\varepsilon') \quad \exp(2\int_t^\varepsilon M_x(s)ds) < C\delta(t)$$

Conversely let's assume that Φ is D-bounded (with any weight). Then $(P_1), (P_3)$ and

$$(P_2') \qquad \exists C_2' \quad s.t. \quad \forall x \in A, \quad \forall t \in (0,\varepsilon') \quad \int_t^\varepsilon m_x(s)ds < C_2$$

are verified.

Proof : Let $\Phi : (M, g') \to (B, g)$ be the mct induced by ξ and ε and let's assume that $(P_1), (P_2), (P_3)$ hold. Let Ξ be the flow of ξ ; we consider as in Lemma I-6 the maps $\Xi_t = \Xi|_{E\times\{\varepsilon - t\}}$. Hence, if $m = [z,t] \in M$ and $v = \hat{v}(t) + \lambda\frac{d}{dt} \in T_m M (\hat{v} \in T_z E, \lambda \in \mathbf{R})$, we have

$$(1) \qquad d_m\Phi(v) = d_z\Xi_t\hat{v} - \lambda\xi_{\Phi(m)} = d_z\Xi_t P_z\hat{v} + d_z\Xi_t N_z\hat{v} - \lambda\xi_{\Phi(m)}$$

Hence(to distinguish the norms relative to g and g' we write $\|\ \|_g$ and $\|\ \|_g'$ respectively):

$$(2) \qquad \|d_m\Phi(v)\|_g \leq \|d_z\Xi_t P_z\hat{v}\|_g + \|d_z\Xi_t N_z\hat{v}\|_g + |\lambda|\|\xi_{\Phi(m)}\|_g$$

Let us assume that $\|v\|_{g'}^2 = \delta(t)\|P_z\hat{v}\|_g^2 + \|N_z\hat{v}\|_g^2 + \lambda^2 = 1$; this implies that

$$(3) \qquad \|P_z\hat{v}\|_g^2 \leq \delta(t)^{-1}, \|N_z\hat{v}\|_g^2 \leq 1, |\lambda| \leq 1$$

By (P_2), $\exp(2\int_t^\varepsilon M_x(s)ds)$ is bounded positive on $A \times (0,\varepsilon')$ and the set $\Delta' = \{\delta \in \Delta; \delta \text{ fulfils } (*)\}$ is not empty. Let $\delta \in \Delta'$; we apply the right inequality of $(*)$ of Lemma I-7 with the positions $x = \pi(z)$, $W = F_z$, $f = \rho|_{F_z}$, $\xi = \xi|_{F_z}$ and $\alpha = 0$, $\beta = \varepsilon'$. Then, from (P_2) and (3),

$$(4) \qquad \|d_z\Xi_t P_z\hat{v}\|_g \leq \exp(\int_t^\varepsilon M_x(s)ds)\|P_z\hat{v}\|_g \leq C$$

Moreover, we get from (P_3), NA1 and (3) that

$$(5) \qquad C_\pi \geq \|d_z\pi N_z\hat{v}\|_g = \|d_{\Phi(m)}\pi d_z\Xi_t N_z\hat{v}\|_g \geq \|d_z\Xi_t N_z\hat{v}\|_g C_3$$

Putting together (P_1), (2), (4) and (5) that $\forall m \in M$

$$(6) \qquad \|d_m\Phi\| \leq C + C_3^{-1}C_\pi + C_1 = C_\Phi$$

Let's suppose now Φ D-bounded. Then (P_1) is verified, since $\xi = d\Phi(\frac{d}{dt})$. Let $x = \Phi([z,t]) \in B \setminus A$. If $v \in \mathcal{H}_x$, there is $v' \in \mathcal{K}_z^\perp$ such that $v = d_z\Xi_t v'$; then $\|v\| \le C_\Phi \|v'\|$ and, by our hypothesis

$$(7) \quad \|d_x\pi v\|_g/\|v\|_g = \|d_z\pi v'\|_g/\|v\|_g \ge \|d_z\pi v'\|_g/C_\Phi\|v'\|_g \ge C'_\pi/C_\Phi = C_3 \,,$$

which gives (P_3).

As for (P'_2), it is a consequence of the left inequality in $(*)$ of Lemma I-7.

Property (P_3) is quite difficult to handle with since it involves explicitly the flow of ξ. Using Lemma I-7 in a more direct way we may eliminate it ; of course the new condition (P_4) is stronger than (P_2).

2. Theorem. Let's asume that (P_1) of Th. III-1 and

$$(P_4) \qquad \exists C_4 \ \text{ s.t. } \ \forall t \in (0,\varepsilon') \int_t^\varepsilon M_\xi(s)ds < C_4$$

Then the mct of W at X induced by ξ, ε is D-bounded of any $\delta \in \Delta$ s.t.

$$(*') \qquad \forall t \in (0,\varepsilon') \ \ \exp(2\int_t^\varepsilon M_\xi(s)ds) < C\delta(t) \quad \text{for some} \quad C > 0$$

Proof. It is enough to follow Th.III-1's proof without decomposing $\hat{v}$ and apply directly Lemma 1-7 with $f = \rho$.

Of course, there is a corresponding necessary condition (P'_4) involving m_ξ.

If we can define a radial field ξ in term of the control data, the above conditions become "intrisic conditions". One nice thing should be to have $\nabla\rho$ controlled but this is false in general. On the other hand, let us assume that depth$X = 1$. Then $E(A,t)$ and K_x are smooth for any x, t and $\nabla(\rho|_{F_X})_X$ (projection of $\nabla\rho_x$ on T_xF_x) gives a field ζ in $\mathcal{X}_w(B) \cap \mathcal{X}_c(B \setminus A))$ by setting $\zeta_A = 0$. Let $n_x = \zeta_x/\|\zeta_x\|$ be the unit normal to K_x in F_x and $\nu = -n/\|\zeta\|$. We have for $v \in T_xK_x$: $q^\nu(v)_x = -\|\zeta\|^{-1}g_x(\mathcal{L}_x(v),v)_X$, where $\mathcal{L}_x$ is the self-adjoint endomorphism $\mathcal{L}_x$ of $(\mathcal{K}_x, g_x)$ given by $v \to (D_vn)_x$ and which is called the shape operator of K_x in F_x. Then we may state (P_2) in term of the principal curvatures of the fibers of ϕ considered as hypersurfaces of the fibers of π. In fact, if $\dim(W,x) = p$, $\dim X = q$, let $c_i(x)$, $i = 1,\ldots,p-q-1$, be the principal curvatures of K_x in F_x at x ; then $c_i(x) = -\|\zeta_x\|^{-1}\lambda_i(x)$. , being λ_i the eigenvalues of Q^ν. We define for $(x,t) \in A \times (0,\varepsilon')$ $\ X_x(t) = inf_{z\in K_y} c_{p-q-1}(z)$ and $\kappa_x(t) = inf_{z\in K_y} c_i(z)$, where $\phi(y) = (x,t)$, and we introduce instead of NA1 and NA2 a different "natural assumption"

NA3. The map $\rho : (B,g) \to (0,\varepsilon')$ is D-bounded with constant C_ρ.

Then we deduce from Th.III-1 :

3. Theorem. Let depth$X = 1$ and assume that :
(I_1) $0 < C_1 \leq \|\zeta_x\|_g$ for $x \in B \setminus A$,
(I_2) $\exists C_2$ s.t. $\forall x \in A, \forall t \in (0, \varepsilon')$ $\int_t^\varepsilon \chi_x(s)ds > C_2$,
(I_3) (P_3) holds for ν.

Then the mct of W at X induced by ξ, ε is D-bounded of any $\delta \in \Delta$ s.t.

$$(*'') \qquad \exists C > 0 \quad \forall x \in A, \forall t \in (0, \varepsilon') \qquad \exp\left(2C_\rho^{-1} \int_t^\varepsilon \chi_x(s)ds\right) < C\delta(t)$$

Conversely let's assume that Φ is D-bounded (with any weight). Then $(I_1), (I_3)$ and

$$(I_2') \qquad \exists C_2' \text{ s.t. } \forall x \in A, \forall t \in (0, \varepsilon') \qquad \int_t^\varepsilon \kappa_x(s)ds > C_2$$

are verified.

Proof Evidently $(I_1) \Rightarrow (P_1)$. Now, by NA3 and (I_1),

$$-C_1^{-1} c_{p-q-1}(x) \leq \lambda_{p-q-1}(x) \leq -C_\rho^{-1} c_{p-q-1}(x) \ ;$$

hence $(I_2) \Leftrightarrow (P_2)$ for ν.

Since it is easy to see that NA1, NA2 are fulfilled for depth$X = 1$ (E is relatively compact and π is smooth on $\bar{E}$), we get our thesis from (I_3). The converse assertion is analogous.

4. Remarks. If depth$X = 1$, there are two cases in which our hypothesis are simpler.
i) If $\dim X = 0$, we have the so called *isolated singularity* ; is this case (P_3) holds trivially.
ii) If $\dim X = \dim W - 1$, (P_3) doesn't depend on the flow, being $\mathcal{K}^\perp = TE$.

In section IV we shall essentially examine such kind of singularities.

We observe at this point that, if W is compact, to have f.g.d. we need less than D-bounded. In fact

5. Proposition. Let (W, g) be compact. If any $x \in W$ admits a nbd. which is the image of a mct induced by a radial field fulfilling only (P_1), then W has f.g.d.

Proof Let $\Phi : M \to B(A, \varepsilon') = B$ a mct at $X \in \mathcal{S}_W$ induced by ξ and satisfying (P_1). Up to restriction, we may assume that $\varepsilon' < +\infty$ and that A

has f.g.d. (with smooth arcs). If $x_1 = \Phi([z_1,t_1])$, $x_2 = \Phi([z_2,t_2]) \in B$, let for $i = 1,2$, $\gamma_i : I \to M$ be be $\gamma_i(u) = [z_i, ut_i]$; now $\exists \gamma' : I \to A$ joining $\pi_X(z_1)$ with $\pi_X(z_2)$ and with length $< C'$, where $C' > 0$ is independent from z_1, z_2. Then the arc γ obtained connecting γ_1, γ' and γ_2 has length $< C' + 2\varepsilon'$. It is easy to prove now from our hypothesis that the arc $\Phi \circ \gamma$ has length $< C_1(C' + 2\varepsilon')$ ($\Phi|_A = Id$ and $d(\Phi \circ \gamma_1)/dt = \xi$ for $i = 1,2$). The thesis follows from compactness arguments.

IV THE EMBEDDED CASE

Let N be an n-dimensional smooth manifold : a locally closed subset $W \subseteq N$ endowed with a stratification $\mathcal{S}_W$ formed by smooth submanifolds of N is called a stratified subset of N. It is well known that b-regularity condition implies the existence of a control data system for $\mathcal{S}_W$ which makes W a stratified space. Recently, K. Bekka introduced incidence conditions weaker than (b)-regularity giving a stratified structure (see [B]). In particular, $(a) + (\delta)$-regularity maintains some of the metric properties of b-regular stratifications. In fact :

1. Lemma. If $\mathcal{S}_W$ is $(a) + (\delta)$-regular, we have a control data system with the following properties :

i) $\forall X \in \mathcal{S}_W$, the map ϕ_X is defined on a nbd. V_X of X in N and it is smooth on $V_X \setminus X$; moreover π_X is smooth on V_X (Observe that there are proper tubes for the top dimensional strata too, in this case).

ii) Let $X \in \mathcal{S}_W$, $\dim X = q$. $\forall x \in X$, $\exists$ open nbd. A of x in X, $\varepsilon \in (0, r_X)$ and an embedding $\tau : B(A,\varepsilon) \to \mathbf{R}^n$ s.t. $\tau(x) = 0$, $\tau(A) \subseteq \{u \in \mathbf{R}^n; u_j = 0, j > q\}$, $\tau \circ \pi_X \circ \tau^{-1}(u) = (u_1, \ldots, u_q)$, $\rho_X \circ \tau^{-1}(u) = \left(\sum_{j>q} u_j^2\right)^{1/2}$ and $\tau(B(A,\varepsilon) \cap W)$ is $(a) + (\delta)$-regular.

It is direct consequence of Lemma IV-1 that NA1, NA2, NA3 of sec.III hold in the $(a) + (\delta)$-regular case. Moreover, if depth$X = 1$, we have (I_1) as well.

From now on, W will be a closed stratified subset of N, $\mathcal{S}_W$ will be $(a) + (\delta)$-regular and we shall assume fixed a control data system for $\mathcal{S}_W$ with the properties i), ii) of Lemma IV-1. Moreover g will be a riemannian metric on N and we shall consider on W the stratified metric (named g too) given by restriction.

At first, using Remark III-4 ii) and Lemma IV-1 ii), we may find again a result proved in $[F_1]$ in the b-regular case and in $[B_1]$ in the $(a) + (\delta)$-regular one :

2. Proposition. If $W \subseteq (N, g)$ has strata in dimension $\dim W$ and $\dim W - 1$ only, then it is D-bounded trivial. Hence, by Th. II-9, it has 1.f.v..

Actually in [B] and $[F_1]$ was proved that W was v-regular, which is stronger than 1.f.v.; this depends on the fact that the trivializations used there turns out to be Lipschitz maps too.

In general, we may always find for W radial fields fulfilling (P_1). In fact

3. Theorem. Let $X \in \mathcal{S}_W$ with depth$X > 0$, $A \subseteq X$ connected, relatively compact and open and $\varepsilon \in (0, r_X)$. Then there is a radial field $\xi \in \mathcal{X}(B(A, \varepsilon'))$ fulfilling (P_1).

Proof Let $B(A, \varepsilon) = B$ and let $\nu \in \mathcal{X}(B)$ be defined as

$$\nu_x = -\nabla(\rho_X|_{F_x \cap W})_x / \|\nabla(\rho_X|_{F_x \cap W})_x\|_g^2$$

if $x \in B \setminus A$ and $\nu_A = 0$. Then, by Lemma IV-1, ν fulfils (P_1) and is a ϕ_X-lift of $(0, \frac{d}{dt})$ but is not in general controlled : then ν is to be changed to get a controlled field without loosing the other properties. We use downward induction on $k =$ depth$Y$, $Y \in \mathcal{S}_B$. If $k \geq$ depth$X - 1$, we set $\xi_Y = \nu_Y$. Let now $k <$ depth$X - 1$ and assume that ξ_Z is defined for any $Z \in \mathcal{S}_B$ with depth$Z > k$. We build ξ_Y by a recursive procedure ; for the sake of simplicity we assume there is only one chain $A = Z_0 < Z_1 < \cdots < Z_m < Y$ between A and Y. For any i, $0 \leq i \leq m$, let $Z^1 = \cup_{j<1} Z_j$; we build a extension ξ_1 of $\xi|_{Z^1}$ to $Z^1 \cup Y$ which has bounded norm on $Z^i \setminus A$. If $i = 0$, $\xi_0 = \nu|_{Y \cup A}$. We assume that ξ_{i-1} is defined and we set $Z_1 = Z$. Let $\hat{\xi}$ be a controlled ϕ_Z-lift of $(\xi_Z, 0)$ on B_Z ; by $a + \delta$-regularity and Lemma IV-1, we can find a smooth dim Z-plane field $y \to H_y$ on $B_Z \cap Y$ s.t.

1) $T_y Y = H_y \oplus \ker d_y(\pi_Z|_Y)$,
2) $d_y \rho_Z H_y = 0$ and
3) $\lim_{y \to z} H_y = T_z Z$ for any $z \in Z$.

By 2) we may decompose $\hat{\xi} = \xi' + \xi''$, with $\xi'_y \in H_y$. It is not difficult to see that, in force of 1), 3) and Lemma IV-1, ξ' extends C^0 to Z with ξ_Z and ξ' is radial. By continuity, $\forall z \in Z$ $\exists \mu(z) > 0$ s.t. $|\|\xi'_y\|_g - \|\xi_{Z,z}\|_g| < 1$ if dist$(y, z) < \mu(z)$ and we may take $\mu \in C^\infty(Z)$.

If $y \in B_Z \cap Y$, let $\lambda(z) = \alpha(\text{dist}(y, \pi_Z(y))/\mu(\pi_Z(y)))$, where $\alpha \in C^\infty(\mathbf{R})$, $\alpha \geq 0$ and $\alpha(t) \equiv 1$ if $t \geq 1$, $\alpha(t) \equiv 0$ if $t \leq \frac{1}{2}$.

Then the field $\xi_{1,y} = (1 - \lambda(y))\xi'_y + \lambda(y)\xi_{i-1,y}$ is the one we looked for at the end we set $\xi_Y = \xi_m$.

We have to observe that the existence of controlled continuous fields was at first established by Shiota ([Sh]). Very recently K. Bekka gave a result of equivalence between c-regularity and extendibilty of controlled continuous fields ($[B_2]$). As a consequence of Prop. III-5 and Th. IV-2 we get

4. Theorem. If (W, g) is compact, then it has f.g.d..

This theorem was proved in [BT] for a wider class of sets but without requiring that the joining arcs were stratified.

We give now an application in the isolated singularity case. Let $N = \mathbf{R}^n$, g be the euclidean metric and let W be a closed stratified subset of $\mathbf{R}^n$ with an isolated singularity in 0. Assume that W admits a D-bounded mct Φ of weight $\delta = t^2$ at 0 induced by ξ and ε ; we may suppose $\varepsilon = 1$. Then $E \subseteq S^{n-1}$, and $(\mathbf{R}^n, g)$ is isometric to $(C(S^{n-1}), g^\delta)$ and we may consider $C^\varepsilon(E) \subseteq \mathbf{R}^n$ with the induced metric. Let $k(t) = \varepsilon^{-1}(\varepsilon + 1 - t)t^2$ and $\psi : (C^\varepsilon(E), g^\delta) \to (C^\varepsilon(E), g^\delta)$, $\psi([z,t]) = [z, k(t)]$; then $\|d_m\psi\|$ goes to 0 as t does and, if we set $\Phi' = \Phi \circ \psi$, we get that Φ' is a C^1 map with $d_0\Phi' = 0$. Given a smooth triangulation of E, $\tau : K \to E$, we have the associated triangulation of the cone, $\tau' : K' \to C^\varepsilon(E)$; then $\Phi' \circ \tau' : K' \to B = \mathrm{Im}\Phi$ is a C^1-triangulation (on any q-simplex σ it extends to a C^1 map on a nbd. of σ in the q-plane generated by σ). By an observation of K. Bekka, if W is c-regular, then the triangulation is c-regular.

In a more specific situation, we can sketch a method to check (P_2). Let $f \in C^0(\mathbf{R}^n) \cap C^2(\mathbf{R}^n \setminus \{0\})$ s.t. $f(0) = 0$. Let $W = \{f = 0\}$ and assume that

a) $\forall x \in X = W \setminus \{0\}$, $d_x f \neq 0$.

b) If θ_x is the angle between x and ∇f_x, $\exists \alpha \in (0,1)$ s.t. $\forall x \in X$, $\|x\| < 1$, $\cos\theta_x < \alpha$ (namely the couple $(\{0\}, X)$ is a $(a) + (\delta)$-regular).

Then W is a stratified subset with an isolated singularity and $\rho(x) = \|x\|$ gives the stratified structure. Let $E_t = \{x \in W; \|x\| = t\} = W \cap S_t^{n-1}$, and let, for x near 0, ν_x be the field defined previously (here $\nabla(\rho|_X)_x = P_{T_xX}$, $(\frac{X}{\|X\|})$. We get for $x \in X$, $\|x\| = t$ and $v \in T_xE_t$.

$$q^\nu(D_v\nu, v)_x = (sen\theta_x)^{-2}((\cos\theta_x)D_v\frac{\nabla f}{\|\nabla f\|} - \frac{1}{\|x\|}\|v\|^2) ;$$

thus we can extimate our quadratic form by means of the principal curvatures of X in $\mathbf{R}^n$. In fact, fix $x \in X$; on T_xX the form $q = q_x^\nu$ is associated to the symmetric operator $aL_x + b.Id$, where $a = (sen\theta_x)^{-2}cos\theta_x$, $b = (sen\theta_x)^{-2}\|x\|$ and $\mathcal{L}_x$ is the shape operator of X. Then the eigenvalues of q are $\mu_i = ac_i + b$, $i = 1, \ldots n-1$, where the c_i's are the principal curvatures of X at x ordered so that $\mu_1 \leq \ldots \leq \mu_{n-1}$. When we restrict q to $T_xE_t = T_xX \cap T_xS_t^{n-1}$, we have a symmetric operator with eigenvalues $\lambda_1 \leq \ldots \leq \lambda_{n-2}$. By the extremal properties of eigenvalues on the unit ball, operator we get $\lambda_{n-2} \leq \mu_{n-1}$ and $\mu_1 \leq \lambda_1$. Obviously this estimate is not the best possible.

An open question is : are (sub/semi-)analytic sets D-bounded ? We recall that subanalytic sets have not only l.f.v., but they also admits a density in any point ([KR]).

References

[B] K. Bekka *Sur les propriétés topologiques et métriques des espaces stratifiés Thèse* Université de Paris-Sud, Centre d'Orsay (1988).

[BT] K. Bekka, D. Trotman *Propriétés métriques de familles Φ-radiales de sous-variétés différentiables C.R.A.S.* t. 305 s.I, 389-392 (1987).

[Bu] H. Busemann *The geometry of geodesics Academic Press* (1955).

[C] J. Cheeger *On Hodge theory of Riemannian Pseudomanifolds Proc. Symp. Pure Maths. A.M.S.* XXXVI, (1980).

[D] J.C. Darchen *Integration sur un ensemble stratifié* (1987).

[F_1] M. Ferrarotti *Volume on stratified sets Annali di Mat. pura ed appl.* (IV) Vol. CXLIV, 183-201 (1986).

[F_2] M. Ferrarotti *Some results about integration on regular stratified sets Annali di Mat. pura ed appl.* (IV) Vol. CL (1988), 263-279.

[F_3] M. Ferrarotti *Differential forms and metric contractibility of stratified spaces Pub. Dip. Mat. Univ. Pisa* 267 (dicembre 1988).

[GW] C. Gibson, K. Wirthmüller, E. Loojienga, A. DuPlessis *Topological stability of smooth mappings Lec. Notes in Maths.* 552, (1976) 193-200 (1978).

[KR] K. Kurdyka - G. Raby. *Densité des ensembles sous-analytiques. Ann. Inst. Fourier* 39 (1989).

[M] J. Mather *Notes on topological stability. Preprint, Harward* (1970).

[Mo] G. Monti-Bragadin *Abstract riemannian stratifications Pac.Jour.Maths.* 133, *n*°1 (1988).

[Ms] T. Mostowski *Lipschitz equisingularity Dissertationes Math.* 243 (1985).

[S] M.H. Schwartz *Classes caractéristiques définies par une stratification d'une variété analytique complexe C.R.A.S.* t. 260 (1965).

[Sh] M. Shiota *Piecewise linearization of real analytic functions Pub. R.I.M.S. Kyoto Univ.* 20 (1984).

[T] R. Thom *Ensembles et morphismes stratifiés. Bull. A.M.S.* 75 (1969).

[V] A. Verona *Stratified Mappings-Structure and Triangulability Lec. Notes in Maths.* 1102, (1984).

[W] H. Winkelkamper *Estimating* $\|d\phi^t\|$ *for unit vector fields whose orbits are geodesics J. Diff. Geom.* 31 (1990).

Massimo Ferrarotti
Dipartemento di matematica
Università di Pisa
Via F. Buonarroti 2
56100 Pisa - Italy

MODULI FOR SINGULARITIES

Gert-Martin Greuel
Universität Kaiserslautern
Fachbereich Mathematik
Erwin-Schrödinger-Straße
D-6750 Kaiserslautern

Gerhard Pfister
Humboldt-Universität zu Berlin
Fachbereich Mathematik
Unter den Linden 6
D-1086 Berlin

Introduction

The aim of this article is to report on the authors' recent methods and results about moduli spaces for curve singularities and for modules over the local ring of a fixed curve singularity. We emphasize especially the general concept which lies behind these constructions. Therefore, the article might be useful to the reader who wishes to have the leading ideas and the main steps of the proofs explained without going into all the details. We also calculate explicit examples (for singularities and for modules) which illustrate the general theorems.

The construction of moduli spaces for certain objects means a **geometric classification** of the objects with respect to some equivalence relation. This is adequate, in particular, when it becomes too complicated to give a complete classification through (parametrized) normal forms. The basic concept, which stems from Mumford's Geometric Invariant Theory, is that of a **coarse moduli space**. Such a coarse moduli space for singularities respectively for modules over the local ring of a fixed singularity consists of a complex space or an algebraic variety M such that the points of M correspond in a unique way to equivalence classes of singularities or modules (with certain invariants fixed). Moreover, it is required that (flat) families of singularities or modules correspond to subvarieties of M. This means that the structure of M reflects neighbouring relations between objects which are given by small deformations.

There exists also the concept of a fine moduli space M (where it is required that a universal family over M exists, from which any other family can be induced by a unique base change), but such a fine moduli space does, in general, not exist without imposing some extra rigidifying structure on the objects to be classified. Nevertheless, in many cases of interest one has first to find some fine moduli space from which the coarse one can be constructed.

The construction of moduli spaces for objects of projective geometry has a long and interesting tradition and is still a very active field of research. In his famous paper on Abelian Functions [Ri], Riemann states: "Die allgemeine Fläche vom Geschlecht g hängt von $3g-3$ stetig veränderlichen Größen ab, welche die Moduln dieser Klasse genannt werden sollen." This means that $3g-3$ is the dimension of the moduli space of Riemann surfaces of genus g. Note that it makes perfect sense to talk about dimensions, even without having a precise definition of a moduli space.

Precise algebraic definitions of different concepts of moduli spaces were given 1965 by Mumford in his celebrated book "Geometric Invariant Theory" [MuF], where he also applied this to Riemann surfaces. According to Mumford, one has to fix the functor, for which a moduli space is to be constructed:

- First of all fix the equivalence relation. For subvarieties of a fixed $\mathbf{P}^n$ it could be just equality (leading to the Hilbert scheme) or isomorphism of varieties. For hypersurface singularities it could be right equivalence of the defining function or contact equivalence, that is isomorphism of the germ defined by the function (which turns out to be much more complicated than right equivalence and which will be our concern). For modules, it will be isomorphism of modules.
- Define the notion of a (flat) family of objects over some category of base spaces, together with the notion of base change.
- The equivalence relation has to be extended to families. Note that, in general, there exist different possible extensions, which may lead to different concepts of a moduli space.

We shall not discuss the last two items in this paper, but would just like to mention the following. If one wants to classify singularities with respect to isomorphism of their complete local rings, so-called algebroid singularities, or modules over complete local rings, one can consider families over complex germs respectively over the spectrum of a complete local ring, or over base spaces, which are algebraic varieties of finite type over the ground field. In the latter case, one has to be careful with base change, since the fibres are still formal objects along a section. The necessary foundational material for such families of modules was developed in [GP 2].

After fixing the functor, a general method for constructing a moduli space is the following:

1. Fix some rough invariants (for Riemann surfaces, for instance, the genus, for isolated singularities, for instance, the Tjurina number τ and the Milnor number μ).

2. Construct an algebraic family $X \to T$ with finite dimensional base space T, which contains all isomorphism classes to be classified. This is sometimes called a parametrizing **bounded family** of the problem. This family should be locally versal. That is, for any family $Y \to S$ and $s \in S$ there is a neighbourhood U of s in S such that $Y|U \to U$ can be induced from $X \to T$ by some base change. In singularity theory this can be the versal deformation of the "worst" object (if such a worst object exists). This will be our approach. Of course, one then has to prove that indeed all objects which we want to classify, and not only those which are close to the worst object, are contained in T.

3. In general T will contain many equivalent objects and one tries to interpret the equivalence classes as orbits of the action of a Lie group or an algebraic group G acting on T. If one starts with a versal deformation $X \to T$, the kernel of the Kodaira–Spencer map is a (infinite dimensional) Lie algebra $\mathcal{L}$ and the integral manifolds of $\mathcal{L}$ give rise to trivial subfamilies. In our cases, we can show that each equivalence class coincides with an integral manifold. Moreover, it is possible to find a finite dimensional Lie subalgebra $L \subset \mathcal{L}$ with the same integral manifolds. $G = exp\, L$ is the algebraic group with the desired properties. The topological space T/G may be considered as a classifying space but, in general, it is not an algebraic variety or a complex space.

4. (a) Describe a big open subset $T_0 \subset T$ such that T_0/G exists as complex space or algebraic variety. The objects of T_0 may be called generic, or stable.

   (b) Better, find a stratification $T = \cup T_i$ of T into locally closed G–invariant subvarieties T_i such that the closure $\bar{T_i}$ is the union of strata and such that T_i/G exists.

5. Find fine invariants of the objects to be classified, such that at least the reduction of each stratum is characterized exactly by fixing these invariants.

So far, this somewhat schematic description applies to moduli problems in global algebraic geometry as well as to singularities. Roughly speaking, one has the following analogy between the local and global situation:

- Moduli spaces for singularities (local) versus moduli spaces for projective varieties (global).
- Moduli spaces for modules over the local ring of a fixed singularity (local) versus moduli spaces for vector bundles on a fixed projective variety (global).

But the basic fundamental difference between the local and the global situation is that the group G occurring in 3), for the global case, is mostly reductive and hence has a finitely generated ring of invariant algebraic functions. Moreover, Mumford's theory of stable points [MuF] gives a powerful criterion (the Hilbert–Mumford criterion) for the existence of a geometric quotient. In the local case, the groups are almost never reductive, but either solvable or even unipotent, or the product of such a group with a reductive group. Consequently, the ring of invariants can be infinitely generated. It seems that the first really practicable criterion for the existence of a geometric quotient for actions of unipotent groups was developed in [GP 1]. At least for us, this criterion plays the same basic role as the Hilbert–Mumford criterion in the global case. Note, however, that the criterion of [GP 1] gives a complete stratification, whilst the Hilbert–Mumford criterion gives only the big open stratum of stable points.

The construction of moduli spaces for singularities has been carried out only for few cases. The first paper aiming at this goal seems to be that of Ebey [Eb], who carried out step 1) to 3) above for reduced curve singularities. But, in his lecture notes, Zariski [Za] was the first to actually construct moduli spaces. Fixing the equisingularity type of a plane curve singularity, Zariski studied the problem of moduli from the parametric point of view as well as from the point of view of deformations. Many basic ideas about moduli for singularities go back to him. He was able to compute several special examples just using power series methods, in particular, he carried out step 1) to 4a) and 5) for singularities with equation $x^m + y^{m+1}$. Laudal and Pfister [LaP] developed a quite general frame for the construction of moduli spaces for arbitrary singularities. The starting point is the semiuniversal deformation of the "worst" object to be classified and the existence of a moduli stratum (before considered by Palamodov [Pa] in the complex analytic category) in the base space of the semiuniversal deformation. They proved that it has an algebraic representative under certain assumptions, which hold, for instance, for isolated hypersurface singularities. In this case, the moduli stratum is just the τ–constant stratum. Moreover, they solve steps 1) — 4a) and 5) for plane curve singularities of the form $x^p + y^q$ by this approach (with respect to contact equivalence). An explicit formula for the dimension of this moduli space is given in [BGM].

Note that the τ–constant stratum is, in general, not a moduli space, not even for families over complex germs respectively over the spectrum of a complete local ring. Usually there is a finite group which has to be divided out. For more details see [LaP], further references are given in [LaP] and in the survey article [Gr].

In this paper we shall discuss three special moduli problems where all steps 1) — 5) have been carried out.

First we explain in some detail the classification of irreducible plane curve singularities with semigroup $\langle p, q \rangle$. In this case we extend the results of [LaP] in from the generic case to a complete stratification of the semiuniversal μ–constant deformation by fixing a certain Hilbert function of the Tjurina algebra. This extension to lower dimensional strata was possible because of the results of [GP 1]. Subsequently, the authors were able to extend these results to semi–Brieskorn singularities of arbitrary dimension [GP 3]. We report on the results of [LuP] on plane curve singularities with semigroup $\langle 2p, 2q, 2pq + d \rangle$, which is somewhat exceptional since no further stratification was necessary. Moreover, we state some results of [GP 2] on torsion free modules of rank 1 over the local ring of an irreducible curve singularity. The original approach in [GP 2] was different, via "sandwiching" of modules, but we show that this case also fits into the general method via deformation theory of the "worst" object. As to all this a basic ingredient is the criterion for the existence of a geometric quotient of a unipotent group action which has to be discussed first.

We give an outline of the arguments in all three cases and explain the main steps of the constructions. For complete proofs we refer to [LaP], [LuP], [GP 1] and [GP 2].

1 Geometric quotients of unipotent group actions

Let K be a field of characteristic 0.
Let G be an algebraic group acting algebraically on an algebraic variety X. If Y is an algebraic variety and $\pi : X \to Y$ a morphism then $\pi : X \to Y$ is called

a **geometric quotient**, if

1. π is surjective and open

2. $(\pi_* \mathcal{O}_X)^G = \mathcal{O}_Y$

3. π is an orbit map, i.e. the fibres of π are orbits of G.

If a geometric quotient exists it is uniquely determined and we just say that X/G exists.

By a general result of Rosenlicht which holds for arbitrary algebraic groups there exists an open dense G-stable subset $U \subset X$ such that U/G exists if X is reduced. But U is not uniquely determined and it is not at all clear how to construct such an open subset.

If G is reductive and $X = Spec\,A$, A of finite type over K, then A^G is of finite type over K and $X \to Spec\,A^G$ is a geometric quotient iff all orbits are closed and have the same dimension. The Hilbert Mumford criterion for "stable" points ([MuF]) is the basic tool for the construction of moduli spaces in global algebraic geometry.

In singularity theory the groups are almost never reductive. In our applications the groups are unipotent. Furthermore it may happen that the ring of invariants A^G is not of finite type. But even if A^G is of finite type and if all orbits have the same dimension (they are closed since G is unipotent) it may happen that X/G does not exist.

An analysis of "bad" examples suggested the following definition of stability which we proposed in [GP 1].

Definition: Let G be a unipotent algebraic group, $Z = Spec\,A$ an affine G-variety and $X \subset Z$ open and G-stable. Let $\pi : X \to Y := Spec\,A^G$ be the canonical map. A point $x \in X$ is called **stable** under the action of G **with respect to** A **(or with respect to** Z**)** if the following holds:
There exists an $f \in A^G$ such that $x \in X_f = \{y \in X, f(y) \neq 0\}$ and $\pi : X_f \to Y_f = Spec\,A_f^G$ is open and an orbit map.
If $X = Z = Spec\,A$ we call a point stable with respect to A just **stable**.

Let $X^s(A)$ denote the set of stable points of X (under G with respect to A).

Proposition 1.1 *1. $X^s(A)$ is open and G-stable*

2. *$X^s(A)/G$ exists and is a quasiaffine algebraic variety*

3. *If $V \subset Spec\, A^G$ is open, $U = \pi^{-1}(V)$ and $\pi : U \to V$ is a geometric quotient then $U \subset X^s(A)$*

4. *If X is reduced then $X^s(A)$ is dense in X.*

The aim of this chapter is to describe effective criteria for stability, i.e. to give sufficient conditions for the existence of a geometric quotient in terms of given coordinates of X and a given representation of G in $Aut(X)$. These criteria are easier to formulate in terms of the Lie algebra of G.
Let $L = \text{Lie }(G)$. If G is unipotent then L is nilpotent and the representation of $G \to Aut_K(A)$ induces a commutative diagramme:

$$\begin{array}{ccc} G & \to & Aut_K(A) \\ exp \uparrow & & \uparrow \; exp \\ L & \to & Der_K^{nil}(A) \end{array}$$

Here $Der_K^{nil}(A)$ is the set of K-linear nilpotent derivations δ of A (δ is nilpotent if for any $a \in A$ there is an $n(a)$ such that $\delta^{n(a)}(a) = 0$).

The best results are obtained for free actions or for abelian L. In these cases we obtain necessary and sufficient conditions for the existence of a locally trivial geometric quotient.

Definition: A geometric quotient $\pi : X \to Y$ is **locally trivial** if an open covering $\{V_i\}_{i\in I}$ of Y and $n_i \geq 0$ exist, such that $\pi^{-1}(V_i) \cong V_i \times A_K^{n_i}$ over V_i where A_k^n denotes the n-dimensional affine space over K.

We use the following notations:
Let $L \subseteq Der_K^{nil}(A)$ be a nilpotent Lie-algebra and $d : A \to Hom_K(L, A)$ the differential defined by $da(\delta) = \delta(a)$. If $B \subseteq A$ is a subalgebra then $\int B := \{a \in A \mid \delta(a) \in B \text{ for all } \delta \in L\}$.

Theorem 1.2 *Let A be a reduced noetherian K-algebra and $L \subseteq Der_K^{nil}(A)$ be a finite dimensional abelian Lie algebra. The following conditions are equivalent:*

1. *There exists an open subset $U \subset Spec\, A^L$ such that $Spec\, A \to U$ is a locally trivial geometric quotient.*

2. *$AdA = Ad\int A^L$ and $Hom_K(L, A)/AdA$ is flat over A.*

2'. *There are $x_1, \dots, x_n \in A, \delta_1, \dots, \delta_m \in L$ such that*

- *rank* $(\delta_i(x_j))$ *is locally constant and equal to the orbit dimension of the action of* L.
- $d\delta_i(x_j) = 0$ *for all* i, j.

3. There is a filtration $F^\bullet(A)$ *such that*

- $0 = F^{-1}(A) \subset F^0(A) \subset F^1(A) \subset \ldots$ *and* $A = \cup_{i \in \mathbf{Z}} F^i(A)$,
- $\delta F^i(A) \subseteq F^{i-1}(A)$ *for all* $i \in \mathbf{Z}$ *and* $\delta \in L$,
- $Hom_K(L, A)/AdF^i(A)$ *is flat over* A *for all* i.

3'. There are $x_1, \ldots, x_n \in A, \delta_1, \ldots, \delta_m \in L$ *and* $i_1, \ldots, i_k \in \{1, \ldots, n\}$ *such that*

- $1 \leq i_1 < i_2 < \ldots < i_k = n$
- $E(s) := rank(\delta_i(x_j))_{j \leq i_s}$ *is locally constant and* $E(k)$ *is the orbit dimension of the action of* L
- $d\delta_i(x_\ell) \in \sum_{\nu \leq i_{r-1}} Adx_\nu$ *for all* i *and* $\ell \leq i_r$.

4. $Spec\,A = \cup_{f \in S} D(f), S \subseteq A^L$ *and for* $f \in S$ *there is a sub-Lie algebra* $L^{(f)} \subseteq L$ *such that*

- $L^{(f)} \otimes_K A_f = L \otimes_K A_f$
- $H^1(L^{(f)}, A_f) = 0$

Proof: (1) implies (2) and (4) is proved in [GP 1] Theorem 4.1.
(2') resp. (3') is the same as (2) resp. (3) expressed in coordinates.
(2) implies (3) using the filtration defined by $F^0(A) = A^L, F^i(A) = \int F^{i-1}(A)$.
(4) implies (1) by theorem 3.10 of [GP 1].
(3) implies (1) follows from theorem 4.7 in [GP 1].

Corollary 1.3 *Let* A *and* L *be as in the theorem and let* $F^\bullet(A)$ *be a filtration of* A *such that* $\delta F^i(A) \subseteq F^{i-1}(A)$ *for all* i *and all* $\delta \in L$. *Let* $Spec\,A = \cup U_\alpha$ *be the flattening stratification of* $Spec\,A$ *defined by the* A-*modules* $Hom_K(L, A)/AdF^i(A)$. *Then* U_α *is invariant under the action of* L *and* $U_\alpha \to U_\alpha/L$ *is a geometric quotient.*

If the Lie algebra L is nilpotent but no longer abelian we need some extra conditions on a central series of L to obtain a stratification as in the corollary:

So, let A be a noetherian K-algebra and $L \subseteq Der_K^{nil} A$ a finite dimensional nilpotent Lie algebra. Suppose that $A = \cup_{i \in \mathbf{Z}} F^i(A)$ has a filtration

$$F^\bullet : 0 = F^{-1}(A) \subset F^0(A) \subset F^1(A) \subset \ldots$$

by subvector spaces $F^i(A)$ such that

(F) $\quad \delta F^i(A) \subseteq F^{i-1}(A)$ for all $i \in \mathbf{Z}$ and all $\delta \in L$.

Assume, furthermore, that

$$Z_\bullet : L = Z_0(L) \supseteq Z_1(L) \supseteq \ldots \supseteq Z_\ell(L) \supseteq Z_{\ell+1}(L) = 0$$

is filtered by sub Lie algebras $Z_j(L)$ such that

(Z) $\quad [L, Z_j(L)] \supseteq Z_{j+1}(L)$ for all $j \in \mathbf{Z}$

The filtration $Z_\bullet$ of L induces projections

$$\pi_j : Hom_K(L, A) \to Hom_K(Z_j(L), A).$$

For a point $t \in Spec\, A$ with residue field $\kappa(t)$ let

$$r_i(t) := dim_{\kappa(t)} AdF^i(A) \otimes_A \kappa(t) \qquad i = 1, \ldots, k,$$

$$k \text{ minimal such that } AdF^k(A) = AdA$$

$$s_j(t) := dim_{\kappa(t)} \pi_j(AdA) \otimes_A \kappa(t) \qquad j = 1, \ldots, \ell$$

($s_j(t)$ is the orbit dimension of $Z_j(L)$ at t).

Let $Spec\, A = \cup U_\alpha$ be the flattening stratification of the modules

$$Hom_K(L, A)/AdF^i(A), \qquad i = 1, \ldots, k$$

and

$$Hom_K(Z_j(L), A)/\pi_j(AdA), \qquad j = 1, \ldots, \ell.$$

Theorem 1.4 *U_α is invariant and admits a locally trivial geometric quotient with respect to the action of L.*

Remarks:

1. The functions $r_i(t)$ and $s_i(t)$ are constant along U_α.

2. Let $x_1, \ldots, x_n \in A$, $\delta_1, \ldots, \delta_m \in L$ satisfying the following properties:

   - there are $\nu_1, \ldots \nu_k$, $0 \leq \nu_1 < \ldots < \nu_k = n$, such that $dx_1, \ldots, dx_{\nu_i}$ generate the A-module $AdF^i(A)$;

- there are $\mu_o, \dots, \mu_\ell$, $1 = \mu_o < \mu_1 < \dots < \mu_\ell$ such that $\delta_{\mu_j}, \dots, \delta_m \in Z_j(L)$ and $Z_j(L) \subseteq \sum_{i \geq \mu_j} A\delta_i$. Then

$$rank(\delta_\alpha(x_\beta)(t))_{\beta \leq \nu_i} = r_i(t),\ i = 1, \dots, k$$

$$rank(\delta_\alpha(x_\beta)(t))_{\alpha \geq \mu_j} = s_j(t),\ j = 1, \dots, \ell.$$

Hence the U_α are a defined set theoretically by fixing

$$rank(\delta_\alpha(x_\beta)(t))_{\beta \leq \nu_i}, i = 1, \dots, k$$

and

$$rank(\delta_\alpha(x_\beta)(t))_{\alpha \geq \mu_j}, j = 1, \dots, \ell$$

But notice that the U_α carry a unique, not necessarily reduced, analytic structure with respect to the flattening property and which is defined by the corresponding subminors.

The key lemma to prove these theorems is the following:

Proposition 1.5 *Let A be a commutative K-algebra, $\delta_1, \dots, \delta_n \in Der_K^{nil}(A)$ and $x_1, \dots, x_n \in A$ satisfying the following properties:*

1. $[\delta_i, \delta_j] \in \sum_{\nu=1}^{n} A\delta_\nu$
2. *$det(\delta_i(x_j))$ is a unit in A*
3. *For any $k = 1, \dots, n$ and any k-minor M of the first k columns of $(\delta_i(x_j))$ we have*

$$\underline{\delta}(M) \in \sum_{\nu<k} A\underline{\delta}(x_\nu)$$

(with the conventions $x_0 = 0$ and $\underline{\delta} = \begin{pmatrix} \delta_1 \\ \vdots \\ \delta_n \end{pmatrix}$).

Let $L \subseteq \sum_{\nu=1}^{n} A\delta_\nu$ be any K-Lie algebra with $\delta_1, \dots, \delta_n \in L$, then $A^L[x_1, \dots, x_n] = A$ and $x_1, \dots, x_n$ are algebraically independent over A^L.

The proposition implies that $Spec\,A \to Spec\,A^L$ is a (trivial) geometric quotient with fibre K^n. In particular, $A^L \cong A/(x_1, \dots, x_n)$ is of finite type over K if A is of finite type and every point of $Spec\,A$ is stable.
The proof of this proposition is done by induction on n. The condition (3)

guaranties that the elements of the first column of $(\delta_i(x_j))$ *are already in* A^L *and furthermore this property can be kept during the induction.*

Condition (3) is satisfied in our application, even a stronger one:

3'. $\underline{\delta}(\delta_j(x_k)) \in \sum_{\nu<k} A\underline{\delta}(x_\nu) \qquad k = 1, \ldots, n$

i.e. the derivative-vector of each element of the matrix $(\delta_i(x_j))$ *is an A-linear combination of earlier columns.*

We conjecture that in case of a free action the condition (3) of the proposition can be omitted.
Conjecture: Let $L \subseteq Der_K^{nil}(A)$ be a nilpotent Lie algebra of dimension n and $\delta_1 \ldots, \delta_n \in L, x_1, \ldots, x_n \in A$ such that $det(\delta_i(x_j))$ is a unit. Then there are $y_1, \ldots, y_n \in A$ such that $A = A^L[y_1, \ldots, y_n]$ (equivalently $H^1(L, A) = 0$).

Remark: If we would require in the conjecture $x_i = y_i, i = 1, \ldots, n$ then this conjecture is equivalent to the Jacobian Umkehrproblem.

2 A moduli space for plane curve singularities with semigroup $\langle p, q \rangle$

We assume that $K = \mathbf{C}$. We follow the advice of the introduction:

1. The "worst" object is the singularity defined by $x^p + y^q$. A versal deformation of $x^p + y^q$ fixing the semigroup $\langle p, q \rangle, p < q$ and $gcd(p, q) = 1$, is given by
$$F(x, y, \underline{T}) = x^p + y^q + \sum_{(i,j)\in B} T_{iq+jp-pq} x^i y^j,$$
where
$$B = \{(i, j) \mid iq + jp > pq, i \leq p - 2, j \leq q - 2\}.$$
Let $\underline{T} = \{T_{iq+jp-pq}\}_{(i,j)\in B}$, $X = Spec\,\mathbf{C}[\underline{T}][[x, y]]/F,$, $F_t \in \mathbf{C}\ [[x, y]]$ given by $F_t(x, y) = F(x, y, t),\ t \in T = Spec\,\mathbf{C}[\underline{T}]$. Then the family $X \to T$ has the following properties:

   1.1 $X \to T$ is a versal deformation of $Spec\,\mathbf{C}[[x, y]]/x^p + y^q$ fixing the semigroup $\langle p, q \rangle$.

   1.2 Every plane curve singularity with semigroup $\langle p, q \rangle$ is represented in this family, i.e. there is a $t \in T$ such that the given singularity is isomorphic to $X_t = Spec\,\mathbf{C}[[x, y]]/(F_t)$.

2. The Kodaira-Spencer map of the family $X \to T$ is given by

$$\rho : Der_{\mathbf{C}}\, \mathbf{C}[\underline{T}] \longrightarrow \mathbf{C}[\underline{T}][[x,y]]/\left(F, \frac{\partial F}{\partial x}, \frac{\partial F}{\partial y}\right)$$

$$\rho(\delta) = \text{ class } (\delta F) = \text{ class } \left(\sum_{(i,j)\in B} \delta(T_{iq+jp-pq}) x^i y^j \right).$$

The kernel of the Kodaira-Spencer map is a Lie algebra $\mathcal{L}$ which is a finitely generated $\mathbf{C}[\underline{T}]$-module and has the following property:

2.1 For $t, t' \in T$ the singularities X_t and $X_{t'}$ are isomorphic iff t and t' are in the same integral manifold of $\mathcal{L}$, i.e. $T/\mathcal{L}$ is a classifying space for all singularities with semigroup $\langle p, q\rangle$.
T has a natural $\mathbf{C}^*$-action defined by $deg\, T_\alpha = -\alpha$. This $\mathbf{C}^*$-action is induced by the $\mathbf{C}^*$-action on $\mathbf{C}\,[[x,y]]/x^p + y^q$ given by $deg\, x = q$ and $deg\, y = p$ in order to keep F homogeneous. The induced grading of $\mathcal{L} \subseteq Der_{\mathbf{C}}\mathbf{C}[\underline{T}]$ is defined by $deg \frac{\partial}{\partial T_\alpha} = \alpha$. One can show that $\mathcal{L}$ is generated as $\mathbf{C}[\underline{T}]$-module by homogeneous vector fields $\{\delta_\alpha\}$ with the following properties:

2.2 There are homogeneous vector fields $\delta_{aq+bp} \in \mathcal{L}$ for $(a,b) \in B^{\checkmark} := \{(p-2-i, q-2-j) \mid (i,j) \in B\}$ such that

- $\{\delta_{aq+bp}\}_{(a,b)\in B^{\checkmark}}$ generate $\mathcal{L}$ as $\mathbf{C}[\underline{T}]$-module
- $deg\, \delta_\alpha = \alpha$
- $\delta_\alpha(T_\beta) = \delta_{\beta^{\checkmark}}(T_{\alpha^{\checkmark}})$, where $a^{\checkmark} := pq - 2p - 2q - a$ for $a \in \mathbf{Z}$
- $[\delta_\alpha, \delta_\beta] \in \sum_{\nu \geq \alpha+\beta} \mathbf{C}[\underline{T}]\delta_\nu$

Remark: $\mathbf{C}[\underline{T}]\,[[x,y]]/(\frac{\partial F}{\partial x}/\frac{\partial F}{\partial y})$ is a free $\mathbf{C}[\underline{T}]$-module of rank μ. The multiplication by F defines an endomorphism of this module. The module admits a basis $\{u_\alpha\}$ represented by quasihomogeneous polynomials of degree α such that

$$u_\alpha F = \sum_\beta \delta_\alpha(T_\beta) \cdot u_{\beta+pq}.$$

This determines the Lie algebra L_0 generated by $\delta_\alpha = \sum_\beta \delta_\alpha(T_\beta)\partial/\partial T_\beta$ and the action of L_0 on T. For concrete examples, one has to compute the matrix of the endomorphism of $\mathbf{C}[\underline{T}]\,[[x,y]]/(\frac{\partial F}{\partial x}/\frac{\partial F}{\partial y})$ given by multiplication with F (with respect to the basis $\{u_\alpha\}$). For this purpose there exists a fast algorithm which has also been implemented (cf. [LaP], appendix with B. Martin).

Let L_0 be the Lie algebra generated by $\{\delta_{aq+bp}\}_{(a,b)\in B^{\checkmark}}$ as Lie algebra and $L = [L_0, L_0]$, then L_0 is finite dimensional and solvable and L is nilpotent $L_0/L \cong \mathbf{C}\delta_0, \delta_0$ is the Euler vector field. Because the $\mathbf{C}[\underline{T}]$-generators $\{\delta_\alpha\}$ of $\mathcal{L}$ are in L_0, L_0 and $\mathcal{L}$ have the same integral manifolds which are the orbits of the action of L_0. This implies that the action of the kernel of the Kodaira-Spencer map $\mathcal{L}$ is induced by the action of the algebraic group $G_0 := exp\, L_0$ with the same quotient ring of invariants $\mathbf{C}[I]^{\mathcal{L}} = \mathbf{C}[I]^{G_0}$.

3. The grading of $\mathbf{C}[\underline{T}]$ induces for each $a \geq 0$ a filtration $F_a^\bullet(\mathbf{C}[\underline{T}])$, where $F_a^i(\mathbf{C}[\underline{T}])$ is the $\mathbf{C}$-vector space generated by all quasihomogeneous polynomials of degree $\geq -(a+ip), p \geq a \geq 0$. Similarly, we get filtrations $H_a^\bullet = H_a^\bullet(\mathbf{C}\,[[x,y]])$ on $\mathbf{C}\,[[x,y]]$ by defining H_a^n to be the ideal generated by all quasihomogeneous polynomials of degree $\geq a+np$. The **Hilbert-function of the Tjurina algebra** of X_t, $\mathbf{C}\,[[x,y]]/(F_t, \partial F_t/\partial x, \partial F_t/\partial y)$, with respect to $H_a^\bullet$ is by definition the function $\tau_a^\bullet(t)$,

$$n \mapsto \tau_a^n(t) := dim_{\mathbf{C}}\mathbf{C}\,[[x,y]]/(F_t, \partial F_t/\partial x, \partial F_t/\partial y, H_a^n)$$

Notice that $\tau_a^n(t) = \tau(X_t)$, the Tjurina number of X_t if n is big and $\tau_a^n(t) = dim_{\mathbf{C}}\mathbf{C}\,[[x,y]]/H_a^n$ (hence independent of t) if n is small.

Remark: We introduced the filtrations $F_a^\bullet$ and $H_a^\bullet$ for different a because the general theory works for arbitrary a but in some cases a good choice of a gives bigger strata (cf. Theorem 2.1(3) and the examples at the end of this section). There is only a finite range of n such that $\tau_a^n(t)$ can vary with t. We usually identify $\tau_a^\bullet$ with the finite tuple of values which might vary with t. Moreover, if $\mu \in \mathbf{N}$, we also write μ for the constant function on $\mathbf{N}$.

If $\delta \in L$ is a homogeneous vector field then $deg\,\delta \geq p$. This implies that $\delta F_a^i \subseteq F_a^{i-1}$ for all $\delta \in L$.
Let k be minimal such that $dF_a^k(\mathbf{C}[\underline{T}])$ generates $\mathbf{C}[\underline{T}]d\mathbf{C}[\underline{T}]$ over $\mathbf{C}[\underline{T}]$ and consider the following filtration of L induced by the filtration of $\mathbf{C}[\underline{T}]$:

$$L = Z_1^a(L) \supset Z_2^a(L) \supset \ldots \supset Z_k^a(L) \supset Z_{k+1}^a(L) = \{0\},$$

$Z_i^a(L)$:= the Lie algebra generated by $\{\delta_\alpha\}_{\alpha\in S_i}$, where $S_i = \{\alpha \mid T_{\alpha^{\checkmark}} \in F_a^{k-i}, \alpha \neq 0\}$. Since $deg\,\delta_\alpha = \alpha \geq p$ if $\alpha \neq 0$ we obtain $[L, Z_i^a(L)] \subseteq Z_{i+1}^a(L)$.

Some pictures might be helpful (see next page)
The monomials $x^iy^j, (i,j)$ any point of the rectangle $\{i \leq p-2, j \leq q-2\}$, are a $\mathbf{C}$-basis of the Tjurina algebra $\mathbf{C}\,[[x,y]]/(x^p+y^q)$ of X_0. The monomials in the shaded region $B = \{(i,j) \mid iq+jp > pq, i \leq p-2, j \leq q-2\}$ correspond

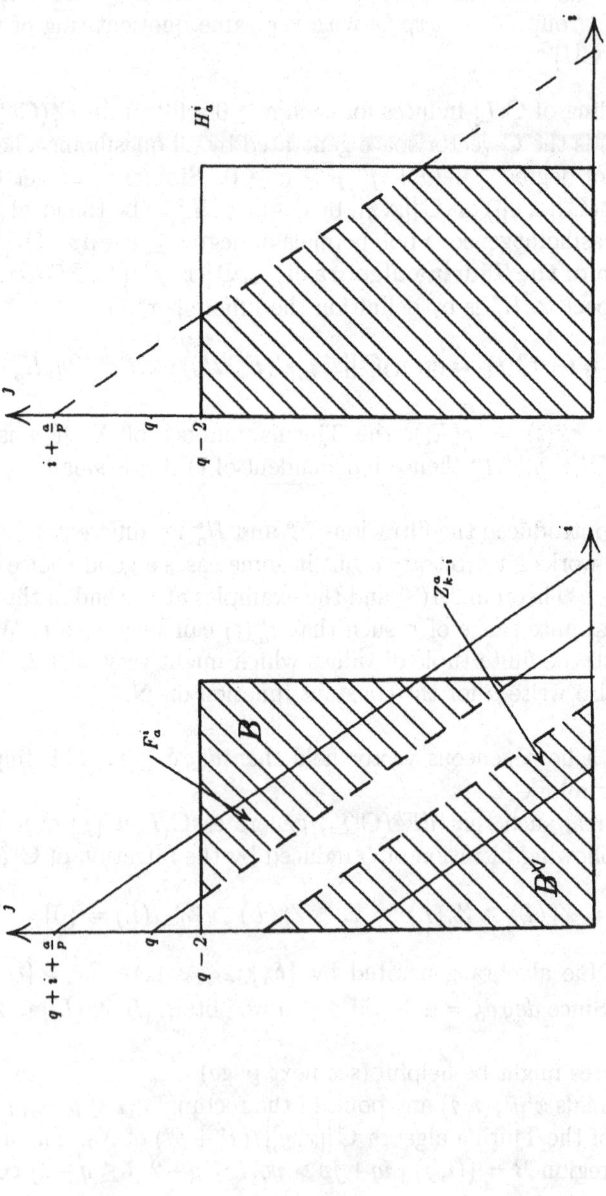

j
i + a/p
q
q − 2
H^i_a
p − 2
p
i
q + i + a/p
F^i_a
B
Z^a_{k−i}
B^∨

to deformations of X_0 with fixed semigroup $\langle p,q\rangle$. They occur as coefficients of the parameters T_α of the versal base space T for such deformations. The T_α are indexed by $\alpha = ip + jp - pq$ if $x^i y^j$ is the coefficient of T_α (this is unique since $gcd(p,q) = 1$). Hence the (increasing) filtration of $\mathbf{C}[\underline{T}]$, $F^{\bullet}_{a}$, is generated by those T_α such that the coefficients $x^i y^j$ belong to a strip as indicated above. Note that the degree of the T_α decreases at most by p $(= min\{p,q\})$ if we go from F^i_a to F^{i+1}_a, the different choices of $a, 0 \le a \le p$, means just a shift of the starting point.

$B^{\checkmark} = \{(p-2-i, q-2-j) \mid (i,j) \in B\}$ is just the mirror of B at the centre of the rectangle. The vector fields δ_α which generate the Lie algebra L are indexed by the weights $\alpha = aq + bp$ of the points $(a,b) \in B^{\checkmark}$. The (decreasing) filtration $Z^a_{\bullet}$ of L is given by dual strips, indexed in a complementary manner: $\delta_\alpha \in Z^a_i \Longleftrightarrow T_{\alpha^{\checkmark}} \in F^{k-i}_a, \alpha^{\checkmark} = pq - 2p - 2q - \alpha$, i.e. Z^a_{k-i} is the mirror image of F^i_a.

The second picture shows the (decreasing) filtration $H^{\bullet}_a$ of $\mathbf{C}[[x,y]]$, H^i_a is generated by all monomials above and on the dotted line. Hence $\tau^i_a(t)$ is the number of monomials in the shaded region which are linear independent modulo $(F_t, \partial F_t/\partial x,$
$\partial F_t/\partial y, H^i_a)$, for $t = 0$ these are all monomials.

Let $\{U^a_\alpha\}$ now be the flattening stratification on $T = Spec\,\mathbf{C}[\underline{T}]$ corresponding to $F^{\bullet}_a$ and $Z^a_{\bullet}$. Notice that each U^a_α is a locally closed, not necessarily reduced subvariety of T.

Now we can apply Theorem 1.4 and obtain that $U^a_\alpha \longrightarrow U^a_\alpha/L$ is a geometric quotient. Moreover, $L_0/L \cong \mathbf{C}^*$ acts on U^a_α/L and $U^a_\alpha/\mathcal{L} = U^a_\alpha/L_0$ is a geometric quotient of U^a_α by L_0. For $t \in T$ define $\underline{e}^a_{(t)} = (e^a_0(t), \ldots, e^a_k(t)) \in \mathbf{N}^{k+1}$ by

$$e^a_i(t) = rank(\delta_\alpha(T_\beta)(t))_{\substack{\beta \le a+ip \\ \alpha \ge 0}},\, i = 0, \ldots, k.$$

Theorem 2.1 *Let T be the base space of the versal deformation with fixed semigroup of $Spec\,\mathbf{C}[[x,y]]/(x^p + y^q)$ and $\{U^a_\alpha\}_\alpha$ the stratification of T defined above. The following holds:*

1. *$\underline{e}^a$ is constant on U^a_α and takes different values for different α. The scheme structure of U^a_α is defined by the corresponding minors of $(\delta_\alpha(T_\beta))$. Moreover, $e^a_i(t) = \mu(X_t) - \tau^{q+i}_a(t)$. In particular, $e_k(t) = \mu(X_t) - \tau(X_t)$, where $\mu(X_t) = \mu = (p-1)(q-1)$ is the Milnor number and $\tau(X_t)$ the Tjurina number of the curve singularity $X_t =$ Spec $\mathbf{C}[[x,y]]/(F_t)$.*

2. *Let $\underline{e} = (e_1, \ldots, e_k) \in \mathbf{N}^{k+1}$ and let $U^a_{\underline{e}}$ denote the unique stratum such $\underline{e}^a(t) = \underline{e}$ for $t \in U^a_{\underline{e}}$ and assume that $U^a_{\underline{e}}$ is not empty. The geometric quotient $U^a_{\underline{e}}/\mathcal{L}$ is quasiaffine and of finite type over $\mathbf{C}$. It is a coarse moduli space for the functor which associates to any complex space germ S the set of isomorphism classes of flat families (with section) over S of plane curve singularities with fixed semigroup $\langle p, q\rangle$ and fixed Hilbert function $\tau^\bullet_a(t) = \mu - \underline{e}$ of the Tjurina algebra.*

3. *Let $T_{\tau_{min}}$ be the open dense subset of T defined by singularities with minimal Tjurina number τ_{min}. Then there exists an a such that $T_{\tau_{min}} = U^a_\alpha$ for a suitable α. In particular, the geometric quotient $T_{\tau_{min}}/\mathcal{L}$ exists and is a coarse moduli space for curves with semigroup $\langle p, q\rangle$ and Tjurinanumber τ_{min}. $T_{\tau_{min}}/\mathcal{L}$ is locally isomorphic to an open subset of a weighted projective space.*

Proof:

1. $t \in U^a_\alpha$ iff $rank\,(\delta_\alpha(T_\beta)(t))_{\substack{\beta \le a+ip \\ \alpha > 0}} =: r_i(t)$ ($\alpha = 0$ excluded) and $rank\,(\delta_\alpha(T_\beta)(t))_{\alpha \in S_i} =: s_i(t)$ are constant (remark 2 after Theorem 1.4). But since $\delta_\alpha(T_\beta) = \delta_{\beta^\checkmark}(T_\alpha^\checkmark)$ ($\alpha = 0$ included) we have $s_k = e_0, \ldots, s_1 = e_{k-1}$ and $r_i = max\{0,\ e_i - 1\}$. It is also clear that the scheme structure required by the flattening property is given by the minors. For $t \in U^a_\alpha$ consider the induced $\mathbf{C}$-base $\{u_i(t)\}$ of $\mathbf{C}\,[[x,\,y]]\,/\,(\frac{\partial F_t}{\partial x}, \frac{\partial F_t}{\partial y})$, $u_i(t)F(x,y,t) = \sum_j \delta_i(T_j)(t)u_{j+pq}(t)$ (as in the remark 2.3). This implies by definition of $\tau^\bullet_a(t), e_i(t) = \mu(X_t) - \tau^{q+i}_a(t)$.

2. This follows from the fact that $U^a_{\underline{e}}$ is locally a versal family for singularities with fixed semigroup and Hilbert function and that $U^a_{\underline{e}}/\mathcal{L}$ is a geometric quotient.

3. Is proved in [LaP].

Remark: We can actually enlarge the functor for which $\mathcal{U}^a_{\underline{e}}/\mathcal{L}$ is a coarse moduli space. Namely, we can extend the category of base spaces from germs to complex spaces S for which $H^1(S, \mathbf{Z}/d\mathbf{Z}) = 0$ (compare [GP 3]).

Example: $p = 5, q = 11$

1. The versal deformation of $x^5 + y^{11}$ fixing the semigroup $\langle 5, 11\rangle$ is given by

$$F(x,y,\underline{T}) = \\ x^5 + y^{11} + T_1\,x\,y^9 + T_2\,x^2\,y^7 + T_3\,x^3\,y^5 + T_7\,x^2\,y^8 + \\ T_8\,x^3\,y^6 + T_{12}\,x^2\,y^9 + T_{13}\,x^3\,y^7 + T_{18}\,x^3\,y^8 + T_{23}\,x^3\,y^9$$

$$B = \{(1,9),(2,7),(3,5),(2,8),(3,6),(2,9),(3,7),(3,8),(3,9)\}$$

$$B^{\checkmark} = \{(0,0),(0,1),(0,2),(1,0),(0,3),(1,1),(0,4),(1,2),(2,0)\}$$

2. The following vector fields generate the kernel $\mathcal{L}$ of the Kodaira-Spencer map (these can be computed using the algorithm given in [LaP]).

$$\begin{aligned}
\delta_0 &= T_1\tfrac{\partial}{\partial T_1} + 2T_2\tfrac{\partial}{\partial T_2} + 3T_3\tfrac{\partial}{\partial T_3} + \ldots + 23T_{23}\tfrac{\partial}{\partial T_{23}} \\
\delta_5 &= A\tfrac{\partial}{\partial T_7} + B\tfrac{\partial}{\partial T_8} + C\tfrac{\partial}{\partial T_{12}} + D\tfrac{\partial}{\partial T_{13}} E\tfrac{\partial}{\partial T_{18}} + 18T_{18}\tfrac{\partial}{\partial T_{23}} \\
\delta_{10} &= 2T_2\tfrac{\partial}{\partial T_{12}} + (3T_3 + T_1T_2)\tfrac{\partial}{\partial T_{13}} + D\tfrac{\partial}{\partial T_{18}} + 13T_{13}\tfrac{\partial}{\partial T_{23}} \\
\delta_{11} &= T_1\tfrac{\partial}{\partial T_{12}} + 2T_2\tfrac{\partial}{\partial T_{13}} + C\tfrac{\partial}{\partial T_{18}} + 12T_{12}\tfrac{\partial}{\partial_{23}} \\
\delta_{15} &= B\tfrac{\partial}{\partial T_{18}} + 8T_8\tfrac{\partial}{\partial T_{23}} \\
\delta_{16} &= A\tfrac{\partial}{\partial T_{18}} + 7T_7\tfrac{\partial}{\partial T_{23}} \\
\delta_{20} &= 3T_3\tfrac{\partial}{\partial T_{23}} \\
\delta_{21} &= 2T_2\tfrac{\partial}{\partial T_{23}} \\
\delta_{22} &= T_1\tfrac{\partial}{\partial T_{23}}
\end{aligned}$$

with $A = 2T_2 - \frac{9}{11}T_1^2$, $B = 3T_3 - \frac{7}{11}T_1T_2$, $C = 7T_7 + \frac{3}{11}T_1T_3^2$,
$D = 8T_8 - \frac{8}{11}T_1T_7 + \frac{2}{11}T_1^2T_3^2$,
$E = 13T_{,3} - \frac{117}{11}T_1T_{12} + \frac{3}{11}T_1^2T_3T_8 + \frac{55}{11^2}T_1T_2T_3T_7 + \frac{7}{5\cdot 11^3}T_1^3T_2^2T_3^2$.
For the filtration F_a^i of $\mathbf{C}[\underline{T}]$ we choose $a = 5$, i.e. $F_a^i = \langle T_\alpha \mid \alpha \leq 5(1+i)\rangle$ (we omit the index a) $F^0 = \langle T_1, T_2, T_3\rangle, F^1 = \langle T_1, T_2, T_3, T_7, T_8\rangle, F^2 = \langle T_1, T_2, T_3, T_7, T_8, T_{12}, T_{13}\rangle$, $F^3 = \langle T_1, T_2, T_3, T_7, T_8, T_{12}, T_{13}, T_{18}\rangle, F^4 = \langle T_1, \ldots, T_{23}\rangle$.
The minimial i such that dF^i generates $\mathbf{C}[\underline{T}]\, d\,\mathbf{C}[\underline{T}]$ is 4, hence $k = 4$.
The stratification $\{U_\alpha\}$ is given by fixing the rank of $(\delta_\alpha(T_\beta))_{\substack{\beta \leq 5(i+1) \\ \alpha \geq 0}}$ for $i = 0, \ldots, 4$. Calculation shows:

$$
\begin{aligned}
U_1 &= \{t \in Spec\,\mathbf{C}\,[\underline{T}] \mid rank(\delta_\alpha(T_\beta)(t)) = 6\} = \{t \mid \tau(X_t) = 34\} \\
&= \{t \mid \underline{e}(t) = (1,2,4,5,6)\} = \{t \mid 4t_2^2 - 3t_1t_3 + t_1^2t_2 \neq 0\} \\
&\quad \text{with } e_i(t) = \mu(x_t) - \tau_a^{q+i}(t). \text{We have } U_1 = T_{\tau_{min}}. \\
U_2 &= \{t \mid \underline{e}(t) = (1,2,3,4,5)\} \\
&= \{t \mid 4t_2^2 - 3t_1t_3 - t_1^2t_2 = 0 \text{ and } A(t) \neq 0 \text{ or } B(t) \neq 0\} \\
U_3 &= \{t \mid \underline{e}(t) = (1,1,3,4,5)\} \\
&= \{t \mid 4t_2^2 - 3t_1t_3 - t_1^2t_2 = A(t) = B(t) = 0 \text{ and } D(t) \\
&\quad (2t_2C(t) - t_1D(t) - C(t)(C(t)(3t_3 + t_1t_2) - 2t_2D(t)) \neq 0\} \\
U_4 &= \{t \mid \underline{e}(t) = (1,1,2,3,4)\} \\
&= \{t \mid A(t) = B(t) = t_1(9t_1C(t) - 11D(t)) = 0 \text{ and} \\
&\quad C(t)^2 - t_1E(t) \neq 0 \text{ or } D(t)^2 - (\tfrac{9}{11})^2t_1^3E(t) \neq 0\} \\
U_5 &= \{t \mid \underline{e}(t) = (1,1,2,2,3)\} \\
&= \{t \mid A(t) = B(t) = C(t)^2 - t_1E(t) = D(t) - \tfrac{9}{11}t_1C(t) = 0 \\
&\quad \text{and } t_1 \neq 0\} \\
U_6 &= \{t \mid \underline{e}(t) = (0,0,1,2,3)\} \\
&= \{t \mid t_1 = t_2 = t_3 = t_7 = t_8 = 0, t_{13} \neq 0\} \\
U_7 &= \{t \mid \underline{e}(t) = (0,0,1,1,2)\} \\
&= \{t \mid t_1 = \ldots = t_8 = 0, t_{13} = 0 \text{ and } t_{12} \neq 0\} \\
U_8 &= \{t \mid \underline{e}(t) = (0,0,0,1,2)\} \\
&= \{t \mid t_1 = \ldots = t_{13} = 0 \text{ and } t_{18} \neq 0\} \\
U_9 &= \{t \mid \underline{e}(t) = (0,0,0,0,1)\} \\
&= \{t \mid t_1 = \ldots = t_{18} = 0 \text{ and } t_{23} \neq 0\} \\
U_{10} &= \{t \mid \underline{e}(t) = (0,0,0,0,0)\} \\
&= \{0\} \\
U_1/\mathcal{L} &= D(2T_2A - T_1B) \subsetneq \text{Proj } \mathbf{C}[T_1T_2T_3, y] = \mathbf{P}^3_{(1:2:3:10)}, \\
&\quad y = AT_8 - BT_7
\end{aligned}
$$

Conclusion: The space of plane curves with semigroup $\langle 5, 11\rangle$ is stratified into ten strata $U_1, \ldots, U_{10}$, corresponding to the different values of the Hilbert function $\tau_5^\bullet$ of the Tjurina algebra; U_1 is the τ_{min}- and U_{10} the τ_{max}-stratum. The quotients $U_i/\mathcal{L}$ exist and are a coarse moduli space for such singularities with corresponding fixed value of $\tau_5^\bullet$.

Remarks:

1. It is not always possible to choose $a = p$ for the filtration to obtain $T_{\tau_{min}}$ as one stratum in the corresponding stratification. In the case

$p = 13, q = 36$ we have to choose $a = 9$ (cf. [LaP]).

2. In the $< 5, 11 >$-example we have $U_2 Cup U_3 = \{t \mid rank(\delta_\alpha(T_\beta))(t) = 5\} = \{t \mid \tau(X_t) = 35\}$.
   The geometric quotient $U_2 Cup U_3/\mathcal{L}$ does not exist (cf. [LaP]), i.e. fixing τ is not enough, it is necessary to work with a finer stratification.

3 A moduli space for irreducible plane curve singularities with semigroup $\langle 2p, 2q, 2pq + d\rangle$

1. The "worst" object is the singularity defined by $(x^p + y^q)^2 + x^\alpha y^\beta, \alpha q + \beta p = 2pq + d, \alpha < p$. A versal deformation of $(x^p + y^q)^2 + x^\alpha y^\beta$, fixing the semigroup $\langle 2p, 2q, 2pq + d\rangle, p < q, gcd(p, q) = 1$ and $d\, odd$, is given by

$$\begin{aligned} F(x, y, \underline{H}, \underline{W}) &= (x^p + y^q + \textstyle\sum_{(i,j)\in B_0} H_{iq+jp-pq} x^i y^j)^2 + x^\alpha y^\beta \\ &+ \textstyle\sum_{(i,j)\in B_1} W_{iq+jp-2pq} x^i y^j, \end{aligned}$$

$$B_0 = \{(i, j), iq + jq > pq, i \leq p - 2, j \leq q - 2\},$$

$$B_1 = \{(i, j), iq + jp > 2pq + d, i < p, j < \delta\}$$

$$Cup\{(i, j), iq + jp > 2pq + d, i < \gamma, j < \delta + q\}$$

with γ, δ defined by $\gamma < p, \gamma q + \delta p = 3pq - q - p + d$. Let $\underline{H} = \{H_{iq+jp-qp}\}_{(i,j)\in B_0}, \underline{W} = \{W_{iq+jp-2pq}\}_{(i,j)\in B_1}$,
$T := Spec\, \mathbf{C}[\underline{H}, \underline{W}], X := Spec\, \mathbf{C}[\underline{H}, \underline{W}][[x, y]]/F$. The family $X \to T$ has the following properties:

   1.1 $X \to T$ is a versal deformation of $Spec\, \mathbf{C}[[x, y]]/(x^p + y^q)^2 + x^\alpha y^\beta$ fixing the semigroup $\langle 2p, 2q, 2pq + d\rangle$.

   1.2 Every plane curve singularity with semigroup $\langle 2p, 2q, 2pq + d\rangle$ is represented in this family.

   1.3 The group μ_d of $d{-}th$ roots of unity acts on T via $F(\lambda^q x, \lambda^p y, \underline{h}, \underline{w}) = \lambda^{2pq} F(\lambda, y, \lambda \circ \underline{h}, \lambda \circ \underline{w})$ for $\lambda \in \mu_d$.

   1.4 $dim\, T = 2(p - 1)(q - 1) - p - q + 2 + \left[\frac{q}{p}\right]$.

2. The Kodaira-Spencer map of the family $X \to T$ is given by
   $\rho : Der_{\mathbf{C}} \mathbf{C}[\underline{H}, \underline{W}] \longrightarrow \mathbf{C}[\underline{H}, \underline{W}][[x, y]]/(F, \frac{\partial F}{\partial x}, \frac{\partial F}{\partial y})$
   $\rho(\delta) = class\,(\delta F)$
   $= class\,(2(x^p + y^q + \sum_{(i,j)\in B_0} H_{iq+jp-pq} x^i y^j) \sum_{(i,j)\in B_0} \delta H_{iq+jp-pq} x^i y^j$
   $+ \sum_{(i,j)\in B_1} \delta W_{iq+jp-2pq} x^i y^j)$.

The kernel of the Kodaira-Spencer map is a Lie algebra $\mathcal{L}$ which is a finitely generated $\mathbf{C}[\underline{H}, \underline{W}]$-module and has the following property:

2.1 For t, t' the singularities X_t and $X_{t'}$ are isomorphic iff for a suitable $\lambda \in \mu_d$ $\lambda \circ t$ and t' are in the same integral manifold of $\mathcal{L}$, i.e. $T/\mathcal{L}/\mu_d$ is a classifiying space for all singularities with semigroup $\langle 2p, 2q, 2pq + d\rangle$.

2.2 It is always possible for $(\alpha, b) \neq (0,0)$ to obtain a unique decomposition

$$x^a y^b F = (x^p + y^q + \sum_{(i,j)\in B_0} H_{iq+jp-pq} x^i y^j) \sum_{(i,j)\in B_0} E^{qa+pb}_{iq+jp-pq} x^i y^j$$
$$+ \sum_{(i,j)\in B_1} D^{qa+pb}_{iq+jp-2pq} x^i y^j mod(\frac{\partial F}{\partial x}, \frac{\partial F}{\partial y}).$$

This defines vector fields δ_{aq+pb} of the kernel of the Kodaira-Spencer map by $\delta_{qa+pb}(H_s) = \frac{1}{2}E^{qa+pb}_s$ and $\delta_{qa+pb}(W_s) = D^{qa+pb}_s$. $\mathcal{L}$ is generated as $\mathbf{C}[\underline{H}, \underline{W}]$-module by the vector fields $\{\delta_s\}$.
These vector fields have the following properties:

- $[\delta_\ell, \delta_m] \in \sum_{s>l+m} \mathbf{C}[\underline{H}, \underline{W}]\delta_s$
- $\delta_\ell = 0$ if $l > 2pq - 2p - 2q$
- $\delta_\ell(W_m) = 0$ if $m < l + d$
- if $(\alpha + a, \beta + b)$ or $(\alpha + a - p, \beta + b - q) \in B_1$ then we have that $\delta_{aq+bp}(W_{d+aq+bp}) = -\frac{d}{2pq}$.

Now it is not difficult to see that

- $aq+bp < pq-q$ implies $(\alpha+a, \beta+b)$ or $(\alpha+a-p, \beta+b+q) \in B_1$
- $(i,j) \in B_1$ and $iq+jp \geq 3pq+d-q$ implies $(i,j) = (\alpha+a, \beta+b)$ or $(i,j) = (\alpha + a - p, \beta + b + q)$ for a suitable (a,b).

From these properties we deduce that already

$$\{\delta_\ell\}_{\ell\in L}, L = \{\ell = aq+bp, (\alpha+a, \beta+b) \,\text{or}\, (\alpha+a-p, \beta+b-q) \in B_1\}$$

generates $\mathcal{L}$ as $\mathbf{C}[\underline{H}, \underline{W}]$-module. Now for $\ell \in L$ we know that $\delta_\ell(W_{\ell+d}) = -\frac{d}{2pq}$ and $\delta_\ell(W_m) = 0$ if $m < \ell + d$. This implies that $\{\delta_\ell\}_{\ell\in L}$ and $\{W_{\ell+d}\}_{\ell\in L}$ satisfy the properties of proposition 1.5.
This implies that $T \to T/\mathcal{L}$ is a geometric quotient and in particular $T/\mathcal{L} \cong Spec\,\mathbf{C}[\underline{H}, \underline{W'}]$ with $\underline{W'} = \{W_{iq+jp-2pq}\}_{(i,j)\in B_2}, B_2 = \{(i,j) \in B_1, iq + jp - 2pq - d \notin L\}$.
Notice that $dim\, T/\mathcal{L} = (p-2)(q-2) + [\frac{q}{p}] - 1$. We obtain as in section 2:

Theorem 3.1 $\mathbf{C}^{(p-2)(q-2)+[\frac{q}{p}]-1}/\mu_d$ *is a coarse moduli space for families of plane curve singularities with semi-group* $\langle 2p, 2q, 2pq+d\rangle$*. The Tjurina number of all these singularities is constant and equal to* $\tau = \mu - (p-1)(q-1)$*, where* $\mu = (2p-1)(2q-1)+d$.

For details of the proof see [LuP]. Note that we did not need to stratify in this case. The remark after Theorem 2.1 applies also to Theorem 3.1.

4 A moduli space for torsion free modules of rank 1 over the local ring of an irreducible curve singularity

Let R be the local ring of an irreducible curve singularity and $\bar{R} = \mathbf{C}[[t]]$ the normalization of R. Let $Mod(R)$ be the category of torsion free rank 1 R-modules, c the conductor of $R, \delta = \delta(R) = dim_{\mathbf{C}}\bar{R}/R$ be the δ-invariant, and $\bar{M} = M \otimes_R \bar{R}/\text{torsion}$.

Lemma 4.1 1. *Any* $M \in Mod(R)$ *is isomorphic to some fractional ideal* M' *such that* $R \subset M' \subset \bar{R}$ *and* $dim_{\mathbf{C}}\bar{R}/M' = dim_{\mathbf{C}}\bar{M}/M$.

2. *Let* $M, M' \subset \bar{R}$ *be two fractional ideals with* $dim_{\mathbf{C}}\bar{R}/M = dim_{\mathbf{C}}\bar{R}/M'$*. Then* $M \simeq M'$ *iff there is* $u \in \bar{R}^*$ *such that* $uM = M'$.

3. *For any* $M \in Mod(R), M \subset \bar{R}$ *and* $dim\,\bar{R}/M = d$ *we have* $t^{d+\delta}\bar{R} \subseteq M$.

For a proof cf. [GP 2]; it is easy, use $M \subset M \otimes_R Quot(R) \cong \mathbf{C}[[t]][t^{-1}]$.

Definition: Let $M \in Mod(R)$ and $R \subset M' \subset \bar{R}$ such that $M \simeq M'$. We define $\Gamma(M) := \{v(m'), m' \in M'\}$, ($v$ the valuation of $\bar{R}$) and $\delta(M)$ as the number of gaps in $\Gamma(M)$.

Remark: $\Gamma(M)$ does not depend on the choice of M'. $\Gamma(M)$ is a $\Gamma(R)$ set. $\delta(M) = dim_{\mathbf{C}}\bar{M}/M$. In analogy to singularities, we may consider $\Gamma(M)$ as the "topological type" of M.

The aim is now to classify all torsion free rank 1 modules with fixed value set Γ. Let $R_c = \mathbf{C}[[t^c, t^{c+1}, \ldots]]$ and $M \in Mod(R)$ then $M \in Mod(R_c)$. We first solve the problem for $Mod(R_c)$. It is easy to see that R^* acts on the space which parametrizes objects of $Mod(R_c)$. The fixed point scheme will then be the solution for $Mod(R)$ since these points correspond to R_c modules which are also R-modules.

Let $\Gamma = \{\gamma_0, \gamma_1, \ldots, \gamma_k, c, c+1, \ldots\}, 0 = \gamma_0 < \gamma_1 < \ldots < \gamma_k < c$.

1. The "worst" object in $Mod_\Gamma(R_c)$ is the monomial module $M_0 = \sum_{i=0}^{k} t^{\gamma_i} R_c + t^c \bar{R}$. A versal deformation of M_0 fixing Γ is the $\mathbf{C}[\underline{\lambda}][[t^c, t^{c+1}, \ldots]]$-module

$$\mathcal{M}_\Gamma = \sum_{i=0}^{k} m_i \mathbf{C}[\underline{\lambda}][[t^c, t^{c+1}, \ldots]] + t^c \mathbf{C}[\underline{\lambda}][[t]] \subset \mathbf{C}[\underline{\lambda}][[t]],$$

with

$$\begin{aligned} \underline{\lambda} &= \{\lambda_{ij}\}_{(i,j)\in I} \\ I &= \{(i,j), 0 \le i \le k, j > 0, j + \gamma_i \notin \Gamma\} \\ m_i &= t^{\gamma_i} + \textstyle\sum_{j+\gamma_i \notin \Gamma} \lambda_{ij} t^{j+\gamma_i} \end{aligned}$$

Let $T = Spec\,\mathbf{C}[\underline{\lambda}]$ and $\mathfrak{X} = Spec\,\mathbf{C}[\underline{\lambda}][[t^c, t^{c+1}, \ldots]]$ be the trivial deformation of R_c. $\mathcal{M}_\Gamma$ is a coherent $\mathcal{O}_{\mathfrak{X}}$-module, which is flat over T and its restriction M_t to the fibre $\mathfrak{X}_t = Spec\,R_c, t \in T$, is an element of $Mod\,(R_c)$. One shows:

1.1 $\mathcal{M}_\Gamma$ is a versal deformation of M_0 fixing the value set Γ.

1.2 Every R_c-module with value set Γ is represented in this family, i.e. there is a $t \in T$ such that M_t is isomorphic to the given module.

2. The Kodaira-Spencer map of the family $\mathcal{M}_\Gamma$ is given by

$$\rho : Der_{\mathbf{C}}\mathbf{C}[\underline{\lambda}] \to Ext^1_{\mathbf{C}[\underline{\lambda}][[t^c,t^{c+1},\ldots]]}(\mathcal{M}_\Gamma, \mathcal{M}_\Gamma).$$

Because $\mathcal{M}_\Gamma \subset \mathbf{C}[\underline{\lambda}][[t]]$ is embedded it factors through the Kodaira-Spencer map of the embedded family:

$$Der_{\mathbf{C}}\mathbf{C}[\underline{\lambda}] \to Hom_{\mathbf{C}[\underline{\lambda}][[t^c,t^{c+1},\ldots]]}(\mathcal{M}_\Gamma, \mathbf{C}[\underline{\lambda}][[t]]/\mathcal{M}_\Gamma),\ \delta \mapsto \varphi_\delta$$

and $\varphi_\delta(m) =$class of $\delta(m)$, where δ is lifted to $\mathbf{C}[\underline{\lambda}][[t]]$ by $\delta(t) = 0$.

2.1 Now one can prove that the kernel of the Kodaira-Spencer map $\mathcal{L}$ is the Lie algebra generated as $\mathbf{C}[\underline{\lambda}]$-module by the vectorfields $\{\delta_\ell\}_{1\le\ell\le c-1}$ $\delta_\ell = \sum_{(i,j)\in I} h_{\ell,i,j} \frac{\partial}{\partial\lambda_{ij}}$

with $h_{\ell,i,j} = \lambda_{i,j-\ell} - \sum_{\mathbf{N}u=i+1}^{k} \lambda_{i,\gamma_{\mathbf{N}u}-\gamma_i-\ell}\lambda_{\mathbf{N}u,j+\gamma_i-\gamma_{\mathbf{N}u}}$
(with the convention $\lambda_{ij} = 0$ if $j < 0$, $\lambda_{i,0} = 1$).

Remark:

(1) $h_{\ell,i,j} = 0$ if $\ell > j$

(2) $h_{j,i,j} = 1$ if $(i,j) \in I$.

Furthermore $L := \sum_{\ell=1}^{c-1} \mathbf{C}\delta_\ell$ is an abelian Lie algebra.

2.2 It is not difficult to see that for $t, t' \in T$ the modules M_t and $M_{t'}$ are isomorphic iff they are in the same integral manifold of $\mathcal{L}$, i.e. in the same orbit under the action of L.

2.3 $\mathbf{C}[\underline{\lambda}]$ admits a $\mathbf{C}^*$-action defined by $deg\, \lambda_{ij} = j$. The vector fields δ_ℓ are homogeneous of degree-ℓ. Let $\Gamma_0 \subset \Gamma$ be the maximal semigroup acting on Γ and $a = mult(\Gamma_0 \backslash \{0\})$. For the sub Lie algebras $L^{(0)} := \sum_{i \geq a} \mathbf{C}\delta_i$ and $L^{(1)} := \sum_{i<a} \mathbf{C}\delta_i$, we have $H^1(L^{(1)}, \mathbf{C}[\underline{\lambda}]) = 0$ and $H^1(L^{(1)}, \mathbf{C}[\underline{\lambda}]^{L^{(0)}}) = 0$.

3. Consider now the filtration $F^i(\mathbf{C}[\underline{\lambda}]) :=$ the $\mathbf{C}$-vector space generated by all quasihomogeneous polynomials of degree less than $(i+1)a$. If $\delta \in L^{(0)}$ then $\delta F^i \subseteq F^{i-1}$. Let $T = Spec\,\mathbf{C}[\underline{\lambda}] = CupU_\alpha$ be the flattening stratification of the $\mathbf{C}[\underline{\lambda}]$-modules $Hom_{\mathbf{C}}(L^{(0)}, \mathbf{C}[\underline{\lambda}])/\mathbf{C}[\underline{\lambda}]dF^i\mathbf{C}[\underline{\lambda}]$. We may apply corollary 1.3 and obtain that $U_\alpha \to U_\alpha/L^{(0)}$ is a geometric quotient. Since $H^1(L^{(0)}, \mathbf{C}[\underline{\lambda}]^{L^{(0)}}) = 0$ we obtain that $U_\alpha \to U_\alpha/L$ is a geometric quotient.

Remark: For $t \in T$ let $E(t)(n) := rank\,(\delta_\ell(\lambda_{ij})(t))_{j<n}$, then $E(t)(\mathbf{N}ua)$ is constant along U_α for all $\mathbf{N}u$. More precisely, let d be maximal such that $d \notin \Gamma$ and $\underline{r} = (r_1, \ldots, r_{[\frac{d}{a}]})$ such that $E(t_0)(\mathbf{N}ua) = r_{\mathbf{N}u-1}, \mathbf{N}u = 2, \ldots, [\frac{d}{a}]+1$ for some $t_0 \in U_\alpha$ then

$$t \in U_\alpha \iff E(t)(\mathbf{N}ua) = r_{\mathbf{N}u-1}, \quad \mathbf{N}u = 2, \ldots, \left[\frac{d}{a}\right] + 1$$

Let $\underline{E}(t) := (E(t)(2a), \ldots E(t)([\frac{d}{a}]+1)a))$. We also write $U_{\underline{r}}$ instead of U_α. The invariants $\underline{E}(t)$ which describe the stratification $\{U_{\underline{r}}\}$ can be interpreted as follows:

Remark: For $M \in Mod(R)$, the ring $End_R(M)$ dominates R, we have $R \subset End_R(M) \subset \bar{R}$, and M is an $End_R(M)$-module. We assume that $M \subset \bar{R}$ and define a filtration $End^\bullet(M)$ of $\bar{R}$ by

$$End^n(M) := \{g \in \bar{R} \mid g\, M_{n,i} \subset M_{n,i} \text{ for all } i\}$$

where $M_{n,i} = M Cap \mathbf{C}[[t^i, t^{i+1}, \ldots,]] + \mathbf{C}[[t^{n+i}, t^{n+i+1}, \ldots]]$. $End^n(M)$ is independent of the embedding of M into $\bar{R}$. The function $E(t)$ has the following interpretation:

$$E(t)(n) = dim_{\mathbf{C}} \bar{R}/End^n(M_t),$$

where M_t is the module corresponding to $t \in T$, i.e. $E(t)$ is the Hilbert function of $\bar{R}/End_R(M)$ with respect to the filtration $End^\bullet(M)$.

Theorem 4.2 *Let $R_c = \mathbf{C}[[t^c, t^{c+1}, \ldots]], \underline{r} \in \mathbf{Z}^k, k = [\frac{d}{a}]$, and $U_{\underline{r}}$ the stratum such that $\underline{E}(t) = \underline{r}$ for $t \in U_{\underline{r}}$. The geometric quotient $U_{\underline{r}}/L$ exists, is a quasi-affine algebraic scheme and a coarse moduli space for flat families of torsion free R_c-modules with fixed value set Γ and fixed Hilbert function of $\bar{R}_c/End(M)$ with respect to the filtration $End^\bullet(M)$. The same holds for the local ring R of an arbitrary irreducible curve singularity if we replace $U_{\underline{r}}$ by the fixed point scheme $U_{\underline{r}}^{R^*}$ of $U_{\underline{r}}$ with respect to the natural action of R^*.*

Proof: For the proof that the geometric quotient $U_{\underline{R}}/L$ exists, we use the criteria of chapter 1. To see that $U_{\underline{R}}/L$ indeed is a coarse moduli space, also for families over non-reduced base spaces, many details have to be checked and one has to be very careful about the functors defining families of such modules. First of all, one has to generalize the notion of Γ- and E-constant families to non-reduced base spaces. This can finally be achieved by requiring that certain sheaves are flat over the base. Then one has to prove that the strata $U_{\underline{r}}$, together with the family of modules which can be constructed over it, is, in each of its points, a representative of a versal Γ- and E-constant deformation of the corresponding module. Last, but not least, some foundational work has to be done on families of objects which are algebraic along the base but formal resp. analytic along the fibres.

For a complete proof see [GP 2].

Example: We construct in detail the stratification $\{U_{\underline{r}}\}$ of the space of torsion free R_c-modules of rank 1 with fixed value set Γ, and for each stratum we determine the quotient for the example

$$R = R_c = \mathbf{C}[[t^8, t^9, \ldots]], c = 8,$$
$$\Gamma = \Gamma_0 = \{0, 2, 4, 6, 8, 9, \ldots\}, k = 3.$$

Recall that Γ_0 is the maximal semigroup contained in the set Γ, $a =$ smallest non-gap ($\neq 0$) of $\Gamma_0, d =$ biggest gap of Γ and $k = [\frac{d}{a}]$.

We have:

$$I = \{(0,1), (1,1, (2,1), (3,1), (0,3), (1,3), (2,3), (0,5), (1,5), (0,7)\}.$$

The matrix $H(\underline{\lambda}) = (h_{\ell,i,j}(\underline{\lambda}))$ of the coefficients of the vector fields $\delta_1, \ldots, \delta_7$ is:

$$\begin{pmatrix} 1\ldots1 & -\lambda_{11}\lambda_{01} & -\lambda_{21}\lambda_{11} & -\lambda_{31}\lambda_{21} & -\lambda_{21}\lambda_{03}-\lambda_{12}\lambda_{01} & -\lambda_{31}\lambda_{13}-\lambda_{23}\lambda_{11} & * \\ & \lambda_{01}-\lambda_{11} & \lambda_{11}-\lambda_{21} & \lambda_{21}-\lambda_{31} & \lambda_{03}-\lambda_{13} & \lambda_{13}-\lambda_{23} & * \\ & 1 & 1 & 1 & -\lambda_{21}\lambda_{01} & -\lambda_{31}\lambda_{11} & * \\ & & & & \lambda_{01}-\lambda_{21} & \lambda_{11}-\lambda_{31} & * \\ & & & & 1 & 1 & \lambda_{01}-\lambda_{31} \\ & & & & & & 1 \end{pmatrix}$$

We have $a = 2, d = 7$.

Let $T = Spec\, \mathbf{C}[\underline{\lambda}] = \mathbf{C}upU_{\underline{r}}$ be the stratification constructed before. Then for $\underline{r} \in \{(3,5,6), (2,4,5), (3,4,5), (2,3,4)\}$ we have $U_{\underline{r}} \neq \emptyset$.

$$\begin{aligned} U_{\Gamma,(3,5,6)} &= \{\lambda, \lambda_{01} - \lambda_{11} - \lambda_{21} + \lambda_{31} \neq 0\} \\ U_{\Gamma,(3,4,5)} &= \{\lambda, \lambda_{01} - \lambda_{11} - \lambda_{21} + \lambda_{31} = 0, 2\lambda_{11} - \lambda_{01} - \lambda_{21} \neq 0\} \\ U_{\Gamma,(2,4,5)} &= \{\lambda, \lambda_{01} - \lambda_{11} - \lambda_{21} + \lambda_{31} = 2\lambda_{11} - \lambda_{01} - \lambda_{21} = 0, \\ &\quad 2\lambda_{31} - \lambda_{03} - \lambda_{23} + (\lambda_{01} - \lambda_{11})(\lambda_{11}\lambda_{31} - \lambda_{01}\lambda_{21}) \neq 0\} \\ U_{\Gamma,(2,3,4)} &= \{\lambda, \lambda_{01} - \lambda_{11} - \lambda_{21} + \lambda_{31} = 2\lambda_{11} - \lambda_{01} - \lambda_{21} = 0, \\ &\quad 2\lambda_{13} - \lambda_{03} - \lambda_{23} + (\lambda_{01} - \lambda_{11})(\lambda_{11}\lambda_{31} - \lambda_{01}\lambda_{21}) = 0\} \end{aligned}$$

$$\begin{aligned} \text{let } L_1 &:= \mathbf{C}\delta_1 + \mathbf{C}\delta_3 + \mathbf{C}\delta_5 + \mathbf{C}\delta_7, \text{ then} \\ \mathbf{C}[\underline{\lambda}] &= \mathbf{C}[\underline{\lambda}]^{L_1}[\lambda_{01}, \lambda_{03}, \lambda_{05}, \lambda_{07}] \\ \text{and} & \\ \mathbf{C}[\underline{\lambda}]^{L_1} &= \mathbf{C}[\bar{\lambda}_{11}, \bar{\lambda}_{21}, \bar{\lambda}_{31}, \bar{\lambda}_{13}, \bar{\lambda}_{23}, \bar{\lambda}_{15}] \\ \bar{\lambda}_{11} &= \lambda_{11} - \lambda_{01} \\ \bar{\lambda}_{21} &= \lambda_{21} - \lambda_{01} \\ \bar{\lambda}_{31} &= \lambda_{31} - \lambda_{01} \\ \bar{\lambda}_{13} &= \lambda_{13} - \lambda_{03} - \lambda_{01}(\lambda_{11}\lambda_{01} - \lambda_{11}\lambda_{21}) + \tfrac{1}{2}\lambda_{01}^2(\lambda_{01} - \lambda_{21}) \\ \bar{\lambda}_{23} &= \lambda_{23} - \lambda_{03} - \lambda_{01}(\lambda_{11}\lambda_{01} - \lambda_{21}\lambda_{31}) + \tfrac{1}{2}\lambda_{01}^2(\lambda_{11} + \lambda_{01} - \lambda_{21} - \lambda_{31}) \\ \bar{\lambda}_{15} &= \lambda_{15} - \lambda_{05} - (\lambda_{01}\lambda_{21} - \lambda_{11}\lambda_{31})\lambda_{03} + \lambda_{01}(\lambda_{31} - \lambda_{01})(\lambda_{13} - \lambda_{03}) \\ &\quad + \lambda_{01}\lambda_{11}(\lambda_{23} - \lambda_{03}) - \tfrac{1}{2}\lambda_{01}^2(\lambda_{23} - \lambda_{03}) \\ &\quad + \text{polynomial in } \lambda_{01}, \lambda_{11}, \lambda_{21}, \lambda_{31}. \end{aligned}$$

We have

$$\begin{aligned}\delta_2 \mid \mathbf{C}[\underline{\lambda}]^{L_1} &= (2\bar\lambda_{11}-\bar\lambda_{21})\tfrac{\partial}{\partial\bar\lambda_{13}} + (\bar\lambda_{11}-\bar\lambda_{31}+\bar\lambda_{21})\tfrac{\partial}{\partial\bar\lambda_{23}} \\ &\quad +(2\bar\lambda_{13}-\bar\lambda_{23}-\bar\lambda_{11}^2\bar\lambda_{31})\tfrac{\partial}{\partial\bar\lambda_{15}} \\ \delta_4 \mid \mathbf{C}[\underline{\lambda}]^{L_1} &= (\bar\lambda_{11}-\bar\lambda_{31}+\bar\lambda_{21})\tfrac{\partial}{\partial\bar\lambda_{15}} \\ \delta_6 \mid \mathbf{C}[\underline{\lambda}]^{L_1} &= 0\end{aligned}$$

Let $\bar U_{\Gamma,\underline{r}} = \varphi(U_{\Gamma,\underline{r}}), \varphi : U_\Gamma \to U_\Gamma/L_1 = Spec\,\mathbf{C}[\bar{\underline{\lambda}}]^{L_1}$ be the quotient map and $L_0 = K\delta_2 + K\delta_4$. Then

$$\begin{aligned}\bar U_{\Gamma,(3,5,6)} &= \{\bar\lambda, \bar\lambda_{11}-\bar\lambda_{31}+\bar\lambda_{21}\neq 0\} \\ \bar U_{\Gamma,(3,4,5)} &= \{\bar\lambda, \bar\lambda_{11}-\bar\lambda_{31}+\bar\lambda_{21}=0, 2\bar\lambda_{11}-\bar\lambda_{21}\neq 0\} \\ \bar U_{\Gamma,(2,4,5)} &= \{\bar\lambda, \bar\lambda_{11}-\bar\lambda_{31}+\bar\lambda_{21}=2\bar\lambda_{11}-\bar\lambda_{21}=0, \\ &\quad 2\bar\lambda_{13}-\bar\lambda_{23}-\bar\lambda_{11}^2\bar\lambda_{31}\neq 0\} \\ \bar U_{\Gamma,(2,3,4)} &= \{\bar\lambda, \bar\lambda_{11}-\bar\lambda_{31}+\bar\lambda_{21}=2\bar\lambda_{11}-\bar\lambda_{21}= \\ &\quad 2\bar\lambda_{31}-\bar\lambda_{23}-\bar\lambda_{11}^2\bar\lambda_{31}=0\} \\ U_{\Gamma,(3,5,6)}/L &= \bar U_{\Gamma,(3,5,6)}/L_0 \\ &= Spec\,\mathbf{C}\,[\bar\lambda_{11},\bar\lambda_{21},\bar\lambda_{31},\bar\lambda_{13}(\bar\lambda_{11}-\bar\lambda_{31}+\bar\lambda_{21}) \\ &\quad -\bar\lambda_{23}(2\bar\lambda_{11}-\bar\lambda_{21})]_g \\ g &:= \bar\lambda_{11}-\bar\lambda_{31}+\bar\lambda_{21} \\ U_{\Gamma,(3,4,5)}/L &= \bar U_{\Gamma,(3,4,5)}/L_0, \\ &= Spec\,\mathbf{C}[\bar\lambda_{11},\bar\lambda_{21},\bar\lambda_{23},\bar\lambda_{15}(2\bar\lambda_{11}-\bar\lambda_{21})-\bar\lambda_{13}(2\bar\lambda_{13}-\bar\lambda_{23}-\bar\lambda_{11}^2 \\ &\quad (\bar\lambda_{11}+\bar\lambda_{21}))+\bar\lambda_{13}^2]_h \\ h &= 2\bar\lambda_{11}-\bar\lambda_{21} \\ U_{\Gamma,(2,4,5)}/L &= \bar U_{\Gamma,(2,4,5)}/L_0 \\ &= Spec\,\mathbf{C}\,[\bar\lambda_{21},\bar\lambda_{13},\bar\lambda_{23}]_{2\bar\lambda_{13}-\bar\lambda_{23}-3\bar\lambda_{21}^3} \\ U_{\Gamma,(2,3,4)}/L &= \bar U_{\Gamma,(2,3,4)} = Spec\,\mathbf{C}[\bar\lambda_{11},\bar\lambda_{13},\bar\lambda_{15}]\end{aligned}$$

Remark: Let $V := U_{\Gamma,(2,4,5)} \mathbf{C}up U_{\Gamma,(3,4,5)}$. Then

$$V = \{\underline{\lambda} \in U_\Gamma \mid \text{ orbit dimension at } \lambda is 5\}$$

and even $\Gamma(End_R(M_\lambda)) = \{0,2,4,8,9,\dots\}$ is constant on V. It can be shown that the geometric quotient V/L does not exist, neither in the algebraic nor in the analytic category.

Conclusion: The space of torsion free R-modules, $R = \mathbf{C}[[t^8, t^9, \ldots]]$ is stratified into four strata, corresponding to the four different values of the Hilbert function of $\bar{R}/End_R(M)$. The quotients of these strata by L are coarse moduli spaces for R-modules (torsion free, rank 1) with value set Γ and Hilbert function the corresponding value. On the union of two of these strata the orbit dimension of L is constant but the quotient does not exist.

References

[BGM] BrianCcon, J.; Granger, M.; Maisonobe, Ph.: Le nombre de modules du germe de courbe plane $x^a + y^b = 0$. Math. Ann. **279**, 535–551 (1988).

[Ca] Casas, E.: Moduli of algebroid plane curves. Lecture Notes in Math., Vol. 961, 32–83 (1983).

[De] Delorme, C.: Sur les modules des singularités des courbes planes. Bull. Soc. Math. France **106**, 417–446 (1978).

[DG] Demazure, M.; Gabriel, P.: "Groupes Algébriques". Masson & Cie, North Holland, Amsterdam 1970.

[DR] Dixmier, J.; Raynaud, M.: Sur le quotient d'une variété algébrique par un groupe algébrique. Adv. Math. Suppl. Stud. **7A**, 327–344 (1981).

[Eb] Ebey, S.; The classification of singular points of algebraic curves. Trans. Amer. Math. Soc. **118**, 454–471 (1965).

[GP 1] Greuel, G.-M.; Pfister, G.: Geometric quotients of unipotent group actions. To be published in Proc. Lond. Math. Soc. (1993).

[GP 2] Greuel, G.-M.; Pfister, G.: Moduli spaces for torsion free modules on curve singularities I. J. Algebraic Geometry **2** 81–135 (1993).

[GP 3] Greuel, G.M.; Pfister, G.: On moduli spaces of semiquasihomogeneous singularities. Preprint, Kaiserslautern 1993.

[Gr] Greuel, G.-M.: Deformation und Klassifikation von Singularitäten und Moduln. Jber. d. Dt. Math.–Verein., Jubiläumstagung 1990, 177-238, B.G. Teubner (1992).

[Gra] Granger, M.: Sur une espace de modules de germe du courbe plane. Bull. Soc. Math. France 2è sér. **103** (1979).

[LaP] Laudal, O.A.; Pfister, G.: "Local Moduli and Singularities". Lecture Notes in Math., Vol. 1310 (1988).

[LMP] Laudal, O.A.; Martin, B.; Pfister, G. Moduli of plane curve singularities with C^*-action. Banach Center Publ. **40** (1988).

[LuP] Luengo, I.; Pfister, G.: Normal forms and moduli spaces of curve singularities with semigroup $\langle 2p, 2q, 2pq+d\rangle$. Compositio Math. **76**, 247-264 (1990).

[Ma] Martin, B.: Moduli space of binary semi–homogeneous functions over the complex field. Preprint (1990).

[MuF] Mumford, D.; Fogarty, J.: "Geometric Invariant Theory". (Second, enlarged edition.) Ergeb. Math. Grenzgeb., Bd. 34. Berlin-Heidelberg-New York: Springer 1982.

[Pa] Palamodov, V.P.; Moduli in versal deformations of complex spaces. In: "Variétés Analytiques Complexes", Nice 1977 (Eds.: Hervier, Y.; Hirschowitz, A.). Lecture Notes in Math., Vol. 683, pp. 74–115. Berlin–Heidelberg–New York: Springer 1978.

[PS] Pfister, G.; Steenbrink, J.H.M.: Reduced Hilbert scheme for irreducible curve singularities. Preprint, Nijmegen 1990.

[Ri] Riemann, B.: Theorie der Abel'schen Funktionen. J. reine angew. Math. **54**, 115-155 (1857).

[Ro] Rosenlicht, M.: A remark on quotient spaces. An. Acad. Brasil. Ciênc. **35**, 487–489 (1963).

[Sch] Schönemann, H.: Normalformen und Modulräume von Kurvensingularitäten mit der Halbgruppe $np, nq, npq+d$. Dissertation, Berlin 1990.

[Za] Zariski, O.: Le problème des modules pour les branches planes (redigè par F. Kmety et M. Merle avec un appendice de B. Teissier). Pub. du Centre de Mathématiques de l'Ecole Polytechnique, Paris 1976.

CONORMAL SPACE AND JACOBIAN MODULES
A short dictionary

J.-P. HENRY
Centre de Mathématiques
École Polytechnique
91128 PALAISEAU Cedex
Unité associée au C.N.R.S n°169

and

M. MERLE
Laboratoire de Mathématiques
Université de Nice–Sophia Antipolis
Parc Valrose. B.P. 71
06108 NICE Cedex 2
Unité associée au C.N.R.S n°168

1 Introduction and historical ramblings

The present work comes after a long series of papers on equisingularity and stratification of analytic complex varieties. It is motivated by a recent attempt of T. Gaffney to extract, from inside the latest results, the essence of the old intuitions. He proceeds towards this goal by introducing in the equisingularity field a new mathematical tool : the integral closure of modules.

In his famous [Cargese] paper, Teissier considered the following situation: a family of (germs) of hypersurfaces in $\mathbf{C}^n$ with isolated singularities is given, together with a smooth subspace Y of the total space of the family X. He studied several kinds of equisingularity of this family

1. The Whitney equisingularity of X along Y, which insures, by Thom-Mather first isotopy lemma, the topological triviality.

   It is easy to see that Whitney equisingularity is implied by saying certain derivatives are in the integral closure of the product of the jacobian ideal by the ideal of Y (this happened to be the condition **w** of Verdier)

2. The equidimensionality above Y of the exceptional divisor of the blowing up of the product of the jacobian ideal by the ideal of Y

3. The constancy of the different Milnor numbers of the plane sections of the fibres along Y

Teissier proved the equivalence of 2) and 3) and that these two implied 1). The inverse proposition, conjectured by him, was proved by Joel Briançon and Jean-Paul Speder in their thesis ([B-S]) in 1975, thus achieving the completion of the dreamed dictionary in case of families of isolated singularities for analytic complex hypersurfaces.
Most of these results on families of isolated singularities of hypersurfaces have been successfully generalized both in absolute ([H-M 2], [T 2]) or relative case ([H-M-S]), except that there was no purely algebraic translation in terms of equations (integral dependance) of Whitney (or Verdier) conditions. It was the merit of T. Gaffney to provide such a translation by introducing in this field a new tool, integral dependance on module and multiplicities of modules, and using it for families of isolated singularities of complete intersections : the jacobian **module** playing the role devoted in the old days to the jacobian ideal.
We will see now in the following paragraphs that there are not really two ways of generalizing the ancient results but only two ways of describing them.
In Teissier's Cargese paper, algebra could seem, via extensive use of multiplicities and integral closures, to supersede geometry. In the general case geometry took its revenge and the conormal space was first considered in preceding papers by the authors or by Bernard Teissier himself, a few years before the introduction in that field, by Terence Gaffney, of the Buchsbaum-Rees multiplicity of the jacobian module : the jacobian module is (using an adequate word of Teissier) just an "avatar" of the conormal space ; more precisely we show here that the conormal space is the geometric space associated to the jacobian module, the **Proj** of this module, and the related multiplicities are degrees of the divisor already studied in [H-M 2].
Therefore this paper is also a generalization to submodules of a free module of the result of Ramanujam on ideals ([Ra]).
We may hope that, with a new way of phrasing things, new ideas and new results will emerge. This is already true here as results on constancy of multiplicities in relation with modules are not restricted to the jacobian module. In particular, we want to study, in a forthcoming work, applications to systems of differential forms.

A first version of this paper contained some unjustified and/or presumptuous assertions about cohomology groups (in sections 2.2 and 4.2), S. Kleiman warned us about them, and we thank him for his remarks. He also informed us that, in collaboration with Anders Thorup, he has written a paper [K-T] on a geometric theory of Buchsbaum-Rim multiplicity. This paper certainly overlaps our work.
This work has been exposed during one session (in Luminy) of the Hawaï-Luminy seminar in 1990.
The redaction has been revised using also useful remarks of the referees.

2 The projective cone of a sheaf of modules

2.1 Definitions

Let X be an analytic space, and $\mathcal{M}$ a coherent $\mathcal{O}_X$ module.
The module $\mathcal{M}$ has a presentation

$$\mathcal{O}_X^p \xrightarrow{\psi} \mathcal{O}_X^k \longrightarrow \mathcal{M} \longrightarrow 0$$

Associated to this presentation, we have several objects, algebraic and geometric, which are closely related:

- The image of the map $\varphi = {}^t\psi$ which we call $\mathcal{K}$. The symmetric algebra $\mathrm{Sym}(\mathcal{K})$ of the submodule $\mathcal{K}$ of $\mathcal{O}_X^p$, and its image $S(\mathcal{K})$ in $\mathrm{Sym}(\mathcal{O}_X^p)$. We shall also consider the reduced algebra associated to $S(\mathcal{K})$, namely the quotient by its nilpotent part.

- The projective cone of $\mathcal{K}$, which we will denote by $\mathbf{P}(\mathcal{K})$ and call (as in [F]), the projective cone associated to the cone $\mathbf{Spec}S(\mathcal{K})$. We will also define the reduced associated space $\mathbf{P}(\mathcal{K})_{\mathrm{red}}$.

  When $\mathcal{K}$ is the sheaf of sections of a vector bundle K, then the projective cone $\mathbf{P}(\mathcal{K})$ is the projective bundle $\mathbf{P}\check{K}$ on the dual bundle $\check{K}$.

On $\mathbf{P}(\mathcal{K})$ we have the tautological sheaf $\mathcal{O}(1)$ and the associated graded algebra $\Gamma_* = \bigoplus_{n\geq 0} H^0(\mathbf{P}S(\mathcal{K}), \mathcal{O}(n))$. We have canonical surjective morphisms from $\mathrm{Sym}(\mathcal{K})$ to $S(\mathcal{K})$ and from $S(\mathcal{K})$ to Γ_*.
Following [EGA III] 2.3, we get that the map $\mathrm{Sym}(\mathcal{K}) \longrightarrow \Gamma_*$ is a **TN**-isomorphism, which means that there exists an integer n_0 such that, for $n \geq n_0$ the map

(2.1.0.1) $$\mathrm{Sym}(\mathcal{K})(n) \longrightarrow \Gamma_*(n)$$

is an isomorphism.
We are going to define local additive invariants of X and $\mathcal{M}$.

Hence we will assume X is an irreducible germ of analytic space at a point x. In the case where X is not locally irreducible at x, the invariants have to depend only on the restriction of $\mathcal{K}$ to the components of maximal dimension of X.
Under this assumption, the algebra $S(\mathcal{K})$ is reduced (as a subalgebra of $S(\mathcal{O}^p)$ which is reduced) and the morphism $S(\mathcal{K}) \longrightarrow \Gamma_*$ is thence an isomorphism, using the property 2.1.0.1. Hence, the following is devoted to the study of numerical invariants of Γ_*.
As X is irreducible, we have an open set $\mathcal{U}$, everywhere dense, where the module $\mathcal{K}$ has maximal rank r. The complement of this set is the singular locus of the map φ which is sometimes called the *singular locus* of $\mathcal{K}$.
We are going to show that the Hilbert invariants of the quotient $S(\mathcal{O}^p)/S(\mathcal{K})$ are related to geometrical invariants of $\mathbf{P}(\mathcal{K})$ and eventually we will prove that they depend only on $\mathcal{M}$ and not on the given presentation.

2.2 Conormal space

Example 2.2.1 (The sheaf of relative differentials of a map)

Let $f : X \longrightarrow S$ be a germ of analytic mapping. One can define the sheaf Ω^1_f of relative differentials by the exact sequence (First exact sequence in [Ha] p. 173) :

$$f^*(\Omega^1_S) \longrightarrow \Omega^1_X \longrightarrow \Omega^1_f \longrightarrow 0.$$

When we have a (relative) local embedding of f (in this case we shall often call 0 the closed point of $\mathbf{C}^k \times S$),

$$\begin{array}{ccc} X & \subset & \mathbf{C}^k \times S \\ \downarrow f & & \downarrow pr_2 \\ S & = & S \end{array}$$

such as f is induced by the second projection, we get a well-known exact sequence (Second exact sequence in [Ha] p. 173):

$$(2.2.1.1) \qquad \mathcal{I}_X/(\mathcal{I}_X)^2 \longrightarrow \Omega^1_{\mathbf{C}^k\times S/S} \otimes \mathcal{O}_X \longrightarrow \Omega^1_f \longrightarrow 0$$

which induces a presentation of Ω^1_f :

$$\mathcal{O}^p_X \xrightarrow{\ {}^t\varphi\ } \mathcal{O}^k_X \longrightarrow \Omega^1_f \longrightarrow 0.$$

Dualizing, we get

$$\mathcal{O}^k_X \xrightarrow{\ \varphi\ } \mathcal{O}^p_X$$

Let $\mathcal{K}$ the image of φ. On the smooth part of f —that is, on the open set of X where f is smooth—, the sheaf $\check{\mathcal{K}}$ is the sheaf of sections of $\check{K}$, the (relative) conormal bundle to f, and $\mathbf{P}(\mathcal{K})$ is nothing but the projective bundle $\mathbf{P}\check{K}$.

To be more explicit, if $(g_1, \ldots, g_p)$ are a system of generators of the ideal $\mathcal{I}_X$, and $(x_1, \ldots, x_k)$ a system of coordinates on $\mathbf{C}^k$, then the matrix of φ is the jacobian matrix:

$$\left(\frac{\partial g_j}{\partial x_i}\right)_{1 \leq i \leq k, 1 \leq j \leq p}$$

and the *jacobian module* is the image $\mathcal{K}$ of φ in $\mathcal{O}_X^p$.

Proposition 2.2.2 *Assume X is reduced of pure dimension and the singular part of f is nowhere dense in X. Then $\mathbf{P}(\mathcal{K})$ is the (relative) conormal space of the map f.*

Proof : The only thing to prove is that $\mathbf{P}(\mathcal{K})$ is the closure of its restriction to the smooth points of f. This is true if any homogeneous function of degree ν on $\mathbf{P}(\mathcal{K})$ (by definition, an element of $S(\mathcal{K})_\nu$) which is zero above the smooth part of f is zero. The hypothesis on the singular locus of f makes this clear. ▪

2.2.3 Remark

In the general case, $\mathbf{P}(\mathcal{K})$ —or the associated reduced space— could be a candidate for the relative conormal space of f.

Example 2.2.4 (Local complete intersections morphisms)

Assume in the preceding example that f is a flat, surjective morphism onto S and $f^{-1}(0)$ is a complete intersection in $\mathbf{C}^k$. In this particular case, the map ψ fits in a short exact sequence

$$\mathcal{I}_X/\mathcal{I}_X^2 \xrightarrow{\psi} \Omega^1_{\mathbf{C}^k \times S/S} \otimes \mathcal{O}_X \longrightarrow \Omega^1_f \longrightarrow 0$$

so the map φ fits in

$$0 \to \mathrm{Hom}_{\mathcal{O}_X}(\Omega^1_f, \mathcal{O}_X) \to$$

$$\mathrm{Hom}_{\mathcal{O}_X}(\Omega^1_{\mathbf{C}^k \times S/S} \otimes \mathcal{O}_X, \mathcal{O}_X) \xrightarrow{\varphi} \mathrm{Hom}_{\mathcal{O}_X}(\mathcal{I}_X/\mathcal{I}_X^2, \mathcal{O}_X) \to T^1_{\mathcal{O}_X}(X/S) \to 0$$

and is a presentation of the module $T^1_{\mathcal{O}_X}(X/S)$.

Example 2.2.5 (Local complete intersections)

Let $X \subset \mathbf{C}^k$ a regular embedding given by a regular sequence $f_1, \ldots, f_p$ in $\mathcal{O}_{\mathbf{C}^k}$. Let $\mathcal{M} = \Omega^1_X$ be the sheaf of differentials on X and take the presentation

$$\mathcal{I}_X/(\mathcal{I}_X)^2 \xrightarrow{\psi} \Omega^1_{\mathbf{C}^k} \otimes \mathcal{O}_X \longrightarrow \Omega^1_X \longrightarrow 0$$

The image $\mathcal{K} = Im(\varphi)$ of the transpose map of ψ is the $\mathcal{O}_X$-module generated by the columns of the jacobian matrix.

Example 2.2.6 (Conormal space of a system of 1-forms) (T. Suwa)

Let us start with the situation of example 2.2.1. If we have a system ω of p relative 1-forms which generate a submodule of Ω^1_f, we can consider the quotient module Ω^1_ω.
Taking representatives of the generators of ω in $\Omega^1_{\mathbf{C}^k \times S/S} \otimes \mathcal{O}_X$, we are able, using the exact sequence 2.2.1.1 to give a presentation

$$\mathcal{O}_X^p \xrightarrow{\psi} \Omega^1_{\mathbf{C}^k \times S/S} \otimes \mathcal{O}_X \longrightarrow \Omega^1_\omega \to 0$$

and by the transpose map φ, we get a submodule $\mathcal{K}$ of $\mathcal{O}_X^p$. The projective cone $\mathbf{P}(\mathcal{K})$ is a subspace of the projectivized cotangent bundle $\mathbf{P}T^*(\mathbf{C}^k \times S/S)$.
We have an assertion similar to 2.2.2. If U is an open set everywhere dense in X where Ω^1_ω is locally free, then $\mathbf{P}(\mathcal{K})$ is the closure of the set of points in $\mathbf{P}T^*(\mathbf{C}^k \times S/S)$ where the canonical 1-form is in the module $\check{\mathcal{K}}$.
In the particular case of a full submodule of $\Omega^1_{\mathbf{C}^k}$, T. Suwa [Su] defined this way the unfolding of a system of 1-forms.

2.3 Using a compact divisor

It is generally convenient to blow-up in $\mathbf{P}(\mathcal{K})$ the inverse image of the maximal ideal $\mathbf{m}$. By this procedure, which can be traced back at least to [Ra] and [Cargese], we get a *compact* divisor D on the blown up space $\mathcal{W}$. On this compact divisor we are able to compute degrees and to obtain local discrete invariants of $\mathcal{K}$ at 0, whatever the dimension of the singular locus of $\mathcal{K}$ is.

If X is embedded in $\mathbf{C}^n$ (that is if $\mathbf{m}$ has a system of n generators), this blow-up provides an exact sequence of sheaves:

$$0 \to \mathcal{O}_\mathcal{W} \otimes \mathcal{O}_{\mathbf{P}^{n-1}}(1) \longrightarrow \mathcal{O}_\mathcal{W} \longrightarrow \mathcal{O}_D \to 0$$

For the sake of simplicity we shall abbreviate $\mathcal{O}_\mathcal{W} \otimes \mathcal{O}_{\mathbf{P}^{n-1}}(i) \otimes \mathcal{O}_{\mathbf{P}^{k-1}}(r)$ in $\mathcal{O}_\mathcal{W}(i,r)$ and $\mathcal{O}_D \otimes \mathcal{O}_{\mathbf{P}^{n-1}}(i) \otimes \mathcal{O}_{\mathbf{P}^{k-1}}(r)$ in $\mathcal{O}_D(i,r)$
When we tensor by $\mathcal{O}_{\mathbf{P}^{n-1}}(s) \otimes \mathcal{O}_{\mathbf{P}^{k-1}}(r)$ the preceding exact sequence, we get the following:

$$(2.3.0.1)0 \to \mathcal{O}_\mathcal{W}(s,r) \longrightarrow \mathcal{O}_\mathcal{W}(s-1,r) \longrightarrow \mathcal{O}_D(s-1,r) \longrightarrow 0.$$

As the sheaf $\mathcal{O}_{\mathbf{P}^{n-1}}(1) \otimes \mathcal{O}_{\mathbf{P}^{k-1}}(1)$ is ample, there exists an integer r_0 such that for all $r \geq r_0$ and for all $q \geq 1$,

$$H^q(\mathcal{O}_\mathcal{W}(r,r)) = H^q(\mathcal{O}_\mathcal{W}(r_0,r)) = 0.$$

Gluing together for $r \geq s \geq r_0$ the sequences 2.3.0.1 we obtain

$$0 \to \mathcal{O}_{\mathcal{W}}(r,r) \longrightarrow \mathcal{O}_{\mathcal{W}}(r_0,r) \longrightarrow \mathcal{O}_{D_{r_0}}(r_0,r) \longrightarrow 0.$$

where D_{r-r_0} is the Cartier divisor defined in $\mathcal{W}$ by the $r-r_0{}^{\text{th}}$ power of the maximal ideal $\mathbf{m}$. ¿From the last exact sequence, we obtain a long exact sequence of cohomology groups, and, for $r \geq r_0$,

$$H^q(\mathcal{O}_{D_{r-r_0}}(r_0\,,r)) = 0 \qquad \text{for} \quad q \geq 1$$

and

$$\begin{aligned} \ell(H^0(\mathcal{O}_{\mathcal{W}}(r\,,r))/H^0(\mathcal{O}_{\mathcal{W}}(r_0\,,r))) &= h^0(\mathcal{O}_{D_{r_0}}(r_0\,,r)) \\ &= \chi(\mathcal{O}_{D_{r-r_0}}(r_0\,,r)) \end{aligned}$$

Let us recall that $\mathcal{K}_r$ is the part of degree r of the graded algebra $S(\mathcal{K})$, and notice that, because $\mathcal{W}$ is reduced,

$$H^0(\mathcal{O}_{\mathcal{W}}(i\,,r)) \;=\; \mathbf{m}^i\mathcal{K}_r$$

¿From the sequence of inclusions:

$$\mathcal{O}_{\mathcal{W}}(r,r) \longrightarrow \mathcal{O}_{\mathcal{W}}(r-1,r) \longrightarrow \cdots \longrightarrow \mathcal{O}_{\mathcal{W}}(r_0,r)$$

the successive quotients of which are the $\mathcal{O}_D(i,r)$ with $r_0 \leq i \leq r-1$, we get:

$$\chi(\mathcal{O}_{D_{r-r_0}}(r_0\,,r)) = \sum_{i=r_0}^{r-1} \chi(\mathcal{O}_D(i\,,r))$$

and finally:

$$\ell\, \mathbf{m}^{(r_0)}\mathcal{K}_r/\mathbf{m}^{(r)}\mathcal{K}_r = \sum_{i=r_0}^{r-1} \chi(\mathcal{O}_D(i\,,r))$$

Going down to $\mathbf{P}(\mathcal{K})$, we consider the short exact sequence

$$0 \longrightarrow \mathbf{m}^{r_0}\mathcal{O}_{\mathbf{P}(\mathcal{K})}(r) \longrightarrow \mathcal{O}_{\mathbf{P}(\mathcal{K})}(r) \longrightarrow \left(\mathcal{O}_{\mathbf{P}(\mathcal{K})}/\mathbf{m}^{r_0}\right) \otimes \mathcal{O}_{\mathbf{P}(\mathcal{K})}(r) \longrightarrow 0$$

For r big enough, say $r \geq r_1$, the higher cohomology groups of these three sheaves are zero, and we get that the length of the module

$$\mathcal{K}_r/\mathbf{m}^{r_0}\mathcal{K}_r = H^0((\mathcal{O}_{\mathbf{P}(\mathcal{K})}/\mathbf{m}^{r_0}) \otimes \mathcal{O}_{\mathbf{P}(\mathcal{K})}(r))$$

is a numerical polynomial in r of degree equal to the dimension of the subspace of $\mathbf{P}(\mathcal{K})$ defined by $\mathbf{m}$, which is less or equal to N, the dimension of the divisor D.
We want to evaluate this Euler characteristic $\chi(\mathcal{O}_D(i\,,r))$ by using bidegrees of the divisor D which is included in a product of projective spaces $\mathbf{P}^{k-1} \times \mathbf{P}^{n-1}$. In the next section we shall recall a few results that we shall use in the following section for the desired evaluation.

2.4 Snapper's formula

Let us recall in the weaker form which we will use, some results of Snapper and Kleiman ([K],[S]).

Theorem 2.4.1 (Snapper) *Let V be an algebraic variety, $\mathcal{F}$ a coherent sheaf on V, with support of dimension N, and $\mathcal{L}_1, \ldots, \mathcal{L}_t$ invertible sheaves on V. The Euler-Poincaré characteristic $\chi(\mathcal{F} \otimes \mathcal{L}_1{}^{\otimes n_1} \otimes \ldots \otimes \mathcal{L}_t{}^{\otimes n_t})$ is a numerical polynomial in $n_1, \ldots, n_t$, of total degree N.*

Proof : By induction on N, dimension of the support of $\mathcal{F}$. For $N = -1$,($\mathcal{F}$ is 0) the result is trivial. We can replace V by **supp** ($\mathcal{F}$) and work with a sheaf $\mathcal{O}_V$. Let $\mathcal{I} = \mathcal{O}_V(-\mathcal{D})$, the sheaf of ideals which defines the divisor $\mathcal{D}$, such that $\mathcal{L}_1 = \mathcal{O}_V(\mathcal{D})$. The exact sequence

$$0 \longrightarrow \mathcal{I} \longrightarrow \mathcal{O}_V \longrightarrow \mathcal{O}_\mathcal{D} \longrightarrow 0$$

gives after tensoring

$$0 \longrightarrow \mathcal{I} \otimes \mathcal{L}_1{}^{\otimes n_1} \otimes \cdots \otimes \mathcal{L}_t{}^{\otimes n_t} \longrightarrow \mathcal{L}_1{}^{\otimes n_1} \otimes \cdots \otimes \mathcal{L}_t{}^{\otimes n_t}$$
$$\longrightarrow \mathcal{O}_\mathcal{D} \otimes \mathcal{L}_1{}^{\otimes n_1} \otimes \cdots \otimes \mathcal{L}_t{}^{\otimes n_t} \longrightarrow 0$$

as $\mathcal{I} \otimes \mathcal{L}_1 = \mathcal{O}_V(-\mathcal{D}) \otimes \mathcal{O}_V(\mathcal{D}) = \mathcal{O}_V$, we obtain:

$$\chi(\mathcal{L}_1{}^{\otimes n_1} \otimes \cdots \otimes \mathcal{L}_t{}^{\otimes n_t}) - \chi(\mathcal{L}_1{}^{\otimes n_1 - 1} \otimes \cdots \otimes \mathcal{L}_t{}^{\otimes n_t}) = \chi(\mathcal{O}_\mathcal{D} \otimes \mathcal{L}_1{}^{\otimes n_1} \otimes \cdots \otimes \mathcal{L}_t{}^{\otimes n_t})$$

by hypothesis $\mathcal{O}_\mathcal{D}$ has support of dimension $N - 1$ and one concludes in the usual way, using the induction hypothesis. ∎

Definition 2.4.2 (Snapper) *With notations as before, if* **supp** $\mathcal{F}$ *is of dimension at most N, define the intersection number*

$$(\mathcal{F} . \mathcal{L}_1 \ldots \mathcal{L}_N)$$

as the coefficient of $n_1 \ldots n_N$ in the polynomial $\chi(\mathcal{L}_1{}^{\otimes n_1} \otimes \ldots \otimes \mathcal{L}_N{}^{\otimes n_N})$. This number is an integer. If $\dim$ **supp** $\mathcal{F} < N$, *it is clearly zero.*

2.4.3 Remarks

If $D_1, \ldots, D_N$ are the divisors associated to the sheaves $\mathcal{L}_1 \ldots \mathcal{L}_N$, and $\mathcal{F}$ the structure sheaf of a subspace X of V, then $(\mathcal{F} . \mathcal{L}_1 \ldots \mathcal{L}_N)$ is a possible definition for the intersection number:

$$(X . D_1 \ldots D_N)_V$$

having all properties of intersection theory ([F]).

Let us apply this definition to the special case where the sheaf $\mathcal{F}$ is the structure sheaf $\mathcal{O}_V$, and the N divisors are equal :

$$\begin{aligned}\chi(\mathcal{O}_V \otimes \mathcal{L}_1{}^{\otimes n_1} \otimes \cdots \otimes \mathcal{L}_1{}^{\otimes n_N}) &= \chi(\mathcal{O}_V \otimes \mathcal{L}_1{}^{\otimes (n_1+\cdots+n_N)}) \\ &= c.(n_1+\cdots+n_N)^N + \text{lower total degree terms} \\ &= \cdots + (\mathcal{O}_V \,.\, \mathcal{L}_1 \ldots \mathcal{L}_N)\, n_1 \ldots n_N + \cdots\end{aligned}$$

it is easy to conclude that $N!\,c = (\mathcal{O}_V \,.\, \mathcal{L}_1 \ldots \mathcal{L}_N)$ because there are $N!$ terms equal to $n_1 \ldots n_N$ when expanding $(n_1 + \cdots + n_N)^N$.

2.4.4 Remark

Let us introduce the coefficients c_i defined (using Snapper's theorem) by

$$\chi(\mathcal{O}_V \otimes \mathcal{L}_1{}^{\otimes r_1} \otimes \mathcal{L}_2{}^{\otimes r_2}) = \sum_{i=0}^{i=N} c_i r_1{}^i r_2{}^{N-i} + \cdots$$

Let us compute in two different ways the coefficient of $n_1 \ldots n_N$ in the expansion of

$$\chi(\mathcal{O}_V \otimes \mathcal{L}_1{}^{\otimes n_1} \otimes \cdots \otimes \mathcal{L}_1{}^{\otimes n_i} \otimes \mathcal{L}_2{}^{\otimes n_{i+1}} \otimes \cdots \otimes \mathcal{L}_2{}^{\otimes n_N})$$

Using the definition 2.4.2, we find:

$$(\mathcal{O}_V \,.\, \mathcal{L}_1{}^{\otimes i} \mathcal{L}_2{}^{\otimes N-i})$$

now using the definition of the c_i we have just introduced, we can notice that in

$$(n_1 + \cdots + n_i)^i (n_{i+1} + \cdots n_N)^{N-i}$$

there are exactly $i!(N-i)!$ terms equal to $n_1 \cdots n_i n_{i+1} \ldots n_N$ whence the formula:

$$\begin{aligned}\chi(\mathcal{O}_V \otimes \mathcal{L}_1{}^{\otimes r_1} \otimes \mathcal{L}_2{}^{\otimes r_2}) &= \sum_{i=0}^{i=N} (\mathcal{O}_V \,.\, \mathcal{L}_1{}^{\otimes i} \mathcal{L}_2{}^{\otimes N-i}) \frac{r_1{}^i r_2{}^{N-i}}{i!(N-i)!} + \ldots \\ &= \sum_{i=0}^{i=N} (\mathcal{O}_V \,.\, \mathcal{L}_1{}^{\otimes i} \mathcal{L}_2{}^{\otimes N-i}) \binom{r_1+i}{i} \binom{r_2+N-i}{N-i} + \ldots\end{aligned}$$

where the ... stands for terms of lower total degree.

2.5 Formulas on the degrees of the divisor associated to a Module

We are now ready to compute the Euler characteristic of 2.3 by using the Snapper polynomial of remark 2.4.4. We can write, denoting by N the dimension of the divisor D,

$$\begin{aligned}\chi(\mathcal{O}_D \otimes \mathcal{O}_{\mathbf{P}^{n-1}}(i) \otimes \mathcal{O}_{\mathbf{P}^{k-1}}(r)) &= \chi(\mathcal{O}_D(i\,,r)) \\ &= \sum_{l=0}^{l=N}(D\,.\,H_1{}^l\,.\,H_2{}^{N-l})\binom{i+l}{l}\binom{r+N-l}{N-l}+\cdots\end{aligned}$$

where we write $(D\,.\,H_1{}^l\,.\,H_2{}^{N-l})$ for the intersection number of the invertible sheaves $(\mathcal{O}_D\,.\,\mathcal{O}_{\mathbf{P}^{k-1}}(l).\mathcal{O}_{\mathbf{P}^{n-1}}(N-l))$
Remark that some of these terms of higher degree can be zero. Now we want to sum these polynomial for $i = r_0$ to $i = r-1$. Using:

$$\sum_{i=r_0}^{r-1}\binom{i+l}{l} = \binom{l+r}{l+1} - \binom{l+r_0}{l+1}$$

we find

$$\sum_{i=r_0}^{r-1}\chi(\mathcal{O}_D(r\,,i)) \quad = \quad \sum_{l=0}^{l=N}(D\,.\,H_1{}^l\,.\,H_2{}^{N-l})\binom{r+l}{l+1}\binom{N-l+r}{N-l}+\cdots$$

whence

$$\ell(\mathbf{m}^{r_0}\mathcal{K}_r/\mathbf{m}^r\mathcal{K}_r) \quad = \quad \sum_{l=0}^{l=N}\binom{N+1}{l+1}(D\,.\,H_1{}^l\,.\,H_2{}^{N-l})\frac{r^{N+1}}{(N+1)!}+\cdots$$

where the dots stand for terms of lower degree. Reminding that the length $\ell(M_r/\mathbf{m}^{r_0}\mathcal{K}_r)$ is for $r \geq r_1$ a numerical polynomial of degree N, we have proved the

Proposition 2.5.1 *For r big enough, the length $\ell(\mathcal{K}_r/\mathbf{m}^r\mathcal{K}_r)$ is a numerical polynomial in r of degree $N+1$, the coefficient of the leading term $\frac{r^{N+1}}{(N+1)!}$ is equal to*

$$\sum_{l=0}^{l=N}\binom{N+1}{l+1}(D\,.\,H_1{}^l\,.\,H_2{}^{N-l}) \tag{2.5.1.1}$$

Example 2.5.2 (Local complete intersection morphisms)

If $\mathcal{K}$ is the jacobian module (cf 2.2.4), we said that $\mathbf{P}(\mathcal{K})$ is the conormal space C_f. Then the intersection number $(D \,.\, {H_1}^l \,.\, {H_2}^{N-l})$ is the multiplicity of the (general) polar variety of f associated to a (general) plane of codimension $N-l$ in $\mathbf{C}^k$ (cf [H-M-S] 4.4).
We point out the fact that the formula 2.5.1.1 we have obtained is valid without any hypothesis on the singularities of f.

Example 2.5.3 (Teissier [Cargese])

Let $f : \mathbf{C}^k \longrightarrow \mathbf{C}$ be a germ of function with isolated singularity. In this case the polar multiplicities of f are the Milnor numbers of the restrictions of f to (generic) plane sections. As the jacobian module is the jacobian ideal, for r big enough, the length of the quotient $j(f)^r/(\mathbf{m}j(f))^r$ is a polynomial of degree k, the leading term of which is

$$\frac{r^k}{k!}\left(\binom{k}{1}\mu^{(k-1)}(f)+\cdots+\binom{k}{i}\mu^{(k-i)}(f)+\cdots+\mu^{(0)}\right).$$

Note the "missing term" $\mu^{(k)}(f)$ in this formula.

2.6 Geometric interpretation of the formulas

As for the jacobian module, the intersections numbers $(D \,.\, {H_1}^l \,.\, {H_2}^{N-l})$ can be interpreted as multiplicities of some subvarieties of X.
Let us introduce some notations. As the module $\mathcal{K}$ is the image of a linear map

$$\varphi : \mathcal{O}_X^k \longrightarrow \mathcal{O}_X^p$$

we shall write down $\mathcal{K}_p^k$ for $\mathcal{K}$, and $\mathcal{K}_r^s$ for the image of the composed map

$$\mathcal{O}_X^s \xrightarrow{\iota} \mathcal{O}_X^k \xrightarrow{\varphi} \mathcal{O}_X^p \xrightarrow{\varpi} \mathcal{O}_X^r$$

where ι is a generic linear injection and ϖ a generic linear surjection. We shall write down $\mathcal{J}_r^s$ for $\Lambda^r \mathcal{K}_r^s$.

Proposition 2.6.1 *Let r be the rank of the module $\mathcal{K}$.*
For $0 \le l \le N$ the degree $(D \,.\, {H_1}^l \,.\, {H_2}^{N-l})$ is the multiplicity of the subvariety of X equal to

- *X if $N-l<r$,*
- *the closure of the variety (empty or of dimension $l+1$) $V(\mathcal{J}_r^{N-l}) \setminus V(\mathcal{J}_r^k)$, otherwise.*

Proof : This can be worked out using projection formula and general position arguments as in ([H-M-S] 4.4.3 and 4.4.6.). ∎

3 Integral closure of a module

3.1 Reminders on the integral closure of an ideal

Let us first recall the definition and some properties of the integral closure of an ideal (for the most general case consult [B] or [Z-S], and a convenient and detailed exposition in the algebro-geometric landscape see [L-T] or more recently [T2]).

Definition 3.1.1 (Integral closure of an ideal). *Let I be an ideal in a unitary ring R, define the integral closure of the ideal I as the ideal (denoted $\overline{I}$) of all elements h of the ring R satisfying a relation of integral dependance:*

$$h^n + \sum_{i=1}^{n-1} f_{n-i}\, h^i = 0$$

where the coefficients f_k are in the k-th power I^k of the ideal I.

Proposition 3.1.2 (Characterizations of the integral closure). *For two ideals of finite type $I \subset J$ in a noetherian domain R, the following conditions are equivalent :*

1. $\overline{I} = \overline{J}$
2. *For any morphism h of the ring R in a discrete valuation ring A_v the images of the ideals I and J by h coincide.*
3. *For any morphism h of the ring R in a discrete valuation ring $R_v \supset R$ included in K the fraction field of R, the images h^*I of I and h^*J of the ideal J in the ring R_v are equal.*
4. *There exists a faithful R module (or ideal) L of finite type such that $JL = IL$*
5. *There exists an integer N, and a fixed integer r such that for all $q \geq N$ $J^{q+r} \subset I^q$*

The last condition can be rephrased by saying that $\bar{\nu}_I(f) \geq 1$ for all elements $f \in J$ (see [L-T] definition 0.2.3 for a definition of $\bar{\nu}$ and paragraph 2 of loc. cit. for a detailed use of this notion in analytic geometry). If X is a reduced complex analytic space, $\mathcal{O}_X$ the local ring of germs of holomorphic functions near the origin, for a coherent sheaf of ideals $\mathcal{I}$ in $\mathcal{O}_X$ there are at least four convenient criteria characterizing its integral closure (we shall see a fifth one in the next section, but which is not always valid).
The germ at a point x of the integral closure of a sheaf of ideals $\overline{\mathcal{I}}$ depends only on the germ $\mathcal{I}_x$, so the following proposition is within the local frame.

Proposition 3.1.3 (Characterizations of the integral closure in local analytic geometry). *For two ideals $I \subset J$ of the ring $\mathcal{O}_X$ of a germ of analytic space, the following conditions are equivalent*

1. $\bar{I} = \bar{J}$
2. *For any local path in X (image of the germ of the line $\mathbf{C}$ by an analytic morphism), going through the origin $h : \mathbf{C} \longrightarrow X$ the image h^*I of I and h^*J of the ideal J in the ring $\mathbf{C}\{t\}$ are equal.*
3. *For any system $g_1, \ldots g_k$ of generators of I, there exists a representative U of the germ X such that, for any function f of J there exists a constant C and a bounding :*
$$|f(z)| \le C \sup_{1 \le i \le n} |g_i(z)| \quad \text{for all } z \text{ in } U$$
4. *On the normalized blown up space of the ideal I in X, the ideal J is equal to I.*

This interpretation of integral dependance in terms of a bounding of a function by the sup of several others, is precious in translating Verdier condition towards algebra. As for conditions $\mathfrak{a}$ and $\mathfrak{b}$ of Whitney, they are equivalent to some quotient nearing 0 (see also [H-M 3]). This can be expressed by an inequality of the type $\bar{\nu}_I(f) > 1$, (as in [Cargese]) or $f \in \bar{I}^+$ using the notations of [L-T] (p. 38 prop. 4 .4.1.). This is easily seen to be equivalent to the property that for some N

$$f^N \in I^{N+1}$$

(compare to condition 5 in 3.1.2) or $f^N \in \bar{I}^{N+1}$, which provides a convenient "test disc" characterization of this condition. It can also be expressed as

$$f^N \in m I^{N+1}$$

if the ideal I is m-primary.

3.2 Notations and motivations

The notion of integral closure of a module has been developped by David Rees ([R 5]). Recently T. Gaffney has promoted it in the context of isolated singularities of complete intersections (ICIS). He had in mind to use Teissier's work (relating integral closures of jacobian ideals with Verdier's and Whitney's conditions) as a template for the use of the jacobian module in the ICIS case.
Our point of view is different, with the emphasis being on geometry, and we shall introduce this notion of integral closure in a third manner, related

to constructions of section 2, (especially to symmetric algebras) and more convenient for our purpose.
In fact we use the integral closure of ideals to define the integral closure of modules (recall that the integral closure of rings can be used to define integral closure of ideals). We shall then proceed to prove the equivalence of our definition with the definitions of Rees and Gaffney (which generalize the characterizations 2 of 3.1.2 and 3.1.3). The ring R is supposed to be a noetherian domain as in Rees ([R 5]), and we will barely indicate how to extend to the general case; in the geometric landscape it is easy to understand how one can work on the irreducible components of X and apply on each component the criterion.
As in paragraph 2 we shall denote by $S(R^p)$ the symmetric algebra of the R module R^p which is isomorphic to $R[\lambda_1, \ldots, \lambda_p]$, by S_q its graded part of degree q and, for a submodule M of R^p generated by k vectors

$$f_i = \begin{pmatrix} f_{i1} \\ \vdots \\ f_{ip} \end{pmatrix} \quad 1 \leq i \leq k,$$

we denote by $\mathcal{I}(M)$ the ideal generated by $S(M)_1 = M$ in $S(\mathcal{O}^p) := S$. If we are given a system of generators of M, then $\mathcal{I}(M)$ is generated by:

$$(\sum_{j=1}^{p} \lambda_j f_{i,j}) \quad \text{for} \quad 1 \leq i \leq k$$

in $R[\lambda_1, \ldots, \lambda_p]$.
We shall denote by $Frac(R)$ the field of fraction of R.

3.3 Definitions and basic properties

Definition 3.3.1 (Integral closure of a module). *Let R be a noetherian commutative ring without zero divisors and M an R module of finite type. Suppose that M is a submodule of R^p. We define the integral closure of M in R^p denoted by $\overline{M}$ as the module of all elements $g = (g_1, \ldots, g_p)$ of R^p such that*

$$\sum_{j=1}^{p} \lambda_j g_j \in \overline{\mathcal{I}(M)} \,.\, R[\lambda_1, \cdots, \lambda_p]$$

which can also be translated as asking that two functors commute : $\mathcal{I}(\overline{M}) = \overline{\mathcal{I}(M)}$.

Proposition 3.3.2 *The integral closure —as an algebra— of the graded algebra $S(M)$ in the polynomial algebra $S(R^p) = R[\lambda_1, \cdots, \lambda_p]$ is $S(\overline{M})$. The three modules $\overline{S(M)}_1$, $\overline{\mathcal{I}(M)}_1$ and $\overline{M}$ are equal.*

Let us now prove that the previous definition is equivalent to the definition of Rees (loc. cit.).

Proposition 3.3.3 ("Discrete valuation" criterion). *Let R be a noetherian domain and let M be a finitely generated submodule of R^p. For an element $g = (g_1, \ldots, g_p)$ of R^p the following conditions are equivalent:*

1. $g \in \overline{M}$
2. *For any discrete valuation v in $Frac(R)$, such that, for all f in R, $v(f) \geq 0$, g is in $R_v M$*

Proof : We shall first prove that (1) implies (2). The hypothesis is that $g \in \overline{M}$, which means that

$$\sum_{j=1}^{p} \lambda_j g_j \in \overline{\mathcal{I}(M)} \,.\, R[\lambda_1, \cdots, \lambda_p]$$

Now suppose that (2) is not verified. We can find v a discrete valuation on $Frac(R)$ such that, for all f in R, $v(f) \geq 0$, and $g \notin R_v M$. Let $N \supset M$ be the module generated by M and g in R^p. We have $\mathcal{I}(N) \supset \mathcal{I}(M)$ and $\overline{\mathcal{I}(N)} = \overline{\mathcal{I}(M)}$. Let $N_v = R_v N \supset M_v = R_v M$ be the corresponding modules in R_v^p, they are torsion free submodules of finite rank of a free finite rank module on the discrete valuation ring R_v, and are thence free ([B] Algèbre, chapitre 7, paragraphe 4, Théorème 1).
By proposition 4 of loc. cit. we may choose a basis of N_v, $(g_1, \ldots, g_r)$ such that M_v has a basis $(v^{m_1} g_1, \ldots, v^{m_r} g_r)$ where v is a generator of the maximal ideal of R_v, and, using that $N_v \neq M_v$ at least one of the $m_j > 0$ (which we can take to be m_1) with the convention of taking $0 = v^\infty$. We consider $\mathcal{I}(M_v) \subset \mathcal{I}(N_v)$ in the algebra $R_v[\lambda_1, \cdots, \lambda_p]$ and remark that

$$\begin{aligned} \mathcal{I}(M_v) &= \mathcal{I}(M) R_v[\lambda_1, \cdots, \lambda_p] \\ \mathcal{I}(N_v) &= \mathcal{I}(N) R_v[\lambda_1, \cdots, \lambda_p] \end{aligned}$$

The next step is to assign to the $(\lambda_1, \cdots, \lambda_p)$, values $(\alpha_1, \cdots, \alpha_p)$ in R_v^p such that $\sum_i \alpha_i g_{1i} = v^q \neq 0, \sum_i \alpha_i g_{2i} = 0, \ldots, \sum_i \alpha_i g_{ri} = 0$ this provides a morphism $h : R[\lambda_1, \cdots, \lambda_p] \to R_v$ such that the image of $\mathcal{I}(N)$ is generated by $v^q \neq 0$ and the image of $\mathcal{I}(M)$ is generated by $v^{q+m_1} \neq v^q$, which contradicts the hypothesis $\overline{\mathcal{I}(M)} = \overline{\mathcal{I}(N)}$.

Conversely let us prove that (2) implies (1). Suppose that an element $g = (g_1, \ldots, g_p)$ of R^p is such that for all discrete valuation v on $Frac(R)$ positive on R we have $g \in R_v M$. We intend to prove that $g \in \overline{\mathcal{K}}$, that is to say, by definition, that

$$\mathcal{I}(g) := \sum_{j=1}^{p} \lambda_j g_j \in \overline{\mathcal{I}(M)} \,.\, R[\lambda_1, \cdots, \lambda_p]$$

We use the " discrete valuation " criterion (criterion (2) of proposition 3.1.2) to prove that result. Let w be a discrete valuation on $Frac(R)(\lambda_1, \cdots, \lambda_p)$, positive on $R[\lambda_1, \cdots, \lambda_p]$ and let v be the induced discrete valuation on R. We want to test if $\mathcal{I}(g)$ is in $\mathcal{I}(\mathcal{K})R[\lambda_1, \cdots, \lambda_p]_w$ By hypothesis, applying the discrete valuation criterion for modules for the valuation v we know that

$$\begin{aligned}
g &\in R_v M \\
g &= \sum_{i=1}^{k} \frac{a_i}{b_i} f_i \quad \text{with} \quad a_i \quad \text{and} \quad b_i \quad \text{in} \quad R, \quad v(a_i) \geq v(b_i) \\
g_j &= \sum_{i=1}^{k} \frac{a_i}{b_i} f_{i,j} \qquad j \in \{1, \ldots, p\} \\
\sum_{j=1}^{p} \lambda_j g_j &= \sum_{j=1}^{p} \lambda_j \sum_{i=1}^{k} \frac{a_i}{b_i} f_{i,j} \\
&= \sum_{i=1}^{k} \frac{a_i}{b_i} \sum_{j=1}^{p} \lambda_j f_{i,j}
\end{aligned}$$

then the last equality implies that

$$w(\mathcal{I}(g)) \geq \min_{1 \leq i \leq k} w(\mathcal{I}(f_i))$$

and we get the result. ▪

Proposition 3.3.4 *Let R be a noetherian domain and let M be a finitely generated submodule of R^p. For a submodule of R^p, containing M, $M \subset N$, the following conditions are equivalent:*

1. $\overline{N} = \overline{M}$
2. *There exists a non trivial ideal $\mathcal{J} \subset R[\lambda_1, \cdots, \lambda_p]$ such that $\mathcal{J}\mathcal{I}(M) = \mathcal{J}\mathcal{I}(N)$*
3. *There exists a non trivial ideal $J \subset R$ such that $J\,\mathcal{I}(M) = J\,\mathcal{I}(N)$*

Proof : The equivalence of (1) and (2) is a consequence of the criterion 4 in 3.1.2 and of the definition. Clearly (3) is much stronger than (2) because the ideal $\mathcal{J}$ needs not to have a non trivial graded part of degree 0. In fact if (1) is verified, one can take for J in (3) any ideal such that $J.\overline{\Lambda^p(M)} = J.\Lambda^p(M)$, as can be seen by using theorem 1.2 of David Rees (in [R 5]). ▪

The definition we have chosen for the integral dependance on a module is perhaps not so convenient to define a function corresponding to the $\bar{\nu}$, however corresponding to $\bar{\nu}_I(f) > 1$ which was noted $f \in \bar{I}^+$ by Lejeune-Teissier, it is consistent to define $\bar{M}^+$ as the set of f such that there exists an integer N verifying

$$S_1 f^N \subset \mathcal{I}(M)^{N+1}$$

(note the multiplication by S_1 due to graduation).

Proposition 3.3.5 ("Test disc" characterization). *Let X be a reduced complex analytic space, $\mathcal{O}_X$ the local ring of germs of holomorphic functions near the origin. Let $\mathcal{K}$ be a coherent sheaf of modules which is a submodule of $\mathcal{O}_X^p$. For an element $g = (g_1, \ldots, g_p)$ of $\mathcal{O}_X^p$ the following conditions are equivalent:*

1. $g \in \overline{\mathcal{K}}$
2. *For any local path in X (image of the germ at the origin of $\mathbf{C}$ by an analytic morphism), going through the origin $h : \mathbf{C} \longrightarrow X$, the image $h^*g = g \circ h$ of g is included in $h^*\mathcal{K}$*
3. *On $\overline{\overline{\mathcal{W}}}$, the normalized blown-up space of $\mathcal{I}(\mathcal{K})$ in $X \times \mathbf{P}^{k-1}$, the vector g is in the module $\mathcal{O}_{\overline{\overline{\mathcal{W}}}}.\mathcal{K}$.*

Proof : The implication (1) $\Rightarrow$ (2) is a consequence of the same implication in proposition 3.3.2. For the converse the proof is the same, but has to be rephrased, restricting to the valuations given by local paths. ∎

Remark : the equivalence of (2) and (3) is in Gaffney who uses (2) as a definition of the integral closure of a module.

4 Multiplicity of a module

4.1 Notations and motivations

For a graded algebra or graded module $\mathcal{N}$ we shall designate by $\mathcal{N}_\nu$ its graded part of degree ν. If $\mathcal{K}$ is a submodule of $\mathcal{O}_X^p$, we shall write down $\mathcal{K}_\nu$ for the degree ν piece of the symmetric algebra $S(\mathcal{K})$.
Instead of computing the length of $\mathcal{K}_r/\mathbf{m}^r\mathcal{K}_r$ we would like to evaluate the one of $S_r/\mathbf{m}^r\mathcal{K}_r$, but this is infinite as soon as $\ell(S_r/\mathcal{K}_r)$ is not finite. On the contrary, if this colength is finite, as

$$\ell(S_r/\mathbf{m}^r\mathcal{K}_r) = \ell(S_r/\mathcal{K}_r) + \ell(\mathcal{K}_r/\mathbf{m}^r\mathcal{K}_r),$$

using the result of section 2.4 we can try to prove that it is, for r sufficiently big, a numerical polynomial in r. And we can expect a formula of the following type:

$$e(\mathbf{m}\mathcal{K}) = e(\mathcal{K}) + \sum_{i=0}^{N} \binom{N+1}{i} (D \,.\, {H_1}^i {H_2}^{N-i}) \tag{4.1.0.1}$$

on the coefficients of the terms of higher degree in $\frac{r^{N+1}}{(N+1)!}$. It remains first to define $e(\mathcal{K})$ as a suitable invariant associated to an $\mathcal{O}_X$-module $\mathcal{K}$ of finite colength in ${\mathcal{O}_X}^p$, then to prove the expected formula.
We will assume from now on, that $\mathcal{K}$ is an $\mathcal{O}_X$-module of finite colength in $\mathcal{O}_X^p$ and that X is an irreducible analytic space as in section 2.

Definition 4.1.1 (Rees) *Let $\mathcal{K}$ be a finite colength module in $\mathcal{O}_X^p$. The multiplicity of the module $\mathcal{K}$, which we will denote by $e(\mathcal{K})$, is the coefficient of the term of higher degree of the Hilbert function*

$$\nu \longrightarrow \ell(S_\nu/\mathcal{K}_\nu)$$

of the graded quotient algebra $S/S(\mathcal{K})$

4.1.2 Remark

We will see later that this multiplicity is an invariant of the quotient module (of finite length) $\mathcal{O}_X^p/\mathcal{K}$.

4.2 Computation of the multiplicity of a module

We give here new proofs of old results of David Buchsbaum and Dock Rim ([B.R]) on the multiplicity of a module. As we write them in a geometric way, using blowing ups and degree of divisors rather than Koszul complexes our proofs may seem shorter and simpler to the reader; it is at least our hope. When we will have to compare colengths and multiplicities, we will assume the ring $\mathcal{O}_X$ to be Cohen-Macaulay.
As for the multiplicity of an ideal two lemmas are proved, which enable us to calculate multiplicities of modules by restriction simpler cases. If $\mathcal{K}$ is generated by the minimum number of generators $d+p-1$, then this multiplicity will be equal to a colength if the ring $\mathcal{O}_X$ is Cohen-Macaulay.

Lemma 4.2.1 *Let X be an analytic complex space of dimension d, let $\mathcal{K}$ an $\mathcal{O}_X$ module of finite colength in ${\mathcal{O}_X}^p$. Then the inverse image $\mathcal{D}$ of 0 in $\mathbf{P}(\mathcal{K})$ is of dimension $d+p-2$*

Proof : Let us remark that above a generic point of X the fiber of $\mathbf{P}(\mathcal{K})$ is of dimension $p-1$, and $\mathbf{P}(\mathcal{K})$ is consequently of dimension $d+p-1$ and $\mathcal{D}$ is of dimension at most $d+p-2$, so all we have to prove is that it is of dimension at least $d+p-2$.
We shall prove that a linear subspace $\mathbf{P}V$ of codimension $d+p-2$ in $\mathbf{P}^{k-1}$, meets $\mathcal{D}$.

$$\begin{array}{ccccc} \mathbf{P}^{k-1} & \supset & \mathcal{D} & \subset & \mathbf{P}(\mathcal{K}) \\ & & \downarrow & & \downarrow \\ & & 0 & \in & X \end{array}$$

The intersection of $\mathbf{P}V$ with $\mathbf{P}(\mathcal{K})$ is defined by the vanishing of $(d+p-2)$ linear combinations of the generators of $\mathcal{K}$. Its image $\Gamma(V)$ in X is then defined by the ideal of $p{\times}p$ minors of a $(d+p-2){\times}p$ -matrix, of codimension at most $d+p-2-p+1=d-1$ (the codimension is bounded up by that of the *generic* determinantal variety). As X is of dimension d, $\Gamma(V)$ is at least of dimension 1 and is not restricted to $\{0\}$.
This shows that the intersection of the linear subspace $\mathbf{P}V$ of $\mathbf{P}^{k-1}$ with $\mathbf{P}(\mathcal{K})\setminus\mathcal{D}$ is non empty as well as its intersection with $\mathcal{D}$. ∎

4.2.2 Remark

While $\mathbf{P}(\mathcal{K})$ is of dimension $d+p-1$ and $\mathcal{D}$ is of dimension $d+p-2$ there is no reason for $\mathcal{D}$ to be a Cartier divisor, and it may not have locally a unique equation.

Proposition 4.2.3 (Expression of e($\mathcal{K}$)). *With the same hypothesis as in 4.2.1 we consider $\widetilde{\mathcal{W}}$ the blowing-up in $X\times\mathbf{P}^{p-1}$ of the sheaf of ideals $\mathcal{I}(\mathcal{K})$. The divisor of the blowing up $\widetilde{\mathcal{D}}$ is a subspace of dimension N of $\mathbf{P}^{k-1}{\times}\mathbf{P}^{p-1}$, and we shall denote by H_2 (resp. H_3) a generic hyperplane section in the first factor (resp the second factor). The multiplicity of $\mathcal{K}$ is given by the formula:*

$$e(\mathcal{K})=\sum_{l=0}^{N}(\widetilde{\mathcal{D}}.{H_2}^l.{H_3}^{N-l})$$

Proof : We construct an exact sequence associated to the blow-up of $\mathcal{I}(\mathcal{K})$ (which is an homogeneous ideal). Define

$$\mathcal{O}_{\widetilde{\mathcal{W}}}\otimes\mathcal{O}_{\mathbf{P}^{k-1}}(1)\otimes\mathcal{O}_{\mathbf{P}^{p-1}}(-1)\longrightarrow\mathcal{O}_{\widetilde{\mathcal{W}}}$$

as the multiplication by

$$\frac{1}{\lambda_s}\sum_{j=1}^{p}\lambda_j f_{r,j}$$

in the chart of $\mathbf{P}^{k-1}\times\mathbf{P}^{p-1}$ where η_r and λ_s are invertible. We obtain an exact sequence of sheaves

$$0\longrightarrow\mathcal{O}_{\widetilde{\mathcal{W}}}\otimes\mathcal{O}_{\mathbf{P}^{k-1}}(1)\otimes\mathcal{O}_{\mathbf{P}^{p-1}}(-1)\longrightarrow\mathcal{O}_{\widetilde{\mathcal{W}}}\longrightarrow\mathcal{O}_{\widetilde{\mathcal{D}}}\longrightarrow 0$$

Let us abbreviate the tensor product of sheaves $\mathcal{O}_{\widetilde{\mathcal{W}}}\otimes\mathcal{O}_{\mathbf{P}^{k-1}}(\nu)\otimes\mathcal{O}_{\mathbf{P}^{p-1}}(\rho)$ in $\mathcal{O}_{\widetilde{\mathcal{W}}}(\nu,\rho)$ and $\mathcal{O}_{\widetilde{\mathcal{D}}}\otimes\mathcal{O}_{\mathbf{P}^{k-1}}(\nu)\otimes\mathcal{O}_{\mathbf{P}^{p-1}}(\rho)$ in $\mathcal{O}_{\widetilde{\mathcal{D}}}(\nu,\rho)$. If we tensor the exact sequence defining $\widetilde{\mathcal{D}}$ by $\mathcal{O}_{\mathbf{P}^{k-1}}(\nu)\otimes\mathcal{O}_{\mathbf{P}^{p-1}}(\rho)$, we get:

$$0\rightarrow\mathcal{O}_{\widetilde{\mathcal{W}}}(\nu{+}1,\rho-1)\longrightarrow\mathcal{O}_{\widetilde{\mathcal{W}}}(\nu,\rho)\longrightarrow\mathcal{O}_{\widetilde{\mathcal{D}}}(\nu,\rho)\rightarrow 0$$

Remark that we are interested in the sequence of inclusions :

(4.2.3.1) $\mathcal{O}_{\widetilde{\mathcal{W}}}(\nu,0) \hookrightarrow \mathcal{O}_{\widetilde{\mathcal{W}}}(\nu-1,1) \hookrightarrow \cdots \hookrightarrow \mathcal{O}_{\widetilde{\mathcal{W}}}(0,\nu)$

where the successive quotients are the $\mathcal{O}_{\widetilde{\mathcal{D}}}(\nu-i,i)$. As we assume X to be irreducible, the space $\widetilde{\mathcal{W}}$ is reduced, the space of global sections $H^0(\mathcal{O}_{\widetilde{\mathcal{W}}}(\nu-i\,,i))$ is equal to $S_i\mathcal{K}_{\nu-i}$, so the quotient $H^0(\mathcal{O}_{\widetilde{\mathcal{W}}}(0\,,\nu))/H^0(\mathcal{O}_{\widetilde{\mathcal{W}}}(\nu\,,0))$ is the quotient $S_\nu/\mathcal{K}_\nu$ used for the definition of $e(\mathcal{K})$.

We shall write the same kind of proof as in 2.2. As the sheaf $\mathcal{O}_{\mathbf{P}^{k-1}}(1) \otimes \mathcal{O}_{\mathbf{P}^{p-1}}(1)$ is ample, there exists an integer ν_0 such that for all $\nu \geq \nu_0$ $\rho \geq \nu_0$ and for all $q \geq 1$,

$$H^q(\mathcal{O}_{\widetilde{\mathcal{W}}}(\nu\,,\rho)) = 0$$

The sequence (4.2.3.1) of inclusions gives an exact sequence:

$$0 \to \mathcal{O}_{\widetilde{\mathcal{W}}}(\nu,\nu_0) \longrightarrow \mathcal{O}_{\widetilde{\mathcal{W}}}(\nu_0,\nu) \longrightarrow \mathcal{O}_{\widetilde{\mathcal{D}}_{\nu-\nu_0}}(\nu_0,\nu) \longrightarrow 0.$$

where $\widetilde{\mathcal{D}}_{\nu-\nu_0}$ is the Cartier divisor defined in $\widetilde{\mathcal{W}}$ by the $\nu-\nu_0{}^{\text{th}}$ power of the ideal $\mathcal{I}(\mathcal{K})$.

The last short exact sequence gives birth to a long exact sequence of cohomology groups, and, as the higher groups are zero we obtain

$$\begin{aligned}\ell(H^0(\mathcal{O}_{\widetilde{\mathcal{W}}}(\nu_0\,,\nu))/H^0(\mathcal{O}_{\widetilde{\mathcal{W}}}(\nu\,,\nu_0))) &= h^0(\mathcal{O}_{\widetilde{\mathcal{D}}_{\nu-\nu_0}}(\nu_0\,,\nu)) \\ &= \chi(\mathcal{O}_{\widetilde{\mathcal{D}}_{\nu-\nu_0}}(\nu_0\,,\nu))\end{aligned}$$

Using the sequence (4.2.3.1) we get

$$\chi(\mathcal{O}_{\widetilde{\mathcal{D}}_{\nu-\nu_0}}(\nu_0\,,\nu)) = \sum_{i=1+\nu_0}^{\nu} \chi(\mathcal{O}_{\widetilde{\mathcal{D}}}(\nu+\nu_0-i\,,i))$$

and finally:

$$\ell(\,\mathcal{K}_{\nu_0}S_\nu/S_{\nu_0}\mathcal{K}_\nu) \;=\; \sum_{i=1+\nu_0}^{\nu} \chi(\mathcal{O}_{\widetilde{\mathcal{D}}}(\nu+\nu_0-i\,,i)).$$

We now use that $\dim\widetilde{\mathcal{D}} = N$ and apply the Snapper formula of remark 2.4.4 to express the

$$\chi(\mathcal{O}_{\widetilde{\mathcal{D}}}(\nu-i\,,i)) = \sum_{l=0}^{N}(\widetilde{\mathcal{D}}.H_2{}^l.H_3{}^{N-l})\binom{\nu-i+l}{l}\binom{i+N-l}{N-l} + \ldots$$

where the ... stands for terms of lower total degree. Whence,

$$\begin{aligned}\ell(\,\mathcal{K}_{\nu_0}S_\nu/S_{\nu_0}\mathcal{K}_\nu) &= \sum_{i=1+\nu_0}^{\nu} \chi(\mathcal{O}_{\widetilde{\mathcal{D}}}(\nu+\nu_0-i\,,i)) \\ &= \sum_{i=1+\nu_0}^{\nu}\sum_{l=0}^{N}(\widetilde{\mathcal{D}}.H_2{}^l.H_3{}^{N-l})\binom{\nu+\nu_0-i+l}{l}\binom{i+N-l}{N-l} + \ldots\end{aligned}$$

But, computing the coefficient of T^ν in the expansion of the formal power series $(1-T)^{-(l+1)}(1-T)^{-(N-l+1)}$ we find that

$$\sum_{i=0}^{\nu}\binom{\nu-i+l}{l}\binom{i+N-l}{N-l}=\binom{N+1+\nu}{\nu}$$

We obtain this way

$$\sum_{i=1+\nu_0}^{\nu}\binom{\nu+\nu_0-i+l}{l}\binom{i+N-l}{N-l}=\binom{N+1+\nu+\nu_0}{\nu+\nu_0}$$

$$-\sum_{j=1}^{\nu_0}\binom{\nu_0-j+l}{l}\binom{j+\nu+N-l}{N-l}-\sum_{i=0}^{\nu_0}\binom{\nu+\nu_0-i+l}{l}\binom{i+N-l}{N-l}$$

which is equivalent, when ν goes to infinity, to $\nu^{N+1}/(N+1)!$. We have just proved that the length of the quotient module

$$\mathcal{K}_{\nu_0}S_\nu/S_{\nu_0}\mathcal{K}_\nu$$

is, when ν is big enough, a polynomial in ν, the leading term of which is

$$\frac{\nu^{N+1}}{(N+1)!}\sum_{l=0}^{N}(\tilde{\mathcal{D}}.H_2{}^l.H_3{}^{N-l})$$

Let us examine now the two quotients $S_{\nu+\nu_0}/\mathcal{K}_{\nu_0}S_\nu$ and $S_{\nu_0}\mathcal{K}_\nu/\mathcal{K}_{\nu_0+\nu}$. On $X\times\mathbf{P}^{p-1}$ consider the exact sequence

$$0\longrightarrow\mathcal{K}_{\nu_0}\mathcal{O}_{\mathbf{P}^{p-1}}(\nu)\longrightarrow S_{\nu_0}\otimes\mathcal{O}_{\mathbf{P}^{p-1}}(\nu)\longrightarrow(S_{\nu_0}/\mathcal{K}_{\nu_0})\otimes\mathcal{O}_{\mathbf{P}^{p-1}}(\nu)\longrightarrow 0$$

As the sheaf $\mathcal{O}_{\mathbf{P}^{p-1}}(1)$ is ample, we get that, for ν big enough, the module $H^0((S_{\nu_0}/\mathcal{K}_{\nu_0})\otimes\mathcal{O}_{\mathbf{P}^{p-1}}(\nu))$ is equal to the quotient $S_{\nu+\nu_0}/\mathcal{K}_{\nu_0}S_\nu$ and its length is a polynomial in ν of degree equal to the dimension of the support of $(S_{\nu_0}/\mathcal{K}_{\nu_0})$ which is at most N.
In the same way, consider on $\mathbf{P}(\mathcal{K})$ the sequence

$$0\longrightarrow\mathcal{K}_{\nu_0}\mathcal{O}_{\mathbf{P}^{k-1}}(\nu)\longrightarrow S_{\nu_0}\otimes\mathcal{O}_{\mathbf{P}^{k-1}}(\nu)\longrightarrow(S_{\nu_0}/\mathcal{K}_{\nu_0})\otimes\mathcal{O}_{\mathbf{P}^{k-1}}(\nu)\longrightarrow 0$$

The same argument as before shows that the length of $S_{\nu_0}\mathcal{K}_\nu/\mathcal{K}_{\nu_0+\nu}$ is, for ν big enough, a polynomial in ν of degree equal to the dimension of the support of $(S_{\nu_0}/\mathcal{K}_{\nu_0})$ which is at most N.
Putting together these three computations, we get the announced formula for the multiplicity of the module $\mathcal{K}$. ■

Lemma 4.2.4 (Buchsbaum and Rim) (Existence of a parameter matrix). *Let X a space of dimension d and $\mathcal{K}$ an $\mathcal{O}_X$ module of finite colength in $\mathcal{O}_X{}^p$ generated by $f_1,\dots f_k$. There exists a submodule $\mathcal{K}'$ of $\mathcal{K}$ generated by $d+p-1$ linear combinations of $f_1,\dots f_k$, the multiplicity of which is the same as the multiplicity of $\mathcal{K}$.*

Proof : By the lemma 4.2.1 on dimension, we know that $\mathbf{P}(\mathcal{K})$ is of dimension $d+p-1$ and $\mathcal{D}$, the inverse image of 0 is of dimension $d+p-2$, where d stands for the dimension of X. There exists a linear projection π of the complement of a linear subspace of $\mathbf{P}^{k-1}$ onto $\mathbf{P}^{d+p-2}$ defined on the whole $\mathcal{D}$, which induces on $\mathcal{D}$ and on $\mathbf{P}(\mathcal{K})$ a finite morphism.
More precisely, it is possible to insure, by a linear change on the coordinates $\eta_1, \ldots, \eta_{d+p-1}, \eta_{d+p}, \ldots, \eta_k$, that the kernel of the projection is the linear space of codimension $k-d-p$ in $\mathbf{P}^{k-1}$ given by $\eta_{d+p+1} = \ldots = \eta_k = 0$. This linear space does not cut $\mathcal{D}$ (of dimension $d+p-2$). Let us denote by $\mathcal{D}'$ (resp. $\mathbf{P}(\mathcal{K})'$) the image of $\mathcal{D}$ (resp. $\mathbf{P}(\mathcal{K})$) by this morphism.

$$\begin{array}{ccccccc} \mathbf{P}(\mathcal{K}) & \supset & \mathcal{D} & \hookrightarrow & \mathbf{P}^{k-1} & & \\ \downarrow & & \downarrow & \searrow & & \searrow & \\ X & \ni & 0 & & \mathcal{D}' & \hookrightarrow & \mathbf{P}^{d+p-2} \end{array}$$

Let us call $\mathcal{K}'$ the module of global sections

$$\mathcal{K}' = H^0(\mathcal{O}_X \otimes \mathcal{O}_{\mathbf{P}^{d+p-2}})(1) \hookrightarrow H^0(\mathcal{O}_X \otimes \mathcal{O}_{\mathbf{P}^{k-1}}(1)) = \mathcal{K}.$$

The space $\mathbf{P}(\mathcal{K})'$ is the **Proj** of $\mathbf{S}(\mathcal{K}')$. We can blow up in $X \times \mathbf{P}^{p-1}$ the ideal $\mathcal{I}(\mathcal{K})$ (resp. $\mathcal{I}(\mathcal{K}')$, and get the space $\widetilde{\mathcal{W}}$ and the exceptionnal divisor $\widetilde{\mathcal{D}}$ (resp $\widetilde{\mathcal{W}'}$ with exceptional divisor $\widetilde{\mathcal{D}'}$):

$$\begin{array}{ccccccc} & & \widetilde{\mathcal{D}} & \hookrightarrow & \mathbf{P}^{k-1} \times \mathbf{P}^{p-1} & & \\ & \swarrow & \downarrow & & \downarrow & \searrow & \\ \widetilde{\mathcal{D}'} & & \mathcal{D} & \hookrightarrow & \mathbf{P}^{k-1} & & \mathbf{P}^{d+p-2} \times \mathbf{P}^{p-1} \\ \downarrow & \swarrow & & & & \searrow & \downarrow \\ \mathcal{D}' & & & & & & \mathbf{P}^{d+p-2} \end{array}$$

As the morphism $\mathcal{D} \longrightarrow \mathcal{D}'$ is finite, the morphism $\widetilde{\mathcal{D}} \longrightarrow \widetilde{\mathcal{D}'}$ is finite. We can use the preceding proposition on the computation of multiplicity with bidegrees. We shall write H_2 for a generic hyperplane section in $\mathbf{P}^{d+p-2}$ (for example $\eta_1 = 0$) as well as for its inverse image $\pi^* H_2$ (also given by $\eta_1 = 0$) in $\mathbf{P}^{k-1}$,

$$\begin{aligned} e(\mathcal{K}) &= \sum_{l=0}^{N} (\widetilde{\mathcal{D}}.{H_2}^l.{H_3}^{N-l}) \\ e(\mathcal{K}') &= \sum_{l=0}^{N} (\widetilde{\mathcal{D}'}.{H_2}^l.{H_3}^{N-l}) \end{aligned}$$

We can use the projection formula and write

$$\begin{aligned} (\widetilde{\mathcal{D}}.{H_2}^l.{H_3}^{N-l}) &= (\widetilde{\mathcal{D}}.\pi^*{H_2}^l.{H_3}^{N-l}) \\ &= (\pi_*\widetilde{\mathcal{D}}.{H_2}^l.{H_3}^{N-l}) \\ &= (\widetilde{\mathcal{D}'}.{H_2}^l.{H_3}^{N-l}) \end{aligned}$$

therefore $e(\mathcal{K}) = e(\mathcal{K}')$, as announced. ■

Theorem 4.2.5 (Buchsbaum and Rim) *Let X be Cohen-Macaulay of dimension d. Assume $\mathcal{K}$ is an $\mathcal{O}_X$ module generated by $k = d+p-1$ generators, (according to Buchsbaum and Rim, the morphism $f : \mathcal{O}_X^p \longrightarrow \mathcal{O}_X^k$ is given by a parameter matrix), then the following assertions are true:*

1. $e(\mathcal{K}) = \ell(\mathcal{O}_X^p/\mathcal{K})$
2. $e(\mathcal{K}) = \ell(\mathcal{O}_X/\Lambda^p\mathcal{K})$

Proof : Assertion 1 of this theorem is corollary 4.5 (assertion 2) of theorem 4.2 of [B-R] and assertion 2 of this theorem can be deduced from corollary 4.3 of [B-R] p. 223 in the special case where $p = n$ (with the notations of [B-R]) exactly as corollary 4.5 comes from the case $p = 1$ by using corollary 2.6. loc. cit.
Let us give another proof of (2) by induction on p.
Recall the notations of 2.5. We write down $\mathcal{K}_r^s$ for the image of the composed map

$$\mathcal{O}_X^s \xrightarrow{\iota} \mathcal{O}_X^k \xrightarrow{\varphi} \mathcal{O}_X^p \xrightarrow{\varpi} \mathcal{O}_X^r$$

where ι is a generic linear injection and ϖ a generic linear surjection. We shall write down $\mathcal{J}_r^s$ for $\Lambda^r\mathcal{K}_r^s$.
For $p = 1$ the result is well known as it is the classical case of multiplicities of ideals.
By proposition 4.2.3, we can express $e(\mathcal{K}) = e(\mathcal{K}_p^k)$ as sum of degrees of the divisor $\tilde{\mathcal{D}}$:

$$\begin{aligned} e(\mathcal{K}_p^k) &= \sum_{i=0}^{N} (\tilde{\mathcal{D}}.{H_2}^{N-i}.{H_3}^i) \\ e(\mathcal{K}_p^k) &= (\tilde{\mathcal{D}}.{H_2}^N) + \sum_{i=0}^{N-1} ((\tilde{\mathcal{D}}.H_3).{H_2}^{N-1-i}.{H_3}^i) \end{aligned}$$

It is easy to interpret $\sum_{i=0}^{N-1}((\tilde{\mathcal{D}}.H_3).{H_2}^{N-1-i}.{H_3}^i)$ as the multiplicity of a module $\mathcal{K}_{p-1}^k$ and using the recursion hypothesis we can write

$$\begin{aligned} e(\mathcal{K}_p^k) &= (\tilde{\mathcal{D}}.{H_2}^N) + e(\mathcal{K}_{p-1}^k) \\ &= (\tilde{\mathcal{D}}.{H_2}^N) + col(\Lambda^{p-1}\mathcal{K}_{p-1}^{k-1}) \\ &= (\tilde{\mathcal{D}}.{H_2}^N) + col(\mathcal{J}_{p-1}^{k-1}) \end{aligned}$$

We have now only to prove that

$$(\tilde{\mathcal{D}}.{H_2}^N) = col(\mathcal{J}_p^k) - col(\mathcal{J}_{p-1}^{k-1})$$

By a result of Giusti-Henry ([G-H] Theoreme 2.4), this difference of colengths can be computed as

$$\begin{aligned} &col(\mathcal{J}_p^k) - col(\mathcal{J}_{p-1}^{k-1}) \\ &= \; col(\mathcal{J}_p^{k-1}, M(k-p+1,\cdots,k)) - col(\mathcal{J}_p^{k-1}, M(k-p+1,\cdots,k-1)) \end{aligned}$$

In fact, the result of [G-H] is an equality of determinantal cycles in a smooth space where X is embedded.
Intersecting with X we get the equality of colengths, because X is Cohen-Macaulay, as well as the determinantal cycles.
Choosing an appropriate general point on $\widetilde{\mathcal{D}}$ and an adapted chart, we can suppose that we have exactly

$$M(k-p+1,\cdots,k) = f_{p,k}.M(k-p+1,\cdots,k-1)$$

with $f_{p,k}$ defining locally the exceptional divisor, which ends the proof. ∎

Example 4.2.6 Isolated singularity function (Teissier [Cargese])

Let f be an isolated singularity function as in example 2.5.3. We have $\mathcal{K} = j(f)$ and $e(\mathcal{K}) = \mu^{(k)}(f)$. So we get the "complete" formula, like in Teissier's paper

$$e(\mathbf{m}j(f)) = \mu^{(k)}(f) + \binom{k}{1}\mu^{(k-1)}(f) + \cdots + \binom{k}{i}\mu^{(k-i)}(f) + \cdots + \mu^{(0)}.$$

Example 4.2.7 Local complete intersections with isolated singularities (Gaffney [G 3])

Let $X \subset \mathbf{C}^k$ a regular embedding given by a regular sequence $f_1, \ldots, f_p$ in $\mathcal{O}_{\mathbf{C}^k}$. Let $\mathcal{M} = \Omega_X^1$ be the sheaf of differentials on X and take the presentation

$$\mathcal{I}_X/(\mathcal{I}_X)^2 \xrightarrow{\psi} \Omega^1_{\mathbf{C}^k} \otimes \mathcal{O}_X \longrightarrow \Omega^1_X \longrightarrow 0$$

The image $\mathcal{K} = Im(\varphi)$ of the transpose map of ψ is the $\mathcal{O}_X$-module generated by the columns of the jacobian matrix. We have $e(\mathcal{K}) = col(\mathcal{J}_p^{k-1})$ and this last integer is the multiplicity of the discriminant of a generic linear projection from X to $\mathbf{C}$ (this projection is determined by the generic embedding $\mathcal{O}_X^{k-1} \xrightarrow{\iota} \mathcal{O}_X^k$) and is equal to $\mu^k(X) + \mu^{k-1}(X)$, substituting, we get the formula

$$e(\mathbf{m}\mathcal{K}) = \sum_{i=0}^{i=d} \binom{k}{i} \left[\mu^{(k-i)}(X) + \mu^{(k-i-1)}(X)\right]$$

with the usual convention that $\mu^0(X) = 1$.

5 Independance of the presentation

In the beginning, we introduced an $\mathcal{O}_X$-module $\mathcal{M}$ and till now, we have computed invariants of a presentation of $\mathcal{M}$. We have to prove they are independent of the given presentation.
In view of 2.4 and 4.2, we have only to take care of the various exceptionnal divisors we have built in our constructions, for the invariants are some of their multi-degrees.
We present here a proof similar to the one of [H-M-S] (4).
We can restrict ourselves to the case we have two presentations of $\mathcal{M}$, one obtained from the other by a surjective map

$$\begin{array}{ccccccc} \mathcal{O}_X^p & \xrightarrow{\psi} & \mathcal{O}_X^k & \longrightarrow & \mathcal{M} & \longrightarrow & 0 \\ & & \uparrow & & \| & & \\ \mathcal{O}_X^{p'} & \xrightarrow{\psi'} & \mathcal{O}_X^{k'} & \longrightarrow & \mathcal{M} & \longrightarrow & 0 \end{array}$$

The spaces $\mathbf{Proj}(S(\mathcal{K}))$ and $\mathbf{Proj}(S(\mathcal{K}'))$ are different but the divisor $\mathcal{D}'$ is a cone on the divisor $\mathcal{D}$ and their degrees are equal. The same happens for the other divisors D and $\widetilde{\mathcal{D}}$.

6 Relation with equisingularity

In the case of isolated singularity functions, Teissier [Cargese] and Briançon-Speder [B-S] proved that the constancy of all the Milnor numbers in a family is equivalent to Whitney equisingularity.
What about the constancy of all the numbers we considered extracted from the jacobian module, and in particular the multiplicity?
In fact the answer is already known. In a family where all the fibres are isolated singularities, the constancy of the multiplicity of the jacobian module has two consequences

1. the emptyness of the t-dimensional polar variety, where t is the dimension of the parameter space

2. the fact that the singular set of the total space of the family—which is finite onto the parameter space—is unramified on the parameter space.

The converse is also true. This equivalence follows from (equi)multiplicity theory (see [L], [Cargese]).
Let X be an (local) analytic space together with a mapping onto a smooth space S, the parameter space. Assume that the critical locus of $X \longrightarrow S$ is finite and unramified on S, and that the polar variety of dimension $t =$

$dim(S)$ is empty. Then we claim that the multiplicity of the jacobian module of the fibres is constant.
The proof relies on the fact that, for an isolated singularity X_0, the jacobian multiplicity is nothing but the multiplicity of the discriminant of a generic projection projection from X_0 onto $\mathbf{C}$ (this is our example 4.2.8 and is well known for hypersurface singularities). This is necessarily the case as long as one wants a notion of discriminant which is compatible with base change.
To conclude, it is enough to say that the discriminant of a generic (relative) projection from X onto $S \times \mathbf{C}$ induces on each fibre, the discriminant of a projection from the fibre onto $\mathbf{C}$, and that the critical locus is unramified on the parameter space.
When the singularities of the members of the family are no longer isolated ones, that is when the total space has more than two strata, we have to consider not only properties of the bigger stratum but also of the smaller ones to test equisingularity along a smooth subspace. This involves polar varieties of the closure of the smaller strata. In particular, the jacobian module of a member of the family is no longer of finite colength.
Let's end with a question. When we take the situation of example 4.2.6, then a result of Lazzeri says that if we assume the discriminant of a family of isolated singularity hypersurfaces is unramified above the parameter space, then so is the singular locus of the family of functions (in short the jacobian module is equimultiple along the singular set). In what situations does this remain true?

References

[B] N. Bourbaki. *Éléments de Mathématique, Algèbre commutative, chapitre 8 , 9, 10* Masson éditeur, Paris, 1983.

[B-R] Buchsbaum and Rim. *A generalized Koszul complex. II. Depth and multiplicity* Trans. Amer. Math. Soc. **111** (1964) 197-224

[F] W. Fulton. *Intersection Theory* Ergebnisse der Math; 3. Folge, Band **2**. Springer Verlag 1984

[G-H] M. Giusti et J.-P. Henry. *Minorations de nombres de Milnor* Bull. Soc. Math. France 108, **2** (1980), 279-282.

[G 1] T. Gaffney. *Polar Multiplicities and equisingularity of map germs*, to appear in Topology.

[G 2] T. Gaffney. *Integral Closure of modules and Whitney Equisingularity*, preprint, Northeastern University, 1989.

[G 3] T. Gaffney. *Multiplicities and equisingularity of ICIS germs*, Preprint, Northeastern University, 1990.

[EGA III] A. Grothendieck. *Eléments de géométrie algébrique, Chap. III*, Pub. Math. I.H.E.S. **9** 1961.

[Ha] R. Hartshorne *Algebraic Geometry* Graduate Texts in Mathematics **52** Springer-Verlag 1977

[H-M 1] J.-P. Henry et M. Merle. *Limites d'espaces tangents et transversalité des variétés polaires.* Actes de la conférence de la Rabida, Springer, Lecture Notes **961**

[H-M 2] J.-P. Henry et M. Merle. *Limites de normales, conditions de Whitney et éclatement d'Hironaka.* Proc. A.M.S. Summer Institute on singularities, Arcata, 1981.

[H-M 3] J.-P. Henry et M. Merle. *Conditions de régularité et éclatements.* Ann. Inst. Fourier **37**,3 1987 pp 159-190

[H-M-S] J.-P. Henry , M. Merle et C. Sabbah. *Sur la condition de Thom stricte pour un morphisme analytique complexe.* Ann. Scient. Ec. Norm. Sup. 4ème série, **17**, 1984.

[H 1] H. Hironaka. *Normal cones in analytic Whitney Stratifications.* Publ. Math. I.H.E.S., **36**, P.U.F., 1970.

[H 2] H. Hironaka. *Stratifications and Flatness.* in Real and Complex Singularities (Nordic Summer School), Oslo, 1976; Sijthoff and Noordhoff, 1977.

[K] S. Kleiman. *Towards a numerical theory of ampleness* Annals of Math. **84**, 1966, pp 293-344.

[K-T] S. Kleiman and A. Thorup *Geometric theory of Buchsbaum-Rim multiplicity*, to appear in Copenhagen Preprint Series, summer 1991.

[L] J. Lipman. *Reduction, Blowing up and Multiplicities*, Proc. Conf. on Transcendental Methods in Commutative Algebra. George Mason U. Decker 1979

[L-T] M. Lejeune et B. Teissier. *Cloture intégrale des idéaux et équisingularité*, Centre de Mathématiques de l'École Polytechnique, Publications de l'université scientifique et médicale de Grenoble 1974.

[M] S. McAdam. *Asymptotic Prime Divisors*, Lecture Notes in Mathematics **1023**, Springer-Verlag 1983.

[Ra] C.P. Ramanujam *On a Geometric Interpretation of Multiplicity* Inventiones math. 22, 63-67 (1973).

[R 1] D. Rees. *A transform of local rings and a theorem on multiplicities of ideals*, Proc. Cambridge Philos. Soc. 57 (1961), pp. 8-17.

[R 2] D. Rees. *Rings associated with ideals and analytic spread*, Proc. Cambridge Philos. Soc. 89 (1981), pp. 423-432.

[R 3] D. Rees. *Generalizations of reductions and mixed multiplicities*, J. London Math. Soc. (2) 29 (1984) pp. 397-414.

[R 4] D. Rees. *Amao's theorem and Reduction criteria*, J. London Math. Soc. (2) 32 (1985) pp. 404-410.

[R 5] D. Rees *Reduction of modules* , Math. Proc. Camb. Phil. Soc. (1987) **101**, pp. 431-449.

[Sa] P. Samuel. *Some asymptotic properties of ideals*, Annals of Math., Series **2** tome 56.

[S] E. Snapper. *Polynomials associated with divisors*, J. Math. and Mech. **9** 1960, pp 123-129.

[Su] T. Suwa. *Singularities of complex analytic foliations* Proc. A.M.S. Summer Institute on singularities, Arcata, 1981. pp 551-559.

[Cargese] B. Teissier. *Cycles évanescents, sections planes et conditions de Whitney*, Singularités à Cargèse, Astérisque n^o **7-8** , 1973.

[T 2] B. Teissier. *Multiplicités polaires, sections planes et conditions de Whitney*, La Rabida, Lecture Notes in Mathematics **961**, pp. 314-491 , Springer-Verlag 1983.

[Z-S] O. Zariski and P. Samuel. *Commutative Algebra, chapter V , volume I*, Van Nostrand ed. 1958.

Weak Lefschetz and Topological q-Completeness

LUDGER KAUP

Konstanz University

TABLE OF CONTENTS

Abstract Let $\imath : A \hookrightarrow X$ denote the inclusion mapping of a closed complex subspace of a complex analytic space X. The Weak Lefschetz Theorem deals with the question to what extent the induced homomorphisms in homology (with closed supports if X is not compact) or intersection cohomology are bijective. Our answer uses the degree of topological completeness of $X \setminus A$, an invariant that facilitates the reduction of relative to absolute situations. See 5.1 for an absolute and 5.2, 5.11, 5.7, and 8.3 for relative homological results, moreover 6.1 and 8.3 for intersection cohomology.

Résumé Notons $\imath : A \hookrightarrow X$ l'inclusion d'un sous-espace fermé dans un espace analytique complexe X. Le petit théorème de Lefschetz répond à la question : dans quelle mesure les homomorphismes induits par $\imath$ en homologie (à supports fermés si X n'est pas compact) ou en cohomologie d'intersection sont-ils bijectifs? Dans cet article on donne une réponse qui se sert de la proportion dans laquelle le complémentaire $X \setminus A$ est topologiquement q-complet, notion qui est particulièrement convenable pour réduire une situation relative à un cas absolu. Voir 5.1 pour une version absolue, 5.2, 5.11, 5.7, et 8.3 pour des versions relatives en homologie, ainsi que 6.1 et 8.3 pour des résultats en cohomologie d'intersection.

0: INTRODUCTION

"Connectedness" properties in a rather general meaning play an important role in many mathematical problems. That applies not only to an absolute setting, but also to a subspace compared with the ambient space. In particular, if X and Y are closed subspaces of a space Z, how do connectedness properties of X carry over to $X \cap Y$? In complex analysis one has got accustomed to ask such a question when Z is a complex projective space $\mathbb{P}_k$ and the inclusion mapping $X \hookrightarrow \mathbb{P}_k$ is replaced by a more general holomorphic mapping $f: X \to \mathbb{P}_k$: consider a "pull-back diagram" of the form

$$\begin{array}{ccc} A := f^{-1}(Y) & \hookrightarrow & X \\ \downarrow & & \downarrow f \\ Y & \hookrightarrow & \mathbb{P}_k \end{array}$$

For the purpose of the introduction, let X and Y be in addition irreducible and X compact.

0) The simplest question (sometimes called of "Bézout type") is whether $\mathrm{Im}(f)$ meets Y, i.e., whether $f^{-1}(Y) \neq \emptyset$. Obviously, the answer will be "no" in general; nevertheless[1)]

$$f^{-1}(Y) \neq \emptyset \ \text{ if } \ \dim f(X) + \dim Y \geq k.$$

There is a simple criterion in terms of relative cohomology (with coefficients in a fixed ring), which is a standard device to compare connectedness properties:

$$f^{-1}(Y) \neq \emptyset \ \text{ iff } \ H^0(X, f^{-1}(Y)) = 0.$$

1) The next question ("Bertini type") is the following: Is $f^{-1}(Y)$ connected? Again the answer will be "no" in general; consider for instance the intersection of a line and a curve of higher degree in $\mathbb{P}_2$. Nevertheless there is the following result (for a proof, see [FuLa, 4.3]):

$$f^{-1}(Y) \text{ is connected if } \ \dim f(X) + \dim Y \geq k+1.$$

If X in addition is locally irreducible, then $H^1(X, f^{-1}(Y)) = 0$. – As above, the cohomological condition $H^1(X, f^{-1}(Y)) = 0$ guarantees a positive answer, even if X is not locally irreducible; it obviously is a necessary condition if the first Betti number $b_1(X)$ vanishes.

2) One might hope that one could continue in the same way: If $\dim f(X) + \dim Y \geq k+i$ and X is locally irreducible, then $H^i(X, f^{-1}(Y)) = 0$ and thus $H^j(X) = H^j(f^{-1}(Y))$ for $j \leq i-1$. Unfortunately, such a result only holds true in particular situations. Let us give counter-examples under two different

1) Since $f(X)$ and Y are not homologous to zero in $\mathbb{P}_k$, their intersection product does not define the zero class in $H_\bullet(\mathbb{P}_k, \mathbb{Z})$.

aspects: **a)** *Global obstructions.* If Y is a *smooth* surface in $X := \mathbb{P}_4$ such that $H^1(Y, \mathcal{O})$ is not zero, then $H^2(\mathbb{P}_4, Y; \mathbb{Q}) \neq 0$, see 7.5 (for the existence of Y cf. [Ht,]8]). **b)** *Local obstructions.* For $n \geq 2$ let W denote the quotient variety $\mathbb{C}^n/z \sim -z$, which can be realized as a cone in $\mathbb{C}^k$ where $k = \binom{n+1}{2}$. Its projective closure X in $\mathbb{P}_k$ has the affine origin 0 as the only singular point. Let Y be the *hyperplane* $\mathbb{P}_k \setminus \mathbb{C}^k$ in $\mathbb{P}_k$. As W is contractible, the exact sequence for the singular Poincaré - duality homomorphisms yields:

$$0 = H^{2n-2}(W, \mathbb{Z}_2) \to H_2(X, X \cap Y; \mathbb{Z}_2) \to {}_W\mathcal{H}_{2,0}\mathbb{Z}_2 \to H^{2n-1}(W, \mathbb{Z}_2) = 0.$$

Since a typical neighborhood-boundary of the singularity is homeomorphic to the real projective space $\mathbb{P}_{2n-1}(\mathbb{R})$, we obtain, using the local homology ${}_W\mathcal{H}_{2,0}$ of the singular point 0:

$$H^2(X, X \cap Y; \mathbb{Z}_2) \cong H_2(X, X \cap Y; \mathbb{Z}_2) \cong {}_W\mathcal{H}_{2,0}\mathbb{Z}_2 \cong H_1(\mathbb{P}_{2n-1}(\mathbb{R}), \mathbb{Z}_2) \cong \mathbb{Z}_2.$$

For that reason, other conditions for the vanishing of $H^i(X, f^{-1}(Y))$ have to be searched for (which leads to the so called Weak Lefschetz Theorems)! There is an abundance of literature about that problem, see [FuLa] and [Fu] for an interesting survey and numerous references.

In the present article we intend to reduce the "relative problem" concerning f to an absolute problem in terms of X and A only (see 5.1). To that end we need an appropriate measure for the topological position of A in X. That cannot be the degree of "geometrical" completeness of $X \setminus A$; hence, we replace that number by the "topological completeness" (2.1), as in [FiKp$_2$]. That invariant is even well adapted to the relative setup. But then the restriction to $\mathbb{P}_k$ or at least a homogeneous manifold becomes obsolete. Thus we shall deal with a pull-back diagram of the form

$$(D) \qquad \begin{array}{ccc} A := f^{-1}(B) & \subset & X \\ \downarrow & & \downarrow f \\ B & \subset & Z \end{array}$$

where f is a holomorphic mapping of complex spaces and B a closed nonvoid subset of Z.

We want to give an answer to the question of Lefschetz type for general pairs (X, A), though it becomes rather intricate, if one does not only deal with particularly nice singularities like local complete intersections. Our formal approach yields an estimate of the form

$$\dim(X \setminus A) - 1 - \delta,$$

where the defect δ essentially is composed of the degree of topological completeness of

- $Z \setminus B$ (substituting for "positivity"),
- the fibers of f over $Z \setminus B$ (substituting for their dimension),
- the singular locus of $X \setminus A$,

and *local invariants* $lh(X \setminus A)$ depending on the local homology in the singular points, measuring the difference to local complete intersections in homology, respectively, in intersection cohomology for a perversity $\mathbf{p}$, an estimate $\beta \in \mathbb{Z}$, such that $\mathbf{p} \geq \mathbf{m} - 1 - \beta$ (so $\beta = 0$ for the middle perversity $\mathbf{m}$ and the top perversity $\mathbf{t}$).

I would like to express my gratitude to Karl-Heinz Fieseler and Gottfried Barthel for their helpful comments in discussion over this topic. Moreover, my thanks go to the organizers of the stimulating conference, in particular to Jean Paul Brasselet.

1: THE LEFSCHETZ PRINCIPLE

Let X denote a complex space of dimension n and A a closed subset. The inclusion mapping $A \hookrightarrow X$ induces natural homomorphisms in homology, cohomology, intersection cohomology, such as

$$\lambda^j : H^j(X, \mathbb{Z}) \to H^j(A, \mathbb{Z}).$$

We start the discussion of (D) with a motivation that does not include f (i.e., with the absolute case). A standard result about such kind of homomorphisms in *cohomology* is as follows:

1.1 Lefschetz Theorem on Hyperplane Sections (Cohomological Version) *If X is a projective algebraic variety and A a hyperplane section of X such that $X \setminus A$ is an n-dimensional homology manifold, then λ^j is bijective for $j \leq n-2$ and injective for $j = n-1$. Moreover, the homomorphism λ^{n-1} induces an isomorphism when restricted to the torsion subgroups.*

One standard proof consists of two steps:

1) There is an exact cohomology sequence with integer coefficients

$$\ldots \to H^j(X, A) \to H^j(X) \xrightarrow{\lambda^j} H^j(A) \to H^{j+1}(X, A) \to \ldots ;$$

hence, the obstruction groups are $H^j(X, A)$ for $j \leq n$. Moreover, there is an isomorphism

$$H^j(X, A) \cong H^j_{cld(X)|(X\setminus A)}(X \setminus A);$$

hence, we have to analyze the groups $H^j_{cld(X)|(X\setminus A)}(X \setminus A)$ for $j \leq n$. Obviously, $cld(X)|(X \setminus A) = c(X \setminus A)$, since X is compact; moreover, $X \setminus A$ is an orientable homology manifold, so the *Poincaré-duality-homomorphisms*

$$P_{2n-j}(X \setminus A) : H^j_c(X \setminus A) \to H^c_{2n-j}(X \setminus A)$$

are isomorphisms, and we are left with the investigation of $H^c_{2n-j}(X \setminus A)$.

2)As $X \setminus A$ is a Stein space, we may apply the *vanishing* result for the homology of Stein spaces [Kp]. Moreover, $H^c_n(X \setminus A)$ is a free abelian group, which implies the additional property of λ^{n-1} ∎

One might ask, how to include into such a result a more precise estimate in such a simple case as $X \setminus A \cong \mathbf{C}^n$. Actually, the above proof can serve as a pattern for far more general situations, even in order to get the desired stronger result. We proceed as follows:

1) The obstructions, why Poincaré-duality-homomorphisms on $V := X \setminus A$ are not bijective, are global cohomology classes with values in the sheaf $\bigoplus_{j<2n} \mathcal{H}_j R$ with support in $\Sigma(V)$. In our theory, we guarantee the desired vanishing of obstruction modules using an invariant $\mathrm{lh}(V)$ for the local homology sheaves and a global estimate $\mathrm{tc}\,\Sigma(V) \leq \dim \Sigma(V)$ for their supports.

2) The global homological vanishing property is estimated by means of a topological invariant $\mathrm{tc}(V)$, called "topological completeness" of V, which is not positive if V is Stein and which satisfies $\mathrm{tc}(\mathbf{C}^n) = -n$.

If we go over to the relative situation as described in diagram (D), then the estimates of 1) are done with respect to V, while the global aspect of 2) is induced from $W := f(V)$ and the fibers of $f_{|V}$. To that end we shall introduce an invariant $\mathrm{tc}(f_{|V})$, where $\mathrm{tc}(\mathrm{id}_V) = \mathrm{tc}(V)$.

We start our general investigation with homology, using *local homology sheaves* of the form $\mathcal{H}_k\mathcal{L}$, where $\mathcal{L}$ is a locally constant sheaf of R-modules on X over a principal ideal domain R, which is considered to be fixed for the whole theory:

We say that the *Weak Lefschetz in homology* holds true for n_0, (X, A), and coefficients in $\mathcal{L}$ if the Lefschetz homomorphisms

$$(\mathbf{LEF_{ho}}) \qquad \lambda_j : H_j^{cld}(A, \mathcal{L}) \to H_j^{cld}(X, \mathcal{L})$$

are bijective for $j \leq n_0 - 1$, and λ_{n_0} is surjective. We also say for short: $(\mathrm{LEF_{ho}})$ *holds for* n_0.

A very general, but rather formal result is as follows:

1.2 Proposition *If A is HLC and*

$$H^j(X \setminus A, \mathcal{H}_k\mathcal{L}) = 0 \textit{ if } j, k \textit{ satisfy } k - j \leq n_0,$$

then (LEF_{ho}) holds in singular homology with n_0[1].

[1] For an equivalent formulation in the terminology of section 2 see 3.5.

Proof. If A is HLC, then Borel-Moore homology and singular homology of A coincide [Br, V.11.12]). It then suffices to show that the obstruction modules $H_j^{cld}(X,A;\mathcal{L})$ in the exact homology sequence vanish for $j \leq n_0$. That is true since the relative singular Poincaré-duality-homomorphism stems from a convergent spectral sequence [Br, V.8.5]

$$E_2^{i,-k} = H^i(X \setminus A, \mathcal{H}_k\mathcal{L}) \Rightarrow H_{k-i}^{cld}(X \setminus A, \mathcal{L}) \cong H_{k-i}^{cld}(X, A; \mathcal{L}),$$

since $\mathcal{L}$ is locally constant on X ∎

Among other things, *intersection cohomology* can serve as a unified theory for a simultaneous description of homology and cohomology. On the other hand, constructible instead of locally constant coefficients are admissible in that theory, a fact of particular importance for the investigation of mappings! Hence, intersection cohomology provides more general results even for manifolds. For notations see [Bo].

In the discussion of intersection cohomology, let X always be of *pure* dimension n, $\mathbf{X}$ a stratification[2)] of X, and $\mathcal{L}$ a locally constant sheaf of R-modules that is defined outside the singular locus $\Sigma := \Sigma(\mathbf{X})$. Then, for every perversity $\mathbf{p}$, the Deligne complex ${}_{\mathbf{X}}\mathcal{P}^{\bullet}_{\mathbf{p}}\mathcal{L}$ provides an exact sequence in hypercohomology:

$$1.3 \quad .. \to \mathbb{H}^j_{c(X)|X\setminus A}(X, {}_{\mathbf{X}}\mathcal{P}^{\bullet}_{\mathbf{p}}\mathcal{L}) \to \mathbb{H}^j_c(X, {}_{\mathbf{X}}\mathcal{P}^{\bullet}_{\mathbf{p}}\mathcal{L}) \xrightarrow{\lambda^j_{\mathbf{p}}} \mathbb{H}^j_{c\cap A}(A, {}_{\mathbf{X}}\mathcal{P}^{\bullet}_{\mathbf{p}}\mathcal{L}_{|A}) \to ..$$

Intersection cohomology includes (Borel-Moore) homology and cohomology of the normalization $\pi : \tilde{X} \to X$, since

$$\mathbb{H}^j_{\varphi}(X, \mathcal{P}^{\bullet}_{\mathbf{o}}\mathcal{L}) \cong H^j_{\varphi}(\tilde{X}, \tilde{\mathcal{L}})$$
$$\mathbb{H}^j_{\varphi}(X, \mathcal{P}^{\bullet}_{\mathbf{t}}\mathcal{L}) \cong H^{\varphi}_{2n-j}(\tilde{X}, \tilde{\mathcal{L}})$$

where we write $\tilde{\mathcal{L}}$ instead of $\pi^*\iota_*\mathcal{L}$ for the inclusion mapping $\iota : X \setminus \Sigma \to X$; if $\mathcal{L}$ is locally constant on the whole space X, then those modules are isomorphic to the corresponding modules in singular homology (resp. cohomology if φ is paracompactifying). If X is locally irreducible[3)], then, obviously, $H^j_{\varphi}(\tilde{X}, \tilde{\mathcal{L}}) \cong H^j_{\varphi}(X, \mathcal{L})$.

For the additional information about the torsion groups in the cohomological Lefschetz Theorem 1.1, we need an extra notion: We call a perversity $\mathbf{p}$ *dualizing on X with respect to a torsion-free sheaf* $\mathcal{L}$ on $X \setminus \Sigma$ if the natural morphism

$$\sigma_{\mathbf{p}} \; : \mathcal{P}^{\bullet}_{\mathbf{p}^*}\mathcal{L} \to \mathcal{D}\mathcal{P}^{\bullet}_{\mathbf{p}}\mathcal{L}^*[-2n]$$

---

2) Unless otherwise stated, we always consider locally normal trivial stratifications, see [Bo, 1.1], [GoMPh]. That holds for instance if Whitney's condition B is satisfied.

3) That assumption can be weakened in the whole theory, corresponding to the j under consideration. For details see [KpFi,] 2] or 5.14.

is a quasi-isomorphism, see [FiKp,] 3]. That condition is satisfied if R is a field, or if $\mathcal{L}$ extends to a locally free sheaf on X and $\mathbf{p} \leq \mathbf{m} - \mathrm{lh}(\tilde{X}) - 1$ or $\mathbf{p} \geq \mathbf{m} + \mathrm{lh}(\tilde{X}) + 1$, see 8.1. Thus we can formulate the condition

(TD) *The sheaf of R-modules $\mathcal{L}$ on $X \setminus (A \cup \Sigma)$ is torsion-free and the perversity $\mathbf{p}$ is dualizing on $X \setminus A$ with respect to $\mathcal{L}$.*

In intersection cohomology, we shall say that the *Weak Lefschetz holds for n_0*, (X, A), perversity $\mathbf{p}$, and coefficients in $\mathcal{L}$ if the "Lefschetz homomorphisms" in (1.3),

$$(\mathbf{LEF_p}) \qquad \lambda^j_{\mathbf{p}} : \mathbb{H}^j_c(X, \mathcal{P}^\bullet_{\mathbf{p}}\mathcal{L}) \to \mathbb{H}^j_c(A, {}_X\mathcal{P}^\bullet_{\mathbf{p}}\mathcal{L}),$$

are bijective for $j \leq n_0 - 1$ and $\lambda^{n_0}_{\mathbf{p}}$ is injective and identifies the torsion parts if $\mathcal{L}$ satisfies condition (TD). We say for short: $(\mathrm{LEF}_{\mathbf{p}})$ *holds for n_0*.

1.4 Proposition *If*

$$\mathbb{H}^j_c(X \setminus A, \mathcal{P}^\bullet_{\mathbf{p}}\mathcal{L}) = 0 \quad \text{for} \quad j \leq n_0,$$
$$\mathbb{H}^{n_0+1}_c(X \setminus A, \mathcal{P}^\bullet_{\mathbf{p}}\mathcal{L}) \quad \text{is a free } R\text{-module if } \mathcal{L} \text{ satisfies condition } (TD),$$

then $(LEF_{\mathbf{p}})$ holds for n_0.

Proof. (1.3) can be read as

$$\to \mathbb{H}^j_c(X \setminus A, \mathcal{P}^\bullet_{\mathbf{p}}\mathcal{L}) \to \mathbb{H}^j_c(X, \mathcal{P}^\bullet_{\mathbf{p}}\mathcal{L}) \xrightarrow{\lambda^j_{\mathbf{p}}} \mathbb{H}^j_c(A, {}_X\mathcal{P}^\bullet_{\mathbf{p}}\mathcal{L}) \to \mathbb{H}^{j+1}_c(X \setminus A, \mathcal{P}^\bullet_{\mathbf{p}}\mathcal{L}) \to$$

which means that the obstruction modules to be considered for $\lambda^j_{\mathbf{p}}$ are

$$\mathbb{H}^j_c(X \setminus A, \mathcal{P}^\bullet_{\mathbf{p}}\mathcal{L}) \quad \blacksquare$$

There is an easy interpretation of the module $\mathbb{H}^j_c(A, {}_X\mathcal{P}^\bullet_{\mathbf{p}}\mathcal{L}_{|A})$ if *A is in general position* (or *generic*) with respect to $\mathbf{X}$. It means that A is an analytic subset of pure codimension $\delta := \mathrm{codim}^{\mathbb{C}}_X A$ and that every point $a \in A$ has an open neighborhood U in X with a stratified homeomorphism $(U, U \cap A) \cong (U \cap A) \times (\mathbb{R}^{2\delta}, 0)$. We then call A a *subpseudomanifold* of X.

It may be interesting to interpret the results for the perversities $\mathbf{o}$ and $\mathbf{t}$ when A is a subpseudomanifold. Then

$(\mathbf{LEF_o})$ means that the Lefschetz homomorphisms

$$\lambda^j_{\mathbf{o}} = \tilde{\lambda}^j : H^j_c(\tilde{X}, \tilde{\mathcal{L}}) \to H^j_c(\tilde{A}, \tilde{\mathcal{L}})$$

are bijective for $j \leq n_0 - 1$, and $\tilde{\lambda}^{n_0}$ is injective (and identifies the torsion parts if $\mathcal{L}$ satisfies condition (TD));

$(\mathbf{LEF_t})$ means for $\mathcal{L}$ locally constant on X that the "Gysin homomorphisms" in singular homology (with compact supports)

$$\lambda_{\mathbf{t}}^j = \tilde{\gamma}_{2n-j} : H_{2n-j}(\tilde{X}, \tilde{\mathcal{L}}) \to H_{2(n-\delta)-j}(\tilde{A}, \tilde{\mathcal{L}})$$

are bijective for $j \leq n_0 - 1$, and $\tilde{\gamma}_{2n-n_0}$ is injective (and identifies the torsion parts if $\mathcal{L}$ satisfies condition (TD)).

In the same way, there is a version in the language of intersection *homology*, see [FiKp$_2$,] 2]. We do not go into details, since Borel Moore homology with local coefficients may be less familiar than the corresponding sheaf cohomology.

2: TOPOLOGICAL COMPLETENESS FOR COMPLEX SPACES

We have to introduce an appropriate notion of completeness for a complex space W of dimension n (we shall apply that to $Z \setminus B, f(X) \setminus B$, and $X \setminus A$). Since intersection cohomology is constructed by means of a stratification, let

$$\mathbf{W} := (\emptyset = W_{-2} \subset W_0 \subset \ldots \subset W_{2n-2} =: \Sigma =: \Sigma(\mathbf{W}) \subset W_{2n} = W)$$

denote a (not necessarily locally normal trivial) complex stratification of W by analytic subsets W_{2i}; the strata $S_{2i} := W_{2i} \setminus W_{2i-2}$ of $\mathbf{W}$ are topological[1)] manifolds; the index $2i$ refers to the topological dimension. We fix a principal ideal domain R. Since, for a sheaf $\mathcal{F}$ of R-modules defined on W and a closed subset B of W,

$$H^\bullet(W, B; \mathcal{F}) \cong H^\bullet_{cld(W)|W\setminus B}(W \setminus B, \mathcal{F}),$$

for the study of relative cohomology it is sufficient to assume that $\mathcal{F}$ is defined on $W \setminus B$ only.

2.1 Definition *Let q be an integer. We call the stratification $\mathbf{W}$ topologically q-complete, if, for every i and every locally constant sheaf of R-modules $\mathcal{F}$ on $W_{2i} \setminus W_{2i-2}$ with finitely generated stalks,*

$$H^j(W_{2i}, W_{2i-2}; \mathcal{F}) = 0, \quad \textit{for } j \geq i + q + 1\,.$$

We denote with $tc(\mathbf{W})$ the minimal such number. Moreover, we set $tc(W) := \min tc(\mathbf{W})$, where $\mathbf{W}$ runs over all locally normal trivial stratifications of W, and tc$(\emptyset) := -\infty$.

That expression stems from the fact that q-complete complex spaces are topologically q-complete in the above sense, see 7.2. Obviously, the topological notion depends on R. One should keep in mind, that tc$(\mathbf{X})$ depends on the

---

1) At least for the homology results, it suffices to assume the S_{2i} to be R-homology-manifolds.

stratification $\mathbf{X}$ and not only on X: on $\mathbf{C}$, the stratification $(\mathbf{C})$ is topologically (-1)-complete, while $(\{0\}, \mathbf{C})$ is topologically 0-complete. In contrast to ordinary completeness, which one might call "geometric q-completeness", topological q-completeness makes sense also for negative values of q. For some applications a slight generalization is useful: the stratification $\mathbf{W}$ is called *topologically q-complete with respect to a fixed sheaf $\mathcal{L}$* if, in the definiton 2.1, the vanishing condition for $i = n$ is not required for every locally constant sheaf $\mathcal{F}$ on $W \setminus \Sigma$ but for $\mathcal{L}$ only.

2.2 Example 1) If the singular locus Σ of the stratification $\mathbf{W}$ is a stratum and both, W and Σ are contractible, then $\mathbf{W} = (\Sigma, W)$ is topologically $(-\dim \Sigma)$-complete with respect to R.

2) For natural numbers $g \geq 2$ and $m \geq 2$ set

$$X := V(\mathbb{P}_{n+1}; \sum_{j=0}^{m} z_j^g) \text{ and } A := V(X; z_m).$$

Then the singular locus of X is $\Sigma := V(\mathbb{P}_{n+1}; z_0, \ldots, z_m)$. The stratification $\mathbf{X} := (\Sigma, X)$ is a Whitney stratification. The space $W := X \setminus A \cong \mathbf{C}^{n+1-m} \times Z$ is of the homotopy type of $Z = V(\mathbf{C}^m; z_0^g + \ldots + z_{m-1}^g - 1)$ and thus of a bouquet of $(m-1)$-spheres. For $m \geq 3$, the complex manifold W is simply connected, so locally constant coefficients are even constant. Then the stratification $W \cap \mathbf{X}$ is topologically $(m - n - 1)$-complete, though "geometrically" only 0-complete. As a consequence, theorems 5.2 and 5.15 apply with $n_0 = 2n - m$ instead of $n - 1$!

3) Let $W \hookrightarrow \mathbf{C}^N$ be a weighted homogeneous subvariety that is an R-homology manifold. Set $X := \overline{W} \subset \mathbb{P}_N$ and $A := X \setminus W$. Then the trivial stratification of W is topologically $(-n)$-complete (remember the last footnote!), which corresponds to the fact that Theorem 5.2 applies with $n_0 = 2n - 1$ and not only with $n - 1$. Nice examples can be found in [BaDi, 2v)]:

For $1 \leq a \leq b - 2$ set

$$X := V(\mathbb{P}_{n+1}; z_0^a z_1^{b-a} + \sum_{j=1}^{n-1} z_j z_{j+1}^{b-1} + z_{n+1}^b).$$

If $\gcd(a, b) = \gcd(a, b - 1) = 1$, then, with $A := V(X; z_1)$, the hypotheses are satisfied (in the end of the proof of Theorem 2 in [BaDi], the lattice L_1 is unimodular). In fact, X and A are examples for nontrivial $\mathbb{Z}$-homology $\mathbb{P}_n$'s.

4) If $\mathbf{W}$ is topologically q-complete and A a closed complex subspace of W that is a union of connected components of strata, then the induced stratification $A \cap \mathbf{W}$ is topologically q-complete [FiKp5, 1.2]. In particular, $\operatorname{tc}(\Sigma \cap \mathbf{W}) \leq \operatorname{tc}(\mathbf{W})$.

5) **W** is topologically n-complete; it is topologically $(n-1)$-complete iff W has no compact irreducible component of dimension n.

6) The stratification **W** is topologically $(n-2)$-complete with respect to a field R iff Σ has no compact irreducible component of dimension $n-1$ and $H^{2n-1}(W,R) = 0 = H^{2n}(W,R)$, see [FiKp$_3$,1.5].

3: REDUCTION OF THE RELATIVE TO THE ABSOLUTE CASE

Let us come back to the relative situation and extend diagram (D) of the introduction to a stratified version, which always exists if $f_{|V}: V \to W$ is proper. Since we are interested also in other cases like linear spaces over W, we introduce a more general technical notion:

$$(D_{strat}) \qquad \begin{array}{ccccc} A := f^{-1}(B) & \subset & X & \supset & (V,\mathbf{V}) \\ \downarrow{\scriptstyle f_{|A}} & & \downarrow{\scriptstyle f} & & \downarrow{\scriptstyle f_{|V}} \\ B & \subset & Z & \supset & (W,\mathbf{W}), \end{array}$$

with $V := X \setminus A$ and $W := f(X) \setminus B$, where **V** and **W** are (locally normal trivial) stratifications such that $f|_V$ is a *stratified homologically proper* mapping.

There a stratified[1)] holomorphic mapping $g: (V,\mathbf{V}) \to (W,\mathbf{W})$ is called *stratified homologically proper* if it satisfies the three conditions below, where we use the following notation: For a connected component C of $V_{2i} \setminus V_{2i-2}$ and a locally constant sheaf of R-modules $\mathcal{F}$ on C with finitely generated stalks let $\overline{C}$ be the closure in V and $\overline{\mathcal{F}}$ the trivial extension by 0 to $\overline{C}$; call g_C the induced mapping $g: \overline{C} \to g(\overline{C})$. Then

(SHP 1) every $g(\overline{C})$ is a (closed) analytic subset;

(SHP 2) for every C and every $\mathcal{F}$, the derived sheaf $R^\bullet g_C(\overline{\mathcal{F}})$ is locally constant on strata with finitely generated stalks;

(SHP 3) every mapping g_C is homologically proper over the maximal stratum of $g(\overline{C})$, i.e. for every $\mathcal{F}$ and every $w \in g(\overline{C}) \setminus \Sigma(g(\mathbf{W} \cap \overline{C}))$,

$$R^\bullet g_C(\overline{\mathcal{F}})_w := \varinjlim_{U \to w} H^\bullet(g_C^{-1}(U), \overline{\mathcal{F}}) \cong H^\bullet(g_C^{-1}(w), \overline{\mathcal{F}}),$$

where U ranges over the open neighborhoods of w in $g(\overline{C})$.

Note that conditions (SHP 1) and (SHP 2) are satisfied if g admits an extension to a proper holomorphic mapping.

1) i.e., there exist decompositions $V = \bigcup_\alpha V_\alpha$, $W = \bigcup_\beta W_\beta$ where V_α and W_β are connected components of strata of **V**, resp. **W** such that, for every α, there exists a β for which f induces a surjective submersion $V_\alpha \to W_\beta$.

3.1 Examples A stratified holomorphic mapping $g : (V, \mathbf{V}) \to (W, \mathbf{W})$ is stratified homologically proper in the following two cases:
1) g is proper, or, more generally,
2) over each connected component D of a stratum of $\mathbf{W}$ the stratified mapping g is topologically locally trivial with a typical (stratified) fiber $F_D \neq \emptyset$ that has finitely generated homology.

In order to reduce the relative case of a mapping $g = f_{|V}$ to the absolute case of a pair of spaces, we provide an estimate for the degree of topological completeness of $\mathbf{V}$ in terms of that of $\mathbf{W}$ and of properties of g. A rough estimate would be the number $\mathrm{tc}(\mathbf{W}) + d$ where d denotes the maximal (complex) dimension of the fibers of $f|_V$. But such easy examples as blowing up a point in $\mathbb{P}_n \setminus \{\text{point}\}$ show that this would not be very satisfying. It is more efficient to consider the fibers over each stratum separately. For a connected component D of a stratum of $\mathbf{W}$ the number $\mathrm{tc}\, g^{-1}(w)$ is independent of the point $w \in D$.

3.2 Definition *Let $g : (V, \mathbf{V}) \to (W, \mathbf{W})$ be a stratified homologically proper surjective holomorphic mapping. Then set*

$$\mathrm{tc}(g; \mathbf{V}, \mathbf{W}) := \max_D (\mathrm{tc}(\mathbf{W} \cap \overline{D}) + \mathrm{tc}\, g^{-1}(w)),$$

where D runs over all connected components of strata of $\mathbf{W}$ and w is an arbitrary point in D. Using all such pairs $(\mathbf{V}, \mathbf{W})$ we set

$$\mathrm{tc}(g) := \min_{(\mathbf{V},\mathbf{W})} \mathrm{tc}(g; \mathbf{V}, \mathbf{W}).$$

Note that
1) $\mathrm{tc}(g; \mathbf{V}, \mathbf{W}) \leq \mathrm{tc}(\mathbf{W}) + \max_w(\dim g^{-1}(w))$;
2) $\mathrm{tc}(\mathrm{id}_V) = \mathrm{tc}(V)$.

In the Bertini-type theorem 9.4 we shall use the condition $\mathrm{tc}(f_{|V}) \leq n - 2$. Even if W is Stein, that need not be satisfied, as can be illustrated by blowing up a regular point. Nevertheless we obtain:

3.3 Example Let V be irreducible and denote by d the minimal fiber-dimension of g. Assume that
1) the d-dimensional fibers $g^{-1}(w)$ satisfy
a) $\mathrm{tc}(g^{-1}(w)) \leq d - 2$, or
b) $\mathrm{tc}(\mathbf{W}) \leq \dim W - 1$ and no $g^{-1}(w)$ has a compact irreducible component, or
c) $\mathrm{tc}(\mathbf{W}) \leq \dim W - 2$;
2) for $k \geq d + 1$ the set D^k has no compact irreducible component of dimension $\dim V - k - 1$, where $D^k := \{w \in W; g^{-1}(w)$ has a compact irreducible component, $\dim g^{-1}(w) \geq k\}$.
Then $\mathrm{tc}(g) \leq \dim V - 2$.

Proof The minimal (and thus generic) fiber-dimension of g satisfies $d = \dim V - \dim W$. For every connected component D of the maximal stratum of $\mathbf{W}$, condition 1) guarantees that $\mathrm{tc}(\overline{D} \cap \mathbf{W}) + \mathrm{tc}(g^{-1}(w)) \leq \dim V - 2$. For fibers of exceptional dimension $k > d$ the set $\{w \in W; \dim g^{-1}(w) \geq k\}$ is of dimension at most $\dim V - k - 1$. Let D be a connected component of a lower dimensional stratum of $\mathbf{W}$. If $D \subset D^k$, then the required estimate holds since $\overline{D}$ has no compact irreducible component of dimension $\dim V - k - 1$; if it is not in D^k, then the corresponding number $\mathrm{tc}(g^{-1}(w))$ is at most $k - 1$ on D ∎

3.4 Reduction Theorem *Let $g : (V, \mathbf{V}) \to (W, \mathbf{W})$ be a stratified homologically proper surjective holomorphic mapping. Then $\mathbf{V}$ is topologically $\mathrm{tc}(g; \mathbf{V}, \mathbf{W})$-complete. In particular,*

$$\mathrm{tc}(V) \leq \mathrm{tc}(g).$$

Proof. Set $t := \mathrm{tc}(g; \mathbf{V}, \mathbf{W})$. For every $i \leq n$ and every locally constant sheaf $\mathcal{F}$ of R-modules on $V_{2i} \setminus V_{2i-2}$ with finitely generated stalks, we have to show that

$$H^j(V_{2i}, V_{2i-2}; \mathcal{F}) = 0 \ \text{if} \ \ j \geq i + 1 + t.$$

We may assume that $i = n$. Since

$$H^j(V, \Sigma(\mathbf{V}); \mathcal{F}) \cong H^j_{\mathrm{cld}(V)|V \setminus \Sigma(\mathbf{V})}(V \setminus \Sigma(\mathbf{V}), \mathcal{F}),$$

we may in addition assume that $V \setminus \Sigma(\mathbf{V})$ is connected; set $\overline{\mathcal{F}}$ for the trivial extension of $\mathcal{F}$ to V. Then $H^j(V, \Sigma(\mathbf{V}); \mathcal{F}) \cong H^j(V, \overline{\mathcal{F}})$, and we have to verify that

(3.4.1) $$H^j(V, \overline{\mathcal{F}}) = 0 \ \text{if} \ \ j \geq i + 1 + t.$$

By condition (SHP 1), the set $g(V)$, which is a union of strata of $\mathbf{W}$ since g is stratified, is closed. Thus 2.2 4) yields

$$\mathrm{tc}(g; \mathbf{V}, \mathbf{W} \cap g(V)) \leq \mathrm{tc}(g; \mathbf{V}, \mathbf{W}) = t;$$

hence, we may assume that g is surjective. The (convergent) Leray spectral sequence is of the form

$$H^a(W, R^b g(\overline{\mathcal{F}})) \Rightarrow H^{a+b}(V, \overline{\mathcal{F}})$$

so that it suffices to verify

(3.4.2). $$H^a(W, R^b g(\overline{\mathcal{F}})) = 0 \ \text{for} \ \ a + b \geq n + 1 + t.$$

Condition (SHP 3) implies for $w \in W \setminus \Sigma(\mathbf{W})$ that

$$R^b g(\overline{\mathcal{F}})_w = 0 \ \text{for} \ \ b > \dim g^{-1}(w) + \mathrm{tc}\, g^{-1}(w),$$

where $\dim g^{-1}(w) \leq n - \dim W$. Conditon (SHP 2) yields that

$$H^a(W, R^\bullet g(\mathcal{F})) = 0 \text{ for } a \geq \dim W + 1 + \mathrm{tc}(\mathbf{W}).$$

As a consequence,

$$H^a(W, R^b g(\mathcal{F})) = 0 \text{ for } a + b \geq n + 1 + \mathrm{tc}(\mathbf{W}) + \max_{w \in W \setminus \Sigma(\mathbf{W})} \mathrm{tc}\, g^{-1}(w).$$

Since $t \geq \mathrm{tc}(\mathbf{W}) + \max_{w \in W \setminus \Sigma(\mathbf{W})} \mathrm{tc}\, g^{-1}(w)$, (3.4.2) is verified ∎

3.5 Remark In Proposition 1.2, it suffices to require the vanishing condition for the pairs $(2\dim V, j)$ and $(k < 2\dim V, j \leq \dim_{\mathbf{C}} \Sigma(V) + \mathrm{tc}(S(V))$ where $\Sigma(V)$ denotes the set of points in which V is not an R-homology manifold.

4:THE INVARIANT $LH(V)$

In homology, the range of validity for the Weak Lefschetz depends on the local R-homology ${}_V\mathcal{H}_\bullet R$ of V. Set $n := \dim V$ and $n_v := \dim V_v$ for $v \in V$. For a characterization of its impact the following invariant seems to be best possible:

$$\mathrm{lh}(V_v) := \mathrm{lh}(V_v, R) := \min\{i; \dim^{\mathbf{C}}(\mathrm{supp}\ ({}_V\mathcal{H}_{n_v+j-i}R))_v \leq j, \forall j\}.$$

Since it may be difficult to compute $\mathrm{lh}(V_v)$ in explicit examples, we present estimates: For a point $v \in V$ we set for the purposes of this article:

$$\mathrm{tab}(V_v) := \begin{cases} 0 & \text{if the germ } V_v \text{ is an } R\text{-homology manifold,} \\ \min & \{r; V_v \text{ is } \textit{homeomorphic} \text{ to a germ } V(\mathbf{C}_0^N; f_1, \ldots, f_{N-n_v+r})\} \end{cases}$$

(the notation stems from the word "topologische Abweichung"). If we replace the word "homeomorphic" in the definition of $\mathrm{tab}(V_v)$ by "isomorphic", then we obtain in an analogous way an analytic invariant $\mathrm{ab}(V_v)$.

The basic estimate among them is ([Kp_2, 0.1]):

$$\mathrm{lh}(V_v) \leq \mathrm{tab}(V_v) \leq \mathrm{ab}(V_v).$$

In particular, all those invariants vanish in case of a local complete intersection, which means that, with respect to weak Lefschetz theorems, complex spaces with that property behave like manifolds.

All those invariants have a global version:

$$\mathrm{lh}(V) := \max_{v \in V} \mathrm{lh}(V_v),\ \mathrm{tab}(V) := \max_{v \in V} \mathrm{tab}(V_v),\ \mathrm{ab}(V) := \max_{v \in V} \mathrm{ab}(V_v).$$

Let us note explicitly some simple properties of $\mathrm{lh}(V)$:

a) $\mathrm{lh}(V_v) \geq 0$ and $\mathrm{lh}(V_v) \geq \mathrm{codim}_{V_v}(B_v)$ if B_v is a lower-dimensional irreducible component of V_v;

b) ${}_V\mathcal{H}_j\mathcal{L}_v$ vanishes for every $j \leq n_v - 1 - \mathrm{lh}(V_v)$;

c) if $\mathbf{V}$ is a stratification of V, then $\mathcal{H}_j\mathcal{L}_{|V\setminus V_{2(j+\mathrm{lh}-n)}} = 0$ for $j \neq 2n$ and $\mathrm{lh} := \mathrm{lh}(V)$.

The last property can be seen as follows: since the stratification $\mathbf{V}$ is locally normal trivial, the sheaves $\mathcal{H}_j\mathcal{L}$ are locally constant on the strata of $\mathbf{V}$, and

$$\dim(\mathrm{supp}\ ({}_V\mathcal{H}_j\mathcal{L})) = \dim(\mathrm{supp}\ ({}_V\mathcal{H}_{n+(j+\mathrm{lh}-n)-\mathrm{lh}}\mathcal{L})) \leq j + \mathrm{lh} - n.$$

While the invariant $\mathrm{ab}(V)$ may take the value $\dim V$ (an explicit example is for instance in [KpKp, 46.3]), that does not happen with $\mathrm{lh}(V)$:

4.1 Proposition *Let V admit a stratification $\mathbf{V}$ such that, for some $i \leq \dim V$,*
1) let D denote the set of reducible points of V; for every $x \in D$ and every irreducible component I_x of V_x, $\dim I_x \cap D_x \neq i-2$;
2) $V_{2(i-3)}$ is empty.
Then

$$\mathrm{lh}(V) \leq \dim V - i.$$

In particular, $\mathrm{lh}(X) \leq \dim X - 1$ for an arbitrary complex space X of positive dimension.

Proof. For every j we show that

$$\dim^{\mathbf{C}} \mathrm{supp}\ \mathcal{H}_{j+i} \leq j,$$

or, equivalently, that

$$\mathcal{H}_{j+i}|_{V\setminus V_{2j}} = 0.$$

Then $\mathrm{lh}(V_v) \leq \dim V_v - i \leq \dim V - i$ for every $v \in V$. It suffices to verify for every $k \geq 1$ that

$$\mathcal{H}_{j+i,x} = 0 \quad \text{for} \quad x \in S_{2j+2k} := V_{2j+2k} \setminus V_{2j+2k-2}.$$

A typical neighborhood of x is of the form $U \cong B \times \mathring{c}L$, where B is an acyclic connected oriented R-homology manifold of real dimension $2j+2k$ and $\mathring{c}L$ the open cone over a suitable compact space L that is a union of compact pseudomanifolds according to the irreducible components of V_x. The Künneth Formula in homology for closed supports and coefficients in R yields

$$\mathcal{H}_{j+i,x} \cong H^{cld}_{j+i}(U) \cong H^{cld}_{(j+i)-(2j+2k)}(\mathring{c}L) = H^{cld}_{i-j-2k}(\mathring{c}L) \cong \tilde{H}_{i-j-2k-1}(L).$$

Thus, only the j's with $j \leq i-3$ may contribute to nonvanishing local homology. By assumption 2), there is no such x for $j \leq i-4$. For $j = i-3$ we are left with $k=1$; we obtain ${}_V\mathcal{H}_{j+i,x} \cong \tilde{H}_0(L) \cong_{\mathring{c}L} \mathcal{H}_{1,x}$, which is zero iff L is connected, or, equivalently, if the set of reducible points of $\mathring{c}L$ intersects no irreducible component of $\mathring{c}L$ in an isolated point (which is certainly true if V_x is irreducible) ∎

4.2 Corollary *If V admits a stratification $\mathbf{V}$ such that V_{2i-4} is empty for some $i \le \dim V$, then $\mathrm{lh}(V) \le n - i$. In particular, supp ${}_V\mathcal{H}_1$ is discrete; it is empty if no point $x \in V$ is isolated in $D_x \cap I_x$* ∎

4.3 Example The quotient variety $V := \mathbf{C}^n/z \sim -z$ in 0.2 b) satisfies

$$\mathrm{lh}(V, \mathbb{Z}) = n - 2 \quad \text{and} \quad \mathrm{lh}(V, \mathbf{Q}) = 0.$$

5: THE GENERAL RESULTS IN HOMOLOGY

Let us come back to the extended diagram ($\mathrm{D_{strat}}$) of section 3 with globally defined coefficients $\mathcal{L}$. In the following version of the weak Lefschetz in homology we take into account that the degree of topological completeness of the "homologically singular" locus of a complex space is often strictly smaller than that of V. Set

$$\Sigma(V) := \{v \in V; V_v \text{ is not an } R-\text{homology manifold}\},$$
$$\mathrm{tc}_V \Sigma(V) := \min_{\mathbf{V}} \mathrm{tc}(\mathbf{V} \cap \Sigma(V)).$$

Then $\mathrm{tc}_V\, \Sigma(V) \le \mathrm{tc}\, V \le \mathrm{tc}(f_{|V})$, by 2.2 4) and 3.4, and all those numbers may take negative values. If V is not of pure dimension, then the "defect" in dimension has to be taken into account. For that reason we define:

$$\mathrm{Lh}_v(V) := \mathrm{Lh}_V(V_v) := (\mathrm{lh}(V_v) + \dim V - \dim V_v) \text{ and } \mathrm{Lh}(V) := \max_{v \in V} \mathrm{Lh}_v(V).$$

We present our main result of Weak Lefschetz type in a version that facilitates a comparison with central results of [GoMPh$_3$]. The set

$$A^k := \{v \in V; \mathrm{Lh}_v\, V \ge k\}$$

turns out to be a closed set (since $v \mapsto \max_j\{\dim_v \mathrm{supp}\, \mathcal{H}_j - j\}$ is upper semicontinuous), which is a union of connected components of strata for every not necessarily locally normal trivial stratification which makes the sheaf ${}_V\mathcal{H}_\bullet\mathcal{L}$ locally constant on strata. Let $\mathbf{A}$ be the coarsest such stratification, set

$$\mathbf{A}^k := \mathbf{A} \cap A^k \text{ and } t_k := \mathrm{tc}(\mathbf{A}^k);$$

hence, $t_k = -\infty$ if A^k is empty.

5.1 Theorem (Weak Lefschetz in Homology) *For a closed subset A of a complex space X, (LEF_{ho}) holds with*

$$n_0 = \dim(X \setminus A) - 1 - \max_k (k + t_k).$$

5.2 Corollary *In the situation of* (D_{strat}), (LEF_{ho}) *holds for* (X, A) *with*

$$n_1 = \dim V - 1 - \max\bigl(\mathrm{tc}(f_{|V}), \mathrm{tc}_V\Sigma(V) + \mathrm{Lh}(V)\bigr).$$

Proof of 5.2. We show that

(5.2.1)

$$n_1 = n - 1 - \max\bigl(\mathrm{tc}(f_{|V}), \mathrm{tc}_V\Sigma(V) + \mathrm{Lh}(V)\bigr) \le n - 1 - \max_k(k + t_k) = n_0$$

or, equivalently, $\max_k(k + t_k) \le \max\bigl(\mathrm{tc}(f_{|V}), \mathrm{tc}_V\Sigma(V) + \mathrm{Lh}(V)\bigr)$. For $k = 0$, 3.4 yields because of [FiKp₅, 1.2] that $t_0 = \mathrm{tc}(\mathbf{A}^0) \le \mathrm{tc}(V) \le \mathrm{tc}(f_{|V})$. For $k \ge 1$, the closed subspaces A^k of V are unions of connected components of strata of $\mathbf{V} \cap \Sigma(\mathbf{V})$ for every (locally normal trivial) stratification $\mathbf{V}$ of V. Thus $\mathrm{tc}(\mathbf{A}^k) \le \mathrm{tc}_V\,\Sigma(V)$. On the other hand, A^k is empty and thus $t_k = -\infty$ unless $k \le \mathrm{Lh}(V)$ ■ ■

Proof of 5.1. By 1.2, we have to show that

$$H^j(V,\ {}_V\mathcal{H}_{n+b}\mathcal{L}) = 0 \ \text{ for } \ n + b - j \le n_0. \tag{5.1.1}$$

Let us fix a $k \ge 0$, we may write b in the form $b = l - k$ with some integer l. We claim that

$$H^j(A^k, A^{k+r};\ {}_V\mathcal{H}_{n+l-k}\mathcal{L}) = 0 \ \text{ for } \ j \ge 1 + l + \max_{i\ge 0}(i + t_{k+i}),\ r \ge 1. \tag{5.1.2}$$

For a proof we proceed by decreasing induction over k. First of all, A^n is empty, by 4.1. Let us make the hypothesis that (5.1.2) holds for $k+1$. There is an exact sequence

$$H^j(A^k, A^{k+1}; {}_V\mathcal{H}_{n+l-k}\mathcal{L}) \to H^j(A^k, A^{k+r};\ {}_V\mathcal{H}_{n+l-k}\mathcal{L}) \to H^j(A^{k+1}, A^{k+r}; {}_V\mathcal{H}_{n+l-k}\mathcal{L})$$

By induction hypothesis,

$$H^j(A^{k+1}, A^{k+r};\ {}_V\mathcal{H}_{n+l-k}\mathcal{L}) = 0 \ \text{ for } \ j \ge 1 + l + \max_{i\ge 1}(i + t_{k+i}).$$

Hence, it suffices to consider the case $r = 1$. Set $T_i := \overline{\mathrm{supp}\ {}_V\mathcal{H}_i\mathcal{L}}$. Since the basic formal properties of t_k as discussed in [FiKp₃,]] 1 – 3] only depend on the fact that the coefficient sheaves are locally constant on strata (i.e., the stratification need not be locally normal trivial), we obtain

$$H^j(A^k, A^{k+1};\ {}_V\mathcal{H}_{n+l-k}\mathcal{L}) = 0 \ \text{ for } \ j \ge 1 + \dim(A^k \cap T_{n+l-k}) + \mathrm{tc}(\mathbf{A}^k \cap T_{n+l-k}).$$

By 2.2 4), we know that $\mathrm{tc}(\mathbf{A}^k \cap T_{n+l-k}) \le \mathrm{tc}(\mathbf{A}^k) = t_k$. On the other hand, with $D^k := A^k \setminus A^{k+1}$, we may replace $\dim(A^k \cap T_{n+l-k})$ by

$$\dim(D^k \cap T_{n+l-k}) = \max_{v \in D^k \cap T_{n+l-k}} \dim(D^k \cap T_{n+l-k})_v,$$

since we may replace ${}_V\mathcal{H}_{n+l-k}\mathcal{L}_{|A^{k+1}\cap T_{n+l-k}}$ by 0. For every point $v \in D^k \cap T_{n+l-k}$, we obtain

$$\dim(D^k \cap T_{n+l-k})_v = \dim(D^k \cap T_{n_v+l-(k-n+n_v)})_v \leq l, \tag{5.1.3}$$

since

$$D^k = \{v \in V; \operatorname{lh} V_v = k - (n - n_v)\}$$

and since, for all points $v \in D^k$,

$$\dim(\operatorname{supp}\ {}_V\mathcal{H}_{n_v+l-(k-(n-n_v))}\mathcal{L})_v \ \leq \ l.$$

In particular,

$$H^j(A^k, A^{k+1};\ {}_V\mathcal{H}_{n+l-k}\mathcal{L}) = 0 \ \text{ for } \ j \geq 1 + l + t_k,$$

and (5.1.2) follows.

With $r = n$ and $k = 0$, so that $b = l$, we obtain

$$H^j(V,\ {}_V\mathcal{H}_{n+b}\mathcal{L}) = 0 \ \text{ for } \ j \geq 1 + b + \max_{i \geq 0}(i + t_i),$$

and (5.1.1) follows ∎

The impact of the "non-complete-intersection" singularities on the invariant n_0, provided by $\max_{k\geq 1}(k + t_k)$, may be difficult to compute. For that reason we want to indicate further estimates in a simple example:

5.3 Example Assume that $\operatorname{lh}(V) = 1$. Then
1) $1 + t_1 \leq n - 1$;
2) $1 + t_1 \leq n - 2$ if the dimension of the set of reducible points of V is at most $n - 2$ and if $\Sigma(V)$ does not include a compact analytic set of dimension $n - 2$.

Let us prove the second statement: It is easy to see that n is at least 3. Since A^1 has no compact irreducible component of dimension $n - 2$, it suffices to show that $\dim A^1 \leq n - 2$. In the notation of the proof of 5.1 we have $A^1 = \bigcup_{i\leq n} T_{n-1+i}$, so we may verify that $T_{n-1+i} \subset V_{2(n-2)}$, resp., $\mathcal{H}_{n-1+i}$ vanishes on $V \setminus V_{2(n-2)}$ for every $i \leq n$. We use the notation of the proof of 4.1: First of all, S_{2n-1} is empty. For $x \in S_{2(n-1)}$ we obtain

$$\mathcal{H}_{n-1+i,x} \cong \tilde{H}_{i-n}(L).$$

For $i < n$ the latter module is obviously zero, for $i = n$ that holds as L is connected, i.e., since $x \in S_{2(n-1)}$ is an irreducible point of V ∎

We want to compare 5.1 with the homological version of the corresponding basic relative *homotopy*[1)] Lefschetz results in [GoMPh$_3$], which have been proved by means of stratified Morse theory. There the authors distinguish between the case of a finite mapping and that of a smooth variety. We first formulate the following special case of 5.1 (where, by definition, $\dim \emptyset = -\infty$):

5.4 Corollary *Assume that V in $(D_{\rm strat})$ is an analytic subset of a connected manifold M. Call $t(k)$ the dimension of the subset B^k of all points $v \in V$ such that the (set-theoretic) subgerm V_v of M_v cannot be defined by less than k equations. Then $(LEF_{\rm ho})$ holds for (X, A) with*

$$n_0 = \dim M - 1 - \max_k \Big(k + \inf(t(k), \mathrm{tc}(f_{|V}))\Big).$$

Proof. We use the notations of the proof of 5.1. By [Kp$_2$, 0.1], we obtain $A^k \subset B^{c+k}$, where $c := \dim M - \dim V$. Hence, $t_k \le \dim A^k \le \dim B^{c+k} = t(c+k)$. For every (locally normal trivial) stratification $\mathbf{V}$ of V, the set A^k is a union of connected components of strata; in particular, $\mathbf{t}_k \le \mathrm{tc}(V) \le \mathrm{tc}(f_{|V})$. As a consequence, we find

$$\dim M - 1 - \max_l (l + \inf(t(l), \mathrm{tc}(f_{|V})) \le \dim V - 1 - \max_k (k + t_k),$$

and 5.1 yields the result ■

In the particular case that f is finite, Z a projective space, B a linear subspace of Z, and $\mathcal{L}$ constant, 5.4 is analogue to the result with compact supports in [GoMPh$_3$, II Th.1.2].

Let us now come to the comparison in the smooth case.

5.5 Corollary *If V in (D_{strat}) is an R-homology manifold of pure dimension, then $(LEF_{\rm ho})$ holds for (X, A) with*

$$n_1 = n - 1 - \max_D (\mathrm{tc}(\mathbf{W} \cap \overline{D}) + \mathrm{tc}\, f^{-1}(w)),$$

where D runs over the connected components of strata of $\mathbf{W}$ and w is an arbitrary point in D.

For the comparison of n_1 with the bound in [GoMPh$_3$, part II, Th. 1.1], let us formulate their result in the particular case of a proper mapping using our terminology: *let B be a linear subspace of codimension c in a projective*

1) Of course, homotopy results imply analogue vanishing statements in homology with compact supports, while the converse is true if X and $f^{-1}B$ are connected and $\pi_1(f^{-1}B)$ acts trivially on the groups $\pi_j(X, f^{-1}B)$ under consideration.

space $Z = \mathbb{P}_N$, X *a connected manifold, and assume* f *to be proper. Set* $\varphi(k) := \dim\{w \in W; \dim f^{-1}(w) = k\}$. *Then*

$$\pi_j(X, f^{-1}B) = \{e\} \text{ for } j \leq n - 1 - \max_k\Big\{2k - (n - \varphi(k)) + \inf(\varphi(k), c - 1)\Big\}.$$

We have to liken $\mathrm{tc}(f_{|V})$ with $I(f) := \max_k\big\{2k - (n - \varphi(k)) + \inf(\varphi(k), c-1)\big\}$. Since $\mathrm{tc}\, f^{-1}(w) = \dim f^{-1}(w)$, an upper bound for both numbers is

$$\max_i(\inf(i, c-1) + \max_{w \in S_{2i}} \mathrm{tc}\, f^{-1}(w)) \;=\; \max_k(\inf(\varphi(k), c-1) + k) =: II(f).$$

Let δ denote the generic (and thus minimal) dimension of a fiber of f. Since $\varphi(\delta) + \delta = n$, we obtain $I(f) = n = II(f)$ if $\varphi(\delta) \leq c - 1$. If $f : V \to f(V)$ has constant fiber dimension (e.g., if f is *open*) and $\varphi(\delta) \geq c$, then $I(f) = II(f) = \delta + c - 1$. If, in addition, the stratification $\mathbf{W}$ in the definition of $\mathrm{tc}(f_{|V})$ has no compact stratum, then 5.5 provides a strictly sharper result. Let us consider an example of a nonopen mapping: if X is the blow up of $\mathbb{P}_n$ in a linear subspace $\mathbb{P}_l$, then one finds (here we exceptionally set $\dim \emptyset = -1$)

$$I(f) = \max(n - 2 - l + \min(l, c-1), c - 1)$$
$$\mathrm{tc}(f_{|V}) = \max(n - 2 - \dim(\mathbb{P}_l \cap \mathbb{P}_{n-c}), c - 1).$$

Hence, it depends on $\dim(\mathbb{P}_l \cap \mathbb{P}_{n-c})$, which estimate is the better one.

Since our version of the Weak Lefschetz is not restricted to the class of compact spaces, it includes a rather precise version of a vanishing result:

5.6 Corollary (Vanishing Theorem for the Nonstandard Supports) *Assume that the complex space* V *is paracompact. Let* $\mathcal{F}, \mathcal{I}$, *and* M *be locally constant sheaves of* R*-modules on* V *with* $\mathcal{I}$ *invertible and* M *constant. Then*

$$H_j^{cld}(V, \mathcal{F}) = 0 = H_c^j(V, \mathcal{I} \otimes_R M) \;\text{ for }\; j \leq n_0$$

for n_0 *as in 5.1. Moreover,* $H_c^{n_0+1}(V, R)$ *is a free* R*-module if* R *is slender.*

Proof. In homology the statement is just the particular case $A = \emptyset$ of 5.1. For the cohomology one may apply the Universal Coefficient Formulas V.10.4 and II.18.3 in [Br] and the fact, that a Whitehead module of countable rank over a slender principal ideal domain is free ▪

As an application we obtain a local version of the Weak Lefschetz Theorem – for reasons of simplicity we restrict our formulation to the absolute case; we need the corresponding local invariants in a point $a \in A$. Using small open neighborhoods U of a in X set

$$\mathrm{tc}^a(X \setminus A) := \min_U \mathrm{tc}(U \setminus A), \quad \mathrm{tc}^a_{X\setminus A}\, \Sigma(X \setminus A) := \min_U \mathrm{tc}_{U\setminus A}\, \Sigma(U \setminus A),$$

$$\mathrm{lh}^a(X \setminus A) := \min_U \mathrm{Lh}(U \setminus A).$$

The minimum is realized if U is the open simplicial star in an appropriate triangulation of a stratified neighborhood of a.

5.7 Corollary (Local Weak Lefschetz Theorem) *In local homology, the Weak Lefschetz holds in $a \in A$ with*

$$n_1 = n_a - 1 - \max(\mathrm{tc}^a(X \setminus A), \mathrm{tc}^a_{X\setminus A}\Sigma(X \setminus A) + \mathrm{Lh}^a(X \setminus A)).$$

Proof. The exact sequence in local homology

$$\ldots \to {}_A\mathcal{H}_j\mathcal{L}_a \xrightarrow{\lambda_j} {}_X\mathcal{H}_j\mathcal{L}_a \to {}_{(X,A)}\mathcal{H}_j\mathcal{L}_a \to {}_A\mathcal{H}_{j-1}\mathcal{L}_a \to \ldots$$

can be written by means of the open simplicial star U in an appropriate triangulation of a stratified neighborhood of a in the following way:

$$\ldots \to H_j^{cld}(U \cap A, \mathcal{L}) \xrightarrow{\lambda_j(U)} H_j^{cld}(U, \mathcal{L}) \to H_j^{cld}(U \setminus A, \mathcal{L}) \to H_{j-1}^{cld}(U \cap A, \mathcal{L}) \to \ldots$$

Hence, since $n_1 \le n_0$, it suffices to apply 5.6 with $V = U \cap A$ ∎

While $\mathrm{lh}^a(X \setminus A)$ is independent of the points in A, that is not true for $\mathrm{tc}^a(X \setminus A)$ and $\mathrm{tc}^a_{X\setminus A}\Sigma(X \setminus A)$. For applications we thus mention in the spirit of the subsequent section on (geometrical) completeness:

5.8 Examples a) If A_a can be defined as a subset of X_a by $r+1$ holomorphic equations, then $\mathrm{tc}^a_{X\setminus A}\Sigma(X \setminus A) \le \mathrm{tc}^a(X \setminus A) \le r$, see 8.3.

b) If the generalized normal bundle $\mathcal{N}_X(A)$ of A in X is Finsler-positive, then

$$\mathrm{tc}^a_{X\setminus A}\Sigma(X \setminus A) \le \mathrm{tc}^a(X \setminus A) \le \dim \mathcal{N}_X(A)_a - 1.$$

The proof is immediate with a result of Fritzsche [Fr$_2$, Th. 2.2], which holds not only in the compact, but also in the local case ∎

There is a funny companion result to 5.2 (and 5.1) in the following particular situation of (D): Let us assume that f is a (closed) embedding, $\mathcal{L}$ locally constant on Z and Z admits a locally normal trivial stratification $\mathbf{Z}$ such that $X \setminus B$ is a union of lower dimenional connected components of strata of $\mathbf{Z} \setminus B$. Then (D) is of the form *0

$$(S) \qquad \begin{array}{ccc} X \cap B & \subset & X \\ \downarrow & & \downarrow \\ B & \subset & Z \end{array}$$

which induces natural "horizontal" and "vertical" homomorphisms in homology

$$h_j : H_j^{cld}(B, B \cap X; \mathcal{L}) \to H_j^{cld}(Z, X; \mathcal{L}),$$
$$v_j : H_j^{cld}(X, X \cap B; \mathcal{L}) \to H_j^{cld}(Z, B; \mathcal{L}).$$

5.9 Corollary *Assume that* $\mathbf{Z} \setminus B$ *is topologically* q*-complete. Then the homomorphism* h_j *is*
bijective for $j \leq \dim(V)-1-\max(q, \mathrm{tc}\,\Sigma(\mathbf{Z}\setminus B)+\mathrm{Lh}(Z\setminus B), \mathrm{tc}\,\Sigma(\mathbf{V})+\mathrm{Lh}(V))$,
surjective for $j = \dim(V)-\max(q, \mathrm{tc}\,\Sigma(\mathbf{Z}\setminus B)+\mathrm{Lh}(Z\setminus B), \mathrm{tc}\,\Sigma(\mathbf{V})+\mathrm{Lh}(V))$.

Proof. As in [Ok, Lemma 3], one can use the long exact homology sequences associated to the triples $(Z, X, X\cap B)$ and $(Z, B, X\cap B)$ in order to see that we only have to verify the corresponding properties for v_j instead of h_j. Those follow immediately from 5.2, applied to id_Z and $X \hookrightarrow Z$, since $\dim X < \dim Z$ and since V is topologically q-complete, by 2.2 4) ∎

5A: Supplement for Isolated Bad Singularities.

If more detailed information about stratification and local homology of V are available, then we obtain stronger results. In particular, some of the additional obstruction modules can be described explicitly if the "worst" singularities of V are isolated. The simplest case is that of a locally irreducible V. For a slight generalization let $\pi : \tilde{V} \to V$ denote the normalization mapping and set $\tilde{\mathcal{L}} := \pi^{-1}\mathcal{L}$.

5.10 Remark Let R be a field, $\mathcal{L}$ invertible (i.e., locally constant of rank 1), and V of pure dimension n. Then the Poincaré homomorphism

$$P_j^c(V,\mathcal{L}) \circ \pi^{2n-j} : H_c^{2n-j}(V,\mathcal{L}) \to H_j^c(V,\mathcal{L})$$

is injective (surjective, bijective) iff the Poincaré homomorphism

$$P_{2n-j}^{cld}(V,\mathcal{L}) \circ \pi^j : H^j(V,\mathcal{L}) \to H_{2n-j}^{cld}(V,\mathcal{L})$$

is surjective (injective, bijective).

The proof is a generalization of [Kp$_4$, Satz 4.1]. As a consequence, $P_1^\varphi(V,\mathcal{L})$ is surjective if V is locally irreducible (even if only the homomorphism π^1 : $H^1(V,\mathcal{L}) \to H^1(\tilde{V},\tilde{\mathcal{L}})$ is injective). That extends to $P_i^\varphi(V,\mathcal{L})$, if ${}_V\mathcal{H}_{2n-1}R = \ldots = {}_V\mathcal{H}_{2n-i+1}R = 0$ and $\pi^1, \ldots, \pi^i$ are injective, by [Kp$_4$, Diagramm 4.10]. Hence, it is easy to derive the following result:

5.11 Corollary *Let* V *be locally irreducible and of pure dimension* n*. If* R *is a field,* $\mathcal{L}$ *invertible,* M *an* R*-vectorspace, and* ${}_V\mathcal{H}_jR = 0$ *for* $n+\mathrm{tc}(f_{|V})+2 \leq j \leq 2n-1$*, then (LEF$_{\mathrm{ho}}$) holds for* $\mathcal{L}\otimes_R M$ *with*

$$n_1 = n-1-\mathrm{tc}(f_{|V}) \quad \blacksquare$$

Let us come to a more general result in the locally reducible case, where we allow particularly nasty singularities in isolated points. More generally, denote by V_{cc} the following set: let $\mathbf{A}$ be the coarsest (possibly not locally normal trivial) stratification of V such that all sheaves $\mathcal{H}_jR$ are locally constant on strata; then V_{cc} is the union of all contractible closed connected components of strata of $\Sigma(\mathbf{A})$. In particular, the sheaves $\mathcal{H}_j R$ are acyclic on V_{cc}.

In order to avoid to many technicalities, we shall only deal with the case of a pure-dimensional space V.

5.12 Lemma *Let V be of pure dimension n. For a natural number $k \leq 2n-1$, assume that*
a) $k \leq n - 1 - \mathrm{lh}(V \setminus V_{cc}) - \mathrm{tc}_V\, \Sigma(V)$.
Then there exists an exact "Poincaré-Duality-Sequence" of R-modules

$$H^{2n-k}(\tilde{V}, \tilde{\mathcal{L}}) \xrightarrow{P_k} H_k^{cld}(V, \mathcal{L}) \xrightarrow{\alpha_k} H^0(V, \mathcal{H}_k\mathcal{L}) \xrightarrow{\gamma_k} H^{2n-k+1}(\tilde{V}, \tilde{\mathcal{L}}) \to \dots \ .$$

Proof. Set $\mathrm{lh} := \mathrm{lh}(V \setminus V_{cc})$. In the situation of 5.12 it is easy to verify the following conditions for a stratification of $\Sigma(V)$:

a′) if $1 \leq j \leq (2n-1)$, then $\dim_{\mathbb{C}} V_{2(j+k+\mathrm{lh}-n)} + \mathrm{tc}(V_{2(j+k+\mathrm{lh}-n)}) + 1 \ \leq \ j$.

Let us fix an $i \leq k$. For the following discussion we refer to [Kp₃, Satz 2.2 - 2.4]. The homomorphism α_i is well defined. For the existence of γ_i : $H^0(V, \mathcal{H}_i\mathcal{L}) \to H^{2n-i+1}(\tilde{V}, \tilde{\mathcal{L}})$ and the exactness of the above sequence, we have to verify that

$$H^l(V, \mathcal{H}_{i+l}\mathcal{L}) = H^{l+1}(V, \mathcal{H}_{i+l}\mathcal{L}) = 0 \ \text{ for } \ 1 \leq l \leq (2n-1) - i.$$

The basic estimate is

$$\mathrm{supp}\ \mathcal{H}_j\mathcal{L} \ \subset \ V_{2\min(n-1, j+\mathrm{lh}-n)} \cup V_{cc} \quad \text{for } j \leq 2n-1.$$

Since $l \geq 1$ and the locally constant sheaves $\mathcal{H}_j\mathcal{L}$ are acyclic on V_{cc}, we may ignore the set $V_{cc} \setminus V_{2\min(n-1, j+\mathrm{lh}-n)}$ in the proof of the vanishing statement. Then $i \leq k$ and a′) yield

$$l \geq i - k + l \geq \dim V_{2(i+l+\mathrm{lh}-n)} + \mathrm{tc}(V_{2(i+l+\mathrm{lh}-n)}) + 1 \ .$$

Thus we obtain for the l's under consideration

$$H^l(V, \mathcal{H}_{i+l}\mathcal{L}) = H^l(V_{2(i+l+\mathrm{lh}-n)}, \mathcal{H}_{i+l}\mathcal{L}) = 0 \ \blacksquare$$

Let us specify that in some particularly easy situations:

5.13 Examples 1) If V is outside a discrete subset topologically a local complete intersection, then the exact sequence in 5.12 holds for every $k \leq n - 1 - \mathrm{tc}\,\Sigma(V)$.
2) If the singular set of V is one-dimensional without compact irreducible component, then $k = n - 1 - \mathrm{lh}(V \setminus V_{cc})$ is admissible.
3) For $X := V(\mathbb{P}_{n+1}; \sum_{j=0}^{m} z_j^g)$ as in 2.2 2) and $V := X \setminus V(X; z_{n+1})$ one obtains $V_{cc} = \Sigma(V)$. Hence, $\mathrm{lh}(V \setminus V_{cc}) = -\infty$, and 5.12 applies with

$k = 2n - 1$. It is known from the general theory, that P_{2n-1} is in addition injective. The only obstruction groups for $R = \mathbb{Z}$ in the long exact Poincaré duality sequence for nonisolated singularities are $H^0(V, \mathcal{H}_{2n-m}\mathbb{Z}) \cong \mathbb{Z}^b \oplus T$ and $H^0(V, \mathcal{H}_{2n-m+1}\mathbb{Z}) \cong \mathbb{Z}^b$, where $b := (g-1)g^{-1}((g-1)^m - (-1)^m)$ and T is cyclic of order g for m even and 0 otherwise.

In order to obtain results independent of the normalization mapping $\pi : \tilde{V} \to V$, we mention:

5.14 Lemma *Denote by N the analytic closure of the set of reducible points of V. Then*

$$\pi^i : H^i(V, \mathcal{L}) \to H^i(\tilde{V}, \tilde{\mathcal{L}}) \text{ is} \begin{cases} \textit{surjective} & \text{for } i \geq \dim N + \operatorname{tc}(N) + 1, \\ \textit{bijective} & \text{for } i \geq \dim N + \operatorname{tc}(N) + 2. \end{cases}$$

Proof. For $\tilde{N} := \pi^{-1}N$ there is a commutative diagram with exact rows

$$\begin{array}{ccccccc} H^j_{\mathrm{cld}|V\setminus N}(V \setminus N, \mathcal{L}) & \cong & H^j(V, N; \mathcal{L}) & \to & H^j(V, \mathcal{L}) & \to & H^j(N, \mathcal{L}) \\ \downarrow\cong & & \downarrow & & \downarrow\pi^j & & \downarrow \\ H^j_{\mathrm{cld}|\tilde{V}\setminus\tilde{N}}(\tilde{V} \setminus \tilde{N}, \tilde{\mathcal{L}}) & \cong & H^j(\tilde{V}, \tilde{N}; \tilde{\mathcal{L}}) & \to & H^j(\tilde{V}, \tilde{\mathcal{L}}) & \to & H^j(\tilde{N}, \tilde{\mathcal{L}}) , \end{array}$$

where the first vertical arrow is an isomorphism since π is proper. For a stratification $\mathbf{N}$ of N and $\tilde{\mathbf{N}} := \pi^{-1}\mathbf{N}$, the induced finite mapping $\pi : \tilde{N} \to N$ is stratified. Since $\operatorname{tc}(\tilde{\mathbf{N}}) \leq \operatorname{tc}(\pi; \tilde{\mathbf{N}}, \mathbf{N}) \leq \operatorname{tc}(\mathbf{N})$ by 3.4, $H^j(N, \mathcal{L}) = 0 = H^j(\tilde{N}, \tilde{\mathcal{L}})$ for $j \geq \dim N + \operatorname{tc}(N) + 1$ ∎.

We can combine those results to a

5.15 Weak Homology Lefschetz with Local Obstructions *For fixed $k \leq 2n - 1$ and a pure-dimensional $V = X \setminus A$ assume that*
b) $k \leq n - 1 - \max(\operatorname{tc}(f_{|V}), \operatorname{lh}(V \setminus V_{cc}) + \operatorname{tc}_V \Sigma(V))$.
Then there exists an exact sequence of R-modules

$$H^{cld}_k(A, \mathcal{L}) \xrightarrow{\lambda_k} H^{cld}_k(X, \mathcal{L}) \to H^0(X \setminus A; \mathcal{H}_k\mathcal{L}) \to H^{cld}_{k-1}(A, \mathcal{L}) \xrightarrow{\lambda_{k-1}} \cdots .$$

Proof. We use the exact homology sequence associated to the pair (X, A). It suffices, to show for $i \leq k$ that we may replace $H^{cld}_i(X, A; \mathcal{L})$ by $H^0(X \setminus A; \mathcal{H}_i\mathcal{L})$. To that end we apply the exact sequence associated to V as established in 5.12. There the modules $H^{2n-i}(\tilde{V}, \tilde{\mathcal{L}})$ vanish by 5.14, since $2n - i \geq 2n - k \geq n + 1 + \operatorname{tc}_V(N)$ by the estimate a") in the proof of 5.12, and $2n - i \geq 2n - k \geq n + 1 + \operatorname{tc}(f_{|V})$ by assumtion b) and 3.4 ∎

5.16 Examples 1) Assume that, outside a discrete subset, V is topologically a complete intersection. Then the long exact sequence in 5.15 exists with $k := n - 1 - \operatorname{tc}(f_{|V})$.
2) Assume that $V = X \setminus A$ is constructed from a topologically 0-complete

space U with $\mathrm{lh}(U) = 0$ by identifying two different regular points. Then V is topologically 0-complete and $\mathrm{lh}(V) = n-1$. Hence, 5.2 shows that the corresponding Lefschetz homomorphism λ_0 is surjective, while example 1) provides in addition that λ_1 is injective, $\lambda_2, \ldots, \lambda_{n-2}$ are bijective, and λ_{n-1} is surjective.

3) For the contractible, locally irreducible variety $V := \mathbf{C}^n/z \sim -z$ as in 0.2 b), $L := \mathcal{L}_{\bar{0}}$, and $A := X \cap B$, we obtain from 5.12 as obstruction modules for every $j \neq 2n$

$$H_j(X, A; \mathcal{L}) \cong H^0(V, \mathcal{H}_j\mathcal{L}) \cong \mathcal{H}_{j,\bar{0}}\mathcal{L} \cong \tilde{H}_{j-1}(\mathbb{P}_{2n-1}(\mathbb{R}), L),$$

where the last R-module is known to be

$$\tilde{H}_{j-1}(\mathbb{P}_{2n-1}(\mathbb{R}), L \cong \begin{cases} L & \text{if } j = 2n \\ L/2L & \text{if } j \text{ is even and } 2 \leq j \leq 2n-2 \\ \mathrm{Ker}(L \xrightarrow{\cdot 2} L) & \text{if } j \text{ is odd and } 3 \leq j \leq 2n-1 \\ 0 & \text{otherwise.} \end{cases}$$

6: THE GENERAL RESULTS IN INTERSECTION COHOMOLOGY

Let us continue in the relative situation as described in the extended diagram ($\mathrm{D}_{\mathbf{strat}}$) of section 3. In the discussion of the Weak Lefschetz in *intersection cohomology*, we always make the following additional assumptions:

a) The space X is of pure dimension n, and the stratification $\mathbf{X}$ induces the stratification $\mathbf{V}$;

b) the stratification $\mathbf{W}$ is induced by a stratification $\mathbf{Z}$ of Z;

c) the sheaf $\mathcal{L}$ is locally constant on $X \setminus \Sigma(\mathbf{X})$.

If we do not assume that X is compact, then we need[1)] some mild finiteness condition **(CF)**:

1) the ring R is a field, *or*

2) the stratification $\mathbf{V}$ extends to a compact space $\overline{V}$ in which V is a Zariski-dense open subset, *or*

3) $R = \mathbb{Z}$ (more generally, R is slender, i.e., not a local ring that is complete in its Krull topology), and

a) every stratum of $\mathbf{V}$ has only finitely many connected components, *or*

b) the locally constant sheaf $\mathcal{L}$ on $V \backslash \Sigma(\mathbf{V})$ admits a locally constant extension to V (more generally: $\mathcal{L}$ splits locally along $\Sigma(\mathbf{V})$, i.e., every $x \in V$ has an open neighborhood U in V such that $\mathcal{L}|_{U \backslash \Sigma}$ is a direct sum of its torsion part and its torsion-free part) and $\mathbf{p}$ is dualizing with respect to $\mathcal{L}$.

Quasi-isomorphic perversities[2)] provide isomorphic intersection cohomology. Among all perversities quasi-isomorphic to a given $\mathbf{p}$ there exists a maximal

---

1) Otherwise, one would have to replace $\mathrm{tc}\,\Sigma(\mathbf{V}) + \beta$ by $\mathrm{tc}\,\Sigma(\mathbf{V}) + \beta + 1$ in 6.1, see [FiKp, 4.4].

2) We write $\mathbf{p} \cong \mathbf{q}$ with respect to $\mathcal{L}$ if there exists a quasi-isomorphism $\mathcal{P}^{\bullet}_{\mathbf{p}}\mathcal{L} \cong \mathcal{P}^{\bullet}_{\mathbf{q}}\mathcal{L}$.

element **up** with respect to V:

$$\mathbf{up} := \mathbf{u}_V\mathbf{p} := \max\{\mathbf{q}; \mathbf{q} \cong \mathbf{p} \text{ on } V\}.$$

6.1 Theorem (Weak Lefschetz in Intersection Cohomology) *Assume in diagram (D_{strat}) that (CF) holds for the pair (X, A). If $\mathbf{p}$ is a perversity such that $\mathbf{up} \geq \mathbf{m} - 1 - \beta$ on V, then ($LEF_{\mathbf{p}}$) holds with $n_1 = \dim X - 1 - \max(\mathrm{tc}(f|_V), \mathrm{tc}\,\Sigma(\mathbf{V}) + \beta)$.*

For X an algebraic variety, A a hyperplane section, integer coefficients, and $\mathbf{p} \geq \mathbf{m} - 1$, that result is already in [GoMPh₂,7.1]; for B a linear subspace of $\mathbb{P}_N$ see [GoMPh₃, II.6.10].

Proof. For an application of [FiKp, 4.4, 4.5], we have to verify that

$$H^j_{\mathrm{cld}(V_{2i})|(V_{2i}\setminus V_{2i-2})}(V_{2i} \setminus V_{2i-2}, \mathcal{F}) = 0, \; j \geq i + \mathrm{tc}(f|_V) + 1$$

for every locally constant sheaf of R-modules $\mathcal{F}$ with finitely generated stalks, where we use the notations of the proof of 3.4. Thus, the claim follows from 3.4 ∎

The number $\mathrm{tc}\,\Sigma(\mathbf{V})$ in 6.1 is at most $\mathrm{tc}(f_{|V})$, but it will be smaller in general:

6.2 Example For $Z = X = \mathbb{P}_5$, $B = A = \mathbb{P}_2$, choose any stratification $\mathbf{V}$ of $X \setminus A$ such that $\Sigma(\mathbf{V})$ is a compact curve. Then $\Sigma(\mathbf{V})$ is topologically 1-complete, while V even topologically is only 2-complete ∎

There is an application analogous to 5.9 in the situation of the diagram (S): If the subspaces are in "good position", then there are homomorphisms

$$h^j_{\mathbf{p}} : \mathbb{IH}^j_c(Z, X; \mathcal{P}^\bullet_{\mathbf{p}}\mathcal{L}) \to \mathbb{IH}^j_c(B, B \cap X; \mathcal{P}^\bullet_{\mathbf{p}}\mathcal{L}),$$

$$v^j_{\mathbf{p}} : \mathbb{IH}^j_c(Z, B; \mathcal{P}^\bullet_{\mathbf{p}}\mathcal{L}) \to \mathbb{IH}^j_c(X, X \cap B; \mathcal{P}^\bullet_{\mathbf{p}}\mathcal{L}).$$

6.3 Corollary *Let $\mathcal{L}$ be locally constant outside the singular set of the stratification $\mathbf{Z}$, and let $B \setminus X$ (resp. $X \setminus B$) be in general position with respect to $\mathbf{Z} \setminus X$ (resp. $\mathbf{Z} \setminus B$). Assume that $\mathbf{Z} \setminus B$ and $\mathbf{X} \setminus B$ are topologically q-complete and that their singular locus is topologically s-complete. If $\mathbf{p}$ is a perversity such that $\mathbf{up} \geq \mathbf{m} - 1 - \beta$ on $Z \setminus (X \cap B)$, then the homomorphism $h^j_{\mathbf{p}}$ is*
bijective for $j \leq \dim(X \setminus B) - 1 - q - \max(q, s + \beta)$,
injective for $j = \dim(X \setminus B) - q - \max(q, s + \beta)$.

Proof. Since $B \setminus X$ and $X \setminus B$ are subpseudomanifolds, the homomorphisms $h^j_{\mathbf{p}}$ and $v^j_{\mathbf{p}}$ exist. We now dualize [Ok, Lemma 3] and then follow the lines of the proof of 5.9 ∎

For an application of the above theory to a Hard Lefschetz Theorem in intersection cohomology it is useful to formulate the analogue of 6.1 for the *Gysin homomorphisms*

$$\gamma^j : \mathbb{H}^{2\dim A-j}(A, \mathcal{P}^\bullet_{\mathbf{p}}\mathcal{L}) \to \mathbb{H}^{2\dim X-j}(X, \mathcal{P}^\bullet_{\mathbf{p}}\mathcal{L}).$$

Set $\mathbf{lp} := \mathbf{l}_V\mathbf{p} := \min\{\mathbf{q}; \mathbf{q} \cong \mathbf{p} \text{ on } V\}$.

6.4 Remark (Gysin homomorphism) In the situation of diagram ($\mathrm{D_{strat}}$), assume that $\mathbf{p}$ satisfies $\mathbf{lp} \le \mathbf{m} + 1 + \beta$ on V. Then γ^{j-1} is bijective and γ^j surjective for $j \le \dim X - 1 - \max(\mathrm{tc}(f|_V), \mathrm{tc}\,\Sigma(\mathbf{V}) + \beta)$.

The proof is immediate with [FiKp₂, 2.1 i)] and 3.4 ∎

7: Q-COMPLETE SPACES

We now collect some known results for our main examples: A complex space W is called *(geometrically) q-complete* if there exists a continuous function $h : W \to \mathbb{R}$ which is bounded from below, proper on connected components, and q-convex (i.e., for every $u \in W$, there exists an embedding of germs $W_u \hookrightarrow \mathbf{C}^N_0$ and an extension $\psi \in C^2(\mathbf{C}^N_0, \mathbb{R})$ of the germ h_u such that the Levi form of ψ has at most q nonpositive eigenvalues). — In [AnGr], [Ok], [Pe],..., such a complex space W is called $(q+1)$-complete. So we emphasize that in our notation Stein spaces are precisely the 0-complete spaces. — A q-complete complex space is paracompact, since every connected component is countable at infinity.

7.1 Examples 1) If W is a union of $q+1$ open Stein subspaces, then W is q-complete.

2) If $W = W_1 \cup W_2$ where the W_i are q_i-complete open subspaces, then W is (q_1+q_2+1)-complete and $W_1 \cap W_2$ is (q_1+q_2)-complete [Pe, Satz 2.2].

3) If A is an r-complete analytic subset of W and $W \setminus A$ is q-complete, then W is $(q+r+1)$-complete ([Pe, 6.2]).

4) If W is q-complete and A an analytic subset of W locally defined by $r+1$ holomorphic equations, then $W \setminus A$ is $q+r$-complete [Pe, Kor. of Satz 3.2].

5) Let a holomorphic mapping $f : E \to W$ be a stratified topological fiber bundle with typical fiber F. If W is q-complete and F is r-complete, then E in general will not be $(q+r)$-complete, though it is topologically $(q+r)$-complete, by 3.4.

6) If W is a Stein manifold, A a closed subspace with normal "bundle" $\mathcal{N}$, then $W \setminus A$ is $(q-1)$-complete with $q := \sup_{a \in A} \dim \mathcal{N}_a$ (see [Bl] for this and more general statements).

The following basic result is a consequence of [Ha], see [FiKp₃,] 2 ii)]:

7.2 Theorem *Every stratification* **W** *of a q-complete*[1] *complex space is topologically q-complete.*

The results in [Ha] use Morse theory for manifolds with boundary for the description of the homotopy type. In our applications to homology and intersection cohomology no further use of Morse theory is needed.

Because of 7.2, we now list some classes of examples of q-complete space, where we begin with general spaces and then go over to quasi-projective varieties.

The complex space W is called *q-complete with corners*, if the exhaustion function h in the above definition of a q-complete space is locally only a maximum of finitely many q-convex functions.

For an analytic subset A of X, denote by $\operatorname{codim}_+ A = \operatorname{codim}_+^X A$ the minimal number i such that $\min_{a\in A} \dim A_a \geq \min_{x\in X} \dim X_x - i$; denote the diagonal with $\Delta_X \hookrightarrow X \times X$.

7.3 Theorem *Let X be a compact complex space and r an integer such that the space $X \times X \setminus \Delta_X$ is $(\dim X + r)$-complete with corners. If A is an analytic subset of X, then $X \setminus A$ is q-complete with $q = \dim X - [\frac{\dim X}{r + \operatorname{codim}_+ A}]$.*

Proof. By Theorem 6 in [Pe₂] the space $X \setminus A$ is $(r + \operatorname{codim}_+ A)$-complete with corners; hence [DiFo, Th.2] implies that it is $(\dim X - [\frac{\dim X}{r+\operatorname{codim}_+ A}])$-complete ∎

If X is a compact complex manifold and its tangent bundle globally generated (e.g., if X is a homogeneous space) and r-ample, then the hypothesis in 7.3 is satisfied. Hence, $r = 0$ for $\mathbb{P}_n$, and $r = 1$ for a Hopf manifold. For generalized flag manifolds G/P, where G is a connected semisimple complex Lie group and P a parabolic subgroup (with only finitely many connected components) of G, see [Gn] for a computation of the corresponding r's. A nice and systematic description can be found in [Fr].

We now go over to the particular situation of algebraic subsets of some $\mathbb{P}_k$. A consequence of [DiFo] and [Pe₂] is the following result:

7.4 Theorem *Let X and B denote algebraic subsets of $\mathbb{P}_k$. Then $X \setminus B$ is*

---

[1] *There is a particularly nice example in [Fr] for the fact that q-convexity would not be sufficient: Let A be the image in $\mathbb{P}_n$ of $\mathbb{P}_a$ under the Veronese embedding of degree $d \geq 2$, where $n = \binom{a+d}{a} - 1$. Then $\mathbb{P}_n \setminus A$ is $(n-a-1)$-convex, but not topologically $(n-3)$-complete.*

q-complete with

$$q = \dim X - [\frac{\dim X}{\operatorname{codim}_+^{\mathbb{P}_k} B}].$$

The bound $\dim X - [\frac{\dim X}{\operatorname{codim}_+ B}]$ in 7.4 cannot be improved without additional conditions:

7.5 Example Let B be a smooth surface in $X := \mathbb{P}_4$ with $H^1(B, \mathcal{O}) \neq 0$ (e.g., a quintic elliptic scroll, see [Ht,] 8]). Since B is a compact Kähler manifold, $b_1(B) = 2 \cdot \dim H^1(B, \mathcal{O}) \neq 0$, see [KiBaKp, Th. 1 on page 19]. Thus $H^1(B, \mathbb{Z})$ does not vanish, so $H^2(\mathbb{P}_4, B; \mathbb{Z})$ is not the zero-group. By 7.4, $\mathbb{P}_4 \setminus B$ is 2-complete. If it were (at least topologically) 1-complete, then 6.1 would imply that $H^2(\mathbb{P}_4, B; \mathbb{Z}) = 0$ ∎

While 7.4 is independent of the type of singularities, the next statement takes into account the singular structure of $X \cap B$ compared to that of X.

The simplest case is that of a *global* complete intersection[1]:

7.6 Example If a subset A of a variety $X \hookrightarrow \mathbb{P}_k$ can be defined by at most $r+1$ homogeneous equations, then the complex space $X \setminus A$ is r-complete.

Proof. If the defining homogeneous equations are $f_0, \ldots, f_r$, then

$$X \setminus A = \bigcup_{j=0}^{r} [N(X; f_0, \ldots, f_{j-1}) \setminus N(X; f_0, \ldots, f_j)].$$

Since each of those sets is Stein, the union is r-complete according to 7.1 3). For a more elementary proof see [Fr] ∎

If the number of defining equations for A exceeds considerably the codimension of A, then 7.6 provides a rather poor estimate. Thus it is advisable to look for different criteria.

In the discussion of an analytic subset A of a complex space X that is not a local complete intersection, we use the following notion: For a coherent ideal sheaf $\mathcal{I} \subset_X \mathcal{O}$ with zero set $N(\mathcal{I}) = A$, we denote by cg $\mathcal{I}$ the maximal corank of the stalks, i.e., $\operatorname{cg} \mathcal{I} := \max_a(\operatorname{cg}_{_X\mathcal{O}_a} \mathcal{I}_a; a \in A)$. Since we are not interested in the particular choice of the ideal $\mathcal{I}$, we define:

$$\operatorname{cg}_X A := \min_{\mathcal{I}} \{\operatorname{cg} \mathcal{I}; N(\mathcal{I}) = A\}.$$

The next basic example is the following [Pe, 5.2]:

[1] In this article, local and global complete intersections are always understood in a set-theoretic sense.

7.7 Theorem *Let X and B be algebraic subsets of $\mathbb{P}_k$. Then $X \setminus B$ is q-complete with* $q = \operatorname{codim}_{\mathbb{P}_k} B + \operatorname{cg}_X X \cap B - 2$.

We say that an analytic subset A of X *differs (globally) by*

$$\operatorname{ab}_X(A) := \operatorname{cg}_X A - \operatorname{codim}_X A$$

from a local complete intersection in X (from "Abweichung"). If X is smooth and A of constant codimension, then $\operatorname{ab}_X(A)$ is independent of X, so it is just $\operatorname{ab}(A)$.

There is a particularly interesting case, which generalizes the well known result for local complete intersections (i.e., $\operatorname{ab}(B) = 0$):

7.8 Corollary *For $B \hookrightarrow \mathbb{P}_k$ the manifold $\mathbb{P}_k \setminus B$ is q-complete with*

$$q = 2(\operatorname{codim} B - 1) + \operatorname{ab}_{\mathbb{P}_k}(B) \quad \blacksquare$$

In a comparison of 7.4 and 7.7, the first result is more promising in the case of wild singularities, the second one for mild singularities. Let us recall two

7.9 Examples 1) If B is an algebraic surface in $\mathbb{P}_4$, then $\mathbb{P}_4 \setminus B$ is 2-complete by 7.4; if B is *not* a local complete intersection, then 7.7 only provides 3-completeness.
2) If B is an algebraic subset of dimension 3 in $\mathbb{P}_5$ which is a local complete intersection, then $\mathbb{P}_5 \setminus B$ is 2-complete by 7.8, while 7.4 only provides 3-completeness.

Theorem 7.7 is only a special case of [Pe, 5.2]: First of all, the invariant codim B can be improved. Denote by KB a projective cone over B in $\mathbb{P}_k$ and set $K^{r+1}B := KK^rB$. Then the "coning number"

$$c(X, B) := \min\{r; \exists K^rB \supset X\}$$

satisfies $c(X, B) \leq \operatorname{codim} B$, since $K^{\operatorname{codim} B}B = \mathbb{P}_k$ for an appropriate choice of conings. The next observation is that the points with large $\operatorname{ab}_x(X)$ may lie in a rather small analytic subset B of $X \cap B$. The precise result [Pe, 5.2] reads as

7.7 bis Theorem. *Let X and B be algebraic subsets of $\mathbb{P}_k$ and $A = X \cap B$. Then $X \setminus B$ is q-complete with*

$$q = \min_{B \text{ analytic}} \{\dim B - 1 + \max(\operatorname{cg}_X A, \operatorname{cg}_{X\setminus B}(A \setminus B) + c(X, B))\}.$$

Denote $m_{A,B} := \max(\operatorname{codim}_X A + \operatorname{ab}_X A, \operatorname{codim}_X A + \operatorname{ab}_{X\setminus B}(A \setminus B) + c(X, B))$, the following estimate :

$$\begin{aligned}&\min_B\{\dim B - 1 + \max(\mathrm{cg}_X A, \mathrm{cg}_{X\setminus B}(A\setminus B) + c(X,B))\}\\ &\leq \min_B\{\dim B - 1 + m_{A,B}\}\\ &= \mathrm{codim}_X A - 1 + \min_B\{\dim B + \max(\mathrm{ab}_X A, \mathrm{ab}_{X\setminus B}(A\setminus B) + c(X,B))\},\end{aligned}$$

provides an improved version of 7.8:

Example If the algebraic subset B of $\mathbb{P}_k$ is a local complete intersection outside a nonempty finite subset of B only, then $\mathbb{P}_k \setminus B$ is $(\mathrm{codim}\, B - 1 + \max(\mathrm{ab}(B), \mathrm{codim}\, B))$-complete, which is at most the bound $2(\mathrm{codim}\, B - 1) + \mathrm{ab}(B)$ of 7.8 ■

8: COMPARISON OF HOMOLOGY AND INTERSECTION COHOMOLOGY

It is not difficult to stick together the results of the previous sections. Before we can interpret the situation in intersection cohomology, we have to comment on the comparison of different perversities and their intersection cohomology on a complex space W of pure dimension $n > 0$ (where R is fixed as always):

It is the middle perversity, $\mathbf{m}: 2j \mapsto j-1$, that is often the easiest to handle, since, for field coefficients K, intersection cohomology in the middle perversity satisfies Poincaré duality. If the singular set Σ of the stratification $\mathbf{W}$ is empty, then intersection cohomology is independent of the perversity. In the general situation, the closer the perversity is to $\mathbf{m}$, the easier the theory tends to be.

Let $\tilde{W}$ denote the normalization of W, then we can compare perversities (see [FiKp₄, 1.4]):

8.1 Proposition *If* $\mathbf{p}$ *is an arbitrary perversity and* $\mathcal{L}$ *locally constant on* W*, then*

$$\mathbf{up} \geq \mathbf{m} - \mathrm{lh}(\tilde{W}) - 1.$$

With the results of sections 5 and 6, Proposition 8.1 yields

8.2 Comparison Table *In the situation of* (D_{strat}) *set* $q := \mathrm{tc}(f_{|V})$. *Then* (LEF_*) *holds with*

$$n_0 = \dim(X \setminus A) - 1 - \delta,$$

where the defect δ *given by the table :*

$*$	δ
$\mathbf{up} \geq \mathbf{m}-1-\beta$ (*CF*) *holds*	$\max(q, \operatorname{tc}\Sigma(\mathbf{V})+\beta)$
$\mathbf{p} \cong \mathbf{o}$ (*CF*) *holds* $\mathcal{L}$ *loc.const. on* W	$\max(q, \operatorname{tc}\Sigma(\mathbf{V})+\operatorname{lh}\tilde{V})$
$\mathbf{p} \cong \mathbf{m}, \mathbf{t}$ (*CF*) *holds*	q
ho $\mathcal{L}$ *loc.const. on* X	$\max(q, \operatorname{tc}_V\Sigma(V)+\operatorname{Lh} V)$

We still want to comment on our starting point in section one, Theorem 1.1. The first way to generalize it was to consider the simultaneous intersection of X with several hypersurfaces in the ambient projective space. We present a more general version that holds even if X is not quasi-projective.

8.3 Generalized Weak Lefschetz for a Multiple Intersection *Assume that*
1) *Z is q-complete and B an analytic subset locally defined by $r - q + 1$ holomorphic functions,* or
2) *$B = \bigcap_{j=0}^{r} B_j$, where the B_j's are closed subsets of Z such that $Z \setminus B_j$ is Stein.*
Then
(ho) *If $\mathcal{L}$ be locally constant on Z, then the Weak Lefschetz (LEF_{ho}) holds for (Z, B) with*

$$n_0 = \dim(Z \setminus B) - 1 - \max(r, \operatorname{tc}_W \Sigma(W) + \operatorname{Lh}(W)).$$

(p) *If Z is of pure dimension, the pair (Z, B) satisfies condition (CF), and* **p** *is a perversity such that* $\mathbf{up} \geq \mathbf{m} - 1 - \beta$ *on W, then the Weak Lefschetz ($LEF_{\mathbf{p}}$) holds for (Z, B) with*

$$n_0 = \dim Z - 1 - \max(r, \operatorname{tc}\Sigma(\mathbf{W}) + \beta).$$

The proof is immediate with the following rather technical

8.4 Remark *Assume that $W := Z \setminus B = \bigcup_{j=0}^{r} W_j$ is a union of open, topologically q_j-complete subspaces W_j such that, for every $k + 1$ of them,*

$\mathbf{W} \cap W_{j_0} \cap \ldots \cap W_{j_k}$ is topologically $(\sum_{i=0}^{k} q_{j_i})$-complete .
(ho) *If $\mathcal{L}$ is locally constant on Z, then (LEF_{ho}) holds with*

$$n_0 = \dim(Z \setminus B) - 1 - \max(r + \sum_{j=0}^{r} q_j, \mathrm{tc}_W \Sigma(W) + \mathrm{Lh}(W)).$$

(p) *If Z is of pure dimension, the pair (Z, B) satisfies condition (CF), and* $\mathbf{p}$ *is a perversity such that* $\mathbf{up} \geq \mathbf{m} - 1 - \beta$ *on W, then ($LEF_{\mathbf{p}}$) holds with*

$$n_0 = \dim Z - 1 - \max(r + \sum_{j=0}^{r} q_j, \mathrm{tc}\,\Sigma(\mathbf{W}) + \beta).$$

Proof. As a consequence of a Mayer-Vietoris argument, the difference $Z \setminus B$ is topologically $(r + \sum_{j=0}^{r} q_j)$-complete; see 7.1 4). Hence, the claims follow from 5.2 and 6.1 ▪

Let us give some examples and extensions:

8.5 Applications Projective case. If Z is a projective algebraic variety and each B_j in case 2) a hypersurface section of Z, then $Z \setminus B_j$ is Stein; hence, 8.3 (ho) provides a generalized homological version of 1.1 and of 7.6. For $\beta = 0 = r$, $Z \hookrightarrow \mathbb{P}_k$ and $\mathcal{L} = \mathbb{Z}$, 8.3 (**p**) has been proved in [GoMPh$_2$, 7.1]; note that $\mathrm{tc}(\mathbf{W}) \leq r$.

Stein case. If Z is a Stein space and B_j the zero set of an invertible coherent $\mathcal{O}_Z$-ideal, e.g., $B_j = N(Z; f_j)$ for some $f_j \in \mathcal{O}(Z)$, then $Z \setminus B_j$ is Stein [GrRe,]V.1 Satz 5] and 8.3 2) applies.

Excision. If (Z, B) is as in 8.3 2), then, for C closed in Z, the pair $(Z \setminus C, B \setminus C)$ satisfies the hypothesis of 8.3. Thus the projective case generalizes to special quasi-projective situations.

Relative case. In diagram ($\mathrm{D_{strat}}$) assume that the assumption in 8.3 is satisfied for the pair (Z, B). Then 8.3 (ho) and (**p**) hold for (X, A) with r replaced by $\mathrm{tc}(f_{|V})$, which is at most $r + \max_{w \in W} \dim f^{-1}(w)$.

9: RESULTS OF BÉZOUT OR BERTINI TYPE

We add some further comments and specify the general theory of H^0 and H^1 for a closed subset A of the stratified complex space $(X, \mathbf{X})$ of dimension n.

9.1 Remark *The closed subset A intersects every compact irreducible component of dimension n of X iff $X \setminus A$ is topologically $(n-1)$-complete with respect to R.*

The proof is obvious, since topological $(n-1)$-completeness means that $X \setminus A$ has no compact irreducible component of dimension n, see 2.2 5) ▪

Let us come back to diagram (D_{strat}).

9.2 Corollary *Assume that X is compact and a union of connected components of strata of $\mathbf{Z}$. Let the closed subset B of Z be such that $\mathbf{Z} \setminus B$ is topologically $(q-1)$-complete. Then B intersects every irreducible component of X that is of dimension at least q.*

Proof. We may assume that X is irreducible and of dimension at least q. By 2.2 4), the induced stratification on $X \setminus B$ is topologically $(q-1)$-complete. Thus, 9.1 yields that $X \cap B$ is not empty ∎

In the particular case that X is smooth and Z a connected *homogeneous* manifold, the above corollary has been proved in [L] under the essentially weaker assumption that $Z \setminus B$ be $(\dim X - 1)$-convex (by [Pe$_2$, Th.1], if $Z \setminus B$ is q-convex with corners, it is even q-complete with corners). His result answered in the affirmative a question of Hartshorne [Ht$_2$, III. Conj. 4.5] for Z homogeneous. For further comments and applications see [Fr].

Since connected components can be discovered by lower-dimensional homology only if they are compact, let us denote by X_c the union of the compact irreducible components of X. By 2.2 6), the completeness condition in the following result is satisfied if $\dim V_c < n$, $\dim(\Sigma V)_c < n-1$, and if $H^{2n-1}(V, R) = 0$.

9.3 Proposition *Let X be of pure dimension and B be analytic. Then the intersection of A with every compact irreducible component of X is connected and nonvoid if* $\mathrm{tc}(f_{|V}) \leq \dim X - 2$.

Proof. Note that the above statement generalizes to the case where $f_{|V}$ is topologically $(\dim X - 2)$-complete in a reasonable sense. We leave the details to the reader. - We use homology with coefficients in R. As the fields of fractions $Q(R)$ of R is a flat R-module, the stratification $\mathbf{X} \setminus A$ turns out to be $(\dim X - 2)$-complete with respect to $Q(R)$ as well. Hence, we may assume that R is a field. By 3.4, $\mathrm{tc}(\mathbf{V}) \leq \mathrm{tc}(f_{|V}) \leq n-2$. Denote by $\pi : \tilde{X} \to X$ the normalisation mapping. Obviously, $(\pi^{-1}(A))_c = \pi^{-1}(A_c)$. The number of connected components of X_c is given by $H_0(X_c) \cong H_0^{cld}(X)$. By 9.1, the lines of the following commutative diagram are exact:

$$\begin{array}{ccccc} H_1^{cld}(\tilde{X}, \pi^{-1}A) & \to & H_0^{cld}(\pi^{-1}A) = H_0(\pi^{-1}A_c) & \twoheadrightarrow & H_0(\tilde{X}_c) = H_0^{cld}(\tilde{X}) \\ & & \downarrow \chi & & \downarrow \\ & & H_0^{cld}(A) = H_0(A_c) & \to & H_0(X_c) = H_0^{cld}(X) \end{array}$$

Since $\tilde{X}$ is locally irreducible and $\widetilde{\mathbf{X} \setminus A}$ topologically $(n-2)$-complete, by 5.11, the vector space $H_1^{cld}(\widetilde{X \setminus A}) \cong H_1^{cld}(\tilde{X}, \pi^{-1}A)$ vanishes by 5.10. The vertical homomorphisms are induced by finite surjections and thus surjective.

Every irreducible component of X determines one connected component of $\tilde{X}$, wich, in turn, characterizes one connected component of $\pi^{-1}(A)$, which has a connected image in A ∎

9.4 Corollary(Bertini) *Let X be of pure dimension and A an analytic subset in general position. Then A has the same number of compact irreducible components as X if $\mathbf{X}\setminus A$ is topologically $(\dim X - 2)$-complete with respect to R. If, moreover, X is irreducible at every point, then $H^1_c(X,R) \to H^1_c(A,R)$ is injective.*

Proof. Since A is a subpseudomanifold, we may use intersection cohomology. By 6.1, the obstruction group $\mathbb{H}^1_c(X \setminus A; \mathcal{P}^\bullet_{\mathrm{m}}R)$ vanishes; hence, the natural homomorphism

$$H^0_c(\tilde{X},R) \cong \mathbb{H}^0_c(X,\mathcal{P}^\bullet_{\mathrm{m}}R) \to \mathbb{H}^0_c(A,\mathcal{P}^\bullet_{\mathrm{m}}R) \cong H^0_c(\pi^{-1}A,R)$$

is an isomorphism. Since $H^0(\tilde{X}_c,R) \cong H^0_c(\tilde{X},R)$ and since, in the same way, $H^0_c(\pi^{-1}A,R) \cong H^0(\pi^{-1}A_c,R)$, the claim for the irreducible components follows immediately. Moreover, the homomorphism

$$\mathbb{H}^1_c(X,\mathcal{P}^\bullet_{\mathrm{m}}R) \to \mathbb{H}^1_c(A,\mathcal{P}^\bullet_{\mathrm{m}}R)$$

is injective. Since, for X irreducible at every point, the comparison homomorphism $H^1_c(X,R) \to H^1_c(A,R)$ is injective, also the second claim is correct ∎

References

[AnGr] Andreotti, Aldo and Hans Grauert: Théorèmes de finitude pour la cohomologie des espaces complexes. Bull. Soc. Math. France 90, 193 – 259 (1962)

[BaDi] Barthel, Gottfried, and Alexandru Dimca: On Complex Projective Hypersurfaces which are Homology - $\mathbb{P}_n$'s. Max - Planck - Institut Bonn, MPI 89/53 (1989)

[Bo] Borel, Armand et al.: Intersection Cohomology. Birkhäuser, Boston, Basel, Stuttgart (1984)

[Bl] Ballico, E.: Complements of analytic subvarieties and q-complete spaces. Lincei Rend. Sc. fis. mat. re.nat. 71, 60 – 65 (1981)

[Br] Bredon, Glen E.: Sheaf Theory. McGraw Hill, New York (1967)

[DiFo] Diederich, Klas, and J.E. Fornaess: Smoothing q-convex Functions in the Singular Case. Math. Ann. 273, 665 – 671 (1986)

[FiKp] Fieseler, Karl-Heinz and Ludger Kaup: Vanishing Theorems for the

Intersection Homology of Stein Spaces. Math. Zeitschr. 197, 153 - 176 (1988)

[FiKp$_2$] Fieseler, Karl-Heinz and Ludger Kaup: Theorems of Lefschetz Type in Intersection Homology I. The Hyperplane Section Theorem. Rev. Roum. Math. Pures Appl. 33, 175 - 195 (1988)

[FiKp$_3$] Fieseler, Karl-Heinz and Ludger Kaup: Intersection Cohomology of q-complete complex spaces. Proc. Conf. Algebraic Geometry Berlin 1985, 83 - 104. Leipzig: Teubner 1986

[FiKp$_4$] Fieseler, Karl-Heinz and Ludger Kaup: On the hard Lefschetz Theorem in intersection homology for complex varieties with isolated singularities. Aequ. Math. 34, 240 - 263 (1987)

[Fr] Fritzsche, Klaus: Pseudoconvexity Properties of Complements of Analytic Subvarieties. Math. Ann. 230, 107 - 122 (1977)

[Fr] Fritzsche, Klaus: Convexity of Complements of Analytic Subvarieties. Manuscript (1990)

[Fu] Fulton, William: On the Topology of Algebraic Varieties. Proc. of Symp. in Pure Math. 46, 15 - 46 (1987)

[FuLa] Fulton, William and Robert Lazarsfeld: Connectivity and its applications in algebraic geometry. Algebraic Geometry. LNM 862, 26 - 92. Berlin Heidelberg New York: Springer 1981

[Gn] Goldstein, N.: Ampleness and Connectedness in complex G/P. Trans. AMS 274, 361 - 373 (1982)

[GoMPh] Goresky, Mark and Robert MacPherson: Intersection Homology Theory. Topology 19, 135 - 162 (1980)

[GoMPh$_2$] Goresky, Mark and Robert MacPherson: Intersection homology II. Invent. Math. 71, 77 - 129 (1983)

[GoMPh$_3$] Goresky, Mark and Robert MacPherson: Stratified morse theory. Springer Verlag Berlin, Heidelberg, New York 1988

[GrRe] Grauert, Hans, Reinhold Remmert: Theorie der Steinschen Räume. Springer Verlag Berlin, Heidelberg, New York 1977

[Ha] Hamm, Helmut: Zum Homotopietyp Steinscher Räume. J. Reine Angew. Math. 338, 121 - 135 (1983)

[Ht] Hartshorne, Robin: Cohomological dimension of algebraic varietes. Ann. Math. 88, 403 - 450 (1968)

[Ht$_2$] Hartshorne, Robin: Ample Subvarieties of Algebraic Varieties. LNM 156, Berlin-Heidelberg-New York (1970)

[KiBaKp] Kilambi, Srinivasacharyulu, Gottfried Barthel and Ludger Kaup: Sur la topologie des surfaces complexes compactes. SMS vol. 80. Montréal: Les Presses de l'Université de Montréal (1982)

[Kp] Kaup, Ludger: Eine topologische Eigenschaft Steinscher Räume. Nachr. Akad. Wiss. Göttingen Math. Phys. Klasse 8, 213 - 224 (1966)

[Kp$_2$] Kaup, Ludger: Exakte Sequenzen für globale und lokale Poincaré-Homomorphismen. In: Real and Complex Singularities, Oslo 1976, P. Holm editor; 267 - 296 Sijthoff & Nordhoff Int. Publ. 1976

[Kp$_3$] Kaup, Ludger: Poincaré Dualität für Räume mit Normalisierung. Ann. Sc. Norm. Sup. di Pisa, Classe de Sc. 26, 1 - 31 (1972)

[Kp$_4$] Kaup, Ludger: Zur Homologie projektiv algebraischer Varietäten. Ann. Sc. Norm. Sup. di Pisa, Classe de Sc. 26, 479 - 513 (1972)

[KpFi] Kaup, Ludger and Karl-Heinz Fieseler: Singular Poincaré Duality and Intersection Homology. In: Proceedings of the 1984 Vancouver Conference in Algebraic Geometry. AMS, Providence (1986)

[Lü] Lübke, Martin: Beweis einer Vermutung von Hartshorne für den Fall homogener Mannigfaltigkeiten. J. Reine Angew. Math. 316, 215 - 220 (1980)

[Ok] Okonek, Christian: Barth - Lefschetz Theorems for Singular Spaces. J. Reine Angew. Math. 374, 24 - 38 (1987)

[Pe] Peternell, Matthias: Algebraische Varietäten und q-vollständige komplexe Räume. Math. Z. 200, 547 - 581 (1989)

[Pe$_2$] Peternell, Matthias: Continuous q-convex exhaustion functions. Invent. Math. 85, 249 - 262 (1986)

Fakultät für Mathematik der Universität
Postfach 5560, D 203
D-78434 Konstanz (Germany)
e-mail address: makaup@nyx.uni-konstanz.de

VOLUMES AND LATTICE POINTS-
PROOF OF A CONJECTURE OF L. EHRENPREIS

BEN LICHTIN

Introduction.

A classical result, cf. [La-1], for a positive definite quadratic form Q on $\mathbb{R}^n, n \geq 5$, states the following. For $t > 0$ set

$$\mathcal{N}_Q(t) = \#\{m \in \mathbb{Z}^n : Q(m) \leq t\}$$
$$\mathcal{V}_Q(t) = \int_{Q \leq t} dx_1 \cdots dx_n .$$

Then

$$\mathcal{N}_Q(t) - \mathcal{V}_Q(t) = O(\mathcal{V}_Q(t)/t^{\frac{n}{n+1}}) \ \ as \ t \to \infty. \tag{0.1}$$

Furthermore, it is simple to see that $\mathcal{V}_Q(t) = At^{n/2}$, where

$$A = \int_{Q \leq 1} dx_1 \cdots dx_n.$$

This is to be understood as a significant improvement over the trivial error estimate of $O(\mathcal{V}_Q(t)/t^{\frac{1}{2}})$, obtained solely because of the homogeneity of Q and compactness of $\{Q \leq 1\}$.

The argument of Landau relies heavily upon the functional equation satisfied by the quadratic theta function. Thus, there is a natural extension of this lattice point problem for which the methods used to prove (0.1) do not apply. Moreover, even the elementary argument that gives the error $O(t^{\frac{n-1}{2}})$ does not seem to extend to the following "weighted lattice point problem".

Let φ denote a rational function whose denominator is never zero on $\mathbf{R}^n$. Set

$$\mathcal{N}_Q(t, \varphi) = \sum_{\{m \in \mathbf{Z}^n : Q(m) \leq t\}} \varphi(m)$$
$$\mathcal{V}_Q(t, \varphi) = \int_{Q \leq t} \varphi dx_1 \cdots dx_n.$$

Supported in part by NSA grant MDA904-91-H-0002

Problem. Describe the behavior of $\mathcal{N}_Q(t,\varphi) - \mathcal{V}_Q(t,\varphi)$. In particular, show the existence of $\theta > 0$ so that

$$\mathcal{N}_Q(t,\varphi) - \mathcal{V}_Q(t,\varphi) = O(\mathcal{V}_Q(t,\varphi)/t^\theta) \quad \text{as } t \to \infty.$$

This is a special example of a much more general problem that one can hope to study. Given an unbounded semialgebraic set $\mathcal{S} \subsetneq \mathbf{R}^n$ with positive Lebesgue measure, polynomial P that is proper on $\mathcal{S}$, and rational function φ, defined on $\mathcal{S}$, define

$$N_P(t,\varphi,\mathcal{S}) = \sum_{\{m \in \mathbf{Z}^n \cap \mathcal{S} : |P(m)| \le t\}} \varphi(m)$$

$$V_P(t,\varphi,\mathcal{S}) = \int_{\{|P| \le t\} \cap \mathcal{S}} \varphi \, dx_1 \cdots dx_n \, .$$

Problem. Analyze the asymptotic of $N_P(t,\varphi,\mathcal{S}) - V_P(t,\varphi,\mathcal{S})$ as $t \to \infty$.

This general problem appears to be difficult. On the other hand, if $\mathcal{S}$ is a "box" like $[1,\infty)^n$ then some success has been achieved. The underlying reason for this is the existence of an integral representation for $N_P(t,\varphi,\mathcal{S})$ as the Mellin transform of a meromorphic function whose first pole and growth at infinity are understood reasonably well. The first significant result along these lines was given by Mahler when P satisfied an ellipticity condition on $[0,\infty)^n$, and φ was a quotient of polynomials, each of which was elliptic on $[0,\infty)^n$.

One says that $P \in \mathbf{R}[x_1,\ldots,x_n]$ is elliptic on $[0,\infty)^n$ if the top degree term appearing in P is not zero on $[0,\infty)^n - \{(0,\ldots,0)\}$. Let d denote the degree of P and δ the difference of degrees of the numerator and denominator of φ, assuming their gcd equals 1. Define

$$N_P(t,\varphi) = \sum_{\{m \in \mathbf{N}^n : |P(m)| \le t\}} \varphi(m)$$

$$V_P(t,\varphi) = \int_{\{|P| \le t\} \cap [1,\infty)^n} \varphi \, dx_1 \cdots dx_n \, .$$

Mahler [Ma] investigated the growth in t of $N_P(t,\varphi) - V_P(t,\varphi)$. Assuming ellipticity of numerator and denominator of φ, he showed the existence of $A, \theta > 0$ such that

$$N_P(t,\varphi) = At^{\frac{n+\delta}{d}}(1 + O(t^{-\theta})) \quad \textit{and} \quad V_P(t,\varphi) = At^{\frac{n+\delta}{d}}(1 + O(t^{-\theta}))\, .$$

Thus, one concludes the existence of $\theta > 0$ so that

$$N_P(t,\varphi) - V_P(t,\varphi) = O(V_P(t,\varphi)/t^\theta). \tag{0.2}$$

Moreover, an estimate for the best possible θ is given by $\theta \geq 1/nd$.

If however, one is interested *only* in the standard lattice point problem, in which $\varphi \equiv 1$, then an elementary argument will show (0.2). Whenever, $\varphi \equiv 1$, one denotes $N_P(t,\varphi)$ resp. $V_P(t,\varphi)$ by $N_P(t)$ resp. $V_P(t)$.

Recently, Sargos [Sa-1] addressed these questions for polynomials satisfying a nondegeneracy condition on $[1,\infty)^n$. This is formulated in terms of the Newton polyhedron of P at infinity. Recall that if S denotes the support of P then the Newton polyhedron at infinity is the boundary of the convex hull of $\bigcup_{I\in S}(I - \mathbf{R}^n_+)$. Given P, its polyhedron Γ_∞, and the set of vertices $\mathcal{V}$ of Γ_∞, one first defines the polynomial $P_{\Gamma\infty}$ by

$$P_{\Gamma\infty}(x_1,\ldots,x_n) = \sum_{(i_1,\ldots,i_n)\in\mathcal{V}} x_1^{i_1}\cdots x_n^{i_n}.$$

P is said to be nondegenerate with respect to Γ_∞, if there exists $C > 0$ such that

$$P(x_1,\ldots,x_n) > CP_{\Gamma\infty}(x_1,\ldots,x_n) \quad \text{for all } (x_1,\ldots,x_n)\in[1,\infty)^n.$$

Evidently, this condition, which generalizes the property that P has positive coefficients, is considerably weaker than ellipticity.

No elementary argument appears to be available to establish (0.2) when $\varphi \equiv 1$ and P is nondegenerate. Indeed, it is not in general true that $N_P(t)$ and $V_P(t)$ agree up to a strictly lower order in t. For example, if $a \neq b$ are positive integers, then the polynomial $P = x_1^a x_2^b$ is nondegenerate but $N_P(t)$ and $V_P(t)$ do not have the same dominant term as $t\to\infty$, as a simple calculation will verify.

The first result in [ibid] determines the precise rate of growth of $N_P(t), V_P(t)$, assuming nondegeneracy over $[1,\infty)^n$. Sargos showed:

THEOREM. *There are numbers ρ_0,ν, expressible in terms of the polyhedron of P at infinity, and positive numbers A, B such that*

$$N_P(t) = At^{\rho_0}log^\nu t\,(1+O(1/logt)),$$
$$\text{and}\quad V_P(t) = Bt^{\rho_0}log^\nu t\,(1+O(1/logt)).$$

The constants A, B are given by explicit expressions. Moreover, the values of ρ_0 and ν are expressed in terms of Γ_∞.

The second major result of [ibid] showed:

THEOREM. *Assume that the diagonal intersects the Newton polyhedron at infinity of $P(x_1, \ldots, x_n)$ in compact faces only. Further, assume that P is nondegenerate with respect to this polyhedron. Then there exists $\theta > 0$ so that*

$$N_P(t) - V_P(t) = O(t^{\rho_0 - \theta}). \tag{0.3}$$

An extension of (0.3), involving weights determined by polynomials with positive coefficients, is also proved in [ibid]. Presumably, if φ is a rational function whose numerator and denominator are nondegenerate with respect to their polyhedra at infinity, (0.3) continues to hold. However, it is not yet known if one can give an "elementary" argument of (0.3) when $\varphi \equiv 1$. These would be interesting questions to answer.

The goal of this paper is to investigate this subject for a third class of polynomials. These are polynomials hypoelliptic on $[1, \infty)^n$, cf. Section 1 for a definition. Elliptic polynomials on $[0, \infty)^n$ are certainly hypoelliptic on $[1, \infty)^n$. However, the converse is definitely not true. On the other hand, the precise relation between the nondegeneracy condition used in [Sa-1] and hypoellipticity is not yet completely understood. It would be interesting to characterize precisely the nondegenerate polynomials which must also be hypoelliptic on $[1, \infty)^n$, or on $[a, \infty)^n$ for some $a \in (0, 1)$. To this end, it is useful to observe that Hörmander [Hö, ch. 11] found a class of polynomials that were hypoelliptic on $[1, \infty)^n$ but degenerate with respect to their polyhedra at infinity.

As with elliptic polynomials, one can give an elementary argument that proves

THEOREM A. *If P is hypoelliptic on $[1, \infty)^n$, then there exists $\theta > 0$ such that*

$$N_P(t) - V_P(t) = O(V_P(t)/t^\theta) \quad \text{as } t \to \infty.$$

This is given at the end of Section 2. The author expresses his appreciation to Prof. V. I. Arnold for emphasizing the probable existence of a *simple* argument that proves Theorem A. Estimates for the smallest possible rate of growth of the error term, expressed in terms of the geometry of P would be very interesting to know, cf. [CdV], [Ra], [Va-1]. In particular, they would probably enable one to refine the estimates given in [Sa-2], by incorporating the geometry of the level sets of P more explicitly into the error estimation.

However, now assume that φ is a rational function whose numerator and denominator, in reduced form, are both hypoelliptic on $[1,\infty)^n$. Defining $N_P(t,\varphi)$, $V_P(t,\varphi)$ as above, it is natural to ask how well $N_P(t,\varphi)$ approximates $V_P(t,\varphi)$ (or vice versa) as $t \to \infty$. For the analysis of this weighted lattice point problem, the simple argument, used in the proof of Theorem A, no longer suffices, and one apparently needs to use a subtler analytical argument. To formulate the answer given here, one should first recall the

THEOREM. *There exist $\rho_1(\varphi), \lambda_1(\varphi), \theta > 0$ and nonzero polynomials $A_1(\varphi,u)$, $B_1(\varphi,u) \in \mathbf{R}[u]$ so that*

$$N_P(t,\varphi) = t^{\rho_1(\varphi)} A_1(\varphi, logt)\,(1+O(t^{-\theta})) \qquad [Li-3]\,,$$
$$V_P(t,\varphi) = t^{\lambda_1(\varphi)} B_1(\varphi, logt)\,(1+O(t^{-\theta})) \qquad [Ni]\,.$$

The first main result of this paper is

THEOREM B (CONJECTURE OF EHRENPREIS). *If P is hypoelliptic on $[1,\infty)^n$, and φ is a quotient of hypoelliptic polynomials on $[1,\infty)^n$, then*

$$\rho_1(\varphi) = \lambda_1(\varphi) \quad \text{and} \quad A_1(\varphi,u) = B_1(\varphi,u).$$

Set $\rho(\varphi)$ to be the number denoted $\rho_1(\varphi), \lambda_1(\varphi)$. One concludes immediately

COROLLARY. *There exists $\theta > 0$ such that*

$$N_P(t,\varphi) - V_P(t,\varphi) = O(t^{\rho(\varphi)-\theta}) \text{ as } t \to \infty.$$

Theorems A,B answer a question conjectured by Prof. Ehrenpreis to the author in the spring 1990. As noted below, this question has significance beyond the context of lattice point problems involving one polynomial.

There are two immediate, but also interesting, applications of Theorems A, B. Theorem A provides an alternative method for deriving the asymptotic for the number of eigenvalues at most t for a self-adjoint extension of a hypoelliptic PsDO. Theorem B allows one to extend the results of [Li-2] to all hypoelliptic polynomials that are, in addition, "tame" on $\mathbb{C}^n$ in the sense of [Br]. Thus, the main term in the asymptotic for $N_P(t, \varphi)$ is again shown to be a "cohomological invariant" whenever φ is a hypoelliptic polynomial on $[1, \infty)^n$. This leads to further evidence of a topological "local-global" principle that describes the contribution to $N_P(t, \varphi)$ of the singularities of P in $\mathbb{C}^n$. This principle may be viewed as an analogue of an arithmetic local-global principle that determines when the singular series, arising in the Hardy-Littlewood-Vinogradov analysis (cf. [Dav]) of the counts $\#\{P = t\} \cap \mathbb{N}^n$ as $t \to \infty$, possesses a well-defined rate of growth as $t \to \infty$. A third application is discussed below.

An outline of this paper follows. Sections 1-3 briefly recall some needed technical preliminaries for the proof of Theorem B. In addition to the proof of Theorem A, Section 2 also includes the statement of a Theorem C, from which Theorem B follows immediately. Section 4 then gives the proof of Theorem 4.1 which implies Theorem C as an immediate corollary. Section 5 discusses the application to the analysis of the distribution of the eigenvalues for a hypoelliptic PsDO. Section 6 discusses the extension of [Li-2] referred to above.

Because the methods used in this paper combine general analytic and geometric techniques, they provide a convenient framework to help analyze the asymptotics of a "simultaneous" lattice point problem not previously addressed in the literature. Section 7 of this article proves an analogue of Theorem 4.1, and therefore of Theorem C, when one treats $k > 1$ hypoelliptic polynomials at the same time. Some very interesting questions arise in the context of this problem, for which ideas from the analytic theory of singularities appear essential. The main results of the section are contained in Theorems 7.8 and Theorem D.

Theorem D is of basic importance and should be considered as the second main result of this paper. It determines a geometric criterion, formulated precisely in Theorem 7.14, which if satisfied, implies that the estimates, given in [Li-5, secs. 6,7] for a simultaneous lattice point problem, can indeed be sharpened to precise and effectively determined asymptotics. Accomplishing this would amount to a considerable extension of a well known theorem of Landau for Dirichlet series in one complex variable with positive coefficients (cf. 7.20).

It appears to be difficult to show that the criterion holds in general. However, evidence, that is currently being accumulated, supports the conjecture that the criteron is satisfied in general. 7.18 gives the most general result that can be proved so far. As the reader will conclude, considerable room for improvement remains.

A method of analysis, needed to determine when the conditions in 7.14 hold, involves a natural extension of the techniques used to find the largest pole of the generalized functions $P_{\pm}^{s}$ in terms of data arising from a resolution of singularities, cf. [Ig], [Li-7], [Lo-1]. In the k variable case, an important goal of the analysis is to determine explicitly a certain Newton polyhedron, contained in $\mathbb{R}_{+}^{k}$, that is the boundary of a convex envelope determined by resolution data, cf. 7.7 for the precise definition. This polyhedron is the natural analogue of the "largest pole" for series in one variable.

When the weight function φ is identically 1, [Li-6] will carry out this analysis for classes of *pairs* of hypoelliptic polynomials, whose resolution data enables one to find the precise description of the Newton polygon. With this in hand, the criteria in 7.14 can always be shown to hold. This implies that the vertices appearing in the statement of 7.18 will be "initial vertices" in the sense of 7.12. Such results should therefore be viewed as the first known nontrivial examples of multivariable Tauberian theorems with exact asymptotics [Ga]. Additional examples, of greater interest to number theory, can be obtained by studying local zeta functions in several variables over p−adic fields, cf. [Ig] for the one variable case. This will be discussed in a different article. These applications to several variable Dirichlet series represent a third, and perhaps unexpectedly significant, feature of Ehrenpreis' conjecture. In particular, one sees in this way how Ehrenpreis' conjecture is related to a question asked by Igusa in [ibid, pg. 32], concerning multivariable asymptotic expansions.

The type of problem investigated in Section 7 is also related to the work of Loeser in [Lo-2]. Essentially, Theorem D implies that the asymptotics of the simultaneous lattice point problem, discussed here, reduce to the description of the asymptotics of volume integrals

$$\int_{\{P_1 \le t_1, \dots, P_k \le t_k\} \cap [1,\infty)^n} \varphi \, dx_1 \cdots dx_n,$$

where $P_1, \dots, P_k$ are hypoelliptic polynomials. Thus, it would be quite interesting to extend the results of [ibid] to derive the asymptotics of the corresponding fiber integrals

$$\int_{\{P_1=t_1,\ldots,P_k=t_k\}\cap[1,\infty)^n} \frac{\varphi dx_1\cdots dx_n}{dP_1\cdots dP_k}$$

as $t_1,\ldots,t_k \to \infty$. Undoubtedly, this would lead to a significant extension of Theorem 7.17. However, it still remains unclear whether, by such methods, one could show that the vertices in 7.18 must be initial.

Notation. This paper will use the following notations.

(1) For $A=(A_1,\ldots,A_n)\in \mathbf{R}^n$, one sets $|A| = A_1+\ldots+A_n$.

(2) $s=\sigma+it$ denotes a complex variable. $\mathbf{s}=(s_1,\ldots,s_k)\in \mathbf{C}^k$. It is sometimes useful to write $\mathbf{s}=\boldsymbol{\sigma}+i\mathbf{t}$ to emphasize the real and imaginary parts.

(3) If $x=(x_1,\ldots,x_n)$, then $\|x\| = max\,\{|x_i|\}$.

(4) Set $dx = dx_1\cdots dx_n$.

(5) The interior of a set A is denoted $int(A)$.

(6) For a function $g: X\to \mathbf{R}$ define

$$g_+(x) = \begin{cases} g(x) \text{ if } g(x)>0 \\ 0 \text{ if } g(x)\le 0 \end{cases}$$

$$g_-(x) = \begin{cases} -g(x) \text{ if } g(x)<0 \\ 0 \text{ if } g(x)\ge 0 \end{cases}$$

(7) If φ is a weight function and P is a polynomial,

$$N_P(t,\varphi) = \sum_{\substack{m\in\mathbf{N}^n\\ P(m)\le t}} \varphi(m),$$

$$V_P(t,\varphi) = \int_{\{P\le t\}\cap[1,\infty)^n} \varphi\, dx_1\cdots dx_n\,.$$

When $\varphi\equiv 1$, one denotes these functions as $N_P(t), V_P(t)$.

Section 1. *Some properties of hypoelliptic polynomials*

Recall that $P\in \mathbf{R}[x_1,\ldots,x_n]$ is hypoelliptic on $[1,\infty)^n$ if for all differential monomials D_x^A one has

$$\lim_{\substack{\|x\|\to\infty\\ x\in[1,\infty)^n}} \frac{D_x^A P(x)}{P(x)} = 0\,. \tag{1.1}$$

Hörmander [Hö, ch. 11] has shown that (1.1) is equivalent to

$$\textit{There exist } c, C>0 \textit{ such that for all } x\in[1,\infty)^n \quad \left|\frac{D_x^A P(x)}{P(x)}\right| \le C\|x\|^{-c|A|}\,. \tag{1.2}$$

Further, he also observed that (1.2) is equivalent to the satisfaction of (1.1) for all A with $|A| = 1$.

Definition 1.3. The exponent of hypoellipticity for P is the largest c so that 1.2 holds. It will be denoted c_P in the following. ∎

The consequence of (1.2) of use in this paper is

PROPOSITION 1.4. *There exist $\alpha, C, D > 0$ such that*

$$|P(x_1,\ldots,x_n)| > C(x_1\cdots x_n)^\alpha$$

$$\text{if } (x_1,\ldots,x_n) \in [1,\infty)^n \cap \{\|(x_1,\ldots,x_n)\| \geq D\}. \tag{1.5}$$

The best α so that 1.5 holds is denoted α_P in the following.

Note. One will assume in the rest of the paper, without loss of generality, that P is positive on $[1,\infty)^n \cap \{\|(x_1,\ldots,x_n)\| \geq D\}$. Then

$$\textit{there exists } b \geq 0 \textit{ such that } P + b > 0 \quad \textit{on } [1,\infty)^n. \tag{1.6}$$

Remark 1.7. If it is in fact necessary to use a positive b to insure (1.6), then one will replace P by $P + b$ below. However, for simplicity, this translation of P will continue to be denoted as P throughout the rest of the paper. As a result, one will assume throughout the rest of the article that P is positive on $[1,\infty)^n$.

The effect of the translation will not at all affect the conclusions in Theorems A-D. The reason for this follows from the properties stated in Corollaries 2.2 and 2.4, which have been proved for arbitrary P hypoelliptic on $[1,\infty)^n$, and the easily established statements given below. In these, the notations from 2.2 and 2.4 are used.

$$N_P(t,\varphi) - N_{P+b}(t,\varphi) = N_P(t,\varphi) - N_P(t-b,\varphi) = O(t^{\rho_1(\varphi)-\theta}) \text{ for some } \theta > 0$$
$$V_P(t,\varphi) - V_{P+b}(t,\varphi) = O(t^{\lambda_1(\varphi)-\theta}). \quad ∎$$

A simple calculation will also show the following. Assume that $\varphi = Q/T$, with $gcd(Q,T) = 1$. If Q, T are both hypoelliptic on $[1,\infty)^n$, then for any nonzero index A one has

$$\lim_{\substack{\|x\|\to\infty \\ x\in[1,\infty)^n}} \frac{D_x^A\varphi(x)}{\varphi(x)} = 0\,. \tag{1.8}$$

A fraction $\varphi = Q/T$ with Q, T hypoelliptic polynomials will be called a "hypoelliptic fraction".

Remarks 1.9.

(1) In order to avoid unnecessary complications in the discussion below, one will also assume in the rest of the article that $T(x) \neq 0$ for all $x \in [1,\infty)^n$. The industrious reader will easily be able to modify the arguments in the event that T is allowed to vanish on at most a compact subset of $[1,\infty)^n$. Since Q, T cannot, in general, vanish outside a compact subset of $[1,\infty)^n$, the sign of φ is constant outside such a set. Without loss of generality, one will assume that φ is positive.

(2) The careful reader of [Li-1,3], [Ni], will conclude that the proofs of Corollaries 2.2 and 2.4 extend to include hypoelliptic fractions φ. ∎

Section 2. *A Dirichlet series determined by φ and P*

Given P, φ satisfying the properties discussed in Section 1, define

$$D(s,\varphi) = \sum_{m\in\mathbf{N}^n} \frac{\varphi(m)}{P(m)^s}$$

$$I(s,\varphi) = \int_{[1,\infty)^n} \frac{\varphi}{P^s}\, dx\,.$$

The main results of [Li-1,3,4] and [Ni] established the

THEOREM 2.1.

(i) There exists $B > 0$ such that if $\sigma > B$ then $D(s,\varphi), I(s,\varphi)$ are analytic.

(ii) $D(s,\varphi), I(s,\varphi)$ admit analytic continuations to $\mathbf{C}$ as meromorphic functions with polar locus contained in finitely many arithmetic progressions of rational numbers.

Let $\rho_1(\varphi)$ resp. $\lambda_1(\varphi)$ denote the largest pole of $D(s,\varphi)$ resp. $I(s,\varphi)$.

(iii) There exists $A > 0$ so that for each $\tau > 0$, and $\sigma_1 < \sigma_2 \leq \rho_1(\varphi)$ there exists $C = C(\tau, \sigma_1, \sigma_2)$ such that

$$|D(s,\varphi)| < C(1 + |t|^{A(\rho_1(\varphi)-\sigma)+\tau})$$

if $\sigma \in [\sigma_1, \sigma_2]$ *and* $|t| \geq 1$.

(iv) For any polynomial $B(s)$, *and* $\sigma_1 < \sigma_2$ *there exists* $C > 0$ *such that*

$$|B(s)\, I(s, \varphi)| < C, \text{ for all } \sigma \in [\sigma_1, \sigma_2] \text{ and } |t| \geq 1.$$

A tauberian argument, due to Landau [La-2], uses (i-iii) of (2.1) to prove:

COROLLARY 2.2. *Let* $\rho_1(\varphi) > \rho_2(\varphi) > \cdots > \rho_\ell(\varphi) > \rho_1(\varphi) - \frac{1}{A} \geq \rho_{\ell+1}(\varphi) \cdots$ *be the first* $\ell + 1$ *poles of* $D(s, \varphi)$. *Then there exist nonzero polynomials* $A_1(\varphi, u), \ldots, A_\ell(\varphi, u) \in \mathbb{R}[u]$ *such that*

$$N_P(t, \varphi) = \sum_{i=1}^{\ell} t^{\rho_i(\varphi)} A_i(\varphi, log t) + O_\epsilon(t^{\rho_1(\varphi) - \frac{1}{A} + \epsilon}) \quad \text{as } t \to \infty.$$

Remark 2.3. An estimate for the smallest possible A has been given in two cases. In [Sa-2], Sargos showed that if P has positive coefficients then one can choose $A = deg P$. Moreover, this is an optimal (i.e. the smallest) estimate when taken over all polynomials with positive coefficients. If P is hypoelliptic, then [Li-4] showed that one can choose $A = n\, deg P$. However, this is not an optimal estimate, as Sargos' example indicates. It is not clear what is an optimal estimate for A if the degree of P is at least two. ∎

A standard argument uses (i), (ii), (iv) of (2.1) to give a complete asymptotic expansion for $V_P(t, \varphi)$:

COROLLARY 2.4. *Let* $\lambda_1(\varphi) > \lambda_2(\varphi) > \cdots$ *be the poles of* $I(s, \varphi)$. *Then there exist nonzero polynomials* $B_1(\varphi, u), B_2(\varphi, u), \ldots, \in \mathbb{R}[u]$ *such that*

$$V_P(t, \varphi) = \sum_{i=1}^{\infty} t^{\lambda_i(\varphi)} B_i(\varphi, log t) \quad \text{as } t \to \infty.$$

One next observes the following identities. For b sufficiently large,

$$N_P(t, \varphi) = \frac{1}{2\pi i} \int_{\sigma = b} t^s\, D(s, \varphi) \frac{ds}{s}, \qquad (2.5)$$
$$V_P(t, \varphi) = \frac{1}{2\pi i} \int_{\sigma = b} t^s\, I(s, \varphi) \frac{ds}{s}.$$

Theorem B now follows by combining (2.5), (iii), (iv) of (2.1), a standard application of residue theory, and the following:

THEOREM C.
(i) $\rho_1(\varphi) = \lambda_1(\varphi)$.

Let $\rho(\varphi)$ *denote the common first pole of* $D(s,\varphi), I(s,\varphi)$.

(ii) $D(s,\varphi) - I(s,\varphi)$ *is analytic at* $s = \rho(\varphi)$.

Theorem C, in turn, follows immediately from the proof of Theorem 4.1.

Remark 2.6. Indeed, Theorem C implies that the principal parts of $D(s,\varphi)$, $I(s,\varphi)$ agree at the first pole $s = \rho(\varphi)$. The polynomials $A_1(\varphi,u), B_1(\varphi,u)$, appearing in the statement of Theorem B, are uniquely determined by the principal parts. Thus, Theorem C implies $A_1(\varphi,u) = B_1(\varphi,u)$. ∎

Of course, Theorem A will also follow from Theorem C. However, one can give an elementary argument that uses only (2.2), (2.4) when $\varphi \equiv 1$.

PROOF OF THEOREM A: Let $\epsilon > 0$ and c_P the hypoellipticity exponent of (1.3). Define the sets

$$\mathcal{U}_\epsilon(t) = \{x \in [1,\infty)^n : P(x) \leq t + t^{1-\epsilon c_P}\},$$
$$\mathcal{L}_\epsilon(t) = \{x \in [1,\infty)^n : \|x\| \geq t^\epsilon - \frac{1}{2} \text{ and } P(x) \leq t - t^{1-\epsilon c_P}\},$$
$$\ell_\epsilon(t) = \{x \in [1,\infty)^n : \|x\| \leq t^\epsilon - \frac{1}{2} \text{ and } P(x) \leq t - t^{1-\epsilon c_P}\},$$
$$\mathcal{C}_\epsilon(t) = \bigcup_{\substack{m \in \mathbb{N}^n \\ \|m\| \geq t^\epsilon \\ P(m) \leq t}} C(m),$$

where

$$C(m) = \{x : |x_i - m_i| < 1/2, \text{ for each i}\}.$$

The following is easily verified.

$$vol(\mathcal{L}_\epsilon(t)) + vol(\ell_\epsilon(t)) = vol\left(\{x \in [1,\infty)^n : P(x) \leq t - t^{1-\epsilon c_P}\}\right), \quad (2.7.1)$$

$$vol(\ell_\epsilon(t)) \leq vol\left(\{x \in [1,\infty)^n : \|x\| \leq t^\epsilon - 1/2\}\right) \leq C(t^\epsilon - 1/2)^n$$
$$\text{for some } C > 0. \quad (2.7.2)$$

In (2.4), when $\varphi \equiv 1$, drop the "φ" as an argument of the exponents λ_i and polynomials $B_i(u)$. Then, setting $\beta = \lambda_1 - \lambda_2$, one has

$$\begin{aligned} vol(\mathcal{U}_\epsilon(t)) &= \sum_{i=1}^{\infty} (t + t^{1-\epsilon c_P})^{\lambda_i} B_i\left(log\,(t + t^{1-\epsilon c_P})\right) \\ &= (t + t^{1-\epsilon c_P})^{\lambda_1} B_1\left(log\,(t + t^{1-\epsilon c_P})\right) \\ &\quad + O_\kappa((t + t^{1-\epsilon c_P})^{\lambda_1 - \beta + \kappa}) \\ &= t^{\lambda_1} B_1(log\, t) + O_\kappa(t^{\lambda_1 - \epsilon c_P + \kappa}) + O_\kappa(t^{\lambda_1 - \beta + \kappa}) \end{aligned} \tag{2.8.1}$$

and similarly

$$vol(\mathcal{L}_\epsilon(t)) = t^{\lambda_1} B_1(log\, t) + O_\kappa(t^{\lambda_1 - \epsilon c_P + \kappa}) + O_\kappa(t^{\lambda_1 - \beta + \kappa}). \tag{2.8.2}$$

Property (1.2) of hypoellipticity implies, by means of the Taylor expansion of P around each point m, used in the definition of $\mathcal{C}_\epsilon(t)$, that for all t sufficiently large,

$$\mathcal{L}_\epsilon(t) \subset \mathcal{C}_\epsilon(t) \subset \mathcal{U}_\epsilon(t). \tag{2.9}$$

Moreover,

$$vol(\mathcal{C}_\epsilon(t)) = N_P(t) - \nu_\epsilon(t),$$

where

$$\nu_\epsilon(t) = \#\{m \in [1,\infty)^n \cap \mathbf{N}^n : \|m\| \leq t^\epsilon - \frac{1}{2}, \text{ and } P(m) \leq t\}.$$

Clearly,

$$\nu_\epsilon(t) = O(t^{\epsilon n}).$$

Thus, combining this estimate with (2.7)-(2.9) implies

$$N_P(t) = V_P(t) + O(t^{\epsilon n}) + O_\kappa(t^{\lambda_1 - \epsilon c_P + \kappa}) + O_\kappa(t^{\lambda_1 - \beta + \kappa}).$$

Choosing ϵ so that $\epsilon n < \lambda_1 - \epsilon c_P$ then proves Theorem A. ∎

Section 3. *Sketch of analytic continuation of $D(s, \varphi)$*

For the benefit of the reader a brief sketch is now given of the analytic continuation of $D(s, \varphi)$ that uses the Euler-Maclaurin summation formula. More details may be found by consulting [Li-4, Ma, Lnd]. The key point is that hypoellipticity of P, φ allows this summation method to be used to give an analytic continuation to the entire s plane. For certain purposes, such as proving Theorem 4.1, this seems to give a simpler type of integral representation of the series than that provided by Cauchy residues, as has been used in [L-1,3,5].

To begin, one must introduce some convenient notations.

Notation (3.1).
(1) For any $\ell \in \mathbf{N}$ and $\mathbf{C} = (C_1, \ldots, C_\ell) \in (\mathbf{Z}_+^n)^\ell$ set

$$D_x^{\mathbf{C}}(P)/P^\ell = \prod_{i=1}^{\ell} D_x^{C_i}(P)/P .$$

The reader should note that implicit in this notation is the fact that the exponent of P equals the number of n–tuples comprising the components of $\mathbf{C}$.

(2) For each positive integer k set

$$\mathcal{I}'_k = \{I = (i_1, \ldots, i_n) : i_j < k \text{ for all } j\}$$
$$\mathcal{I}''_k = \{I = (i_1, \ldots, i_n) : i_j \le k \text{ for all } j \text{ and } i_j = k \text{ for some j}\}.$$

(3) For any $I \in \mathcal{I}'_k$ and $\ell \in \mathbf{N}$, define

$$\begin{aligned} \mathcal{M}_\ell(I) = \{(B, \mathbf{C}) \in \mathbf{Z}_+^n \times \mathbf{Z}_+^{n\ell} :\ & \text{if } \mathbf{C} = (C_1, \ldots, C_\ell), \\ & \text{then } |C_i| \ge 1 \text{ for each } i, \text{ and} \\ & B + C_1 + \ldots + C_\ell = I\}, \end{aligned}$$
$$\text{and} \quad \mathcal{M}(I) = \cup_\ell \mathcal{M}_\ell(I).$$

It is convenient below to use the notation $\mathbf{C} \in \mathcal{M}(I)$. This means that $C_1 + \cdots + C_\ell = I$ when $\mathbf{C} = (C_1, \ldots, C_\ell)$.

(4) For $i = 0, 1, 2, \ldots$, set $[-s]_i = (-s)(-s-1)\cdots(-s-i)$. ∎

A simple calculation, left to the reader, shows

LEMMA 3.2. *For each positive integer k, index $I \in \mathcal{I}'_k$ and pair $(B, \mathbf{C}) \in \mathcal{M}(I)$ there exist integers $n_{B,\mathbf{C}}$ such that*

$$D_x^I(\varphi/P^s) = \sum_{(B,\mathbf{C}) \in \mathcal{M}(I)} n_{B,\mathbf{C}} \sum_{\ell=1}^{|\mathbf{C}|} [-s]_{\ell-1}\, D_x^B(\varphi) \cdot \frac{D_x^{\mathbf{C}}(P)}{P^\ell} \cdot \frac{1}{P^s} . \quad ∎$$

Using c_P, c_T from (1.2), α_P from (1.4), and Lemma 3.2, one next sees that the conditions satisfied by P and $\varphi = Q/T$ imply

PROPOSITION 3.3. *Set $N = n \deg Q - \alpha_T$. Then*
(a) For each compact subset K of the halfplane $\sigma > (N+2)/\alpha_P$,

$$|\frac{\varphi}{P^s}| = O_K(\|x\|^{-2}), \quad x \in [1,\infty)^n.$$

(b) For each index I and each compact subset K of the halfplane $\sigma > (N + 2 - c_P|I|)/\alpha_P$

$$|D_x^I(\frac{\varphi}{P^s})| = O_K(\|x\|^{-2}), \quad x \in [1,\infty)^n.$$

PROOF: The straightforward verification is given in [Li-4]. ∎

The Euler-Maclaurin summation formula constructs for each $k = 1, 2, \ldots,$

numbers $c_\ell(k)$, $i = 0, 1, \ldots, k-1$, and a periodic C^∞ function $\sigma_k(u)$, where u denotes a coordinate on $\mathbf{R}$ so that if $f(u+iv)$ is any holomorphic function satisfying the property

$$\lim_{u \to +\infty} f^{(i)}(u) = 0, \quad \text{for each } i = 0, 1, \ldots, \tag{3.4}$$

then

$$\sum_{\nu=1}^{\infty} f(\nu) = \int_1^\infty f(u)du + \sum_{i=0}^{k-1} c_i(k) f^{(i)}(1) + \int_1^\infty \sigma_k(u) f^{(k)}(u)\, du\,.$$

The precise values of the $c_i(k)$ and expressions of σ_k are not needed for this paper. The reader can work out their values by consulting [Lnd, pgs. 75-83].

Set, for each $k = 1, 2, \ldots$ and $i = 0, 1, \ldots, k-1$

$$h_i^{(k)}(u) = c_i(k)$$
$$h_k^{(k)}(u) = \sigma_k(u)\,.$$

Proposition 3.3 implies that (3.4) is satisfied in the interval $[1,\infty)$ of the x_i coordinate plane, for each i, for any function of the form $D_x^I(\varphi/P^s)$, whenever σ is sufficiently large. One can therefore first set $k=1$ and iterate the Euler-Maclaurin summation formula n-times to show

THEOREM 3.5. *If $\sigma > (N+2)/\alpha_P$, and $I = (i_1, \ldots, i_n)$, then*

$$D(s,\varphi) = \sum_{i_1=0}^{1} \cdots \sum_{i_n=0}^{1} \int_{[1,\infty)^n} h_{i_1}^{(1)}(x_1) \cdots h_{i_n}^{(n)}(x_n) D_x^I(\frac{\varphi}{P^s})\, dx$$
$$= I(s,\varphi) + \sum_{\substack{i_1,\ldots,i_n=0 \\ I \neq 0}}^{1} \int_{[1,\infty)^n} h_{i_1}^{(1)}(x_1) \cdots h_{i_n}^{(1)}(x_n) D_x^I(\frac{\varphi}{P^s})\, dx. \quad ∎$$

Moreover, it is clear by Proposition 3.3 that for each $I \neq (0, \ldots, 0)$,

$$\int_{[1,\infty)^n} h_{i_1}^{(1)}(x_1) \cdots h_{i_n}^{(1)}(x_n) D_x^I\left(\frac{\varphi}{P^s}\right) dx \quad \textit{is analytic if } \sigma > \frac{N+2-c_P}{\alpha_P}. \tag{3.6}$$

One can then repeat this procedure $k > 1$ times. In this way one proves

THEOREM 3.7. *If $\sigma > (N+2)/\alpha_P$ then*

$$D(s,\varphi) = \sum_{i_1=0}^{k} \cdots \sum_{i_n=0}^{k} \int_{[1,\infty)^n} h_{i_1}^{(k)}(x_1) \cdots h_{i_n}^{(k)}(x_n) D_x^I\left(\frac{\varphi}{P^s}\right) dx. \tag{1}$$

Thus, there exist constants $c(I)$ for each $I \neq (0, \ldots, 0) \in \mathcal{I}'_k$ so that

$$D(s,\varphi) = I(s,\varphi) + \sum_{I \neq (0,\ldots,0) \in \mathcal{I}'_k} c(I) \int_{[1,\infty)^n} D_x^I\left(\frac{\varphi}{P^s}\right) dx \tag{2}$$
$$+ \sum_{I \in \mathcal{I}''_k} \int_{[1,\infty)^n} h_{i_1}^{(k)}(x_1) \cdots h_{i_n}^{(k)}(x_n) D_x^I\left(\frac{\varphi}{P^s}\right) dx.$$

As in (3.6) one observes that for any $I \in \mathcal{I}''_k$

$$\int_{[1,\infty)^n} h_{i_1}^{(k)}(x_1) \cdots h_{i_n}^{(k)}(x_n) D_x^I\left(\frac{\varphi}{P^s}\right) dx \quad \textit{is analytic if } \sigma > \frac{N+2-kc_P}{\alpha_P}. \tag{3.8}$$

Shown in [Li-4] was the

THEOREM 3.9. *For each $I \in \mathcal{I}'_k$ the function*

$$s \to \int_{[1,\infty)^n} D_x^I(\varphi/P^s)\, dx$$

admits an analytic continuation to $\mathbb{C}$ as a meromorphic function with polar locus contained in finitely many arithmetic progressions of rational numbers.

Thus, (3.8), (3.9) imply

(3.10) *Any pole of $D(s,\varphi)$ in the strip*

$$\frac{N+2-kc_P}{\alpha_P} < \sigma \leq \frac{N+2-(k-1)c_P}{\alpha_P}$$

must be a pole of the analytic continuation of $\int_{[1,\infty)^n} D_x^I(\varphi/P^s)\, dx$ for some $I \in \mathcal{I}'_k$. ∎

Notation. Given $I \in \mathcal{I}'_k$ for some k and any $(B, \mathbf{C}) \in \mathcal{M}(I)$ one sets

$$I(s, B, \mathbf{C}, \varphi) = \int_{[1,\infty)^n} D_x^B(\varphi) \cdot \frac{D_x^{\mathbf{C}} P}{P^\ell} \cdot \frac{1}{P^s}\, dx . \quad ∎$$

Remark 3.11. It is useful to write the integrand of this function as

$$\frac{D_x^B(\varphi)}{\varphi} \cdot \frac{D_x^{\mathbf{C}} P}{P^\ell} \cdot \frac{\varphi}{P^s},$$

since (1.2) allows one to interpret the factor of φ/P^s as a function that vanishes at infinity in $[1,\infty)^n$. ∎

Section 4. *Proof of main result*

¿From the description of the analytic continuation of $D(s, \varphi)$ sketched in Section 3, it is clear that Theorem C is a corollary of the following theorem. Using the notations from Section 3 this is

THEOREM 4.1. *Assume P, Q, T are polynomials hypoelliptic on $[1,\infty)^n$. Set $\varphi = Q/T$. Let k, ℓ be positive integers. Assume $I \in \mathcal{I}'_k$ and $(B, \mathbf{C}) \in \mathcal{M}_\ell(I)$. Then the first pole of $I(s, B, \mathbf{C}, \varphi)$ is strictly smaller than the first pole of $I(s, \varphi)$.*

Preliminary remarks. Since the analysis needed to prove Theorem 4.1 is carried out at infinity, it is first necessary to define the following objects.

Definitions/Notations.

1) The chart at infinity in $(P^1\mathbf{R})^n$ will be denoted $(\mathbf{R}^n, (w_1, \ldots, w_n))$. The hyperplane at infinity $\{w_1 \cdots w_n = 0\}$ is denoted H_∞. The notations $1/w$ resp. dw are used to denote the point $(1/w_1, \ldots, 1/w_n)$ resp. the differential $dw_1 \cdots dw_n$.

2) Define the polynomials $G(w), \psi_1, \psi_2$ by these conditions.

$$R(w) =_{def} \frac{1}{P(1/w)} = \frac{w_1^{M_1} \cdots w_n^{M_n}}{G(w_1, \ldots, w_n)}, \quad w_1, \ldots, w_n \nmid G; \tag{4.2}$$

$$\Phi(w) =_{def} \varphi(1/w) = \frac{\psi_1(w)}{w_1^{m_1} \cdots w_n^{m_n} \psi_2(w)}, \quad gcd(w_1 \cdots w_n \cdot \psi_2, \psi_1) = 1 .$$

3) For $\ell \in \mathbf{N}$ and $(B, \mathbf{C}) \in \mathbf{Z}_+^n \times \mathbf{Z}_+^{n\ell}$, define the rational function $\eta_{B,\mathbf{C}}(w)$

$$\eta_{B,\mathbf{C}}(w) = \frac{D_x^B(\varphi)\, D_x^{\mathbf{C}}(P)}{\varphi\, P^\ell}(1/w)\,.$$

Since T is hypoelliptic and assumed to be positive (for simplicity) on $[1,\infty)^n$, $\eta_{B,\mathbf{C}}$ and Φ are defined on $(0,1]^n$. This suffices for the proof of (4.1). ∎

By the assumptions of hypoellipticity for P, Q, T one concludes from (1.2) the existence of $c', C' > 0$ such that for all $w \in (0,1]^n$ and any $\eta_{B,\mathbf{C}}$ defined as in (4.1), one has

$$|\eta_{B,\mathbf{C}}(w_1, \ldots, w_n)| < C' |w_1 \cdots w_n|^{c'}\,.$$

Thus, one concludes for each $p \in \partial[0,1]^n \cap H_\infty$

$$\lim_{\substack{w \to p \\ w \in (0,1]^n}} \eta_{B,\mathbf{C}}(w) = 0\,. \tag{4.3}$$

Note. It will be convenient in the following to fix a particular pair $B, \mathbf{C}$ and drop the subscript $B, \mathbf{C}$ from η whenever there is no possibility of confusion. ∎

The proof of (4.1) is based upon analyzing the integrands in the statement of the theorem, using a resolution of singularities "at infinity".

There exist a nonsingular real algebraic manifold Y and projective morphism $\pi : Y \to (\mathbf{R}^n, (w_1, \ldots, w_n))$ such that the following properties are satisfied.

(4.4)

i) There exists a divisor $\mathcal{D} \subset Y$ so that $\pi : Y - \mathcal{D} \to \mathbf{R}^n$ is an isomorphism onto its image;

ii) $\mathcal{D}$ is a normally crossing divisor. That is, $\mathcal{D} = \cup \mathcal{D}_\alpha$ where each $\mathcal{D}_\alpha$ is smooth and at each point $p \in \mathcal{D}$ the set of divisors containing p are mutually transverse;

iii) The divisor equal to the zero and polar locus defined by

$$[\prod_{i=1}^n (w_i - 1) \cdot \prod_{i=1}^n w_i \cdot R \cdot \Phi \cdot \eta] \circ \pi$$

has support in $\mathcal{D}$ (so that it too is locally normal crossing);

iv) $(0,1)^n \cap \pi(\mathcal{D}) = \emptyset$.

Thus, $(0,1)^n$ is disjoint from the locus of blowing up determined by π.

Next, one takes an open polydisc U containing $[0,1]^n$ in the chart at infinity and sets

$$X = \pi^{-1}(U)$$
$$D = \mathcal{D} \cap X$$
$$B = \overline{\pi^{-1}(0,1)^n} \cap X .$$

An elementary observation is the

LEMMA 4.5.

i) $\partial B \subset D$.
ii) $B \cap D = \partial B$.

PROOF: (i) follows from (4.4)(iii). To verify (ii), one notes that (4.4)(i,iv) imply

$$\pi^{-1}(0,1)^n \cap D = \emptyset.$$

Moreover, since π is continuous, $\pi^{-1}(0,1)^n$ is open in X and equals $int(B)$. Thus, $B \cap D = \partial B \cap D = \partial B$ by (i). ∎

A second elementary result will also be needed below. For each point $q \in \partial B$ there exists an open neighborhood $\mathcal{U}_q$ and coordinates $(z_1, \ldots, z_n)$, defined in $\mathcal{U}_q$ and centered at q, such that

$$\mathcal{U}_q \cap D \subset \cup_{i=1}^{n} \{z_i = 0\}. \tag{4.6}$$

A "sign distribution" is a function

$$\epsilon : \{1, \ldots, n\} \to \{+, -\}.$$

To each sign distribution one defines an open subset of any $\mathcal{U}_q$ by setting

$$\mathcal{O}_\epsilon = \{z \in \mathcal{U}_q : \epsilon(i) z_i > 0, \text{ for each } i = 1, \ldots, n\}.$$

One notes that the only geometric property of interest possessed by these sets is their disjointness from D.

LEMMA 4.7. *For each $q \in \partial B$ there exists a set $\mathcal{E}_q$ of sign distributions such that*

$$\cup_{\epsilon \in \mathcal{E}_q} \mathcal{O}_\epsilon = int(B) \cap \mathcal{U}_q .$$

PROOF: By (4.4)(i) and (4.5)(i), it is clear that

$$int(B) \cap \mathcal{U}_q \subset \cup_\epsilon \mathcal{O}_\epsilon .$$

Suppose for some ϵ_0 that $int(B) \cap \mathcal{U}_q \cap \mathcal{O}_{\epsilon_0} \neq \emptyset$. Further, suppose that $\mathcal{O}_{\epsilon_0} \not\subset int(B) \cap \mathcal{U}_q$. Then, Lemma (4.5) and (4.6) imply that

$$\mathcal{O}_{\epsilon_0} \cap (int(B) \cap \mathcal{U}_q) \neq \emptyset \quad \text{and } \mathcal{O}_{\epsilon_0} \cap (B^c \cap \mathcal{U}_q) \neq \emptyset$$

$$\text{but} \qquad \mathcal{O}_{\epsilon_0} \cap (\partial B \cap \mathcal{U}_q) = \emptyset .$$

Since $\mathcal{O}_{\epsilon_0}$ is connected this decomposition of $\mathcal{O}_{\epsilon_0}$ into two disjoint open subsets cannot occur. Thus, $\mathcal{O}_{\epsilon_0} \subset int(B) \cap \mathcal{U}_q$. This implies Lemma (4.7). ∎

To each irreducible component D_α of D one defines the following orders.

$$M_\alpha = \text{ord}_{D_\alpha} R \circ \pi =_{\text{def}} \text{ord}_{D_\alpha} (w_1^{M_1} \cdots w_n^{M_n}) \circ \pi - \text{ord}_{D_\alpha} G \circ \pi \tag{4.8}$$

$$m_\alpha = \text{ord}_{D_\alpha} \Phi \circ \pi =_{\text{def}} \text{ord}_{D_\alpha} \psi_1 \circ \pi - \text{ord}_{D_\alpha} (w_1^{m_1} \cdots w_n^{m_n}) \circ \pi - \text{ord}_{D_\alpha} \psi_2 \circ \pi ,$$

$$\kappa_\alpha = \text{ord}_{D_\alpha} \eta \circ \pi ,$$

$$\gamma_\alpha = \text{ord}_{D_\alpha} Jac(\pi) - \text{ord}_{D_\alpha} (w_1^2 \cdots w_n^2) \circ \pi ,$$

where $Jac(\pi)$ denotes the jacobian of π.

To each D_α for which $M_\alpha \neq 0$ define the ratios

$$\rho(D_\alpha) = \frac{-(1 + m_\alpha + \gamma_\alpha)}{M_\alpha}$$

$$\beta(D_\alpha) = \frac{-(1 + m_\alpha + \kappa_\alpha + \gamma_\alpha)}{M_\alpha} .$$

If $M_\alpha = 0$ one sets $\rho(D_\alpha) = \beta(D_\alpha) = -\infty$. The $\rho(D_\alpha)$ resp. $\beta(D_\alpha)$ are possible values for the first pole of $I(s, \varphi)$ resp. $I(s, B, \mathbf{C}, \varphi)$.

Define

$$\rho(\pi) = sup_\alpha \{\rho(D_\alpha)\}, \qquad \beta(\pi) = sup_\alpha \{\beta(D_\alpha)\}. \tag{4.9}$$

Then any pole of $I(s, \varphi)$ is at most $\rho(\pi)$ and any pole of $I(s, B, \mathbf{C}, \varphi)$ is at most $\beta(\pi)$. The key step in the proof of Theorem 4.1 is therefore the proof of the inequality

$$\rho(\pi) > \beta(\pi). \tag{4.10}$$

This will follow immediately from

Lemma 4.11. *Suppose q is a point in ∂B such that $\pi(q) \in H_\infty$. Let D_α be any component of D containing q. Then $\kappa_\alpha > 0$.*

Proof: Assume there exists a point $q \in \partial B$ with $\pi(q) \in H_\infty$ for which $\kappa_{\alpha'} \leq 0$ for some divisor $D_{\alpha'}$ containing q. Let $\mathcal{U}_q$ denote a neighborhood of the point so that (4.6) holds. Assume that coordinates are chosen so that the divisor $D_{\alpha'}$ satisfies the property $D_{\alpha'} \cap \mathcal{U}_q = \{z_1 = 0\}$. There exists at least one sign distribution ϵ so that $\mathcal{O}_\epsilon \subset int(B) \cap \mathcal{U}_q$. Given any point $p = (p_1, p') \in \mathcal{O}_\epsilon$ the path $\mu(t) = (1-t)p + t(0,p'), t \in [0,1)$ is entirely contained in $\mathcal{O}_\epsilon$. By definition, one has that

$$ord_t\,(\eta \circ \pi \circ \mu) = \kappa_{\alpha'}\,.$$

Thus, $\kappa_{\alpha'} \leq 0$ implies

$$\lim_{t \to 0} \eta \circ \pi \circ \mu(t) \neq 0\,.$$

On the other hand, $\mathcal{O}_\epsilon \subset int(B) \cap \mathcal{U}_q$ implies that for all $t > 0$, $\pi \circ \mu(t) \in (0,1)^n$. Moreover, as $t \to 0$, $\pi \circ \mu(t)$ approaches a point in H_∞. Thus, by (4.3) the limit of η along the path $\pi \circ \mu(t)$ must equal 0. So, the point q with the above properties must not exist. This proves the Lemma. ∎

An entirely similar argument that uses (1.4), as expressed in the $(w_1, \ldots, w_n)$ coordinates, shows

Lemma 4.12. *Suppose q is a point of ∂B such that $\pi(q) \in H_\infty$. Let D_α be any component of D containing q. Then $M_\alpha > 0$. Moreover, if $q \in \partial B$ is such that $\pi(q) \notin H_\infty$ then $M_\alpha = 0$ for any component D_α containing q.*

Remark 4.13. Geometrically, Lemma 4.12 says that the strict transform of G is a component of D that is disjoint from B in X. That is, the polar divisor of $R \circ \pi|_X$ cannot intersect B. An entirely similar conclusion holds for the polar locus of $\varphi|_X$. This property is important in describing the polar part of $I(s,\varphi)$ at $s = \rho(\pi)$, as seen in (4.16)ff. ∎

Furnished with these preliminary observations, one can proceed to the

Proof of Theorem 4.1: In light of (4.10), it evidently suffices to show, $\rho(\varphi) = \rho(\pi)$, (cf. Theorem 2.1), that is,

$$\rho(\pi) \textit{ is the first pole of } I(s,\varphi). \tag{4.14}$$

It is clear that one can assume that the sign of φ is constant outside a compact subset of $[1,\infty)^n$. For simplicity, one may therefore assume that φ is positive on all but a compact subset of $[1,\infty)^n$.

One has for $\sigma \gg 1$,

$$\begin{aligned} I(s,\varphi) &= \int_{[0,1]^n} R^s \Phi \, \frac{dw}{w_1^2 \cdots w_n^2} \\ &=_{def} \lim_{\epsilon \to 0} \int_{[\epsilon,1]^n} R^s \Phi \, \frac{dw}{w_1^2 \cdots w_n^2} \\ &= \int_B (R \circ \pi)^s (\Phi \circ \pi) |\pi^* (\frac{dw}{w_1^2 \cdots w_n^2})|, \end{aligned}$$

where $|\pi^*(dw/w_1^2 \cdots w_n^2)|$ denotes a density on X.

Since π is proper and B is a closed subset of the compact set $\pi^{-1}[0,1]^n$, B is also compact. For each $q \in B$ there exists an open neigborhood $\mathcal{U}_q$ so that (4.6) holds iff $q \in \partial B$. The open cover $\{\mathcal{U}_q\}$ of B admits a finite open subcover $\{\mathcal{U}_i\}_{i=1}^N$, where $\mathcal{U}_i$ is centered at q_i. One now takes a finite partition of unity $\{v_c\}$ subordinate to the cover $\{\mathcal{U}_i\}$. Thus, for $\sigma \gg 1$

$$\int_B (R\circ\pi)^s \, (\Phi\circ\pi) \, |\pi^*(\frac{dw}{w_1^2 \cdots w_n^2})| = \sum_c \sum_i \int_{\mathcal{U}_i \cap B} (R\circ\pi)^s \, (\Phi\circ\pi) \, v_c \, |\pi^*(\frac{dw}{w_1^2 \cdots w_n^2})| \tag{4.15}$$

One next fixes an arbitrary $\mathcal{U}_i$. One chooses the coordinates centered at q_i so that

$$\mathcal{U}_i \cap D = \cup_{j=1}^r \{z_j = 0\}.$$

Assume that $\{\epsilon_1, \ldots, \epsilon_{R(i)}\}$ are the sign distributions so that

$\mathcal{O}_{\epsilon_k} \subset int(B) \cap \mathcal{U}_i, k = 1, \ldots, R(i)$. Define for each $j = 1, \ldots, r$

$$\begin{aligned} M_j(i) &= \mathrm{ord}_{D_j} \, (R \circ \pi) \\ m_j(i) &= \mathrm{ord}_{D_j} \, (\Phi \circ \pi) \\ \gamma_j(i) &= \mathrm{ord}_{D_j} \, |\pi^*(dw/w_1^2 \cdots w_n^2)|, \end{aligned}$$

and for each $i = 1, \ldots, N$

$$\nu(i, \rho(\pi)) = \#\{j : \frac{-(1 + m_j(i) + \gamma_j(i))}{M_j(i)} = \rho(\pi)\}.$$

Define

$$\mathcal{J}(\rho(\pi)) = \{i : \nu(i, \rho(\pi)) \geq 1\}.$$

By definition, $\mathcal{J}(\rho(\pi)) \neq \emptyset$. In this regard, one should also note that $r = 0$ is possible. This occurs iff $q_i \in int(B)$. In this case, each $M_j(i) = 0$ and $i \notin \mathcal{J}(\rho(\pi))$.

The Gelfand-Shapiro-Shilov regularization method [G-S] applies to the integral over each open set $\mathcal{O}_{\epsilon_k}, k = 1, \ldots, R(i)$ and $i = 1, \ldots, N$, and thereby gives an analytic continuation to each summand on the right side of (4.15). In particular, if $i \in \mathcal{J}(\rho(\pi))$ then the principal part at $s = \rho(\pi)$ of

$$\int_{\mathcal{U}_i \cap B} (R \circ \pi)^s \, (\Phi \circ \pi) \, v_c \, |\pi^*(\frac{dw}{w_1^2 \cdots w_n^2})|$$

consists of at most $\nu(i, \rho(\pi))$ nonzero terms. The main point is to show that the term of order equal to $\nu(i, \rho(\pi))$ must be *positive.*

Note. When one i is fixed, i and $\rho(\pi)$ are subsequently dropped as the arguments for ν. ∎

After reindexing, if necessary, one may assume that

$$\{j : \frac{-(1 + m_j(i) + \gamma_j(i))}{M_j(i)} = \rho(\pi)\} = \{1, 2, \ldots, \nu\} \subset \{1, \ldots, r\}.$$

One sets $z' = (z_{\nu+1}, \ldots, z_n)$.

Then the contribution from $\mathcal{U}_i$ to the term of order ν in the principal part has the form

$$\sum_c \sum_{k=1}^{R(i)} \int_{\mathcal{U}_i \cap D_1 \ldots \cap D_\nu} (z_{\nu+1})^{\zeta_{\nu+1}}_{\epsilon_k(\nu+1)} \cdots (z_n)^{\zeta_n}_{\epsilon_k(n)} \, g_1(z')^{\rho(\pi)} \, g_2(z') \, v_c(z') \, g_3(z') \, d' \tag{4.16}$$

where the following properties are satisfied:

(1) $\zeta_{\nu+1}, \ldots, \zeta_n > -1$, (cf. [Va-2] where this property was first used for a related problem);

(2) $g_1(z')$ is the restriction to $\cap_{j=1}^{\nu} D_j$ of the quotient of the strict transforms of $w_1^{M_1} \cdots w_n^{M_n}$ and G in $\mathcal{U}_i$;

(3) $g_2(z')$ is the restriction to $\cap_{j=1}^{\nu} D_j$ of the quotient of the strict transforms of ψ_1 and $w_1^{m_1} \cdots w_n^{m_n} \cdot \psi_2$ in $\mathcal{U}_i$;

(4) $g_3(z')$ is the restriction to $\cap_{j=1}^{\nu} D_j$ of the strict transform of the quotient of $|Jac(\pi)|$ with $w_1^2 \cdots w_n^2$ in $\mathcal{U}_i$.

¿From Lemma 4.12 and the assumed positivity of P, φ over $[1,\infty)^n$ (cf. section 1), one concludes that $g_1(z'), g_2(z'), g_3(z')$ are *finite and positive* over the domain of integration in (4.16). Moreover, since $\{v_c\}$ forms a partition of unity, one concludes that the double sum in (4.16) must be positive.

Thus, $\rho(\pi)$ *must* be a pole of $I(s,\varphi)$. Furthermore, any rational number larger than $\rho(\pi)$ could *not* be a pole of $I(s,\varphi)$ since it would be larger than any candidate pole ρ_α used to define $\rho(\pi)$. This proves (4.14) and therefore completes the proof of Theorem 4.1. ∎

Section 5. *Distribution of eigenvalues for hypoelliptic PsDO's*

Let $p(x,\xi)$ denote the symbol of a pseudo differential operator $\mathcal{P}$ on $\mathbf{R}^n$. Let $\overline{\mathcal{P}}$ denote a self-adjoint extension to the Hilbert space $L^2([1,\infty)^n)$. Assume that the spectrum of $\overline{\mathcal{P}}$ is discrete. Denote the eigenvalues as $\lambda_1 \le \lambda_2 \le \ldots$. A standard problem is to understand the behavior as $t \to \infty$ of the spectral function

$$N(t) = \sum_{\lambda_n \le t} 1 .$$

This question has been studied by numerous authors under varying assumptions on $p(x,\xi)$. Of interest here is the behavior of $N(t)$ when $p(x,\xi)$ is hypoelliptic on $[1,\infty)^{2n}$. Robert [R] and Smagin [Sm] (among others, cf. their articles' bibliographies) have shown the following result. In the notation of the Introduction,

THEOREM.

$$N(t) \sim \left(\frac{1}{2\pi}\right)^n V_p(t) .$$

Theorem A shows that the asymptotic of $N(t)$ is determined by a discrete version of $V_p(t)$. That is, one sees immediately, again in the notation of the Introduction,

THEOREM 5.1.

$$N(t) \sim \left(\frac{1}{2\pi}\right)^n N_p(t) \quad \text{as } t \to \infty.$$

Theorem 5.1 appears to be of interest because each $N_p(t)$ can be calculated in polynomial time as a function of t. That is, Proposition 1.3 implies that any lattice point $m \in \mathbb{N}^{2n}$ for which $|p(m)| \leq t$ must be contained inside the part of the hyperboloid given by

$$\{y_1 \cdots y_{2n} < t^{1/\alpha}\} \cap [1, \infty)^{2n}, \tag{5.2}$$

where α denotes an exponent for $p(x, \xi)$ so that (1.4) holds. Since the number of lattice points satisfying (5.2) is $O(t^{2n/\alpha})$, the complexity of determining $N_p(t)$ is clearly polynomial in t. One can therefore determine a reasonable approximation to $N(t)$ for any sufficiently large t in polynomial time by calculating $N_p(t)$, rather than the analytically more difficult function $V_p(t)$. The same properties hold if one replaces $[1, \infty)^{2n}$ by $\mathbb{R}^{2n}$ and insists that $p(x, \xi)$ be hypoelliptic on $\mathbb{R}^{2n}$.

A second interesting feature of Theorem 5.1 can be seen by comparing it with the results of Bochner [Bo], who studied $N(t)$ for a constant coefficient operator $P(D)$ on the n dimensional torus. He essentially showed that if the symbol $p(\xi)$ was hypoelliptic on $\mathbb{R}^n$, then the Dirichlet series

$$\sum_{\{m \in \mathbb{Z}^n : p(m) \neq 0\}} \frac{1}{p(m)^s}$$

determined the spectral function for P. The reason for this was that the *identity* $N(t) = N_p(t)$ follows easily from an explicit description of the eigenfunctions of $P(D)$. As a result, the eigenvalues of $P(D)$ are easily seen to be the values of p at the lattice points m.

Thus, Theorem 5.1 shows that a Dirichlet series of the form

$$\sum_n c_n\, e^{-s \log \eta_n}, \text{ where } 0 < \eta_1 < \eta_2 < \dots,$$

exists *in general* so that

$$N(t) \sim \sum_{\eta_n \leq t} c_n .$$

In particular, this occurs even though the actual eigenvalues *are not* known to be expressible algebraically in terms of the values of p at any set of lattice points contained in $[1, \infty)^{2n}$. This intriguing property does not seem to have been observed earlier.

Section 6. *Cohomological invariance of the main term in $N_P(t, \varphi)$*

Suppose that φ_1, φ_2 are two polynomials that are positive outside a compact subset of $[1,\infty)^n$. By Theorem 2.1 (cf. [Li-3,4]), there exist nonzero real polynomials $A_1(u), A_2(u)$ and rational numbers $\rho(\varphi_1), \rho(\varphi_2)$ such that for some $\tau > 0$

$$N_P(t,\varphi_1) = t^{\rho(\varphi_1)} A_1(logt) + O(t^{\rho(\varphi_1)-\tau})$$
$$N_P(t,\varphi_2) = t^{\rho(\varphi_2)} A_2(logt) + O(t^{\rho(\varphi_2)-\tau}).$$

Denote the dominant term in $N_P(t,\varphi)$ by $\hat{N}_P(t,\varphi)$ below.

A natural question asks:

What conditions upon φ_1, φ_2 can be imposed that insures

$$\hat{N}_P(t,\varphi_1) = \hat{N}_P(t,\varphi_2)?$$

In [Li-2] a cohomological criterion was found that answered this question under conditions a bit more restrictive than were really necessary, in light of Theorem B of this paper. It will suffice here to state the extension of [Li-2, Theorem 4] that can now be made. To do so, first recall the standard constructions of the cohomology fiber bundle for a polynomial mapping on $\mathbb{C}^n$ and the section induced by the Leray residue operation.

According to Verdier [Ve]), for any polynomial mapping $P : \mathbb{C}^n \to \mathbb{C}$, there is a finite set $\Sigma_P \subset \mathbb{C}$ such that $P : \mathbb{C}^n - P^{-1}(\Sigma_P) \to \mathbb{C} - \Sigma_P$ is a locally trivial fibration. Set $\mathbb{C}^* = \mathbb{C} - \Sigma_P$, and define $P^* = P|_{P^{-1}(\mathbb{C}^*)}$. For $t \in \mathbb{C}^*$ set $X_t = P^{-1}(t)$. Let H^{n-1} denote the flat vector bundle on $\mathbb{C}^*$ with fiber at t equal to the finite dimensional vector space $H^{n-1}(X_t, \mathbb{C})$. Let $\mathcal{H}^{n-1} = \mathsf{H}^{n-1} \otimes \mathcal{O}_{\mathbb{C}^*}$ be the sheaf of germs of analytic sections of H^{n-1}. Any rational differential n-form ω determines an analytic section of $\mathcal{H}^{n-1}$, defined as

$$\sigma(\omega) : t \to [\omega/dP|_{X_t}]$$

where $\omega/dP|_{X_t} = \text{Res}\ (\omega/(P-t))|_{X_t}$.

THEOREM 6.1. *Assume the following hypotheses.*

(1) *P is a tame polynomial (cf. [Br]).*
(2) *P, φ_1, φ_2 are hypoelliptic polynomials on $[1,\infty)^n$.*
(3) *For all $t \notin \Sigma_P$ one has*

$$\sigma(\varphi_1 dz)(t) = \sigma(\varphi_2 dz)(t),$$

where φdz is the $(n,0)$ form determined by φ.

Then $\hat{N}_P(t,\varphi_1) = \hat{N}_P(t,\varphi_2)$.

In this sense one can say that the dominant term is a "cohomological invariant". That such a result might be possible arose from studying the work of Cassou-Nogues [Ca-N]. On the other hand, one is still quite far from achieving a result with the precision found in [ibid], which was obtained under the assumption that P was a polynomial of the type studied by Dwork.

Section 7. *The case of several polynomials simultaneously*

The proof of Theorem C via Theorem 4.1 extends easily to a several variable setting. The main result of this section is Theorem D which is an immediate corollary of Theorem 7.8. Theorem D appears to be of particular interest because it is an essential ingredient in the sharpening of the results described in [Li-5]. This is discussed at the end of the section.

Let $P_1, \ldots, P_k \in \mathbb{R}[x_1, \ldots, x_n]$ be hypoelliptic on $[1,\infty)^n$. Let φ be a hypoelliptic fraction, in the sense of Section 1. One assumes, for simplicity, that each $P_i > 0$ and $\varphi > 0$ on $[1,\infty)^n$. Denote the best constants c, α so that (1.2), (1.5) hold for each P_i by c_i, α_i. Define, for $\delta > 0$,

$$\Omega(\delta) = \{\mathbf{s} : \sum_{i=1}^{k} \alpha_i \sigma_i > \delta, \text{ and } \sigma_i > 0, \text{ for each } i\}.$$

One also defines

$$D(\mathbf{s},\varphi) = \sum_{m \in \mathbb{N}^n} \frac{\varphi(m)}{P_1^{s_1}(m) \cdots P_k^{s_k}(m)}.$$

An elementary argument shows

PROPOSITION 7.1. *There exists* $\delta > 0$ *so that* $D(\mathbf{s},\varphi)$ *is absolutely convergent in* $\Omega(\delta)$.

In [Li-5] an analytic continuation of $D(\mathbf{s},\varphi)$ to $\mathbb{C}^k$ was given, using a summatory formula (or integral representation) based upon an iteration of Cauchy residue theory, and a set of k functional equations in $s_1, \ldots, s_k$ deduced from the work of Sabbah [Sb]. In this way, one showed

THEOREM 7.2. *There exist* $\mathbf{b}_i \in \mathbb{Z}_+^k, i = 1, \ldots, M, \beta_1, \ldots, \beta_M \in \mathbb{Z}$ *such that* *the polar divisor of $D(\mathbf{s},\varphi)$ is contained in*

$$\bigcup_{e=0}^{\infty}\bigcup_{i=1}^{M}\{\mathbf{s}:\mathbf{b}_i\cdot\mathbf{s}=\beta_i-e\}.$$

One denotes the polar divisor of $D(\mathbf{s},\varphi)$ by Pol_D.

On the other hand, as observed in [Li-4], as well as Section 4 of this paper, a somewhat sharper set of results can be deduced for hypoelliptic polynomials if one uses the Euler-Maclaurin formula. A similar sharpening is also possible in the several variable setting. Define

$$I(\mathbf{s},\varphi)=\int_{[1,\infty)^n}\frac{\varphi}{P_1^{s_1}\cdots P_k^{s_k}}\,dx\,.$$

To proceed as in Section 3, one must first give the natural extension of Lemma 3.2. Of course, the notations from Section 3 are adopted here.

For $J\in\mathbf{Z}_+^n$, set

$$\mathcal{P}(J)=\{(J_0,\ldots,J_k)\in(\mathbf{Z}_+^n)^{k+1}:J_0+\cdots+J_k=J\}.$$

LEMMA 7.3. *For each index $J\in\mathbf{Z}_+^n$ and element $\mathbf{J}=(J_0,\ldots,J_k)\in\mathcal{P}(J)$, there exist integers $a(\mathbf{J})$ such that*

$$D_x^J\Big(\frac{\varphi}{P_1^{s_1}\cdots P_k^{s_k}}\Big)=\sum_{\mathbf{J}\in\mathcal{P}(J)}a(\mathbf{J})\cdot D_x^{J_0}(\varphi)\prod_{i=1}^{k}D_x^{J_i}\Big(\frac{1}{P_i^{s_i}}\Big)$$

$$=\sum_{\mathbf{J}\in\mathcal{P}(J)}a(\mathbf{J})\cdot\frac{D_x^{J_0}(\varphi)}{\varphi}\prod_{i=1}^{k}\left\{\sum_{\mathbf{C}\in\mathcal{M}(J_i)}\sum_{v=0}^{|\mathbf{C}|-1}[-s_i]_v\frac{D_x^{\mathbf{C}}(P_i)}{P_i^{v+1}}\right\}\frac{\varphi}{P_1^{s_1}\cdots P_k^{s_k}}\,.$$

The hypoellipticity of each P_i and φ shows clearly that

PROPOSITION 7.4. *For each J, the function*

$$\mathbf{s}\to\int_{[1,\infty)^n}D_x^J\Big(\frac{\varphi}{P_1^{s_1}\cdots P_k^{s_k}}\Big)\,dx$$

is analytic if $\mathbf{s}\in\Omega(\delta)$.

The arguments of [Li-4] extend easily to show

THEOREM 7.5. *For each J, the function defined in 7.4 admits an analytic continuation to $\mathbb{C}^k$ as a meromorphic function. Further, there exist elements of $\mathbb{Z}_+^k$, $\mathbf{b}_i(J), i = 1, \ldots, M(J)$, and integers $\beta_1(J), \ldots, \beta_{M(J)}(J)$ such that the polar divisor of this meromorphic function is contained in*

$$\bigcup_{e=0}^{\infty} \bigcup_{i=1}^{M(J)} \{\mathbf{s} : \mathbf{b}_i(J) \cdot \mathbf{s} = \beta_i(J) - e\}.$$

Notation. One denotes the polar divisor of $I(\mathbf{s}, \varphi)$ by Pol_I. ∎

For given $r \in \mathbb{N}$ set

$$\mathcal{I}(r) = \{I = (i_1, \ldots, i_n) : i_j \in [0, r], j = 1, \ldots, n\}.$$

In the notation of (3.1) (with $k = r$), one sees the

THEOREM 7.6.

(1) *If $\mathbf{s} \in \Omega(\delta)$ and $r \geq 1$ then*

$$D(\mathbf{s}, \varphi) = \sum_{I \in \mathcal{I}(r)} \int_{[1,\infty)^n} h_{i_1}^{(r)}(x_1) \cdots h_{i_n}^{(r)}(x_n) \, D_x^I\Big(\frac{\varphi}{P_1^{s_1} \cdots P_k^{s_k}}\Big) \, dx.$$

(2) *There exist constants $c(I)$ for each $I \neq (0, \ldots, 0) \in \mathcal{I}_r'$ so that*

$$D(\mathbf{s}, \varphi) = I(\mathbf{s}, \varphi) + \sum_{I \neq (0,\ldots,0) \in \mathcal{I}_r'} c(I) \int_{[1,\infty)^n} D_x^I\Big(\frac{\varphi}{P_1^{s_1} \cdots P_k^{s_k}}\Big) \, dx$$

$$+ \sum_{I \in \mathcal{I}_r''} \int_{[1,\infty)^n} h_{i_1}^{(r)}(x_1) \cdots h_{i_n}^{(r)}(x_n) \, D_x^I\Big(\frac{\varphi}{P_1^{s_1} \cdots P_k^{s_k}}\Big) \, dx.$$

(3) *Set $c = \min\{c_1, \ldots, c_k\}$. For any $I \in \mathcal{I}_r''$*

$$\int_{[1,\infty)^n} h_{i_1}^{(r)}(x_1) \cdots h_{i_n}^{(r)}(x_n) \, D_x^I\Big(\frac{\varphi}{P_1^{s_1} \cdots P_k^{s_k}}\Big) \, dx \ \text{ is analytic if } \sum_i \alpha_i \sigma_i > \delta - cr.$$

Notation. For any $J_0, J_1, \ldots, J_k \in \mathbb{Z}_+^n$ and $(\mathbf{C}_1, \ldots, \mathbf{C}_k) \in \mathcal{M}(J_1) \times \ldots \times \mathcal{M}(J_k)$, set $\mathcal{C} = (J_0, \mathbf{C}_1, \ldots, \mathbf{C}_k)$. Define

$$I(\mathbf{s}, \mathcal{C}, \varphi) = \int_{[1,\infty)^n} \frac{D_x^{J_0}(\varphi)}{\varphi} \cdot \frac{D_x^{\mathbf{C}_1}(P_1)}{P_1^{v_1}} \cdots \frac{D_x^{\mathbf{C}_k}(P_k)}{P_k^{v_k}} \cdot \frac{\varphi}{P_1^{s_1} \cdots P_k^{s_k}} \, dx. \quad ∎$$

To any meromorphic function on $\mathbb{C}^k$ analytic on a domain like $\Omega(\delta)$ and with polar divisor *pol* contained in a union of hyperplanes as described in (7.2), (7.4), one can associate a Newton polyhedron of *pol*, denoted $\Gamma(pol)$. This is an unbounded subset of $\mathbb{R}^k$ and defined as follows.

Definition 7.7. Suppose

$$pol \subset \bigcup_{e=0}^{\infty} \bigcup_{i=1}^{M} \{b^{(i)} \cdot \mathbf{s} = \beta^{(i)} - e\}.$$

Assume *each* hyperplane $\{b^{(i)} \cdot \mathbf{s} = \beta^{(i)}\}, i = 1, \ldots, M$, is a component of *pol*. Set

$$\mathcal{H} = \bigcap_{i=1}^{M} \{\boldsymbol{\sigma} \in \mathbf{R}^k : b^{(i)} \cdot \boldsymbol{\sigma} \geq \beta^{(i)}, \sigma_1, \ldots, \sigma_k > 0\}$$

and $\quad \Gamma(pol) = \partial\mathcal{H}.$ ∎

Denote the Newton polyhedra of Pol_D resp. Pol_I by Γ_D resp. Γ_I. For simplicity these polyhedra will be called the Newton polyhedra of D resp. I.

The main result of this section is

THEOREM D. *Assume each P_i is hypoelliptic on $[1, \infty)^n$, and φ is a hypoelliptic fraction. Then*

(1) $\Gamma_I = \Gamma_D$.
Denote the common polyhedron in (1) by Γ.

(2) *$D(\mathbf{s}, \varphi) - I(\mathbf{s}, \varphi)$ is analytic at each point of $\Gamma + i\mathbf{R}^k$.*

Remark. One should interpret Theorem D as the several variable analogue of Theorem C of Section 3. In particular, one should think of Γ as the analogue of the "first pole" for a series in one variable. The reader should also note that if $b^{(i')} \cdot \boldsymbol{\sigma} = \beta^{(i')}$ contains a face of Γ, then the only $\mathbf{t} \in \mathbf{R}^k$ for which (2) is a nontrivial assertion are those that satisfy the equation $b^{(i')} \cdot \mathbf{t} = 0$. ∎

Theorem D will follow directly from Theorem 7.6 and the following result.

THEOREM 7.8. *For any $C \neq (0, \ldots, 0)$ the Newton polyhedron of $I(\mathbf{s}, C, \varphi)$ is strictly below Γ_I.*

Remark. By the phrase "strictly below" is meant that the Newton polyhedron of $I(\mathbf{s}, C, \varphi)$ is contained in $\Gamma_I - (\epsilon, \infty)^k$, for some $\epsilon > 0$. ∎

PROOF: One can give a similar proof to that of Theorem 4.1. A sketch of the proof should therefore suffice here. The notation of (4.4) will be used below.

In the following, an arbitrarily given $\mathcal{C} = (J_0, \mathbf{C}_1, \ldots, \mathbf{C}_k) \neq (0, \ldots, 0)$ is fixed throughout the discussion.

As in (4.2), define each R_i by

$$R_i(w) =_{def} \frac{1}{P_i(1/w)} \quad \text{for each } i\,.$$

In addition, define the rational function $\Phi_{\mathcal{C}}$ as

$$\Phi_{\mathcal{C}}(w_1, \ldots, w_n) =_{def} \frac{D_x^{J_0}(\varphi)}{\varphi} \cdot \frac{D_x^{\mathbf{C}_1}(P_1)}{P_1^{v_1}} \cdots \frac{D_x^{\mathbf{C}_k}(P_k)}{P_k^{v_k}}(1/w)\,.$$

One constructs a nonsingular real algebraic manifold Y and proper birational map $\pi : Y \to (\mathbf{R}^n, (w_1, \ldots, w_n))$ such that properties (4.4)(i,ii,iv) are satisfied. On the other hand, one modifies (4.4)(iii) so that the divisor determined by

$$[\prod_{i=1}^{n}(w_i - 1) \cdot \prod_{i=1}^{n} w_i \cdot \prod_{i=1}^{k} R_i \cdot \Phi_{\mathcal{C}}] \circ \pi$$

has support in the normally crossing divisor $\mathcal{D}$. Restricting $\mathcal{D}$ to a divisor D over a polydisc U, as in the proof of (4.1), and setting $D = \cup_\alpha D_\alpha$ to denote its decomposition into irreducible components, one next defines the multiplicities

$$M_\alpha(i) = ord_{D_\alpha}\, R_i \circ \pi$$
$$\kappa_\alpha(\mathcal{C}) = ord_{D_\alpha}\, \Phi_{\mathcal{C}} \circ \pi$$
$$\gamma_\alpha = ord_{D_\alpha} Jac(\pi) - ord_{D_\alpha}(w_1^2 \cdots w_n^2) \circ \pi\,.$$

Additionally, for each component D_α, define the linear form and hyperplane

$$L_\alpha(\boldsymbol{\sigma}) = \sum_{i=1}^{k} M_\alpha(i)\, \sigma_i + \gamma_\alpha\,,$$
$$\mathcal{H}(D_\alpha) = \{\boldsymbol{\sigma} : L_\alpha(\boldsymbol{\sigma}) = -1\,\}\,.$$

Define

$$\mathcal{G} = \bigcap_\alpha \{\boldsymbol{\sigma} : L_\alpha(\boldsymbol{\sigma}) \geq -1 \text{ and for each } i, \sigma_i > 0\}\,,$$

$$\mathcal{G}(\mathcal{C}) = \bigcap_\alpha \{\boldsymbol{\sigma} : L_\alpha(\boldsymbol{\sigma}) + \kappa_\alpha(\mathcal{C}) \geq -1 \text{ and for each } i, \sigma_i > 0\}\,.$$

Define the polyhedra

$$\hat{\Gamma}_I = \partial \mathcal{G} \quad \text{and} \quad \hat{\Gamma}(\mathcal{C}) = \partial \mathcal{G}(\mathcal{C}).$$

$\hat{\Gamma}_I$ satisfies an important (and evident) "convexity" property that is used below:

(7.9) *Assume that the plane $\mathcal{H}(D_\alpha)$ is disjoint from $\hat{\Gamma}_I$. Then $L_\alpha(\boldsymbol{\sigma}) > -1$ whenever $\boldsymbol{\sigma} \in \hat{\Gamma}_I$.*

The hypoellipticity condition satisfied by each P_i implies by (1.2):

(7.10.1) *If q is a point of ∂B such that $\pi(q) \in H_\infty$ and D_α is any component of D containing q, then $\kappa_\alpha(\mathcal{C}) > 0$.*

(7.10.2) *Assume q is a point of ∂B such that $\pi(q) \in H_\infty$. Let D_α be any component of D containing q. Then, $M_\alpha(i) > 0$ for each i. In addition, if $q \in \partial B$ is such that $\pi(q) \notin H_\infty$, then $M_\alpha(i) = 0$ for each i and any component D_α containing q.*

It is clear that (7.10.1) implies:

$\hat{\Gamma}(\mathcal{C})$ *must lie below* $\hat{\Gamma}_I$ *in the sense of the above Remark.*

To complete the proof of Theorem 7.8, it suffices to show that $\hat{\Gamma}_I = \Gamma_I$. This is done by a straightforward adaptation of the analysis used to prove Theorem 4.1. In particular, one reduces to the local situation, as described in (4.16). The fact that $\hat{\Gamma}_I$ is the convex envelope of $\mathcal{G}$, which implies (7.9), is now used to insure that the four properties formulated below (4.16) are satisfied in this new setting. In particular, the exponents $\zeta_{\nu+1}, \ldots, \zeta_n$, appearing in (4.16), are now understood to be functions of $\mathbf{s} = \boldsymbol{\sigma} + i\mathbf{t}$. (7.9) implies that for any $\boldsymbol{\sigma} \in \hat{\Gamma}_I$ one must have $Re(\zeta_{\nu+1}), \ldots, Re(\zeta_n) > -1$. The rest of the argument is entirely similar to that in the proof of Theorem 4.1. Details are left to the reader. ∎

Remark. One can also summarize the content of Theorem 7.8 by saying that a Newton polyhedron, constructed from resolution data, agrees with the Newton polyhedron Γ_I. Evidently, this polyhedron is just $\hat{\Gamma}_I$, defined above. In this way, property (4.14) is seen to be generalized by the equation

$$\hat{\Gamma}_I = \Gamma_I . \quad ∎$$

There is an important analytical application of Theorem D. To formulate this, some preliminary remarks are needed.

Definition 7.11. A component L of Pol_D satisfying the property that $L \cap \mathbb{R}^k$ contains a face of dimension $k-1$ of Γ is called an initial component of Pol_D. A divisor D_α in a resolution of singularities, as constructed in the proof of Theorem 7.8, is called an extremal divisor if the plane $\mathcal{H}(D_\alpha)$ is an initial component of Pol_D. Define

$$\hat{L} = L - \bigcup_{\{L' \neq L : L' \text{ a component of } Pol_D\}} (L \cap L') . \quad \blacksquare$$

Definition 7.12. A vertex of the polyhedron Γ is called an initial vertex of Pol_D. More generally, set

$$\widehat{Pol_D} = Pol_D \cup \{s_1 \cdots s_k = 0\} .$$

The intersection of any k normally crossing components of $\widehat{Pol_D}$ is called a vertex of $\widehat{Pol_D}$. A distinguished vertex of $\widehat{Pol_D}$ is any vertex contained in an initial component of Pol_D. ∎

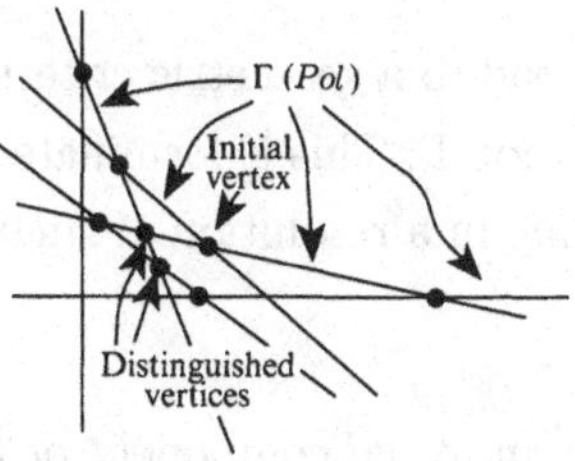

Figure 1

Notation. Define the differential forms

$$\omega_I(t, \mathbf{a}, \mathbf{s}) = t^{\mathbf{a}\cdot\mathbf{s}} I(\mathbf{s}, \varphi) \frac{ds_1 \cdots ds_k}{s_1 \cdots s_k} ,$$

$$\omega_D(t, \mathbf{a}, \mathbf{s}) = t^{\mathbf{a}\cdot\mathbf{s}} D(\mathbf{s}, \varphi) \frac{ds_1 \cdots ds_k}{s_1 \cdots s_k} .$$

Now, for any *initial* component L of Pol_D set

$$\mathcal{R}_L(\omega_D) = Res_L(\omega_D) ,$$

$$\mathcal{R}_L(\omega_I) = Res_L(\omega_I) ,$$

where Res_L is meant the standard Leray residue along the nonsingular variety $\hat{L}$. It is clear that $\mathcal{R}_L(\omega_I), \mathcal{R}_L(\omega_D)$ are analytic on $\hat{L}$ and meromorphic on L. Theorem D evidently implies

THEOREM 7.13. *For any initial component L, one has $\mathcal{R}_L(\omega_D) = \mathcal{R}_L(\omega_I)$ for all $\mathbf{s} \in \hat{L}, t > 0$, and any* $\mathbf{a}$.

As discussed below, the precise description of the polar locus of $\mathcal{R}_L(\omega_D)$ is of great interest. Theorem 7.13 is now seen to be helpful in deriving this description. The reason for this is the existence of a concrete integral representation for $\mathcal{R}_L(\omega_D)$ that is an immediate consequence of the identity in (7.13). Via this representation, one observes that the methods of Gelfand-Shapiro-Shilov, used to determine precisely poles of generalized functions $P_{\pm}^{s}$ for a polynomial P, can now also be applied to describe the polar locus of $\mathcal{R}_L(\omega_D)$, simply by replacing everywhere $D(\mathbf{s}, \varphi)$ by $I(\mathbf{s}, \varphi)$.

Note. For simplicity in the following discussion, one now sets $k = 2$. In this case, one wants to locate *poles* of the residue $(1, 0)$ forms $\mathcal{R}_L(\omega_D)$. The extension to $k > 2$ polynomials is a straightforward, if not slightly complicated exercise, and left to the reader.

The observations above lead to a geometric criterion that locates precisely a pole of $\mathcal{R}_L(\omega_D)$ on the polygon Γ. This is formulated in terms of the geometry of the divisors D_α, appearing in a resolution of singularities, $\pi : Y \to \mathbf{R}^n$, that proves Theorem 7.8.

THEOREM 7.14. *Let L be an initial component of Pol_D and D_α an extremal divisor such that $\mathcal{H}(D_\alpha) = L \cap \mathbf{R}^2$. Let $\mathbf{v}$ be an initial vertex contained in L. Then $\mathbf{v}$ is a pole of $\mathcal{R}_L(\omega_D)$ if there exists an exceptional divisor $D_\beta \subset Y$ such that*

$$D_\beta \cap D_\alpha \cap B \neq \emptyset \tag{7.15.1}$$

$$\text{and} \quad \mathcal{H}(D_\beta) \cap L \cap \mathbf{R}^2 = \{\mathbf{v}\}. \tag{7.15.2}$$

PROOF: This follows immediately from the regularization of the integral which represents the local contribution of $\mathcal{R}_L(\omega_I)$ in each open set $\mathcal{U}_i$, defined in (4.16). The criteria in (7.15) imply there exists an open set $\mathcal{U}_i$, in which is found the point $D_\alpha \cap D_\beta$, such that in each connected component of $\mathcal{U}_i - D$, the integrand determining $\pi^*(\omega_I)$ has the local form

$$\prod_{\ell=1}^{\nu} z_{\ell,\pm}^{L_\alpha(\mathbf{s})} \cdot z_{\nu+1,\pm}^{L_\beta(\mathbf{s})}\, \Omega(\mathbf{s}, z),$$

where $\{z_\nu = 0\} = D_\alpha \cap \mathcal{U}_i$, $\{z_{\nu+1} = 0\} = D_\beta \cap \mathcal{U}_i$, and Ω is an $n-$ form which is

(1) real analytic in z whenever $z \in \mathcal{U}_i - D$,
and
(2) locally integrable if $\boldsymbol{\sigma}$ is contained in an *open* neighborhood of any point of $\Gamma \cap \hat{L}$.

By (7.9), $L_\beta(\boldsymbol{\sigma}) > -1$. The contribution to $\mathcal{R}_L(\omega_I)$ from $\mathcal{U}_i$ will therefore have its first pole at the point $\mathbf{s}$ for which

$$L_\beta(\mathbf{s})\big|_L = -1.$$

(7.15.2) implies that this point must be $\mathbf{s} = \mathbf{v}$. Moreover, the argument given in (4.16)ff. extends easily, cf. the proof of (7.8), and shows that the coefficient of the highest power of $\log t$ in the principal part of $Res_{\mathbf{s}=\mathbf{v}}(\mathcal{R}_L(\omega_I)|_{\mathcal{U}_i})$ is positive. The convexity property (7.9) further implies that in any other open set $\mathcal{U}_j$ used in (4.16), $\mathcal{R}_L(\omega_I)$ is either regular at $\mathbf{s} = \mathbf{v}$ or, also has a pole at $\mathbf{s} = \mathbf{v}$, in which case, the coefficient of the highest power of $\log t$ in $Res_{\mathbf{s}=\mathbf{v}}(\mathcal{R}_L(\omega_I)|_{\mathcal{U}_j})$ is positive. This completes the proof of Theorem 7.14. ∎

Remark. It would be very useful to understand what is the significance of the conditions in (7.15). It is not at all clear what they mean in an intrinsic or geometric sense. ∎

The proof of (7.14) also extends to detect poles of $\mathcal{R}_L(\omega_I)$ in the event that (7.15) is not satisfied. Indeed, to establish the property:

there exists a distinguished vertex $\mathbf{w}$ at which $\mathcal{R}_L(\omega_I)$ has a pole,

it suffices simply to replace $\mathbf{v}$ by $\mathbf{w}$ in (7.15.2).

One can formulate a "simultaneous" lattice point problem whose asymptotic behavior can be precisely determined whenever the criteria in (7.15) are satisfied. Again, for simplicity only, $k = 2$ is assumed. Let $\mathbf{a} = (a_1, a_2) \in (0, \infty)^2$, and $t > 0$. Define

$$N_{\mathbf{P}}(t, \mathbf{a}, \varphi) = \sum_{\{m \in \mathbf{N}^n : P_i(m) \le t^{a_i}, i=1,2\}} \varphi(m).$$

Evidently, one can interpret the family of counts $N_{\mathbf{P}}(t, \mathbf{a}, \varphi)$ as a natural generalization of $N_P(t, \varphi)$ when given 2 polynomials. Thus, the analysis of the asymptotic behavior of $N_{\mathbf{P}}(t, \mathbf{a}, \varphi)$ is a natural problem to understand. Proceeding classically, one can now see why the polygon Γ should be defined so as to lie in $[0, \infty)^2$.

One starts with the following integral formula for $N_{\mathbf{P}}(t, \mathbf{a}, \varphi)$.

Assume that b is a sufficiently large positive number. Then, a 2−fold iteration of Perron's formula [Ti, ch. 9] shows:

$$N_{\mathbf{P}}(t, \mathbf{a}, \varphi) = \frac{1}{(2\pi i)^2} \int_{\sigma_1 = \sigma_2 = b} t^{\mathbf{a} \cdot \mathbf{s}} D(\mathbf{s}, \varphi) \frac{ds_1 ds_2}{s_1 s_2} . \tag{7.16}$$

Thus, ω_D is an evident $(2, 0)$ form to introduce for studying $N_{\mathbf{P}}(t, \mathbf{a}, \varphi)$. The polygon Γ is evidently the polygon of the polar divisor of the meromorphic function appearing in ω_D.

[Li-5] gave a general description of an upper bound for any $N_{\mathbf{P}}(t, \mathbf{a}, \varphi)$ in terms of the geometry of the initial components of Pol_D. In order to sharpen such estimates, in particular, to determine a precise asymptotic for $N_{\mathbf{P}}(t, \mathbf{a}, \varphi)$, it is clear from the analysis in [ibid] that one needs to answer the following:

Question. For each initial component L, what is the polar locus of $\mathcal{R}_L(\omega_D)$?

Based upon the earlier remarks, the most general property that one can prove appears to be the following.

THEOREM 7.17. *For each initial component L there exists a distinguished vertex ν_L such that $\mathcal{R}_L(\omega_D)$ has a pole at $\mathbf{s} = \nu_L$. This vertex is initial if the conditions in (7.15) are satisfied.*

To each initial component L, there corresponds a primitive integral vector $b^{(L)}$ so that $L = \{\mathbf{s} : b^{(L)} \cdot \mathbf{s} = \beta_L\}$. One calls a conical neighborhood of $b^{(L)}$ any (connected) open cone of $(0, \infty)^2$ containing $b^{(L)}$ *in its boundary.* Following the tauberian argument described in [Li-5, sec. 6], one then concludes

COROLLARY 7.18. *For each initial component L there exists a conical neighborhood $\mathcal{U}_L$ of $b^{(L)}$, distinguished vertex ν_L, and nonzero polynomial $A_L(\varphi, u) \in$ $\mathsf{R}[u]$ such that if $\mathbf{a} \in \mathcal{U}_L$ then*

$$N_{\mathbf{P}}(t, \mathbf{a}, \varphi) = t^{\mathbf{a} \cdot \nu_L} A(\varphi, \log t) + O(t^{\mathbf{a} \cdot \nu_L - \theta}) \quad \textit{for some positive } \theta .$$

Thus, there exists a nonempty cone of vectors **a** for which the above asymptotic exists. However, the vertex ν_L will in general "jump" as one crosses the ray containing $b^{(L)}$, as described in detail in [ibid]. On the other hand, the vertex ν_L for which 7.18 holds, will *not* jump as long as one remains inside the conical neighborhood $\mathcal{U}_L$.

As in Section 2, one also has the identity

$$V_{\mathbf{P}}(t, \mathbf{a}, \varphi) =_{def} \int_{\substack{\{P_1 \le t^{a_1}\} \cap \\ \{P_2 \le t^{a_2}\} \cap [1,\infty)^n}} \varphi dx_1 \cdots dx_n$$

$$= \frac{1}{(2\pi i)^2} \int_{\sigma_1 = \sigma_2 = b} t^{\mathbf{a}\cdot\mathbf{s}} I(\mathbf{s}, \varphi) \frac{ds_1 ds_2}{s_1 s_2}.$$

Thus, (7.16) and Theorems D, (7.17) also imply, by a similar argument used to prove (7.18), the

COROLLARY 7.19. *There exists $\theta > 0$ such that if $\mathbf{a} \in \mathcal{U}_L$ then*

$$N_{\mathbf{P}}(t, \mathbf{a}, \varphi) - V_{\mathbf{P}}(t, \mathbf{a}, \varphi) = O(V_{\mathbf{P}}(t, \mathbf{a}, \varphi)/t^{\theta}) \text{ as } t \to \infty.$$

Remarks 7.20.

(1) The knowledeable reader will recognize (7.18) and (7.19) as a generalization of Landau's well known theorem (cf. [Ti, ch. 9]) concerning Dirichlet series with positive coefficients (and moderate growth past the abscissa of convergence).

(2) Although these are still relatively weak results – – – one would like to sharpen the description of the set of **a** for which an asymptotic is known to exist, and also make precise the location of the vertex – – – even the existence of such an asymptotic was not previously known in the general setting of Theorem D. To obtain such precision, one needs to determine exact information about the configuration of divisors, appearing in a resolution of singularities with which one proves Theorem 7.8. This information is a generalization of the type of analysis carried out in [Ig, pg. 87ff].

[Li-6] carries out this analysis in certain cases, that is, when $n = 2$ or when the polynomials satisfy the conditions in the second theorem of Sargos, cf. Introduction. The results in [ibid] support the conjecture that the distinguished vertices appearing in (7.18), should always be vertices of the Newton polygon (or polyhedron when $k > 2$) of Pol_D. Such results have never been able to be proved before due to the lack of a suitable geometric/analytic framework at infinity, which the methods used in the series of paper [Li-1-6] has developed.

(3) Prior to any of this work, it had only been possible to determine the precise asymptotic for $N_{\mathbf{P}}(t, \mathbf{a}, 1)$ when each P_i was elliptic on $[0, \infty)^n$. To state this result, set $degP_i = d_i, i = 1, \ldots, k$, and

$$d_i \mathbf{e}_i = d_i \cdot (0, \ldots, 0, 1, 0, \ldots, 0),$$

where 1 appears in the ith coordinate.

THEOREM 7.21. *Define*

$$\rho(\mathbf{a}) = min_i \{\mathbf{a} \cdot d_i \mathbf{e}_i\}.$$

Then there exist positive numbers A, θ such that

$$N_{\mathbf{P}}(t, \mathbf{a}) = At^{\rho(\mathbf{a})} + O(t^{\rho(\mathbf{a}) - \theta}).$$

The proof of 7.21 is a simple extension of the argument given in [Ma]. This argument however, breaks down once the ellipticity condition is replaced by hypoellipticity.

REFERENCES

Bo S. Bochner, *Zeta functions and Green's functions for linear partial differential operators of elliptic type with constant coefficients*, Ann. of Math. **57** (1953), 32-56.

Br S. A. Broughton, *Milnor Numbers and the Topology of Polynomial Hypersurfaces*, Inv. Math. **92** (1988), 217-241.

Ca-N P. Cassou-Nogues, *Valeurs aux entiers négatifs des sèries de Dirichlet associèes á un polynôme II*, Amer. J. of Math. **106** (1985), 255-299.

CdV Y. Colin de Verdiere, *Nombre de points entiers dans une famille homothetique de domaines de* $\mathbb{R}^n$, Ann. Sci. Ecole Norm. Sup. **10** (1977), 559-575.

Dav H. Davenport, "Analytic methods for Diophantine Equations and Diophantine Inequalities," Univ. of Michigan lecture notes, 1962.

Ga T. Ganelius, "Tauberian Remainder Theorems," Lecture Notes in Mathematics vol 232, Springer-Verlag, 1971.

G-S I. Gelfand and G. Shilov, "Les Distributions t. 1," Dunod, 1972.

Hö L. Hörmander, "Analysis of Linear Partial Differential Operators II," Grundlehren Series vol 257, Springer-Verlag, 1983.

Ig J-I Igusa, "Lectures on Forms of Higher Degree," Tata Institute Lecture Notes Series vol 59, Springer-Verlag, 1978.

La-1 E. Landau, *Zur analytischen Zahlentheorie der definiten quadratischen Formen (Über die Gitterpunkte in einem mehrdimensionalen Ellipsoid)*, Sitzungsb. der Kgl. Preuss. Akad. der Wiss. **31** (1915), 458-476.

La-2 E. Landau, *Über die Anzahl der Gitterpunkte in gewissen Bereichen (Zweite Abhandlung)*, Kgl. Ges. d. Wiss. Nachrichten. Math-Phys. Klasse. (Göttingen) **2** (1915), 209-243.

Li-1 B. Lichtin, *Generalized Dirichlet Series and B-functions*, Comp. Math. **65** (1988), 81–120.

Li-2 B. Lichtin, *Periods and the Asymptotics of a Diophantine problem II*, Canadian J. of Mathematics (to appear).

Li-3 B. Lichtin, *On the Moderate Growth of Generalized Dirichlet Series for Hypoelliptic Polynomials*, Compositio Math. (to appear).

Li-4 B. Lichtin, *The asymptotics of a lattice point problem determined by a hypoelliptic polynomial (to appear)*.

Li-5 B. Lichtin, *The asymptotics of a lattice point problem associated to finitely many polynomials I*, Duke J. of Math. **63** (1991), 139-192.

Li-6 B. Lichtin, *The asymptotics of a lattice point problem associated to finitely many polynomials II (in preparation)*.

Li-7 B. Lichtin, *Some Formulae for Poles of* $|f(x,y)|^s$, Amer. J. of Math. **107** (1985), 139-162.

Lnd E. Lindelöf, "Le Calcul de Rèsidus et ses applications à la Theorie des Fonctions,"

Chelsea Publishing Co., 1947.

Lo-1 F. Loeser, *Fonctions d'Igusa p-adiques et polynomes de Bernstein*, Amer. J. of Math. **109** (1987), 1-22.

Lo-2 F. Loeser, *Fonctions zeta locales d'Igusa à plusieurs variables, intégration dans les fibres, et discriminants*, Ann. Sci. Ecole Norm. Sup. . **22** (1989), 435-472.

Ma K. Mahler, *Über einen Satz von Mellin*, Mathematische Annalen **100** (1928), 384-398.

Ni N. Nillson, *Asymptotic estimates for spectral functions connected with hypoelliptic differential operators*, Arkiv för Math. **5** (1965), 527-540.

Ra B. Randol, *A lattice point problem*, Trans. of AMS. **121** (1966), 257-268.

Ro D. Robert, *Proprietes spectrales d'operateurs pseudo-differential*, Comm. in Partial Differential Equations **3** (1978), 755-826.

Sa-1 P. Sargos, *Series de Dirichlet et Polygones de Newton (to appear)*.

Sa-2 P. Sargos, *Croissance de certaines séries de Dirichlet et applications*, J. für die reine und angewandte Mathematik **367** (1986), 139-154.

Sb C. Sabbah, *Proximité évanescente II. Equations fonctionelles pour plusieurs fonctions analytiques*, Comp. Math. **64** (1987), 213-241.

Sm S. A. Smagin, *Fractional powers of a hypoelliptic operator in* $\mathbf{R}^n$, Soviet math. Dokl. **14** (1973), 585-588.

Ti E. C. Titchmarsh, "The Theory of Functions," Second edition, Oxford University Press, 1939.

Va-1 A. Varcenko, *Number of lattice points in families of homothetic domains in* $\mathbf{R}^n$, Functional Analysis and Applications **17** (1983), 79–83.

Va-2 A. Varcenko, *Newton polyhedra and estimation of oscillating integrals*, Functional Analysis and Applications **10** (1976), 13-38.

Ve J.-L. Verdier, *Stratifications de Whitney et theoreme de Bertini-Sard*, Inv. Math. **36** (1976), 295-312.

Université de Bordeaux, France

Connexions méromorphes

BERNARD MALGRANGE

Université de Grenoble

A la mémoire de Georges de Rham

Ce qui suit expose, sans démonstration, certains des principaux résultats connus sur les connexions méromorphes, en particulier ceux qui ont trait au problème de Riemann. Une partie de ces résultats est classique : voir notamment [De], [Ma 1], [Maj], [Si]. D'autres sont nouveaux (ils avaient été annoncés sous forme conjecturale dans [Ma 2], malheureusement non publié) : leur démonstration paraitra ultérieurement.

1: GÉNÉRALITÉS

Soit X une variété analytique complexe de dimension n, et soit Z une hypersurface de X (= un sous-ensemble analytique fermé, partout de codimension un). On note $\mathcal{O}_X$, resp. Ω^p_X, le faisceau des fonctions, resp. des p-formes, holomorphes sur X. On note encore $\mathcal{O}_X[Z]$ le faisceau des fonctions méromorphes sur X, avec pôles sur Z, et on pose

$$\Omega^p_X[Z] = \Omega^p_X \underset{\mathcal{O}_X}{\otimes} \mathcal{O}_X[Z] \ ;$$

par définition, on a localement

$$\mathcal{O}_X[Z] = \mathcal{O}_X[f^{-1}], \Omega^p_X[Z] = \Omega^p_X[f^{-1}],$$

avec $\{f = 0\}$ une équation locale de Z. Si aucune confusion n'est à craindre, on omettra l'indice X et on écrira $\mathcal{O}, \mathcal{O}[Z]$, *etc*...

De la cohérence de $\mathcal{O}$, on déduit immédiatement celle de $\mathcal{O}[Z]$; on définit alors un $\mathcal{O}[Z]$ module cohérent E de la manière usuelle, comme étant localement le conoyau d'un morphisme $\mathcal{O}[Z]^q \to \mathcal{O}[Z]^p$; il revient au même de demander qu'il existe *localement* un $\mathcal{O}$-module cohérent L tel qu'on ait $E = L \underset{\mathcal{O}}{\otimes} \mathcal{O}[Z]$.

Soit E un $\mathcal{O}[Z]$-module cohérent ; par définition, on appellera *réseau* de E un sous $\mathcal{O}$-module cohérent L tel que la flèche évidente $L \underset{\mathcal{O}}{\otimes} \mathcal{O}[Z] \longrightarrow E$ soit bijective (noter qu'elle est toujours injective, parce que $\mathcal{O}[Z]$ est plat sur $\mathcal{O}$).

L'existence d'un réseau global pour tout $\mathcal{O}[Z]$-module cohérent E est une question ouverte. En particulier, si X provient d'une variété algébrique projective, la question est équivalente à la suivante : E provient-il toujours d'un faisceau algébrique E^{al} ? En effet, si E admet un réseau L, c'est vrai par $GAGA$: L provient d'un L^{al} et E^{al} est le localisé de L^{al} hors de Z ; inversement, si E provient d'un E^{al}, il possède nécessairement un réseau : en effet, E^{al} en possède un, d'après un résultat bien connu de géométrie algébrique.

Soit E un $\mathcal{O}[Z]$-module cohérent ; sur E une connexion se définit de la manière usuelle par la donnée d'un opérateur de dérivation $\nabla \; : E \longrightarrow \Omega^1 \underset{\mathcal{O}}{\otimes} E$ vérifiant les propriétés suivantes :

i) ∇ est **C**-linéaire.

ii) Pour $\varphi \in \mathcal{O}, e \in E$, on a $\nabla(\varphi e) = d\varphi \otimes e + \varphi \nabla e$.

Un tel ∇ se prolonge de la manière habituelle en un "complexe de De Rham"

$$DR(E) \qquad E \xrightarrow{\nabla} \Omega^1 \otimes E \xrightarrow{\nabla} \cdots \xrightarrow{\nabla} \Omega^n \otimes E \longrightarrow 0.$$

Sauf dans la proposition (1.1), on ***supposera toujours dans la suite** E **plat*** (ou : sans courbure), ie *vérifiant* $\nabla^2 = 0$.

D'habitude on restreint l'étude des connexions méromorphes au cas ou E est localement libre, donc provient localement d'un L libre sur $\mathcal{O}$. Ceci n'est pas une restriction essentielle, à cause du résultat suivant :

Proposition (1.1) Soit E un $\mathcal{O}[Z]$-module cohérent muni d'une connexion (non nécessairement plate). Alors sur tout compact $K \subset X$, E est stablement localement libre, ie il existe $F \simeq \mathcal{O}[Z]^p$ tel que $(E \oplus F)/K$ soit localement libre.

Indiquons les grandes lignes de la démonstration :

i) Sur $X - Z$, E est muni d'une connexion holomorphe au sens usuel. Il est alors bien connu (et démontré dans beaucoup d'endroits) que E est localement libre.

ii) Pour tout $a \in Z$, E_a est projectif. Il suffit de démontrer ceci : soit $\mathcal{J}_a$ un idéal de $\mathcal{O}[Z]_a$; alors on a $\mathrm{Ext}^1_{\mathcal{O}[Z]_a}(\mathcal{J}_a, E_a) = 0$. Soit, au voisinage de a, $\mathcal{J}$ un idéal cohérent de $\mathcal{O}[Z]$ qui coincide avec $\mathcal{J}_a$ en a ; il suffit de démontrer qu'on a $\underline{\mathrm{Ext}}^1_{\mathcal{O}[Z]}(\mathcal{J}, E) = 0$. Mais le premier membre est un $\mathcal{O}[Z]$-module cohérent, qui d'après *i)* est à support dans Z. On voit facilement qu'un tel module est nécessairement nul.

iii) Comme $\mathcal{O}_a$ vérifie le théorème des syzygies (= tout $\mathcal{O}_a$ module de type fini a une résolution libre de longueur finie), $\mathcal{O}[Z]_a$ le vérifie aussi par localisation. Mais alors, on voit qu'un module projectif E_a de type fini est stablement libre (raisonner par récurrence sur la longueur d'une résolution de E_a).

La proposition se déduit de là par Borel-Lebesgue.

Remarque (1.2) Si X est projective et si E provient d'un fibré algébrique, toute connexion sur E provient d'une connexion algébrique : ceci se voit en remarquant qu'une connexion est une scission de la suite exacte $0 \longrightarrow E \longrightarrow J^1(E) \longrightarrow \Omega^1 \otimes E \longrightarrow 0$ et en appliquant $GAGA$ à cette dernière suite.

2: LE CAS RÉGULIER

Supposons d'abord qu'on ait $X = D$, un disque de $\mathbf{C}$ défini par $|x| < r$, et qu'on ait $Z = \{0\}$. Soit E une connexion méromorphe sur D, avec pôle sur Z. Quitte à rétrécir D, on peut supposer que E est libre sur $\mathcal{O}_D[Z]$ (en fait, on peut voir qu'il n'est même pas nécessaire de rétrécir D, mais peu importe). Prenant une base (e_j) de E, on a $\nabla e_j = dx \otimes \Sigma m_{ij} e_i$, m_{ij} des sections de $\mathcal{O}[Z]$ ie des fonctions méromorphes sur D avec pôle en 0. Rappelons que (E, ∇) est dit "régulière" ou "à singularité régulière" s'il existe une base de E dans laquelle la matrice de la connexion $M = (m_{ij})$ est à pôle simple. Il revient au même de dire que E admet un réseau L tel qu'on ait

$$\nabla L \subset \frac{1}{x}(\Omega^1 \otimes L).$$

Soient maintenant X et Z comme au §1, et soit E un $\mathcal{O}_X[Z]$-module cohérent muni d'une connexion plate ∇. Par définition, on dit que (E, ∇) est régulier, si pour tout disque D et tout morphisme $f : D \longrightarrow X$ avec $f^{-1}(Z) = \{0\}$, la connexion $(f^*E, f^*\nabla)$ est régulière (cette connexion se définit d'une manière à peu près évidente, que je laisse le lecteur expliciter). A noter que, pour vérifier la régularité, il suffit de tester la propriété précédente sur "suffisamment" de disques : voir un énoncé précis en ce sens dans [De], (on pourra aussi consulter l'exposition qui est donnée dans [Bo]).

Les résultats principaux de Deligne (loc. cit.) sont alors les suivants :

(2.1) Le foncteur "restriction à $X - Z$" : $(E, \nabla) \mapsto (E, \nabla)|X - Z$ est une *équivalence* : connexions plates méromorphes et régulières avec pôles sur $Z \longleftrightarrow$ connexions holomorphes plates sur $X - Z$.

En particulier, si l'on se donne une connexion holomorphe et plate sur $X - Z$, il y a une et une seule manière, à isomorphisme unique près, de l'étendre à X.

Maintenant, la catégorie des connexions holomorphes et plates sur $X - Z$ est équivalente à celle des systèmes locaux de $\mathbf{C}$-vectoriels sur $X - Z$, ou encore, lorsque $X - Z$ est connexe, à celle des représentations sur $\mathbf{C}$ de dimension finie de $\pi_1(X - Z, a)$, a un point de $X - Z$ [dans un sens, associer à $(E, \nabla)|X - Z$ le système local de ses *solutions*, ie les e annulés par ∇ ; dans l'autre, si V est un système local, prendre $E = \mathcal{O} \underset{\mathbf{C}}{\otimes} V$, avec $\nabla(f \otimes v) = df \otimes v, f \in E, v \in V$].

Finalement on a une équivalence dite "de Riemann" (ou suivant les auteurs, "de Riemann-Hilbert") entre les systèmes locaux sur $X - Z$ et les connexions méromorphes et régulières sur X avec pôles sur Z.

(2.2) Si (E, ∇) est régulier, E admet un réseau L, qui est canonique une fois choisi un relèvement τ de la projection $\mathbf{C} \longrightarrow \mathbf{C}/\mathbf{Z}$.

De (2.1) et de *GAGA* résulte alors ceci : si $\check{X}$ est une variété algébrique projective sur $\mathbf{C}$, et $\check{Z}$ une hypersurface de $\check{X}$, la catégorie des systèmes locaux sur $(\check{X} - \check{Z})^{\mathrm{an}}$ est "définissable par voie algébrique", en ce sens qu'elle est équivalente à la catégorie des connexions sur $\check{X}$ avec pôles sur $\check{Z}$ (=les connexions tout court sur $\check{X} - \check{Z}$), *régulières* sur $\check{Z}$. Des arguments de complétion projective et de désingularisation permettent d'appliquer ce résultat à toutes les variétés algébriques lisses et de type fini sur $\mathbf{C}$; *cf* loc. cit.

Comme j'en aurai besoin par la suite, je vais décrire L. Hors de Z, on a nécessairement $L = E$. Sur Z la description se fait en deux temps.

a) Soit Z' la partie lisse de Z ; sur Z', E est localement libre et se décrit ainsi : au voisinage d'un point $a \in Z'$, choisissons des coordonnées locales $(x_1, \ldots, x_n)$ telles que Z' soit défini par $x_1 = 0$. Alors E possède près de a une base $(e_1, \ldots, e_m)$ dans laquelle on ait $\nabla e_j = \frac{dx_1}{x_1} \Sigma c_{ij} e_i, C = (c_{ij}) \in \mathrm{End}(\mathbf{C}^m)$.

La matrice C n'est pas entièrement déterminée : seule est fixée, à conjugaison près, la matrice $\exp 2\pi i C$, qui est la monodromie locale de la connexion. En particulier on peut choisir C de manière que ses valeurs propres appartiennent à l'image du relèvement fixé τ. Alors le réseau engendré (sur $\mathcal{O}$) par $e_1, \ldots, e_m$, est le réseau L considéré (on peut voir que ceci détermine bien un réseau, indépendamment des choix qui ont été faits).

b) Posons maintenant $T = Z - Z'$, et soit i l'injection $X - T \longrightarrow X$; on montre alors que L, qui est défini par *a)* sur $X - T$ admet un prolongement $\mathcal{O}_X$-cohérent. Le prolongement par l'image directe $L = i_*(L|X - T)$ est alors cohérent, en vertu de la proposition suivante, cas particulier d'un résultat de Serre [Se].

Proposition (2.3) Soit T un sous-ensemble analytique (fermé) de codimension ≥ 2 de X, et soit F un $\mathcal{O}$-module localement libre sur $X - T$. Les propriétés suivantes sont équivalentes :

i) F admet un prolongement cohérent.

ii) $i_* F$ est cohérent, avec i l'injection $X - T \longrightarrow X$.

[Pour vérifier que L est bien un réseau, on utilise le fait que la flèche naturelle

$E \longrightarrow i_*i^*E$ est bijective : ce point résulte immédiatement de (1.1) et du théorème de Hartogs].

Les démonstrations de Deligne utilisent une désingularisation de Z, qui permet de se ramener au cas où Z est un diviseur à croisements normaux ; dans ce cas, les démonstrations sont une extension facile du cas classique d'une variable. Plus récemment Mebkhout [Me] a redémontré ces résultats en utilisant seulement la désingularisation des courbes (plus exactement, d'une famille de courbes pour une valeur générique du paramètre). Ceci lui permet d'établir les résultats cherchés modulo un ensemble analytique $S \subset Z$, S de codimension ≥ 2 dans Z (ou ≥ 3 dans X) ; il conclut alors à l'aide du théorème suivant

Théorème (2.4) Dans les mêmes hypothèses qu'en (2.3), soit $S \subset T$, S de codimension ≥ 3 dans X. Alors les propriétés suivantes sont équivalentes :
i) F admet un prolongement cohérent à $X - S$.
ii) F admet un prolongement cohérent à X.

A noter que d'après (2.3) la première condition est locale sur $T - S$.

Ce théorème se voit ainsi : soit j l'injection $X - T \longrightarrow X - S$; si *i)* est satisfait, j_*F est cohérent sur $X - S$. On montre alors facilement que j_*F vérifie les conditions de profondeur de Frisch-Guenot, Siu, Trautmann qui assurent l'existence d'un prolongement cohérent à X. Pour les résultats de ces auteurs, voir par exemple [Do].

(2.5) Pour être complet, je mentionne deux autres résultats importants de l'article de Deligne : d'une part un calcul de la cohomologie de De Rham d'une connexion à singularité régulière ("théorème de comparaison") ; d'autre part un théorème de préservation de la régularité par image directe ("théorème de régularité"). Comme je n'aurai pas à m'en servir dans la suite de cet exposé, je renvoie pour les énoncés précis aux références citées.

3: CONNEXIONS IRRÉGULIÈRES

A l'heure actuelle, leur théorie n'est bien comprise qu'en dimension un, où elle résulte des travaux de nombreux auteurs parmi lesquels Birkhoff, Hukuhara, Turrittin, Sibuya, Jurkat. Une manière commode de l'exposer est fondée sur la notion de "structure de Stokes", due à Deligne ; voir l'exposé qui en est donné dans [Ma 1].

Je vais parler ici directement du cas général (la théorie en dimension un sera un cas particulier de ce que je vais dire). Contrairement au cas régulier, il est nécessaire d'examiner d'abord la théorie formelle, et ensuite les "invariants analytiques" qu'il faut lui rajouter.

(3.1) Le cas formel Gardons X et Z comme au §1, et soit E une connexion méromorphe avec pôles sur Z (on écrit ici E pour (E, ∇)). Pour f méromorphe avec pôles sur Z, on définit la connexion méromorphe $e^f \otimes E$ comme étant le $\mathcal{O}[Z]$ module E muni de la dérivation $\varphi \mapsto \nabla\varphi + df \otimes \varphi, \varphi \in E$. Il sera parfois commode d'écrire ici $e^f \otimes \varphi$ pour φ.

Désignons d'autre part par $\widehat{\mathcal{O}}$ le complété formel de $\mathcal{O}$ le long de Z et posons $\widehat{\mathcal{O}}[Z] = \widehat{\mathcal{O}} \underset{\mathcal{O}}{\otimes} \mathcal{O}[Z], \widehat{E} = E \underset{\mathcal{O}}{\otimes} \widehat{\mathcal{O}}$. Soit alors a un point de la partie lisse Z' de Z ; nous dirons que E admet une *décomposition formelle* au voisinage de a s'il existe des $f_i \in \mathcal{O}[Z]_a$ et des connexions méromorphes *régulières* G_i au voisinage de a telles qu'on ait $\widehat{E} \simeq \oplus e^{f_i} \otimes \widehat{G}_i$ (Bien entendu, il s'agit d'un isomorphisme de connexions et pas seulement de faisceaux).

Dans une telle décomposition, les f_i sont uniquement déterminés modulo $\mathcal{O}_a$; on peut donc, en regroupant les f_i qui diffèrent par une fonction holomorphe, supposer que les $f_i - f_j$ ont effectivement des pôles pour $i \neq j$; alors, les G_i sont uniquement déterminés.

Nous dirons encore que la décomposition est *bonne* si, pour tout $(i, j), i \neq j$, la partie la plus polaire de $f_i - f_j$ ne s'annule pas en a ; l'intérêt de cette hypothèse apparaîtra plus loin.

Le théorème suivant est classique en dimension un ; sa démonstration dans le cas général en est une extension facile.

Théorème (3.1.1) Il existe un ouvert de Zariski Z'' dense dans Z, avec $Z'' \subset Z'$, tel que, au voisinage de tout point $a \in Z''$, E admette, après ramification autour de Z, une bonne décomposition formelle.

L'assertion "après ramification..." signifie ceci : au voisinage de a, on choisit des coordonnées locales $(x_1, \ldots, x_n)$ telles que Z soit défini par $x_1 = 0$; pour $p \geq 1$, on note φ_p l'application $(y, x_2, \ldots, x_n) \mapsto (y^p, x_2, \ldots, x_n)$. Alors il existe p tel que $\varphi_p^* E$ ait une bonne décomposition formelle.

(3.2) Avec les notations du théorème précédent, posons $T = Z - Z''$ (T est donc analytique fermé dans X, de codimension ≥ 2). Le théorème précédent nous donne un réseau L dans $E|X - T$ de la manière suivante :

i) sur $X - Z$, on a $L = E$.

ii) Sur Z'', il suffit de trouver un réseau de $\widehat{E}$ (il est immédiat de vérifier que les réseaux de E et ceux de $\widehat{E}$ se correspondent bijectivement). Si E admet une décomposition formelle $\oplus e^{f_i} \otimes G_i$, il suffit de prendre $\widehat{L} = \oplus e^{f_i} \otimes \widehat{L}_i, \widehat{L}_i$ les réseaux de Deligne des G_i. On vérifie que le même procédé s'applique si E admet une décomposition formelle après ramification, à condition de faire le choix suivant du relèvement τ de $\mathbf{C} \longrightarrow \mathbf{C}/\mathbf{Z}$: on prend celui dont l'image est donnée par $0 \leq \operatorname{Re}\lambda < 1$ (on peut aussi prendre $0 < \operatorname{Re}\lambda \leq 1$).

Notons i l'injection $X - T \longrightarrow X$. On a le résultat suivant qui généralise (2.2).

Théorème (3.2.1) Le faisceau i_*L est cohérent et c'est un réseau de E.

La démonstration sera publiée ultérieurement. Comme au §2, on en déduit les résultats suivants

Corollaire (3.2.2) Si X est projective, toute connexion méromorphe sur X avec pôles sur Z provient d'une connexion algébrique sur $X - Z$.

Corollaire (3.2.3) Soit $S \subset Z$, de codimension ≥ 3 dans X. Toute connexion méromorphe sur $X - S$ avec pôles sur $Z - S$ se prolonge en une connexion méromorphe sur X avec pôles sur Z.

Remarque (3.2.4) : Plus généralement, soit $S \subset X$, analytique fermé de codimension ≥ 2, et tel que $S \cap Z$ soit de codimension ≥ 3. Alors toute connexion sur $X - S$, méromorphe sur $Z - S$, se prolonge à X. En effet le groupe fondamental de $X - (S \cup Z)$ est le même que celui de $X - Z$; donc les systèmes locaux sur ces deux espaces sont en correspondance biunivoque et, par conséquent, E se prolonge en une connexion holomorphe sur $X - Z$; on est alors ramené au corollaire précédent.

(3.3) Structure de Stokes Dans le cas d'une variable, une fois établie la décomposition formelle d'une connexion méromorphe, le travail s'achève essentiellement en analysant les connexions analytiques de formalisé donné ; ceci se fait au moyen d'une étude sectorielle ; voir [Ju], [Si], [Ma 1], et les références citées dans ces articles.

Dans la situation du théorème (3.1.1), il résulte des travaux de Majima [Maj] que la même analyse s'applique au voisinage de Z''. Voici quelques rapides indications.

Soit U un ouvert de X muni d'un système de coordonnées locales dans lequel Z est défini par $x_1 = 0$. Soit $\tilde{U} \xrightarrow{\pi} U$ l'éclaté réel de U le long de Z (ceci revient à remplacer x_1 par les coordonnées polaires (ρ, θ) avec $\rho \geq 0, x_1 = \rho e^{i\theta}$) ; on pose $S = \pi^{-1}(Z)$. Sur S, on considère le faisceau $\mathcal{A}$ défini ainsi : en $\xi = (\theta^0, x_2^0, \cdot, x_n^0)$, $\mathcal{A}_\xi$ est formé par les fonctions f holomorphes en $x_1, \ldots, x_n$ dans un petit secteur $|\theta - \theta^0| < \varepsilon, \rho < \varepsilon, |x_i - x_1^0| < \varepsilon$ et admettant pour $x_1 \longrightarrow 0$ un développement asymptotique $f(x_1, \ldots, x_n) \sim \sum_{p \geq 0} a_p(x_2, \ldots, x_n) x_1^p$, et ceci uniformément par rapport aux $x_i (i \geq 2)$ dans le polydisque $|x_i - x_1^0| < \varepsilon$. Une généralisation facile d'un théorème classique de E. Borel nous dit que l'application "série de Taylor" $f \mapsto Tf = \Sigma a_p x_1^p$ est surjective de $\mathcal{A}_\xi$ sur $\widehat{\mathcal{O}}_a$, avec $a = (0, x_2^0, \ldots, x_n^0) = \pi(\xi)$, $\widehat{\mathcal{O}}$ le complété formel de $\mathcal{O} = \mathcal{O}_X$ le long de Z.

Cela posé, soit E une connexion méromorphe sur U, avec pôles sur $Z \cap U$, admettant une bonne décomposition formelle après ramification ; on se propose de classer les connexions F méromorphes sur Z, définies au voisinage d'un point $a \in Z \cap U$, et munies d'un isomorphisme formel (de connexions, pas seulement de $\widehat{\mathcal{O}}_a$-modules) $\widehat{E}_a \xrightarrow{\widehat{u}} \widehat{F}_a$; la relation d'équivalence est la suivante : $(\widehat{u}, F)$ et $(\widehat{u}', F')$ sont équivalentes si $\widehat{v} = \widehat{u}'\widehat{u}^{-1} : \widehat{F}_a \to \widehat{F}'_a$ est convergent, ie provient d'une flèche $v : F_a \to F'_a$.

Pour cela, on opère ainsi : $\widehat{u}$ définit, au voisinage de a, une section horizontale (=annulée par ∇) de la connexion $\underline{\mathrm{Hom}}(E, F)^\wedge = \widehat{F}^* \otimes \widehat{E}$; en tout point $\xi \in \pi^{-1}(a)$, d'après l'extension du théorème de Borel rappelée plus haut, $\widehat{u}$ se représente par un $u_\xi \in \mathcal{A}_\xi \underset{\mathcal{O}_a}{\otimes} \underline{\mathrm{Hom}}(E, F)_a$; alors, le *théorème des développements asymptotiques nous dit qu'on peut prendre u_ξ annulé par ∇*, *cf* [Maj].

On peut donc trouver un recouvrement ouvert $\{U_i\}$ de $S_a = \pi^{-1}(a)$ et des $u_i \in \Gamma(U_i, \mathcal{A} \otimes \underline{\mathrm{Hom}}(E, F))$, annulés par ∇, et qui représentent $\widehat{u}$. Alors, sur $U_i \cap U_j, u_j^{-1}u_i$ est un morphisme de E dans E, et plus précisément une section horizontale de $\mathcal{A} \otimes \underline{\mathrm{End}} E$ dont le développement asymptotique est égal à l'identité ; désignons par $(\mathcal{A} \otimes \underline{\mathrm{End}} E)^{\nabla, 0}$ le faisceau sur S formé de telles sections (le "∇" pour la condition d'horizontalité, et le "0" pour la condition sur le développement asymptotique) ; ceci est un faisceau de groupes non commutatif pour la loi de composition évidente, et ce qui précède associe à $\widehat{u}$ un élément $\gamma(\widehat{u}) \in H^1(S_a, (\mathcal{A} \otimes \underline{\mathrm{End}} E)^{\nabla, 0})$. Il est immédiat de voir que $(\widehat{u}, F)$ et $(\widehat{u}', F')$ sont équivalents si et seulement si $\gamma(\widehat{u}) = \gamma(\widehat{u}')$. D'où une application injective :

$$\frac{(\text{Connexions } F \text{ munies de } \widehat{E}_a \xrightarrow[\sim]{\widehat{u}} \widehat{F}_a)}{(\text{équivalence})} \xrightarrow{\gamma} H^1(S_a, \mathcal{A} \otimes \underline{\mathrm{End}} E)^{\nabla, 0})$$

En fait, cette application est *bijective* ; la surjectivité se voit en démontrant un théorème de matrices holomorphes inversibles qui est une extension avec paramètres d'un théorème de Birkhoff-Sibuya ; voir [Maj].

On passe de là à une situation globale en montrant que le faisceau (d'ensembles) $a \mapsto H^1(S_a, (\mathcal{A} \otimes \underline{\mathrm{End}} E)^{\nabla, 0})$ est localement constant (la structure de faisceau sur cette "image directe non commutative" est évidente). Cela se voit en reprenant avec l'adjonction de paramètres, ici inoffensifs, les descriptions détaillées données dans le cas d'une variable par [Ju] ou [Lo]. A noter qu'ici et aussi d'ailleurs pour le théorème des développements asymptotiques, l'hypothèse que la décomposition formelle est *bonne* joue un rôle essentiel.

Comme en dimension un, les résultats précédents peuvent s'exprimer sous une forme plus géométrique, en utilisant la notion "structure de Stokes". Cette notion, en fait déjà présente de façon implicite chez Birkhoff, a été dégagée par Deligne ; voir un exposé dans [Ma 1].

Pour ne pas trop allonger l'exposé, je me limiterai à la situation locale, et au cas où il n'y a pas de ramification ; l'extension au cas général n'apporte pas de modification essentielle.

Soit a un point de la partie lisse de Z, et soit U un voisinage ouvert de a (non précisé ; on le restreindra au besoin). Soit, comme ci-dessus, $\widetilde{U} \xrightarrow{\pi} U$ l'éclaté réel de U le long de Z, et soit $S = \pi^{-1}(Z)$. On suppose encore U muni de coordonnées locales, avec $Z \cap U = \{x_1 = 0\}$.

Donnons-nous des fonctions méromorphes $f_i \in \Gamma(U, \mathcal{O}[Z]), i = 1, \dots, p$; on suppose que, pour $i \neq j, f_i - f_j$ a effectivement des pôles, et de plus, que la partie la plus polaire de $f_i - f_j$ ne s'annule pas sur $Z \cap U$. Sur S, on met la relation d'ordre suivante : en $\xi = (\theta^0, x_2^0, \cdot, x_n^0)$, on a $f_i \leq_\xi f_j$ si, dans un petit secteur comme ci-dessus, $|\theta - \theta^0| < \varepsilon, \rho < \varepsilon, |x_i - x_i^0| \leq \varepsilon (i \geq 2), e^{f_i - f_j}$ est bornée ; l'ensemble des ξ où cette relation est satisfaite est un ouvert. Mais ce n'est pas une relation d'ordre total en chaque ξ : si $i \neq j$, il existe des ξ où f_i et f_j sont incomparables.

Ceci étant, une structure de Stokes au voisinage de a, paramétrée par les f_i, consiste en les données suivantes

i) Un système local V de $\mathbf{C}$-vectoriels (de dimension finie) sur $\widetilde{U}$.

ii) Sur S, une famille de sous-faisceaux V^i de V.

On impose en outre la condition suivante :

(*) Pour tout $\xi \in S$, il existe une décomposition (non nécessairement unique) $V_\xi = \oplus W^i$ telle que, pour tout ξ' assez voisin de ξ, on ait $V^i_{\xi'} = \bigoplus_{f_j \leq_{\xi'} f_i} W^j$ (les W^j sont identifiés à des sous espaces de $V_{\xi'}$ grâce à l'isomorphisme canonique $V_{\xi'} \simeq V_\xi$).

La condition précédente entraine en particulier qu'on a $V^j_\xi \subset V^i_\xi$ si $f_j \leq_\xi f_i$.

Soit F une connexion méromorphe avec pôles sur Z au voisinage de a, admettant une bonne décomposition formelle $\widehat{F} \simeq \widehat{E}$, avec $E = \oplus e^{f_i} \otimes G_i, G_i$ à singularités régulières ; on lui associe une structure de Stokes de la manière suivante :

i) On prend pour V les sections horizontales de F.

ii) En $\xi \in S$, l'isomorphisme $\hat{u}$ se réalise par un isomorphisme sectoriel U_ξ comme expliqué plus haut ; si l'on prend pour W_i les sections horizontales de G_i, U_ξ donne un isomorphisme $V_\xi \simeq \oplus W_\xi^i$; on fabrique alors les V^i à partir de (*). Ils ne dépendent pas de U_ξ ni même de $\hat{u}$, car ce sont les sections sectorielles v de V telles que $e^{-f_i}v$ soit à croissance modérée près de Z ; par contre la *décomposition* $V_\xi = \oplus W^i$ dépend effectivement de $\hat{u}$.

On a ainsi fabriqué un foncteur "connexions méromorphes sur Z au voisinage de a, admettant une bonne décomposition formelle "$\mapsto$" structures de Stokes au voisinage de a". On montre à partir des résultats qui précèdent que c'est une *équivalence* ; la démonstration est la même qu'en dimension un. Enfin, comme je l'ai dit plus haut, la même théorie s'applique aussi au cas ramifié, et à la situation globale.

(3.4) Conclusion provisoire? - Elle est fondée sur la remarque suivante : soit E une connexion méromorphe sur X, avec pôles sur Z, et soit $T \subset Z$, analytique fermé de codimension ≥ 2 dans X ; alors toute section de E sur $X - T$ se prolonge à X de manière unique : ceci résulte immédiatement de (1.1) et du théorème de Hartogs.

Soit F une autre connexion méromorphe, avec pôles sur Z ; appliquant ce résultat à $\underline{\mathrm{Hom}}(E, F)$ on trouve le résultat suivant

Proposition (3.4.1) Le foncteur "oubli de T" : (connexions sur X, avec pôles sur Z) $\mapsto$ (connexions sur $X - T$, avec pôles sur $Z - T$) est pleinement fidèle.

De plus, si codim $T \geq 3$, ce foncteur est une équivalence.

La dernière assertion résulte de la première et de (3.2.3).

Maintenant, hors d'un sous-ensemble convenable T de codimension 2, E se décrit de façon géométrique par un système local sur $X - Z$, plus une structure de Stokes sur $Z - T$. Pour achever une description géométrique, il faudrait savoir à quelles conditions une connexion ainsi décrite sur $X - T$ se prolonge à X.

D'après (3.2.3), il suffirait de trouver les conditions d'existence d'un tel prolongement en dehors d'un $S \subset T$, S de codimension ≥ 3 dans X. Un cas crucial à comprendre est celui où $\dim X = 2, T =$ un point. Dans ce cas, les remarquables résultats tout récents de Sabbah [Sa], joints aux travaux de Majima déjà cités devraient permettre d'aborder le problème.

P.S. : M. H. Majima me signale que le théorème (3.1.1) est bien connu de lui ; il figure notamment dans son exposé "Cohomological characterization of regular singularities in several variables" Kyoto, Octobre 1983 (exposé en japonais).

Bibliographie

[Bo] Borel A. & al Algebraic D-modules, Persp. Math. **2** Acad. Press (1987).

[De] Deligne P. Equations différentielles à points singuliers réguliers, Lecture Notes in Math. Springer **163** (1970).

[Do] Douady A. Prolongement de faisceaux analytiques cohérents (travaux de Trautmann, Frisch-Guenot, et Siu) Sém. Bourbaki **366** (1969) 70.

[Ju] Jurkat W. B. Meromorphe Differentialgleichungen. Lecture Notes in Math. Springer **637** (1978).

[Lo] Loday-Richaud M. Classification méromorphe locale des systèmes différentiels linéaires méromorphes,Thèse, Orsay (1991).

[Ma 1] Malgrange B. La classification des connexions irrégulières à une variable Progress in Math., Birkhäuser 37 (1983) 381-399.

[Ma 2] Malgrange B. Meromorphic connections : some problems and conjectures, Proceedings Taniguchi coll. on diff. equations, (non publié) (1987).

[Maj] Majima H. Asymptotic analysis for integrable connections with irregular singular points. Lecture Notes in Math. Springer **1075** (1984).

[Me] Mebkhout Z. Le théorème de comparaison entre cohomologies de de Rham d'une variété analytique complexe et le théorème d'existence de Riemann, Publ. Math. IHES 69 (1989) 47-89.

[Sa] Sabbah C. Connexions méromorphes formelles de deux variables, Manuscrit (1991).

[Si] Sibuya Y. Linear differential equations in the complex domain : problems of analytic continuation, Transl. of Math. Monographs **82** AMS (1990).

Deformations of maps on complete intersections, Damon's $\mathcal{K}_V$-equivalence and bifurcations

David Mond and James Montaldi*

Abstract

A recent result of J. Damon's [4] relates the $\mathcal{A}_e$-versal unfoldings of a map-germ f with the $\mathcal{K}_{D(G)}$-versal unfoldings of an associated map germ which induces f from a stable map G. We extend this result to the case where the source is a complete intersection with an isolated singularity. In a similar vein, we also relate the bifurcation theoretic versal deformation of a bifurcation problem (map-germ) g to the $\mathcal{K}_\Delta$-versal deformation of an associated map germ which induces g from a versal deformation of the organizing centre g_0 of g, where Δ is the bifurcation set of this versal deformation.

The extension of Damon's theorem is used to provide an extension (again to cases where the source is an ICIS) of a result of Damon and Mond relating the discriminant Milnor number of a map to its $\mathcal{A}_e$-codimension.

Introduction

Recently Damon has introduced a generalized version of contact equivalence of map-germs, which he calls $\mathcal{K}_V$-equivalence ([3]; see the background section below for definitions). In [4] he shows how the deformation theory for left-right equivalence of a map-germ $f : \mathbf{C}^n, 0 \to \mathbf{C}^p, 0$ can be understood in terms of the $\mathcal{K}_V$-equivalence and deformation theory of an auxilliary map-germ γ which induces f from a stable unfolding G by a base change, taking as V the discriminant of G (our G and γ replace Damon's F and g):

$$\begin{array}{ccc} \mathbf{C}^{n+q},0 & \xrightarrow{\;G\;} & \mathbf{C}^{p+q},0 \supset V \\ \Big\uparrow i & & \Big\uparrow \gamma \\ \mathbf{C}^n,0 & \xrightarrow{\;f\;} & \mathbf{C}^p,0 \end{array}$$

*Partially supported by the SERC.

where i is the inclusion. This identification of the two deformation theories led to the discovery, by Damon and Mond [5], of a "$\mu \geq \tau$" type theorem, relating the $\mathcal{A}_e$-codimension of a map-germ $\mathbf{C}^n, 0 \to \mathbf{C}^p, 0$ $(n \geq p)$ to the vanishing topology of its discriminant.

In this paper we begin by generalizing Damon's result to include the case where the domain of f is an isolated complete intersection singularity (ICIS) X, giving in the process what we feel to be a clearer proof. This leads immediately to the following generalization of the main theorem of [5] to the case of a map $f : X, 0 \to \mathbf{C}^p$, where X is an n-dimensional ICIS with $n \geq p$ and (n, p) nice-dimensions in the sense of Mather. Let $f_t : X_t \to \mathbf{C}^p$ be a stable perturbation of the pair (X, f), i.e. X_t is a smoothing of X and f_t is a stable map. Then the discriminant of f_t has the homotopy type of a wedge of $(p-1)$-spheres the number μ_Δ of which satisfies

$$\mu_\Delta \geq \mathcal{A}_e\text{-codim}(X, f),$$

with equality if (X, f) is quasihomogeneous. Here $\mathcal{A}_e\text{-codim}(X, f)$ denotes the number of parameters necessary for a versal deformation of (X, f) in the sense defined in Section 1 below. This theorem is proved in Section 2. (A more algebraic proof of Damon's theorem has recently been given by du Plessis, Gaffney and Wilson [15].)

In Section 3, we apply the idea of $\mathcal{K}_V$-equivalence to bifurcation theory. The map-germ $g(x, \lambda)$ (and associated "bifurcation problem" $g(x, \lambda) = 0$) can be obtained from a versal deformation $G(x, u)$ of $g_0(x) = g(x, 0)$, by means of a map γ from λ-space to u-space. We show that the bifurcation-theoretic deformation theory of g is isomorphic to the $\mathcal{K}_V$-deformation theory of γ, where V is the bifurcation set of the deformation G (Theorem 3.1). This provides a theoretical framework for the "path formulation" of bifurcation theory [6, §12(b)]. Combining this with the results of Section 1, we obtain the following extension of a classical theorem due to Martinet (Theorem 3.3): let $V_g = g^{-1}(0)$, and $\pi_g : V_g \to \Lambda$ be the projection (where Λ is λ-space, the parameter space of the deformation). Then the $\mathcal{A}_e$ codimension of the pair (V_g, π_g) (as defined in Section 1) is equal to the bifurcation theoretic codimension of g.

For simplicity, the results in this paper are all proved for the complex analytic category. However, except for those in Section 2, they are all valid in the real analytic category as well.

Acknowledgements We would like to thank Jacques Furter for some useful discussions, especially concerning Section 3. We would also like to take this opportunity to thank J.-P. Brasselet in particular, and also the other organizers, for such an excellent and well-organized conference.

Background

Here we describe briefly three equivalence relations on spaces of map germs and the corresponding deformation theories. In what follows, X and Y denote germs of analytic spaces. If $X \subset E$ (E smooth), Derlog(X) denotes the $\mathcal{O}_E$-submodule of Θ_E consisting of vector fields tangent to X, while Θ_X denotes the $\mathcal{O}_X$-module of vector fields on X. Thus, $\Theta_X \cong \mathrm{Derlog}(X)/I(X)\Theta_E$. For a Cartesian product $X \times Y$ we use $\Theta_{X\times Y/Y}$ to denote the $\mathcal{O}_{X\times Y}$-module of vector fields on $X \times Y$ tangent to the fibres of the Cartesian projection $X \times Y \to Y$.

$\mathcal{A}$- (left-right) equivalence: Two map germs $f, \overline{f} : X \to Y$ are $\mathcal{A}$-equivalent if there are diffeomorphisms ϕ of X and ψ of Y with $\overline{f} = \psi \circ f \circ \phi$. The $\mathcal{O}_Y$-module

$$N\mathcal{A}_e \cdot f := \frac{f^*\Theta_Y}{tf(\Theta_X) + f^{-1}\Theta_Y}$$

is the $\mathcal{A}_e$ normal space for f. When X and Y are smooth, then by the versality theorem ([12], [17]), $\dim_{\mathbb{C}} N\mathcal{A}_e f$ is equal to the number $\mathcal{A}_e$-codim(f) of parameters necessary for an $\mathcal{A}_e$-versal unfolding of f. The same statement with X an ICIS (and Y smooth) is given in Corollary 1.6 below.

$\mathcal{K}_{\mathrm{un}}$-equivalence: Let $g_0 : X \to Y$ be a map germ and let $g, \overline{g} : X \times \Lambda \to Y$ be deformations of g_0. They are $\mathcal{K}_{\mathrm{un}}$-equivalent if there are diffeomorphisms H of $X \times \Lambda$ and h of Λ with $H(x, \lambda) = (H'(x, \lambda), h(\lambda))$ such that

$$g(x, \lambda) = S(x, \lambda).\overline{g}(H(x, \lambda)),$$

with $S(x, \lambda) \in GL_p$. (In bifurcation theory, this is simply called, *equivalence of the bifurcation problems* $g = 0$ and $\overline{g} = 0$. The subscript 'un' is due to Damon [2] and refers to unfoldings.)

The $\mathcal{K}_{\mathrm{un}}$ normal space of g is the $\mathcal{O}_\Lambda$-module

$$N\mathcal{K}_{\mathrm{un}} \cdot g := \frac{g^*\Theta_Y}{tg(\Theta_{X\times\Lambda/\Lambda} + \pi^{-1}\Theta_\Lambda) + g^* m_Y \Theta_Y}$$

where $\pi : X \times \Lambda \to \Lambda$ is the Cartesian projection. (In bifurcation theory, $\dim N\mathcal{K}_{\mathrm{un}} g$ is called the codimension of the bifurcation problem $g = 0$.)

$\mathcal{K}_V$-equivalence: Let V be a subspace of Y. Two map germs $\gamma, \overline{\gamma} : X \to Y$ are said to be $\mathcal{K}_V$-equivalent if there are diffeomorphisms H of $X \times Y$ and h of X with H preserving $X \times V$, and $H(x, y) = (h(x), H'(x, y))$ satisfying

$$H(x, \gamma(x)) = (h(x), \overline{\gamma} \circ h(x)).$$

This group of diffeomorphisms was introduced by J. Damon in [3]. Damon shows that the $\mathcal{K}_V$ normal space for γ is given by

$$N\mathcal{K}_{V,e} \cdot \gamma = \frac{\gamma^*\Theta_Y}{t\gamma(\Theta_X) + \gamma^* \operatorname{Derlog}(V)}.$$

The dimension of $N\mathcal{K}_V\gamma$ is thus the number of parameters necessary for a $\mathcal{K}_{V,e}$-versal deformation of γ.

In the case that $V = \{0\} \subset Y$, then $\mathcal{K}_V$-equivalence reduces to $\mathcal{K}$-equivalence. It is perhaps worth pointing out that in contrast to the case $V = \{0\}$, isomorphism of $\gamma^{-1}(V)$ and $\overline{\gamma}^{-1}(V)$ does not in general imply the $\mathcal{K}_V$-equivalence of γ and $\overline{\gamma}$.

1 Deformations of maps on complete intersections

Let X be a germ of an isolated complete intersection singularity (ICIS), with base point 0. Let $f : X \to P$ be a map-germ, where P is a germ of a smooth space ($P = \mathbf{C}^p, 0$). We say $x \in X$ is a singular point of (X, f) either if x is a singular point of X, or, in the case that $x \in X$ is regular, if f is not a submersion at x. We denote the set of singular points by $\Sigma_{(X,f)}$. We say that f, or (X, f), has *finite singularity type* if the restriction of f to $\Sigma_{(X,f)}$ is finite-to-one.

Let $G : N \to Q$ and $\gamma : P \to Q$ be map germs (where N, P, and Q are germs of smooth spaces, of dimensions n, p and q respectively). Pulling back G by γ gives a space X as the fibre product of G and γ, and a map $f = f_\gamma : X \to P$ which is just the projection to P, giving the following commutative diagram.

$$\begin{array}{ccc} N & \xrightarrow{G} & Q \\ {\scriptstyle i}\uparrow & & \uparrow{\scriptstyle \gamma} \\ X & \xrightarrow{f} & P \end{array}$$

We will say that (X, f) is induced from G by γ (by fibre product).

Recall that the fibre product of G and γ can be defined by

$$X = \{(x, y) \in N \times P \mid G(x) = \gamma(y)\},$$

or equivalently as $(G \times \gamma)^{-1}(\Delta_{Q\times Q})$, where $\Delta_{Q\times Q}$ is the diagonal in $Q \times Q$. It is clear that X is smooth of codimension q in $N \times P$ if and only if γ is transverse to G, and (provided $n + p \geq q$) is an ICIS of codimension q if and only if γ is transverse to G off $(0, 0) \in X$.

The first result of this section asserts that provided X is an ICIS and G is stable then up to isomorphism, any unfolding of $f : X \to P$ can be obtained by deforming γ. Prior to proving this, we show that in fact a large class of pairs (X, f) can be obtained as fibre products.

Lemma 1.1 *Suppose that X is an ICIS and $f : X \to P$ is a map germ of finite singularity type. Then there exist a stable map germ $G : N \to Q$, and an immersion germ $\gamma : P \to Q$, such that f is induced from G by γ (by fibre product).*

Remark The property that (X, f) be of finite singularity type holds in general (e.g. [17], Theorem 5.1). Moreover, a small modification of the argument given in [17] shows that for any given ICIS X the property that $f : X \to P$ is of finite singularity type also holds in general.

PROOF Suppose that $h : E \to \mathbf{C}^k, 0$ is a map-germ with $h^{-1}(0) = X$, and let $\hat{f} : E \to P$ extend f. The hypothesis on (X, f) guarantees that the map-germ $(h, \hat{f}) : E \to (\mathbf{C}^k, 0) \times P$ is of finite singularity type. It then follows by a theorem of Mather that there exists a stable map-germ

$$G : E \times (\mathbf{C}^d, 0) \to (\mathbf{C}^k, 0) \times P \times (\mathbf{C}^d, 0)$$

which is a level preserving unfolding of $(h, \hat{f})$. We can construct G by means of the following algorithmic procedure: choose $\alpha_1, \ldots, \alpha_n \in \Theta(h)$ and $\beta_1, \ldots, \beta_b \in \Theta(\hat{f})$ such that, viewing $\Theta(h)$ and $\Theta(\hat{f})$ as subspaces of $\Theta(h, \hat{f})$, the α_i and β_j together form a $\mathbf{C}$-basis for the quotient

$$\frac{\boldsymbol{m}_E\, \Theta(h, \hat{f})}{T\mathcal{K}_e(h, \hat{f}) \cap \boldsymbol{m}_E\, \Theta(h, \hat{f})}.$$

Let

$$G(x, u) = (h(x) + \sum_{i=1}^{a} u_i \alpha_i(x),\ \hat{f}(x) + \sum_{i=1}^{b} u_{a+j} \beta_{b_j}(x),\ u),$$

where $a + b = d$. Now define $\gamma : P \to (\mathbf{C}^k, 0) \times P \times (\mathbf{C}^d, 0)$ by $\gamma(y) = (0, y, 0)$. Then G is infinitesimally stable (see e.g. [13]) and it is easy to see that γ has the required properties. □

We need to make precise the notions of unfolding, and versal unfolding, of the map $f : X \to P$, and of the pair (X, f).

Definition 1) An *unfolding* of (X, f) *over* a space germ S is a map $F : \mathcal{X} \to P \times S$ together with a flat projection $\pi : \mathcal{X} \to S$, such that if $\pi_2 : P \times S \to S$ is the cartesian projection, then

i) $\pi_2 \circ F = \pi$,
ii) There is an isomorphism $j : X \to \pi^{-1}(0)$, and $F \circ j : X \to P \times \{0\} = P$ is equal to f.
2) The unfoldings $(\mathcal{X}, \pi, F, j)$ and $(\mathcal{X}', \pi', F', j')$ over S are *isomorphic* if there are isomorphisms $\Phi : \mathcal{X} \to \mathcal{X}'$ and $\Psi : P \times S \to P \times S$ over S, such that the diagram below commutes.

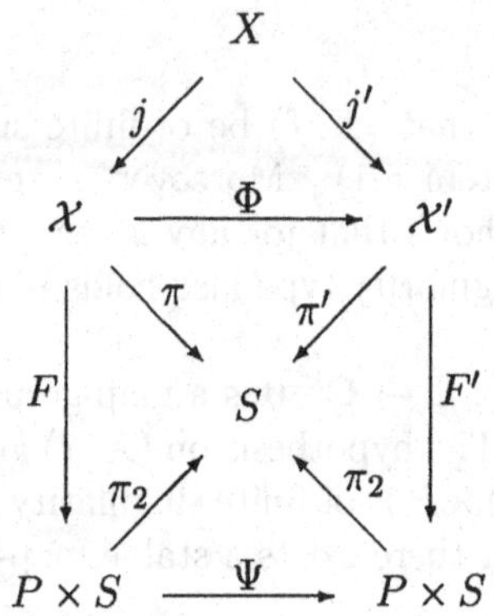

3) If $(\mathcal{X}, \pi, F, j)$ is an unfolding of (X, f) over S, a germ $\rho : T \to S$ induces an unfolding $(X_\rho, \pi_\rho, F_\rho, j_\rho)$ of (X, f) by base change, in the obvious way. The unfolding $(\mathcal{X}, \pi, F, j)$ is *versal* if every unfolding of (X, f) is isomorphic to an unfolding induced from $(\mathcal{X}, \pi, F, j)$ by base change.
4) An *unfolding of* f is an unfolding of (X, f) with $\mathcal{X} = X \times S$, and with $\pi : \mathcal{X} \to S$ being the cartesian projection.

Canonical Construction Suppose that $f : X \to P$ is induced from the map $G : N \to Q$ by $\gamma : P \to Q$ (by fibre product). Associated to any deformation $\Gamma : P \times S \to Q$ of γ there is a fibre product $\mathcal{X} = N \times_Q (P \times S)$, a map $F : \mathcal{X} \to P \times S$ and a natural map $i : X \to \mathcal{X}$ induced by the inclusion $P \times \{0\} \hookrightarrow P \times S$, such that $\pi_2 \circ F \circ i = f$. We make $F : \mathcal{X} \to P \times S$ into an unfolding of f by defining $\pi : \mathcal{X} \to S$ by $\pi = \pi_2 \circ F$.

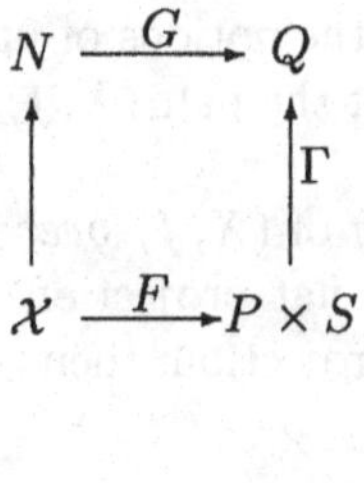

Fibre product is of course unique only up to (unique) isomorphism; in this case such an isomorphism is automatically an isomorphism over S, and thus gives rise to an isomorphism of unfoldings.

Proposition 1.2 *Let X be an* ICIS *and let $f : X \to P$ be a map of finite singularity type. If (X, f) is induced from the stable map $G : N \to Q$ by $\gamma : P \to Q$, and if $(\mathcal{X}, \pi, F, j)$ is an unfolding of (X, f), then there exists a deformation Γ of γ such that $(\mathcal{X}, \pi, F, j)$ is isomorphic to the unfolding obtained from Γ by the canonical construction above.*

PROOF Since X is a complete intersection, a deformation over the base S extends to a deformation over a smooth ambient space of S, so we may assume from the outset that S is smooth. Moreover, any deformation of X can be realised as an embedded deformation, so we may suppose that there is a map $H : E \times S \to \mathbf{C}^k$, with $H(x, s) = h(x) + r(x, s)$ deforming h, such that $\mathcal{X} = H^{-1}(0)$; by extending $F : \mathcal{X} \to P \times S$ to a level-preserving map $\hat{F} : E \times S \to P \times S$, with $\hat{F}(x, s) = (\hat{f}(x) + t(x, s), s)$ (for some extension $\hat{f}$ of f) we obtain an unfolding $(H, \hat{F}) : E \times S \to \mathbf{C}^k \times P \times S$ of the map (h, f), and now we define an unfolding $L : E \times \mathbf{C}^d \times S \to \mathbf{C}^k \times P \times \mathbf{C}^d \times S$ which is the "direct sum" of $(H, \hat{F})$ and G, by

$$L(x, u, s) = \Big(h(x) + \sum_{i=1}^{a} u_i \alpha_i(x) + r(x, s), \hat{f}(x) + \sum_{j=1}^{b} u_{a+j} \beta_j(x) + t(x, v), u, s\Big).$$

Note that $(\mathcal{X}, \pi, F, j)$ is induced from L by the map-germ $\Gamma_0 : P \times S \to \mathbf{C}^k \times P \times \mathbf{C}^d \times S$ defined by $\Gamma_0(y, s) = (0, y, 0, s)$:

$$\begin{array}{ccc} E \times \mathbf{C}^d \times S & \xrightarrow{\quad L \quad} & \mathbf{C}^k \times P \times \mathbf{C}^d \times S \\ {\scriptstyle I_0}\uparrow & & \uparrow{\scriptstyle \Gamma_0} \\ \mathcal{X} & \xrightarrow{\quad F \quad} & P \times S \end{array}$$

where $I_0(x, s) = (x, 0, s)$.

Now L is an unfolding of G, with base S, and as G is stable, L is trivial. Thus, there exist S-level-preserving diffeomorphisms Φ of $E \times \mathbf{C}^d \times S$ and Ψ of $\mathbf{C}^k \times P \times \mathbf{C}^d \times S$, such that the diagram

$$\begin{array}{ccc} E \times \mathbf{C}^d \times S & \xrightarrow{\quad G \times \mathrm{id}_S \quad} & \mathbf{C}^k \times P \times \mathbf{C}^d \times S \\ {\scriptstyle \Phi}\uparrow & & \uparrow{\scriptstyle \Psi} \\ E \times \mathbf{C}^d \times S & \xrightarrow{\quad L \quad} & \mathbf{C}^k \times P \times \mathbf{C}^d \times S \end{array}$$

commutes. Juxtaposing these two diagrams gives the following fibre square.

$$\begin{array}{ccc} E \times \mathbf{C}^d \times S & \xrightarrow{G \times \mathrm{id}_S} & \mathbf{C}^k \times P \times \mathbf{C}^d \times S \\ \uparrow{\scriptstyle \Phi \circ I_0} & & \uparrow{\scriptstyle \Psi \circ \Gamma_0} \\ \mathcal{X} & \xrightarrow{F} & P \times S \end{array}$$

Since I_0, Γ_0, Φ and Ψ are all S-level preserving, if we jettison the trivial S component on the top row, we still have a fibre square

$$\begin{array}{ccc} E \times \mathbf{C}^d & \xrightarrow{G} & \mathbf{C}^k \times P \times \mathbf{C}^d \\ \uparrow{\scriptstyle \phi \circ I_0} & & \uparrow{\scriptstyle \psi \circ \Gamma_0} \\ \mathcal{X} & \xrightarrow{F} & P \times S \end{array}$$

where ϕ and ψ are Φ and Ψ composed with the cartesian projections forgetting S. Now $\Gamma = \psi \circ \Gamma_0$ is a deformation of γ, and the proposition is proved. □

For a map germ G, we denote the discriminant of G by $D(G)$. The main result of this section, generalizing Damon's theorem [4], is

Theorem 1.3 *Let X be an* ICIS *and $f : X \to P$ be of finite singularity type. Suppose that (X, f) is induced from the stable map $G : N \to Q$ by the base-change $\gamma : P \to Q$. Then a $\mathcal{K}_{D(G),e}$-versal deformation $\Gamma_0 : P \times \mathbf{C}^d, 0 \to Q$ of γ induces a versal unfolding F_0 of f. If Γ_0 is miniversal, then so is F_0.*

PROOF We begin by showing that it is enough to prove this result in the case that γ is an immersion.

Define $\tilde{G} : N \times P \to Q \times P$ by $\tilde{G}(x, y) = (G(x), y)$, and the immersion $\tilde{\gamma} : P \to Q \times P$ by $\tilde{\gamma}(y) = (\gamma(y), y)$. Then (X, f) is also induced by $\tilde{\gamma}$ from $\tilde{G}$ (by fibre product). Moreover, $D(\tilde{G}) = D(G) \times P$. It is easy to see that if $\tilde{\Gamma} = (\Gamma, \Gamma')$ is an unfolding of $\tilde{\gamma}$, then $\tilde{\Gamma}$ is $\mathcal{K}_{D(\tilde{G}),e}$-versal if and only if Γ is $\mathcal{K}_{D(G),e}$-versal (see [4, Proposition 1.5]). It follows that it is enough to prove the theorem for γ an immersion. This we now proceed to do.

For the remainder of the proof we follow Damon closely, and give this proof mainly for the sake of completeness. Let $\Gamma_1 : P \times \mathbf{C}^c, 0 \to Q$ be any deformation of γ, and let $\Gamma : P \times \mathbf{C}^d \times \mathbf{C}^c$ be the direct sum of Γ_0 and Γ_1. Each of these deformations of γ induces an unfolding of f by the canonical construction. We represent them as F_0, F_1 and F. Since Γ_1 is obtained from

Γ by restriction, F_1 is obtained from F by restriction. To show that F_0 is versal, it is enough to show that F is a trivial extension of F_0. We show that this follows from the fact that Γ is a $\mathcal{K}_{D(G),e}$-trivial extension of Γ_0, by lifting the vector fields whose integral flows trivialise Γ. As this is done one dimension at a time, we may as well suppose, in order to lighten our notation, that Γ_1 is a 1-parameter deformation of γ.

Now Γ is a $\mathcal{K}_{D(G),e}$-trivial extension of Γ_0, by the $\mathcal{K}_{D(G),e}$-versality of Γ_0. Since $\Gamma \times \mathrm{id}_{\mathbf{C}^d \times \mathbf{C}}$ is an immersion, the infinitesimal criterion for $\mathcal{K}_{D(G),e}$-triviality takes an especially simple form: there exist vector fields $\eta \in \Theta_{\mathbf{C}^d \times \mathbf{C}^c/\mathbf{C}^c}$, $\zeta \in \Theta_{P \times \mathbf{C}^d \times \mathbf{C}^c/\mathbf{C}^d \times \mathbf{C}^c}$ and $\chi \in \mathrm{Derlog}(D(F) \times \mathbf{C}^d \times \mathbf{C}^c)/\mathbf{C}^d \times \mathbf{C}^c$, such that

$$\frac{\partial \Gamma}{\partial v} = -t\Gamma(\zeta + \eta) + \chi \circ (\Gamma \times \mathrm{id}_{\mathbf{C}^d \times \mathbf{C}}) \tag{1.1}$$

(here, as in the remainder of the proof, vector fields defined initially on one factor of a cartesian product are extended trivially to the product).

As χ is tangent to $D(F \times \mathrm{id}_{\mathbf{C}^d \times \mathbf{C}})$, it can be lifted with respect to $F \times \mathrm{id}_{\mathbf{C}^d \times \mathbf{C}}$: there exists $\delta \in \Theta_{N \times \mathbf{C}^d \times \mathbf{C}^c/\mathbf{C}^d \times \mathbf{C}^c}$ such that

$$t(F \times \mathrm{id}_{\mathbf{C}^d \times \mathbf{C}^c})(\delta) = \chi \circ (F \times \mathrm{id}_{\mathbf{C}^d \times \mathbf{C}^c}). \tag{1.2}$$

By the standard construction of fibre products, the spaces $\mathcal{X}_0$ and $\mathcal{X}$, and the maps $F_0 : \mathcal{X}_0 \to P \times \mathbf{C}^d$ and $F : \mathcal{X} \to P \times \mathbf{C}^d \times \mathbf{C}$, may be described as follows: let $H_0 : N \times P \times \mathbf{C}^d \to Q \times Q \times \mathbf{C}^d$ and $H : N \times P \times \mathbf{C}^d \times \mathbf{C} \to Q \times Q \times \mathbf{C}^d \times \mathbf{C}$ be defined by $H_0(x, y, u) = (G(x), \Gamma_0(y, u), u)$, $H(x, y, u, v) = (G(x), \Gamma(y, u, v), u, v)$, and let $\Delta_0 = \mathrm{diag}(Q \times Q) \times \mathbf{C}^d$ and $\Delta = \mathrm{diag}(Q \times Q) \times \mathbf{C}^d \times \mathbf{C}$. Then $\mathcal{X}_0 = H_0^{-1}(\Delta_0)$, $\mathcal{X} = H^{-1}(\Delta)$, F_0 is the restriction of F to $\mathcal{X}_0$, and F is the restriction of F to $\mathcal{X}$.

Write $H(x, y, u, v) = (h(x, y, u, v), u, v)$. Then $\partial h/\partial v = -dH(\delta, \zeta + \eta) + (\chi_i, \chi) \circ H$ by (1.1) and (1.2). Define a vector field $\alpha \in \Theta_{Q \times Q \times \mathbf{C}^d \times \mathbf{C}^c}$ by $\alpha = (\chi, \chi) + \eta + \frac{\partial}{\partial v}$. Then α is tangent to Δ, and if $\xi = \frac{\partial}{\partial v} + \eta + (\delta, \zeta)$, we have

$$tH(\xi) = \alpha \circ H. \tag{1.3}$$

This shows that ξ is tangent to $\mathcal{X}$ at its smooth points, and thus since smooth points are dense in $\mathcal{X}$, the restriction of ξ to $\mathcal{X}$ lies in $\Theta_{\mathcal{X}}$. Now (1.3) restricted to $H^{-1}(\Delta)$, shows that $F : \mathcal{X} \to P \times \mathbf{C}^d \times \mathbf{C}$ is a trivial extension of $F_0 : \mathcal{X}_0 \to P \times \mathbf{C}^d$.

This completes the proof of versality of F_0.

Minimality is proved essentially by reversing the argument: if Γ_0 is miniversal but F_0 is not, then F_0 is a trivial extension of some sub-unfolding F_{-1}. The trivialisation may be lifted to show that Γ_0 is a $\mathcal{K}_{D(G),e}$-trivial extension of a sub-unfolding Γ_{-1}, contradicting the minimality of Γ. □

The unfoldings of $f : X \to P$ we are discussing here simultaneously deform the space X and the map f. Separating these two deformations, one arrives on the one hand at the (well understood) theory of isolated complete intersection singularities, and on the other at the rather less-studied theory of $\mathcal{A}$-equivalence of map-germs defined on singular varieties. In the former, the principal deformation-theoretic invariant is the Tjurina number $\tau(X)$; in the latter, it is the dimension of the quotient space

$$N\mathcal{A}_e f := \frac{f^*\Theta_P}{tf(\Theta_X) + \omega f(\Theta_P)},$$

which we shall refer to as $\mathcal{A}_e$-codim(f). When X is smooth, this reduces to the usual $\mathcal{A}_e$-codimension of f. In the case where X is an ICIS, we shall refer to the dimension of a miniversal unfolding of the pair (X, f) as $\mathcal{A}_e$-codim(X, f).

Theorem 1.4 *With the notation and hypotheses of Theorem 1.3, there is an exact sequence of $\mathcal{O}_P$-modules,*

$$0 \to N\mathcal{A}_e f \to N\mathcal{K}_{D(G),e}\gamma \to T^1_X \to 0.$$

Consequently, $\mathcal{A}_e$-codim$(X, f) = \mathcal{A}_e$-codim$(f) + \tau(X)$.

PROOF Suppose that Γ_0 is a $\mathcal{K}_{D(G),e}$-versal deformation of γ; then the present theorem is proved by identifying, at an infinitesimal level, the sub-deformation of Γ_0 in which X is deformed trivially.

We begin by noting that, as in the proof of Theorem 1.3, we can assume that γ is an immersion; for, with the notation of the proof of Theorem 1.3, it is easy to see that there is a natural isomorphism $N\mathcal{K}_{D(G),e}\gamma \to N\mathcal{K}_{D(\tilde{G}),e}\tilde{\gamma}$.

Let Δ be the diagonal in $Q \times Q$; since $X = (G, \gamma)^{-1}(\Delta)$, an easy calculation shows that

$$T^1_X \cong \frac{(G,\gamma)^*(\Theta_{Q\times Q})}{t(G,\gamma)(\Theta_{N\times P}) + (G,\gamma)^*(\mathrm{Derlog}(\Delta))}.$$

In fact, if z_i and z_i' are coordinates on the two copies of Q, and h is the map on $Q \times Q$ with component functions $z_i - z_i'$, then

$$th : (G,\gamma)^*\Theta_{Q\times Q} \to (h \circ (G,\gamma))^*\Theta_Q$$

induces an isomorphism between $N\mathcal{K}_{\Delta,e}(G,\gamma)$ and $N\mathcal{K}_e(G-\gamma)$.

We shall later make use of the fact that

$$\mathrm{Derlog}(X) = (t(G,\gamma))^{-1}((G,\gamma)^*\,\mathrm{Derlog}(\Delta)).$$

A deformation of γ induces a deformation of X, and thus, at the infinitesimal level there is a map $\phi : \gamma^*(\Theta_Q) \to T^1_X$, which is given by $\phi(\xi) = (0, \xi)$. Recall that

$$N\mathcal{K}_{D(G),e}\gamma = \frac{\gamma^*\Theta_Q}{t\gamma(\Theta_P) + \gamma^*\,\mathrm{Derlog}(D(G))}.$$

Step 1: ϕ passes down to the quotient to give a map $\overline{\phi} : N\mathcal{K}_{D(G)}\gamma \to T^1_X$. To see this, we must show that $t\gamma(\Theta_P) + \gamma^*\operatorname{Derlog}(D(G)) \subset \ker \phi$. First, if $\eta \in \operatorname{Derlog}(D(G))$, there exists $\delta \in \Theta_N$ such that $tG(\delta) = \eta \circ G$, and so $\phi(\eta \circ \gamma) = (0, \eta \circ \gamma) = (\eta, \eta) \circ (G, \gamma) - t(G, \gamma)(\delta, 0)$ which is of course a member of $t(G,\gamma)(\Theta_{N\times P}) + (G,\gamma)^*(\operatorname{Derlog}(\Delta))$; second, if $\zeta \in \Theta_P$, then regarding ζ as an element of $\Theta_{N\times P}$, (which is 0 in the N direction) we have $\phi(t\gamma(\zeta)) = t(G,\gamma)(\zeta)$. Thus, $\overline{\phi}$ is well-defined.

Step 2: $\overline{\phi}$ is an epimorphism. As every deformation of X is induced, up to equivalence of deformations, by an appropriate deformation of γ, by Theorem 1.3, this is immediate. In fact, it is easy to find a direct algebraic proof, avoiding Theorem 1.3.

Step 3: What is the kernel of $\overline{\phi}$? The kernel of ϕ consists of the tangent vector fields to 1-parameter deformations Γ of γ such that the induced unfolding $(\mathcal{X}, S, \pi, F)$ of (X, f) deforms X trivially. Let $f_s : X_s \to P$ be the map obtained by restricting F. If the family $h_s : X \to X_s$ of diffeomorphisms trivialises the deformation of X, then we have a 1- parameter family of maps $f_s \circ h_s : X \to P$, and thus at the infinitesimal level a tangent vector field $\frac{d(f_s \circ h_s)}{ds}|_{s=0} \in f^*(\Theta_P)$. Although the trivialising family h_s is not unique, and thus there is no natural map $\ker(\phi) \to f^*(\Theta_P)$, this construction does enable us to define a map $\psi : \ker(\overline{\phi}) \to N\mathcal{A}_e f$, by $\psi(\xi) = \frac{d(f_s \circ h_s)}{ds}|_{s=0}$. Since $f_s : X_s \to P$ is simply the restriction of the cartesian projection $N \times P \to P$, then in infinitesimal terms ψ is defined as follows: if $\xi \in \gamma^*(\Theta_Q)$, then $\overline{\phi}(\xi + T\mathcal{K}_{D(G),e}\gamma) = 0$ if and only if there exist a vector field $\rho \in \Theta_{N\times P}$ and an element $\beta \in (G,\gamma)^*(\operatorname{Derlog}(\Delta))$ such that

$$(0, \xi) = t(G,\gamma)(\rho) + \beta. \tag{1.4}$$

Write $\rho = \rho_N + \rho_P$, where $\rho_N \in \Theta_{N\times P/P}$ and $\rho_P \in \Theta_{N\times P/N}$. Then the infinitesimal deformation of f induced by $(0,\xi)$ is the class in $N\mathcal{A}_e f$ of $\rho_{P|X}$; that is,

$$\psi(\xi) = \rho_{P|X} + T\mathcal{A}_e f.$$

ψ **is well-defined**: first, the ambiguity in the choice of the solutions ρ, β to (1.4) does not affect the class of $\rho_{P|X}$ in $N\mathcal{A}_e f$; for if $(0,\xi) = t(G,\gamma)(\rho) + \beta = t(G,\gamma)(\rho') + \beta'$, then $t(G,\gamma)(\rho - \rho') \in (G,\gamma)^*(\operatorname{Derlog}(\Delta))$, and so $\rho - \rho' \in \operatorname{Derlog}(X)$ (since $X = (G,\gamma)^{-1}(\Delta)$), and $\rho_{P|X} - \rho'_{P|X} \in tf(\Theta_X)$.

Secondly, if $\xi \in T\mathcal{K}_{D(G),e}\gamma$, then $\rho_{P|X} \in tf(\Theta_X) + f^{-1}(\Theta_P)$; for if $\xi = \eta \circ \gamma$, with $\eta \in \operatorname{Derlog}(D(G))$, then there exists $\delta \in \Theta_N$ such that $tG(\delta) = \eta \circ G$. Thus, (1.4) has a solution with $\rho = -\delta$: $(0,\xi) = -t(G,\gamma)(-\delta) + (\eta,\eta)\circ(G,\gamma)$; here $\rho_P = 0$ and so $\psi(\xi) = 0$. And if $\xi = t\gamma(\alpha)$ with $\alpha \in \Theta_P$, then (1.4) has solution $(0,\xi) = t(G,\gamma)(0,\alpha)$. Here $\rho_P(x,y) = \alpha(y)$, and so $\rho_{P|X} \in f^{-1}(\Theta_P)$.

ψ is 1-1: If (1.4) has solution $(0,\xi) = t(G,\gamma)(\rho_N + \rho_P) + \beta$ with $\rho_{P|X} = tf(\alpha) + \eta \circ f$, then extending $\eta \in \Theta_P$ trivially to $\eta \in \Theta_{N\times P/P}$, and recalling that $tf(\alpha_N + \alpha_P)$ is just α_P, we have that for some α_N, the vector field $\rho_P - \eta + \alpha_N$ is in Derlog(X). As $\text{Derlog}(X) = (t(G,\gamma))^{-1}((G,\gamma)^*\text{Derlog}(\Delta))$, it follows that

$$(0,\xi) = t(G,\gamma)(\rho_N - \alpha_N + \eta) + \tilde{\beta} \tag{1.5}$$

for some $\tilde{\beta} \in (G,\gamma)^*\text{Derlog}(\Delta)$. Write $\sigma_N = \rho_N - \alpha_N$. From (1.5) we deduce that for all $y \in P$ and $x \in G^{-1}(\gamma(y))$,

$$\xi(y) = t\gamma(\eta(y)) - tG(\sigma_N(x, G(x))); \tag{1.6}$$

for since $(x,y) \in X$, the components of $\tilde{\beta}$ in the two copies of Q are equal. Now, we may suppose that γ is an immersion, and at this point it is convenient to regard P as a smooth subvariety of Q. Thus, we regard $t\gamma(\eta) - \xi$ as an element of the restriction $\Theta_{Q|P}$. To complete the proof of the injectivity of ψ, we have to show that $t\gamma(\eta) - \xi$ is the restriction to P of some element of Derlog$(D(G))$; this follows from Lemma 1.5 below.

ψ is onto: Given $\eta \in f^*(\Theta_P)$, choose a 1-parameter deformation $f_t : X \to P$ of f, such that $\frac{df_t}{dt}|_{t=0} = \eta$. By Proposition 1.2, there is a deformation Γ of γ such that the induced unfolding F of f is isomorphic to the unfolding $f_t \times \text{id}_{\mathbf{C}} : X \times \mathbf{C}, 0 \to P \times \mathbf{C}, 0$. Since in the unfolding $f_t \times \text{id}_{\mathbf{C}}$, X is deformed trivially, the same is true in the unfolding F; it follows that $\frac{d\Gamma}{dt}|_{t=0} \in \ker\phi$, and, since F is isomorphic to $f_t \times \text{id}_{\mathbf{C}}$, that $\psi(\frac{d\Gamma}{dt}|_{t=0}) \equiv \eta$ modulo $df(\Theta_X) + f^{-1}(\Theta_P)$. This proves that ψ is onto, and completes the proof of the theorem. □

Lemma 1.5 *Let $P \subset Q$ be a smooth subvariety, and $G : N \to Q$ be stable and such that $P \not\subset D(G)$. Then the sequence*

$$0 \to \text{Derlog}(D(G)) \otimes \mathcal{O}_P \to \Theta_Q \otimes \mathcal{O}_P \to \frac{G^*\Theta_Q \otimes \mathcal{O}_P}{tG(\Theta_N) \otimes \mathcal{O}_P} \to 0$$

is exact, where $\otimes = \otimes_{\mathcal{O}_Q}$.

PROOF ¿From the exact sequence

$$0 \to \text{Derlog}(D(G)) \to \Theta_Q \to \frac{G^*\Theta_Q}{tG(\Theta_N)} \to 0$$

we obtain the exact sequence

$$0 \to \text{Tor}_1\Big(\mathcal{O}_P, \frac{G^*\Theta_Q}{tG(\Theta_N)}\Big) \to \text{Derlog}(D(G)) \otimes \mathcal{O}_P \to$$

$$\to \Theta_Q \otimes \mathcal{O}_P \to \frac{G^*\Theta_Q}{tG(\Theta_N)} \otimes \mathcal{O}_P \to 0,$$

for as Θ_Q is free, $\mathrm{Tor}_1(\mathcal{O}_P, \Theta_Q) = 0$. Here all tensor products are over $\mathcal{O}_Q$, and $\mathrm{Tor}_1(-,-) = \mathrm{Tor}_1^{\mathcal{O}_Q}(-,-)$.

We claim that $\mathrm{Derlog}(D(G)) \otimes \mathcal{O}_P \to \Theta_Q \otimes \mathcal{O}_P$ is injective, for then

$$\mathrm{Tor}_1\Big(\mathcal{O}_P, \frac{G_Q^*}{tG(\Theta_N)}\Big) = 0. \tag{1.7}$$

To prove the claim, it suffices to show that $\mathrm{Derlog}(D(G)) \cap I_P\Theta_Q = I_P\,\mathrm{Derlog}(D(G))$, where I_P is the ideal of functions vanishing on P. This is proved as follows: suppose $v_1, \ldots, v_q$ are free generators of $\mathrm{Derlog}(D(G))$, with $v_i = \sum v_i^j \frac{\partial}{\partial z_j}$. If $v = \sum \alpha_i v_i$, and $v_{|P} = 0$ then by Cramer's rule $\det[v_i^j]\alpha_i \in I_P$ for all i. But $\det[v_i^j]$ is a reduced equation for the irreducible set $D(G)$, and so $\alpha_i \in I_P$, and $v \in \mathrm{Derlog}(D(G))$.

Having established (1.7), the remainder of the proof is simple: tensoring the exact sequence

$$0 \to tG(\Theta_N) \to G^*\Theta_Q \to \frac{G^*\Theta_Q}{tG(\Theta_N)} \to 0$$

over $\mathcal{O}_Q$ with $\mathcal{O}_P$, we obtain the isomorphism

$$\frac{G^*\Theta_Q}{tG(\Theta_N)} \otimes \mathcal{O}_P \cong \frac{G^*\Theta_Q \otimes \mathcal{O}_P}{tG(\Theta_N) \otimes \mathcal{O}_P},$$

whence the desired exact sequence. □

Corollary 1.6 (Infinitesimal versality ⇒ versality) *Let X be an* ICIS *and $f : X \to P$ have finite $\mathcal{A}_e$-codimension. Suppose that the unfolding $F : X \times S \to P \times S$ of f is infinitesimally versal, i.e. that the initial speeds span $N\mathcal{A}_e f$, then F is versal.*

Proof Let H be a versal deformation of h, the defining equation for X, and extend F to $\hat{F}$ on the ambient space of X. Then (H, F) is an unfolding of (h, f) which by Theorems 1.3 and 1.4 is versal. It follows then that F is a versal unfolding of f. □

Remark In [4, Section 3], Damon proves Theorem 1.4 under the assumptions that X is smooth (so that $T_X^1 = 0$) and that f is of finite $\mathcal{A}_e$-codimension. In fact the hypothesis that X be smooth does not greatly simplify the proof; however, the hypothesis of finite codimension of f allows him to avoid the most difficult part of our proof, namely the injectivity of ψ. For from his Lemmas 2.8 and 2.10 (equivalence of versality) (or our Theorem 1.3) one can deduce that $\dim_{\mathbb{C}} N\mathcal{K}_{D(G),e}\gamma = \dim_{\mathbb{C}} N\mathcal{A}_e f$, and thus it is enough to establish the surjectivity of ψ. The proof of Damon's theorem in [15, 3.14] also does not need the assumption that f be of finite $\mathcal{A}_e$-codimension.

2 $\mathcal{A}_e$-codimension and the vanishing topology of the discriminant

In this section we will show how to generalise results of [5] which relate the topology of a stabilisation of an unstable map germ f, to the $\mathcal{A}_e$-codimension of f.

Definition Suppose that $(\mathcal{X}, F, i, \pi)$ is an unfolding of the map-germ $f : X \to P$, with the property that $(\mathcal{X}, \pi)$ is a smoothing of X, and such that for each smooth fibre X_s, the corresponding map $f_s : X_s \to P$ has only stable singularities. We shall call each such map $f_s : X_s \to P$ a *stabilisation* of $f : X \to P$.

If $\mathcal{A}_e$-codimension(X, f) is finite, then all stabilisations of (X, f) are topologically equivalent, for, up to isomorphism, all can be found in a versal unfolding of (X, f), and in the base of the versal unfolding, the parameter values of any two stabilisations can be joined by a path avoiding the bifurcation set; the corresponding family of mappings is locally analytically trivial, and the local analytic trivialisations can be pieced together to give a (global) topological trivialisation.

If Z is a *reduced* analytic space and $f : Z \to P$ is a mapping, then the *discriminant* of f is the set $D(f) = f(\Sigma_{(Z,f)})$, where $\Sigma_{(Z,f)} = Z_{\text{sing}} \cup \Sigma(f_{|Z_{\text{reg}}})$. In [5], the first author and J. Damon proved the following two results.

Theorem 2.1 *Suppose that (n, p) are nice dimensions in the sense of Mather [14], with $n \geq p$, and that f_s is a stabilisation of a map-germ $f : N \to P$ of finite $\mathcal{A}_e$-codimension. Then $D(f_s) \cap B_\epsilon(0)$ has the homotopy type of a wedge of spheres of dimension $p - 1$.*

The number of spheres making up the wedge is independent of the choice of stabilisation, and is called the *discriminant Milnor number* of f, and denoted $\mu_\Delta(f)$.

Theorem 2.2 *Suppose that (n, p) are nice dimensions, with $n \geq p$, and that $f : N \to P$ is a map germ of finite $\mathcal{A}_e$-codimension. Then $\mu_\Delta(f) \geq$ $\mathcal{A}_e$-codim(f), with equality if f is quasihomogeneous.*

The proof of Theorem 2.1 given in [5] relies on a theorem of Lê:

Theorem 2.3 (Lê) *If Y is a germ of a complete intersection of dimension m (not necessarily with isolated singularity) and $h : Y \to \mathbf{C}, 0$ has an isolated singularity at 0 (with respect to the canonical Whitney stratification of Y, in the sense explained in [9]), then the fibres $h^{-1}(t)$ for $t \neq 0$, of a good representative of h, have the homotopy type of a wedge of spheres of dimension m.*

This theorem was announced in [9], and recently a proof was given in [10]; a proof of an equivalent result can be found in [16].

The proof of Theorem 2.1 is then: let f_t, $t \in \mathbf{C}$, be a 1- parameter stabilisation of f, let F be the family $f_t \times \mathrm{id}_{\mathbf{C}}$, let $Y = D(F)$ (which is a hypersurface in $P \times \mathbf{C}$), and let $h : D(F) \to \mathbf{C}$ be the projection to the parameter space. Then (one checks) h has isolated singularity, in the sense of [9]. Since $h^{-1}(t) = D(f_t)$, Theorem 2.1 follows from Theorem 2.3.

Observe now that the same proof shows that the conclusion of Theorem 2.1 still holds if $f_t : X_t \to P$ is a stabilisation of the germ of finite $\mathcal{A}_e$-codimension $f : X \to P$, where X is now an ICIS: namely that $D(f_t) \cap B_\epsilon(0)$ has the homotopy type of a wedge of spheres of dimension $p - 1$. We will denote the number of these spheres by $\mu_\Delta(X, f)$.

Our main theorem is a generalization of Theorem 2.2:

Theorem 2.4 *Let X be an ICIS of dimension n, and let $f : X \to P$ be a map germ of finite $\mathcal{A}_e$-codimension. Suppose that (n, p) are nice dimensions, with $n \geq p$. Then $\mu_\Delta(X, f) \geq \mathcal{A}_e\text{-codim}(X, f)$, with equality if (X, f) is quasihomogeneous.*

(Here, (X, f) is quasihomogeneous if there are so-called good $\mathbf{C}^*$-actions on X and P with respect to which f is equivariant.)

Proof The proof of Theorem 2.2 given in [5] carries over almost verbatim, and here we give no more than a sketch. First, as described in Lemma 1.1, choose $\gamma : P \to Q$ which induces $f : X \to P$ from the stable germ $G : N \to Q$. Then $\mathcal{A}_e\text{-codim}(X, f) = \dim_{\mathbf{C}} N\mathcal{K}_{D(G),e}\gamma$, by Theorem 1.3. Next, introduce the auxilliary module $N\mathcal{K}_{H,e}\gamma$, which is defined as follows: if $H : Q \to \mathbf{C}, 0$ is a reduced equation for the hypersurface $D(G)$, let

$$\mathrm{Derlog}(H) = \{\chi \in \Theta_Q \mid \chi(H) = 0\};$$

then set

$$N\mathcal{K}_{H,e}\gamma = \frac{\gamma^*(\Theta_Q)}{d\gamma(\Theta_P) + \gamma^*(\mathrm{Derlog}(H))}.$$

As shown in [5] Example 3.2, we may, and do, suppose that there exists a vector field χ on Q such that $\chi(H) = H$, from which it follows that

Derlog($D(G)$) splits as a direct sum, Derlog(H) $\oplus \mathcal{O}_Q\{\chi\}$. If G is quasihomogeneous, then we can take the vector field χ to be the Euler vector field generating the $\mathbf{C}^*$-action on Q. If also γ is quasihomogeneous with respect to the same $\mathbf{C}^*$-action on Q, then χ is tangent to the image of γ and it follows that $N\mathcal{K}_{H,e}\gamma = N\mathcal{K}_{D(G),e}\gamma$. This is the case if (X, f) is quasihomogeneous.

Now as G is stable, Derlog($D(G)$) is a free $\mathcal{O}_Q$-module on q generators, and it follows that Derlog(H) is free on $q-1$ generators. Thus, the exact sequence

$$\gamma^*(\text{Derlog}(H)) \oplus \Theta_P \longrightarrow \gamma^*(\Theta_Q) \longrightarrow N\mathcal{K}_{H,e}\gamma \longrightarrow 0$$

defining $N\mathcal{K}_{H,e}\gamma$ in fact gives us a presentation

$$\mathcal{O}_P^{p+q-1} \longrightarrow \mathcal{O}_P^q \rightarrow N\mathcal{K}_{H,e}\gamma \longrightarrow 0.$$

It follows from the form of this presentation that if $\dim_{\mathbf{C}} N\mathcal{K}_{H,e}\gamma < \infty$, then it is conserved in a deformation [5, Corollary 5.5]. In fact, with $(\dim X, p)$ nice dimensions, the finiteness of $\mathcal{A}_e$-codim(X, f) implies that $\dim_{\mathbf{C}} N\mathcal{K}_{H,e}\gamma < \infty$; for if not $\dim \text{supp} N\mathcal{K}_{H,e}\gamma \geq 1$, and hence there exists a parametrized curve σ in $\gamma(P)$, at each point of which P fails to be transverse to the fibres of H. It follows that $H \circ \gamma$ is constant (equal to 0). Now, at all points $y \in \gamma(P) \cap (D(G) \setminus \{0\})$, the map defined by the fibre product of γ and G is stable, and therefore in appropriate coordinates is quasihomogeneous (indeed, quasihomogeneity of all stable maps characterizes the nice dinmensions). It follows that $(N\mathcal{K}_{H,e}\gamma)_y = (N\mathcal{K}_{D(G),e}\gamma)_y = 0$ (see [4, Lemma 3.4]), and thus such a curve σ cannot exist, and $\dim_{\mathbf{C}} N\mathcal{K}_{H,e}\gamma < \infty$.

If γ_t is a $\mathcal{K}_{D(G),e}$-stabilisation of γ, corresponding to a stabilisation of (X, f), then provided $(\dim X, p)$ are nice dimensions, $N\mathcal{K}_{H,e}\gamma_t$ is not supported anywhere on $D(f_t) = \gamma^{-1}(D(G))$, for the reason observed above. If on the other hand $y \notin \gamma_t^{-1}(D(G))$, then $(N\mathcal{K}_{H,e}\gamma_t)_y \cong \mathcal{O}_{P,y}/J_h$, where $h = H \circ \gamma$ is a defining equation for $D(f_t)$ [5, Lemma 5.6]; from this one now concludes, by an argument of Siersma [16], that $\mu_\Delta(X, f)$ is equal to $\dim_{\mathbf{C}} N\mathcal{K}_{H,e}\gamma$. Since it is clear that $\dim_{\mathbf{C}} N\mathcal{K}_{H,e}\gamma \geq \dim_{\mathbf{C}} N\mathcal{K}_{D(G),e}\gamma$, the inequality of the statement of the theorem is proved. Equality when (X, f) is quasihomogeneous follows from the fact that in this case $N\mathcal{K}_{H,e}\gamma = N\mathcal{K}_{D(G),e}$. □

One case of interest is where $p = 1$. In this case $\mu_\Delta(X, f)$ has another topological interpretation, by an argument of Lê and Greuel:

Proposition 2.5 *If $p = 1$, then $\mu_\Delta(X, f) + 1 = \mu(X) + \mu(X')$, where $X' = f^{-1}(0)$.*

PROOF Since $p = 1$, $\mu_\Delta(X, f) + 1$ is just the number of critical values of the non-degenerate function f_t on the Milnor fibre X_t of X. If $h : E \to \mathbf{C}^k$

defines X, then $\mu_\Delta(X, f)+1$ is thus the number of intersection points of the line $\{0\} \times \mathbf{C} \subset \mathbf{C}^k \times \mathbf{C}$ with the discriminant of $h \times f$; hence, as shown for example in [11, §(5.11)], $\mu_\Delta(X, f)+1$ is equal to $\mu(X)+\mu(X')$. □

Corollary 2.6 *If $p = 1$, then*

$$\mu(X)+\mu(X') \geq \tau(X) + \mathcal{A}_e\text{-codim}(f) + 1,$$

with equality if (X, f) is quasihomogeneous. □

Combining this with the theorem of Greuel, that $\tau(X) = \mu(X)$ if X is quasihomogeneous [8], gives:

Corollary 2.7 *If $p = 1$ and (X, f) is quasihomogeneous, then*

$$\mu(X') = \mathcal{A}_e\text{-codim}(f) + 1.$$

□

This fact was already obtained by Bruce and Roberts [1, Proposition 7.7]; their proof also uses Greuel's theorem, together with an explicit description of Θ_X (their Θ_X^-) for X quasihomogeneous. It is also implicit in the paper of Goryunov [7].

3 Bifurcations and $\mathcal{K}_\Delta$-equivalence

Let $g_0 : N \to P$ be a $\mathcal{K}$-finite map-germ, and let $g : N \times \Lambda \to P$ be any deformation of g_0. Then g can be induced from a versal deformation $G : N \times U \to P$ of g_0 by a map-germ $\gamma : \Lambda \to U$. Here Λ and U are germs of smooth spaces. Thus we have the following commutative diagram.

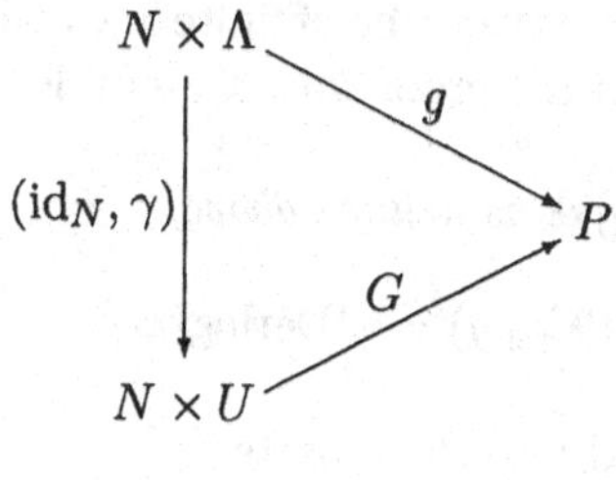

Let $\Delta \subset U$ denote the local bifurcation set of G, that is,

$$\Delta = \{u \in U \mid G_u^{-1}(0) \text{ is not smooth}\}.$$

It is well known that Δ is the discriminant of the projection $\pi_G : V_G \to U$, where V_G is the zero set of G, and π_G is the restriction to V_G of the projection $N \times U \to U$. Let Derlog(Δ) denote the $\mathcal{O}_U$-module of vector fields tangent to Δ. Because G is $\mathcal{K}$-versal, π_G is a stable map [12], so that $v \in \text{Derlog}(\Delta)$ if and only if it is liftable for π_G.

In this section we formalize the close relationship that exists between the deformation g and the map γ which induces it.

For a deformation $g : N \times \Lambda \to P$ of $g_0 : N \to P$, define the *relative* $\mathcal{K}$-tangent space by,

$$T\mathcal{K}_{\text{rel}}.g = tg(\Theta_{N\times\Lambda/\Lambda}) + g^*\boldsymbol{m}_P\Theta_P.$$

It is an $\mathcal{O}_{N\times\Lambda}$-module. Thus $T\mathcal{K}_{\text{un}}.g = T\mathcal{K}_{\text{rel}}.g + t_2g(\Theta_\Lambda)$, where t_2g means differentiating g with respect to the second (λ) variables.

Any map $\gamma : \Lambda \to U$ induces a deformation $g = g_\gamma : N \times \Lambda \to P$. A deformation of γ induces a deformation of g. At the infinitesimal level, this defines a homomorphism of $\mathcal{O}_\Lambda$-modules

$$\begin{aligned} \Psi_\gamma : \Theta(\gamma) &\longrightarrow \Theta(g), \\ \zeta &\longmapsto t_2G(\zeta). \end{aligned}$$

We often omit the subscript γ on Ψ if the context is clear. More explicitly, if $[\lambda \mapsto \zeta(\lambda)] \in \Theta(\gamma)$, then $\Psi(\zeta)$ is the vector field along g given by, $[(x,\lambda) \mapsto d_2G_{(x,\gamma(\lambda))}(\zeta(\lambda))]$. In coordinates, if $G(x,u) = g_0(x) + \sum u_i\phi_i(x)$ then

$$\Psi(\zeta_1(\lambda), \ldots, \zeta_d(\lambda)) = \sum \zeta_i(\lambda)\phi_i(x).$$

Theorem 3.1 *The homomorphism $\Psi = \Psi_\gamma$ introduced above induces an isomorphism of $\mathcal{O}_\Lambda$-modules:*

$$\psi : N\mathcal{K}_\Delta.\gamma \xrightarrow{\simeq} N\mathcal{K}_{\text{un}}.g.$$

Note that we do not assume that γ be of finite $\mathcal{K}_\Delta$ codimension, but only that g_0 be $\mathcal{K}$-finite. We begin the proof with a useful lemma.

Lemma 3.2 *With $g, G, \Psi, \gamma, \Delta$ as defined above,*

$$\Psi_\gamma^{-1}(T\mathcal{K}_{\text{rel}}.g) = \gamma^*\text{Derlog}(\Delta).$$

PROOF The proof is divided into three parts.
(i) We first show that the lemma holds for $g = G$ and $\gamma = \text{id}_U$.

Suppose $\Psi(\eta) \in T\mathcal{K}_{\text{rel}}.G$. Then there exist $\xi \in \Theta_{N\times\Lambda/\Lambda}$ and $\alpha \in G^*\boldsymbol{m}_P\Theta_P$ such that

$$\Psi(\eta) = -tG(\xi) + \alpha.$$

Thus, $tG(\xi, \eta) = \alpha$, whence $(\xi, \eta) \in \Theta_{N \times U}$ is tangent to V_G. Thus η is liftable and hence in Derlog(Δ). To obtain the converse, just reverse the argument.
(ii) We now show that the statement holds if γ is an immersion. Via γ, consider Λ as a subspace of U. Let $\eta \in \Theta(\gamma)$ be such that

$$\Psi(\eta) \in T\mathcal{K}_{\mathrm{rel}}.g = \gamma^{-1} T\mathcal{K}_{\mathrm{rel}}.G$$

Extend η to $\tilde{\eta} \in \Theta_U$. Then $\Psi_{\mathrm{id}}(\tilde{\eta}) \circ \gamma = \Psi_\gamma(\eta)$. Thus,

$$\Psi_{\mathrm{id}}(\tilde{\eta}) \in I_\Lambda.\Theta(G) + T\mathcal{K}_{\mathrm{rel}}.G,$$

where I_Λ is the ideal in $\mathcal{O}_U$ of functions vanishing on Λ. By the preparation theorem, $\Theta(G) = \Psi_{\mathrm{id}}(\Theta_U) + T\mathcal{K}_{\mathrm{rel}}.G$, whence

$$\Psi_{\mathrm{id}}(\tilde{\eta}) \in I_\Lambda.\Psi_{\mathrm{id}}(\Theta_U) + T\mathcal{K}_{\mathrm{rel}}.G.$$

Consequently, modulo $T\mathcal{K}_{\mathrm{rel}}.G$, we can write

$$\Psi_{\mathrm{id}}(\tilde{\eta}) = \sum r_i \Psi_{\mathrm{id}}(e_i)$$

with $r_i \in I_\Lambda$ and $e_i \in \Theta_U$. Let $\eta' = \tilde{\eta} - \sum r_i e_i$. Then $\Psi_{\mathrm{id}}(\eta') \in T\mathcal{K}_{\mathrm{rel}}.G$, so by Part (i), $\eta' \in$ Derlog(Δ). Moreover, η' is also an extension of η, i.e. $\eta = \eta' \circ \gamma$, so $\eta \in \gamma^*$Derlog(Δ) as required.
(iii) This final step reduces the general case to that of γ being an immersion. To this end, define

$$\begin{aligned} \tilde{G} : N \times U \times \Lambda &\longrightarrow P \\ (x, u, u') &\longmapsto G(x, u) \end{aligned}$$

and $\tilde{\gamma} : \Lambda \to U \times \Lambda$ by $\tilde{\gamma}(\lambda) = (\gamma(\lambda), \lambda)$. Now, $\tilde{G}$ is versal because G is, and $\tilde{\Delta} = \Delta \times \Lambda \subset U \times \Lambda$. Since $\tilde{\gamma}$ is an immersion we can apply Part (ii) to get:

$$\Psi_{\tilde{\gamma}}^{-1}(T\mathcal{K}_{\mathrm{rel}}.g) = \tilde{\gamma}^* \,\mathrm{Derlog}(\tilde{\Delta}),$$

since γ and $\tilde{\gamma}$ both induce g. For $\tilde{\eta} \in \Theta(\tilde{\gamma})$ write $\tilde{\eta} = (\eta, \eta')$. Then

$$\Psi_{\tilde{\gamma}}(\tilde{\eta}) = \Psi_\gamma(\eta)$$

as $\tilde{G}$ is independent of u', and so

$$\begin{aligned} \Psi_\gamma(\eta) \in T\mathcal{K}_{\mathrm{rel}}.g &\iff (\eta, \eta') \in \tilde{\gamma}^* \,\mathrm{Derlog}(\tilde{\Delta}) \\ &\iff \eta \in \gamma^* \mathrm{Derlog}(\Delta). \end{aligned}$$

□

PROOF (of Theorem 3.1) We must show that Ψ_γ induces an isomorphism ψ. To do so we must show three things:

(i) $\Psi(T\mathcal{K}_\Delta.\gamma) \subset T\mathcal{K}_{\mathrm{un}}.g$, whence Ψ induces a homomorphism,

$$\psi : N\mathcal{K}_\Delta.\gamma \to N\mathcal{K}_{\mathrm{un}}.g$$

(ii) ψ is injective, and

(iii) ψ is surjective.

The proof is divided into three parts accordingly.

(i) First note that for $\xi \in \Theta_\Lambda$, one has

$$\Psi(t\gamma(\xi)) = t_2G \circ t\gamma(\xi) = t_2g(\xi),$$

so $\Psi(t\gamma(\Theta_\Lambda)) = t_2g(\Theta_\Lambda)$. The remainder of this part follows from the lemma above. Now define

$$\psi : N\mathcal{K}_\Delta.\gamma \to N\mathcal{K}_{\mathrm{un}}.g$$

to be the morphism induced from Ψ.

(ii) Suppose $\Psi(\xi) \in T\mathcal{K}_{\mathrm{un}}.g$. We wish to show that $\xi \in \gamma^*\mathrm{Derlog}(\Delta)$. Write $\Psi(\xi) = \zeta_1 + \zeta_2$, where $\zeta_1 \in T\mathcal{K}_{\mathrm{rel}}.g$ and $\zeta_2 \in t_2g(\Theta_\Lambda)$. Since $\Psi|_{t\gamma(\Theta_\Lambda)} : t\gamma(\Theta_\Lambda) \to t_2g(\Theta_\Lambda)$ is an isomorphism, there is a unique $\xi_2 \in t\gamma(\Theta_\Lambda)$ with $\Psi(\xi_2) = \zeta_2$. Since Ψ is linear, it follows that $\zeta_1 \in \mathrm{image}(\Psi)$, and so by Lemma 3.2, $\xi_1 := \xi - \xi_2 \in \gamma^*\mathrm{Derlog}(\Delta)$.

(iii) By definition, ψ is surjective if and only if,

$$T\mathcal{K}_{\mathrm{un}}.g + t_2G(\Theta(\gamma)) = \Theta(g).$$

Consider the quotient $\Theta(g)/T\mathcal{K}_{\mathrm{un}}.g$. Reducing modulo m_Λ (the maximal ideal in $\mathcal{O}_\Lambda$), this becomes

$$\frac{\Theta(g)}{T\mathcal{K}_{\mathrm{un}}.g + m_\Lambda\Theta(g)} \subset \frac{\Theta(g_0)}{T\mathcal{K}.g_0} = N\mathcal{K}.g_0$$

Now, since G is a versal deformation of g_0 we have that

$$T\mathcal{K}.g_0 + t_2G(U) = \Theta(g_0)$$

so that $t_2G(U)$ spans $N\mathcal{K}.g_0$. Here we have identified U with $\Theta_U/m_U\Theta_U$. It follows from the preparation theorem that $t_2G(\Theta(\gamma))$ spans $\Theta(g)/T\mathcal{K}_{\mathrm{un}}.g$. □

We conclude this article by combining the main theorems of Sections 1 and 3 to refine a classical theorem of Martinet relating stable mappings and versal deformations. First we recall this theorem.

Let $g : N \times \Lambda \to P$ be a regular deformation of g_0 ("regular" means that as a map g is a submersion). Let $V_g = g^{-1}(0)$ and $\pi_g : V_g \to \Lambda$ be the restriction to V_g of the projection $\pi : N \times \Lambda \to \Lambda$. Martinet's Theorem [12] states that g is $\mathcal{K}$-versal if and only if π_g is stable.

Now, lack of versality of g is measured by $N\mathcal{K}_{\text{un}}{\cdot}g$, and lack of stability of π_g is measured by the $\mathcal{A}_e$-codim(V_g, π_g) ($= \dim N\mathcal{A}_e\pi_g$ if g is regular), and it is not unreasonable to expect these two numbers to coincide. Indeed one has,

Theorem 3.3 *Let $g_0 : N \to P$ be a $\mathcal{K}$-finite map germ and let $g : N\times\Lambda \to P$ be a deformation of g_0, then*

$$\mathcal{A}_e\text{-codim}(V_g, \pi_g) = \mathcal{K}_{\text{un}}-\text{codim}(g).$$

If, moreover, g is regular then the following two $\mathcal{O}_\Lambda$-modules are isomorphic:

$$N\mathcal{A}_e.\pi_g \simeq N\mathcal{K}_{\text{un}}.g.$$

Proof Let $G : N \times U \to P$ be a versal deformation of g_0 and $\gamma : \Lambda \to U$ a map which induces g from G. The first part of the theorem follows from combining Theorems 1.3 and 3.1. The second part follows from combining Theorems 1.4 and 3.1 (since $T^1_{V_g} = 0$ if g is regular). □

References

[1] J.W. Bruce & M. Roberts, Critical points of functions on analytic varieties. *Topology* **27** (1988), 57–90.

[2] J. Damon, *The Unfolding and Determinacy Theorems for Subgroups of $\mathcal{A}$ and $\mathcal{K}$*. Memoirs A.M.S. **50**, no. 306 (1984).

[3] J. Damon, Deformations of sections of singularities and Gorenstein surface singularities. *Am. J. Math.* **109** (1987), 695–722.

[4] J. Damon, $\mathcal{A}$-equivalence and the equivalence of sections of images and discriminants. In *Singularity Theory and its Applications, Part I*, Springer Lecture Notes **1462** (1991), 93–121.

[5] J. Damon & D. Mond, $\mathcal{A}$-codimension and the vanishing topology of discriminants. *Invent. math.* **106** (1991), 217–242.

[6] M. Golubitsky & D. Schaeffer, *Singularities and Groups in Bifurcation Theory, Vol. I.* Springer-Verlag, New York etc., 1985.

[7] V.V. Goryunov, Projection and vector fields tangent to the discriminant of a complete intersection. *Func. An. Appl.* **22** (1988), 104–113.

[8] G.-M. Greuel, Dualität in der lokalen Kohomologie isolierter Singularitäten. *Math. Ann.* **250** (1980), 157–173.

[9] Lê Dũng Tráng, Le concept de singularité isolée de fonction analytique. *Adv. Studies in Pure Math.* **8** (1986), 215–227.

[10] Lê Dũng Tráng, Complex analytic functions with isolated singularities. *J. Alg. Geom.* **1** (1992), 83–100.

[11] E.J.N. Looijenga, *Isolated Singular Points on Complete Intersections.* L.M.S. Lecture Notes, **17**, C.U.P., 1984.

[12] J. Martinet, Déploiements versels des applications différentiables et classification des applications stables. Lecture Notes in Math. **535**, Springer, Heidelberg etc., 1976.

[13] J. Mather, Stability of C^∞ mappings, IV: Classification of stable germs by **R**-algebras. *Publ. Math. I.H.E.S.* **37** (1969), 223–248

[14] J. Mather, Stability of C^∞ mappings, VI: The nice dimensions. Springer Lecture Notes **192** (1971), 207–253.

[15] A. du Plessis, T. Gaffney & L. Wilson, Map-germs determined by their discriminant. Preprint, Aarhus, 1991.

[16] D. Siersma, Vanishing cycles and special fibres. In *Singularity Theory and its Applications, Part I*, Springer Lecture Notes **1462** (1991), 292–301.

[17] C.T.C. Wall, Finite determinacy of smooth map-germs. *Bull. L.M.S.* **13** (1981), 481–539.

David Mond
Mathematics Institute

University of Warwick
Coventry CV4 7AL
U.K.

James Montaldi
Institute Non-Linéaire de Nice

1361 route des Lucioles
Sophia Antipolis
06560 Valbonne
France

Cycles évanescents et faisceaux pervers II : cas des courbes planes réductibles

L. Narváez-Macarro*

A la mémoire de Jean-Louis Verdier

TABLE DE MATIÈRES

1991 Mathematics Subject Classification 32S40 (Secondary : 32S60, 32S30).

---

*Supported by DGICYT PB90-0883

Introduction

Soit $(C,0) \subset (\mathbf{C}^2,0)$ un germe de courbe plane complexe et $\boldsymbol{f}: (\mathbf{C}^2,0) \to (\mathbf{C},0)$ une équation réduite. Soit $f: X \to D$ un bon réprésentant de $\boldsymbol{f}$ et $\mathcal{L}$ un système local d'espaces vectoriels sur un corps k dans $X - C$. Il est déterminé par une réprésentation k-linéaire E du groupe fondamental associé G. Dans [26] nous avons calculé, dans les cas où $(C,0)$ est irréductible, le complexe des cycles proches $\mathbf{R}\psi_f(\mathcal{L})$ muni de son automorphisme de monodromie $\mathbf{T}_{\mathcal{L}}$, en fonction de E. En dernière analyse, ce calcul avait un sens précis parce que dans ce cas C est homéomorphe à la droite complexe $\mathbf{C}$ et $\mathbf{R}\psi_f(\mathcal{L})$ est un *faisceau pervers* sur C, stratifié par rapport à $\{0\}$ et $C - \{0\}$ (cf. [12, 18, 6]), et ces faisceaux pervers admettent une description simple et explicite comme objets d'une catégorie convenable de diagrammes de k-espaces vectoriels [9, 11, 22].

Dans ce travail nous traitons le cas général où $(C,0)$ peut être réductible. La plus grande différence avec le cas irréductible se trouve dans la nécessité de travailler simultanément avec plusieurs points de base. Ceci nous amène à considérer dans le §1, non seulement le groupe fondamental de $X - C$ et sa décomposition naturelle comme produit semi-direct, mais aussi les torseurs Δ_i qui lient le point de base aux différentes composantes connexes du bord, munis de l'action de la monodromie. Dans le §2 nous étudions les faisceaux pervers *coniques* sur C, i.e. ceux qui sont stratifiés par rapport à $\{0\}$ et aux composantes connexes de $C - \{0\}$. La description explicite de ces objets est bien connue et découle des résultats généraux sur le recollement des faisceaux pervers [9, 20, 28] (pour un traitement indépendant de ces résultats voir aussi [21]). Mais nous nous plaçons ici dans une situation conique générale et dans un cadre purement topologique. Nous obtenons, à l'aide d'une généralisation du formalisme des cycles évanescents, une description de l'équivalence de catégories Ψ entre la catégorie des faisceaux pervers coniques sur C et une certaine catégorie de diagrammes de k-espaces vectoriels (th. (2.3.4)). Cette description est utilisée de façon essentielle dans la preuve des résultat du paragraphe suivant. La méthode employée pourrait fournir des résultats de la même nature dans le cas où $(C,0)$ est un germe à singularité isolée de dimension ≥ 2 (voir remarque (2.3.7)). Dans le §3 on démontre les résultats principaux de ce travail. A savoir, on décrit le diagramme $\Psi(\mathbf{R}\psi_f(\mathcal{L}))$ et l'automorphisme de monodromie $\Psi(\mathbf{T}_{\mathcal{L}})$ en fonction de E. Pour cela on construit un diagramme de $k[G]$-modules libres de rang fini $\widetilde{\mathcal{P}}_f$, muni d'un automorphisme $\widetilde{t}$, et un isomorphisme fonctoriel

$$\Psi(\mathbf{R}\psi_f(\mathcal{L}), \mathbf{T}_{\mathcal{L}}) \simeq (\mathrm{Hom}_{k[G]}(\widetilde{\mathcal{P}}_f, E), \mathrm{Hom}_{k[G]}(\widetilde{t}, E))$$

(voir th. (3.1.5)). Ceci nous donne en particulier une réprésentation du foncteur exact

$$E \mapsto \Psi(\mathbf{R}\psi_f(\mathcal{L}), \mathbf{T}_{\mathcal{L}}).$$

Le couple $(\widetilde{\mathcal{P}}_f, \widetilde{t})$ est défini à partir des invariants étudiés dans le §1, notamment des torseurs Δ_i. Il contient toute l'information topologique de $f : X \to D$.

Comme application de ce résultat et de la version de Verdier du théorème de recollement des faisceaux pervers [28] on obtient, comme dans [26], une description algébrique des faisceaux pervers sur X stratifiés par rapport à $X - C, C - \{0\}, \{0\}$ (th. (3.2.1)), et une formule pour le cycle caractéristique du complexe d'intersection $IC(\mathfrak{L})$ (cor. (3.2.3)). Ces deux applications ont été obtenues en utilisant d'autres méthodes par Ph. Maisonobe [21].

Je remercie Z. Mebkhout d'avoir relu le manuscrit. Je remercie également le rapporteur qui, grâce à ses commentaires, m'a permis de compléter la première version du texte.

Notations

- ⋆ Si Δ est un ensemble non vide, $\mathcal{S}(\Delta)$ désignera le groupe des bijections de Δ dans lui même.
- ⋆ Si R est un anneau et L est un groupe, on notera $R[L]$ l'algèbre du groupe L à coefficients dans R, et $I(L) \subset R[L]$ l'idéal d'augmentation, i.e. le noyau du morphisme naturel $R[L] \to R$ qui envoie les éléments de L sur 1.
- ⋆ Si R est un anneau et X est un espace topologique, on notera R_X le faisceau d'anneaux constant correspondant et $\mathrm{Mod}_g(R)$ (resp. $\mathrm{Mod}_g(R_X)$) la catégorie abélienne des R-modules (resp. des faisceaux de R-modules, ou R_X-modules) à gauche. Si R est commutatif on notera tout simplement $\mathrm{Mod}(R)$ (resp. $\mathrm{Mod}(R_X)$). Si $R = k$ est un corps, on notera $\mathrm{Vect}(k)$ à la place de $\mathrm{Mod}(k)$.
- ⋆ Les faisceaux sur un espace topologique seront notés par les lettres $\mathcal{E}, \mathcal{F}, \mathfrak{L}, \ldots$
- ⋆ Nous utiliserons les lettres $\boldsymbol{C}, \boldsymbol{F}, \boldsymbol{G}, \ldots$ (resp. $\mathfrak{C}, \mathfrak{F}, \mathfrak{G}, \ldots$) pour désigner les complexes d'objets d'une catégorie abélienne (resp. d'une catégorie abélienne de faisceaux sur un espace topologique). Par exemple, $\boldsymbol{C}$ désignera le complexe $(\{C^n\}_{n\in\mathbf{Z}}, \{d^n_C\}_{n\in\mathbf{Z}})$. De façon analogue, $\boldsymbol{u} : \boldsymbol{C} \to \boldsymbol{F}$ désignera le morphisme $\{u^n\}_{n\in\mathbf{Z}}$. Le symbole • au-dessus d'un objet d'un complexe indique qu'il est placé en dégré 0.
- ⋆ Si $\boldsymbol{C}$ est un complexe d'objets d'une catégorie abélienne, $h^i(\boldsymbol{C})$ désignera le i-ième objet de cohomologie.
- ⋆ Si $\mathcal{A}$ est une catégorie abélienne, $\mathbf{C}(\mathcal{A})$, $\mathbf{K}(\mathcal{A})$, $\mathbf{D}(\mathcal{A})$ désigneront respectivement la catégorie des complexes d'objets de $\mathcal{A}$, la catégorie des

complexes d'objets de $\mathcal{A}$ modulo homotopie, la catégorie dérivée de $\mathcal{A}$. Les sous-catégories pleines dont les objets sont les complexes bornés à gauche seront notées par $\mathbf{C}^+(\mathcal{A})$, $\mathbf{K}^+(\mathcal{A})$, $\mathbf{D}^+(\mathcal{A})$ respectivement. Si $\mathcal{A} = \mathrm{Mod}(R)$ (resp. $\mathcal{A} = \mathrm{Mod}(R_X)$), on notera $\mathbf{C}^*(R)$, $\mathbf{K}^*(R)$, $\mathbf{D}^*(R)$ (resp. $\mathbf{C}^*(R_X)$, $\mathbf{K}^*(R_X)$ $\mathbf{D}^*(R_X)$) à la place de $\mathbf{C}^*(\mathcal{A})$, $\mathbf{K}^*(\mathcal{A})$,$\mathbf{D}^*(\mathcal{A})$ respectivement, pour $* = \emptyset, +$. Si $\boldsymbol{u}$ est un morphisme dans $\mathbf{C}(\mathcal{A})$, $[\boldsymbol{u}]$ désignera sa classe modulo homotopie.

§1 Invariants homotopiques associés à un germe de courbe plane

Soit $(C,0) \subset (\mathbf{C}^2,0)$ un germe de courbe plane complexe eventuellement réductible et $\boldsymbol{f}: (\mathbf{C}^2,0) \to (\mathbf{C},0)$ une équation réduite. Nous allons décrire des objets algébriques qui lui sont attachés (cf. [16, 7, 17, 2, 26]). Ces objets permettent d'extraire l'information du type topologique de $\boldsymbol{f}$ adaptée à notre but, notamment la définition des diagrammes $\mathcal{P}_f$ dans (3.1.2) et $\widetilde{\mathcal{P}}_f$ dans (3.1.4), et les théorèmes (3.1.3) et (3.1.5).

1.1 Réprésentants de Milnor

Soit $(C,0) = \bigcup_{i\in I}(C_i,0)$, $\#I = r$, la décomposition en composantes irréductibles et, pour chaque $i \in I$, soit $\boldsymbol{f}_i: (\mathbf{C}^2,0) \to (\mathbf{C},0)$ une équation réduite de $(C_i,0)$. On a donc $\boldsymbol{f} = \prod \boldsymbol{f}_i$. Soient $f: \overline{X} \to D$ un bon représentant propre de $\boldsymbol{f}$ (cf. [19, 2.B]) et $f_i: \overline{X} \to D$ un représentant de $\boldsymbol{f}_i$, $i \in I$. On peut prendre, par exemple, pour D un disque ouvert de $\mathbf{C}$ centré à l'origine de rayon assez petit, et pour $\overline{X}$ l'intersection d'une boule fermée de Milnor [25] avec $f^{-1}(D)$. Notons X l'intérieur de $\overline{X}$ et ∂X son bord. L'application $f: \overline{X} \to D$ est une fibration C^∞ localement triviale en dehors de l'origine et sa restriction à ∂X est une fibration triviale. En fait, il existe un collier $X^\partial \subset \overline{X}$ du bord ∂X au-dessus duquel la restriction de f est une fibration triviale, et tel que la trace de la fibre de f au-dessus de 0 sur X^∂ coïncide avec la partie lisse de cette fibre $f^{-1}(0) - \{0\}$ (voir [10, exp. XIV, prop. (3.1.5)]). Pour $t \in D$ et $i \in I$ notons

$$\begin{aligned}
\overline{X}_t &:= f^{-1}(t), & \overline{X}^* &:= \overline{X} \cap f^{-1}(D^*), & \partial X^* &:= \partial X \cap \overline{X}^* \\
X^{*\partial} &:= X^\partial \cap \overline{X}^*, & \partial X_t &:= \overline{X}_t \cap \partial X, & X_t^\partial &:= \overline{X}_t \cap X^\partial, \\
\overline{X}_{0,i} &:= f_i^{-1}(0), & \partial X_{0,i} &:= \overline{X}_{0,i} \cap \partial X, & X_{0,i}^\partial &:= \overline{X}_{0,i} \cap X^\partial = \overline{X}_{0,i} - \{0\}.
\end{aligned}$$

Chaque $\partial X_{0,i}$ est homéomorphe au cercle et ∂X_0 est la réunion disjointe des $\partial X_{0,i}$. De plus, $\overline{X}_0$ est homéomorphe au cône sur ∂X_0.

Prenons des points de base $x_i \in X^\partial_{0,i}, i \in I$, $y_0 \in X^{*\partial}$, $t_0 = f(y_0)$ et choisissons une trivialisation $\mathcal{T}: X^\partial_0 \times D \xrightarrow{\sim} X^\partial$ de $f|_{X^\partial}$ qui soit l'identité sur les fibres au dessus de 0. Le collier X^∂ (resp. $X^\partial_{t_0}$) est réunion disjointe des $X^\partial_i := \mathcal{T}(X^\partial_{0,i} \times D)$ (resp. $X^\partial_{t_0,i} := \mathcal{T}(X^\partial_{0,i} \times \{t_0\})$) pour $i \in I$. Notons aussi $X^{*\partial}_i := X^\partial_i \cap \overline{X}^* = \mathcal{T}(X^\partial_{0,i} \times D^*)$.

A partir de la trivialisation $\mathcal{T}$ on obtient un homomorphisme de groupes (cf. [10, exp. XIV, 3.1.3], [19, 2.C]) $\nu_{\mathcal{T}} : \pi_1(D^*, t_0) \to \pi_0(\mathrm{Aut}(\overline{X}_{t_0}; X^\partial_{t_0}))$ et un morphisme, dit de "spécialisation"([10, exp. XIV]), $sp: \overline{X}_{t_0} \longrightarrow \overline{X}_0$, qui identifie chaque $X^\partial_{t_0,i}$ avec $X^\partial_{0,i}$, et qui envoit $\overline{X}_{t_0} - X^\partial_{t_0}$ sur 0.

Soit δ le générateur positif de $\pi_1(D^*, t_0)$ et $T \in \mathrm{Aut}(\overline{X}_{t_0}; X^\partial_{t_0})$ un réprésentant de $\nu_{\mathcal{T}}(\delta)$, i.e. un difféomorphisme caractéristique de la fibration $f|_{\overline{X}^*}$. On a $sp \circ T = sp$.

1.2 Invariants homotopiques

La trivialisation $\mathcal{T}$ et les points de base x_i nous fournissent des points de base $y_i \in X^\partial_{t_0,i}$. Notons :

$$B := \pi_1(D^*, t_0), G := \pi_1(\overline{X}^*, y_0), L := \pi_1(\overline{X}_{t_0}, y_0),$$
$$H_i := \pi_1(X^\partial_{0,i}, x_i) \equiv \pi_1(X^\partial_i, x_i) \equiv \pi_1(X^\partial_i, y_i) \equiv \pi_1(X^\partial_{t_0}, y_i).$$

Les groupes H_i sont cycliques infinis et, grâce à l'orientation complexe, possèdent des génerateurs "positifs", que l'on notera γ_i. Le groupe L est libre de rang μ = nombre de Milnor de f.

Rappelons que la suite exacte d'homotopie de la fibration $f|_{\overline{X}^*}$

$$1 \longrightarrow L \longrightarrow G \longrightarrow B \longrightarrow 1 \tag{1}$$

admet une scission naturelle $\sigma: B \hookrightarrow G$ [26, 1.1.2]. Pour le voir, soit $i_0 \in I$ l'indice tel que $y_0 \in X^\partial_{t_0,i_0} \subset X^\partial_{i_0}$. Considérons la suite exacte d'homotopie de la fibration (triviale) $f|_{X^{*\partial}_{i_0}}$

$$1 \longrightarrow L^\partial = \pi_1(X^\partial_{t_0,i_0}, y_0) \longrightarrow G^\partial = \pi_1(X^{*\partial}_{i_0}, y_0) \longrightarrow B \longrightarrow 1,$$

qui s'injecte dans (1). Comme $f|_{X^\partial_{i_0}}$ est aussi une fibration triviale et D est contractile, l'injection $L^\partial \hookrightarrow \pi_1(X^\partial_{i_0}, y_0)$ est un isomorphisme. Ceci nous donne une rétraction de l'inclusion $L^\partial \hookrightarrow G^\partial$, d'où la section $\sigma: B \hookrightarrow G^\partial \hookrightarrow G$ cherchée.

L'homomorphisme de groupes $\nu_{\mathcal{T}}$ nous donne un autre $\nu: B \to \mathrm{Aut}(L)$ qui est indépendant de $\mathcal{T}$ (cf. [26, 1.1.2]). En fait, σ et ν permettent d'identifier le groupe G avec le produit semi-direct $L \times_\nu B$, car $\nu(\delta)$ coïncide avec la conjugaison par $\sigma(\delta)$ (cf. *loc. cit.*, prop. (1.1.3)).

Voyons maintenant les invariants associés au bord. Rappelons que si H est un groupe, un H-*torseur à gauche* est un ensemble Δ muni d'une action à gauche de H, qui est transitive et pour laquelle le stabilisateur de tout élément est trivial.

Soit Δ_i (resp. $\widetilde{\Delta_i}$) l'ensemble des classes d'homotopie de chemins dans $\overline{X}_{t_0}$ (resp. dans $\overline{X}^*$) allant de y_0 vers y_i, pour $i \in I$. Ces ensembles sont des L-torseurs (resp. G-torseurs) à gauche, et nous permettent de traiter convenablement les problèmes liés au changement des points de base. On a des isomorphismes canoniques

$$\pi_1(\overline{X}_{t_0}, y_i) \simeq \mathrm{Aut}_L(\Delta_i)^{opp} \qquad (\text{resp. } \pi_1(\overline{X}^*, y_i) \simeq \mathrm{Aut}_G(\widetilde{\Delta_i})^{opp}).$$

On a donc des inclusions $H_i \hookrightarrow \mathrm{Aut}_L(\Delta_i)^{opp}, i \in I$. L'homomorphisme $\nu_{\mathcal{T}}$ nous donne des homomorphismes de groupes $\rho_i : B \longrightarrow \mathcal{S}(\Delta_i), i \in I$ indépendants de $\mathcal{T}$ et compatibles avec ν, i.e.

$$\rho_i(b)(l \cdot \alpha) = \nu(b)(l) \cdot \rho_i(b)(\alpha), \quad b \in B, l \in L, \alpha \in \Delta_i, i \in I.$$

Etant donné que les $\nu_{\mathcal{T}}(b)$ laissent invariant le bord, on a aussi

$$\rho_i(b) \circ h = h \circ \rho_i(b), \qquad b \in B, h \in H_i, i \in I. \tag{2}$$

(1.2.1) A partir du représentant $f : (\overline{X}, X^{\partial}) \to D$ du germe $\boldsymbol{f}$, qui est essentiellement unique (cf. [19, (2.9)]), de la trivialisation $\mathcal{T}$ de $f|_{X^\delta}$, qui est aussi essentiellement unique car D est contractile, et du choix des points de base, on a donc les invariants homotopiques suivants

1) $B = \pi_1(D^*, t_0)$, groupe cyclique infini de générateur positif δ,
2) $L = \pi_1(\overline{X}_{t_0}, y_0)$, groupe libre à μ générateurs,
3) $\nu : B \to \mathrm{Aut}(L)$, qui en pratique est déterminé par l'automorphisme de monodromie $\nu(\delta) : L \to L$,
4) $\Delta_i = [y_0, y_i]_{\overline{X}_{t_0}}$, L-torseur à gauche des classes d'homotopie de chemins dans $\overline{X}_{t_0}$ de source y_0 et but y_i, pour $i \in I$,
5) $H_i = \pi_1(X^{\partial}_{0,i}, x_i) \hookrightarrow \mathrm{Aut}_L(\Delta_i)^{opp}$, groupe cyclique infini de générateur positif γ_i, $i \in I$,
6) $\rho_i : B \to \mathcal{S}(\Delta_i)$ homomorphisme de groupes tel que

$$\rho_i(b)(l \cdot \alpha) = \nu(b)(l) \cdot \rho_i(b)(\alpha), \quad b \in B, l \in L, \alpha \in \Delta_i, i \in I,$$
$$\rho_i(b) \circ h = h \circ \rho_i(b), \quad b \in B, h \in H_i, i \in I.$$

(1.2.2) Les invariants précédents nous permettent de récupérer les invariants suivants

a) $G = \pi_1(\overline{X}^*, y_0) \equiv L \times_\nu B$,

b) $L_i = \pi_1(\overline{X}_{t_0}, y_i) \equiv \mathrm{Aut}_L(\Delta_i)^{opp}$, $i \in I$,

c) $\nu_i : B \to \mathrm{Aut}(L_i)$, homomorphisme induit par ν_T :

$$\nu_i(b)(\xi) = \rho_i(b) \circ \xi \circ \rho_i(b)^{-1}$$

pour $b \in B$, $\xi \in \mathrm{Aut}_L(\Delta_i) \equiv L_i$ et $i \in I$,

d) $G_i = \pi_1(\overline{X}^*, y_i) \equiv L_i \times_{\nu_i} B$,

e) $\widetilde{\Delta_i} = [y_0, y_i]_{\overline{X}^*} \equiv \Delta_i \times_{\rho_i} B$, où l'action à gauche de G (resp. à droite de G_i) est donnée par

$$l \cdot (\alpha, b) = (\nu(b)^{-1}(l) \cdot \alpha, b) \qquad b' \cdot (\alpha, b) = (\alpha, b'b)$$

pour $l \in L, b' \in B, (\alpha, b) \in \Delta_i \times_\nu B \equiv G$ (resp.

$$(\alpha, b) \cdot (\xi, b') = (\xi(\rho_i(b')^{-1}(\alpha)), bb')$$

pour $(\alpha, b) \in L \times_\nu B \equiv G$, $(\xi, b') \in \mathrm{Aut}_L(\Delta_i) \times_{\nu_i} B \equiv L_i \times_{\nu_i} B \equiv G_i$ et $i \in I$).

(1.2.3) Remarque. Les ensembles Δ_i sont des $(L; L_i)$-ensembles, i.e. ils portent une action à gauche du groupe L et une action à droite du groupe L_i, compatibles entre elles. Les homomorphismes $\rho_i : B \to \mathcal{S}(\Delta_i)$ sont compatibles à gauche (resp. à droite) avec ν (resp. ν_i). Les ensembles $\Delta_i \times_{\rho_i} B$ définis ci-dessus sont des $(L \times_\nu B; L_i \times_{\nu_i} B)$-ensembles, que l'on appelera *produit semi-direct de Δ_i et B (par rapport à ρ_i)*. Ils sont caractérisés à isomorphisme unique près, par une propriété universelle de l'inclusion $\Delta_i \hookrightarrow \Delta_i \times_{\rho_i} B$ qu'on laisse au lecteur le soin d'énoncer.

Remarquons finalement que, dans le cas où $(C, 0)$ est irréductible, i.e. $\#I = 1$, les bords X_0^∂ et $X_{t_0}^\partial$ ont une seule composante connexe et il est inutile de choisir le seul point de base $y_i, i \in I$ différent de y_0. On a donc $\Delta_i = L = L_i$ et $\rho_i = \nu$. Les invariants 1)-6) ci-dessus deviennent

$$B, \quad L, \quad H \subset L, \quad \nu : B \to \mathrm{Aut}(L; H)$$

(cf. [26, (1.1.5)]).

Dans le cas géneral le point de base y_0 choisi appartient à une de composantes connexes de $X_{t_0}^\partial$, dissons à X_{t_0, i_0}^∂. On aurait pu prendre $y_0 = y_{i_0}$. Ceci sera utile quand'on fait des calculs explicites, mais pour l'instant nous ignorerons ce fait à fin de ne privilegier aucune des composantes irréductibles de $(C, 0)$ et d'avoir des énoncés plus symétriques.

(1.2.4) Remarque. A partir de $\nu : B \to \mathrm{Aut}(L)$ et des $\rho_i : B \to \mathcal{S}(\Delta_i)$ il est possible de récupérer l'action de B sur le groupe $\pi_1(\overline{X}_{t_0}/\star, \star)$ considérée dans [2, page 30, 6)].

§2 Faisceaux pervers coniques

Ce § est destiné à donner une description explicite de la catégorie des faisceaux pervers sur le germe $(C, 0)$, stratifiés par rapport à sa stratification canonique. Ceci s'obtient comme un cas très particulier de la théorie générale dévéloppée par MacPherson-Vilonen dans [20], ou si l'on veut, de la variante de Deligne-Verdier [9, 28] fondée sur le formalisme des cycles évanescents (voir aussi [11, 21, 22]). Nous allons développer une généralisation de ce dernier point de vue dans un contexte purement topologique, que l'on pourra appliquer directement à la situation envisagée. Ceci montrera, encore une fois, comment le formalisme des cycles évanescents est lié au problème de recollement de t-structures [4, 1.4], –indépendamment de la Géométrie Algébrique–, et nous donnera un foncteur quasi-inverse explicite (voir th. (2.3.4)). En plus, ce point de vue peut se généraliser pour traiter le cas où $(C, 0)$ est le germe d'un ensemble analytique complexe à singularité isolée, de dimension ≥ 2 (voir remarque (2.3.7)). Le contenu de ce § est largement inspiré de [10, exp. XIII] et [9].

2.1 t-structures coniques

Soit k un corps et S un "bon"[1] espace topologique compacte, et soient $S_i, i \in I$, ses composantes connexes, avec I fini. On supposera que le revêtement universel de chaque S_i est contractile. Soit Y le cône sur S de sommet 0 et notons $Y^* := Y - \{0\}$, dont les composantes connexes sont les $Y_i^* := S_i \times]0, 1]$. Prenons, pour chaque indice $i \in I$, un point de base $x_i \in S_i \hookrightarrow Y_i^*$ et posons $H_i := \pi_1(S_i, x_i) \equiv \pi_1(Y_i^*, x_i)$. Considérons sur Y la stratification Σ dont les strates sont le sommet 0 et les $Y_i^*, i \in I$, et soit $\mathcal{D} = \mathbf{D}^+_\Sigma(k_Y)$ la sous-catégorie pleine de la catégorie dérivée $\mathbf{D}^+(k_Y)$ dont les objets sont les complexes constructibles par rapport à Σ, i.e. les complexes tels que la restriction de ses faisceaux de cohomologie aux Y_i^* sont des systèmes locaux ($\equiv$ faisceau localement constant de k-espaces vectoriels[2]). Notons $\{0\} \overset{i}{\hookrightarrow} Y \overset{j}{\hookleftarrow} Y^*$ les inclusions, $\mathcal{D}_{\{0\}} = \mathbf{D}^+(k)$ et $\mathcal{D}_{Y^*}$ la sous-catégorie pleine de la catégorie dérivée $\mathbf{D}^+(k_{Y^*})$ dont les objets sont les complexes à cohomologie localement constante. On a les δ-foncteurs

$$\mathcal{D}_{\{0\}} \xrightarrow{i_*} \mathcal{D} \xleftarrow{\mathbf{R}j_*} \mathcal{D}_{Y^*}.$$

On munit la catégorie $\mathcal{D}_{Y^*}$ (resp. $\mathcal{D}_{\{0\}}$) de sa t-structure naturelle [4, ex. 1.3.2] (resp. de sa structure naturelle décalée de $[-1]$), et considérons sur $\mathcal{D}$

[1] p.ex. localement connexe par arcs, localement simplement connexe, localement compacte.

[2] Dans ce § on ne supposera pas que les espaces vectoriels sont de dimension finie, car les résultats traités ne font pas intervenir cette hypothèse.

la t-structure obtenue par recollement de celles de $\mathcal{D}_{Y^*}$ et $\mathcal{D}_{\{0\}}$ [4, 1.4.9], i.e.

$$\mathcal{D}^{\leq 0} = \{\mathcal{K} \in \mathcal{D} \mid h^n(j^{-1}\mathcal{K}) = 0, h^{n+1}(i^{-1}\mathcal{K}) = 0, \forall n > 0\}$$

$$\mathcal{D}^{\geq 0} = \{\mathcal{K} \in \mathcal{D} \mid h^n(j^{-1}\mathcal{K}) = 0, \mathbf{R}^{n+1}\Gamma_0\mathcal{K} = 0, \forall n < 0\}.$$

(2.1.1) Définition. La t-structure précédente $(\mathcal{D}^{\leq 0}, \mathcal{D}^{\geq 0})$ s'appelera *t-structure conique* sur $\mathbf{D}^+_\Sigma(k_Y)$. Son coeur [4, 1.3.1], qui est la sous-catégorie abélienne pleine $\mathcal{D}^{\leq 0} \cap \mathcal{D}^{\geq 0}$ de $\mathcal{D}$, sera noté $\mathrm{Perv}_{con}(Y, k)$, et ses objets seront appelés *k-faisceaux pervers coniques* sur Y.

(2.1.2) Remarque. La t-structure conique sur $\mathbf{D}^+_\Sigma(k_Y)$ coïncide avec la t-structure de perversité $p\colon \Sigma \to \mathbf{Z}$

$$p(\{0\}) = 1, \quad p(Y_i^*) = 0, \ i \in I$$

(voir [4, 2.1]). Si $\mathcal{K}$ est un complexe de $\mathbf{D}^+_\Sigma(k_Y)$, les propriétés suivantes sont équivalentes (cf. *loc. cit.*) :

a) $\mathcal{K}$ est un k-faisceaux pervers conique.

b) $j^{-1}\mathcal{K}$ est concentré en degré 0, $i^{-1}\mathcal{K}$ est concentré en degrés 0 et 1, et $\mathbf{R}^0\Gamma_0\mathcal{K} = \mathbf{R}^0\Gamma_0 h^0(\mathcal{K}) = 0$.

(2.1.3) Proposition. Si $\mathcal{K}$ est un k-faisceau pervers conique sur Y, on a une suite exacte canonique

$$0 \to \mathcal{K}^1 \to \mathcal{K} \to \mathcal{K}^2 \to 0$$

dans $\mathrm{Perv}_{con}(Y, k)$, avec $\mathcal{K}^1$ (resp. $\mathcal{K}^2$) k-faisceau pervers conique sur Y concentré en degré 0 (resp. en degré 1 et supporté par $\{0\}$).

Preuve : Considérons la suite exacte des tronqués

$$0 \to \sigma_{\leq 0}(\mathcal{K}) \to \mathcal{K} \to \sigma_{\geq 1}(\mathcal{K}) \to 0$$

(cf. [13, §7]). Comme $\mathcal{K}$ est concentré en degrés 0 et 1, on a des isomorphismes canoniques dans $\mathbf{D}^+_\Sigma(k_Y)$

$$\sigma_{\leq 0}(\mathcal{K}) \simeq h^0(\mathcal{K}), \quad \sigma_{\geq 1}(\mathcal{K}) \simeq h^1(\mathcal{K})[-1],$$

mais d'après la remarque précédente, $h^0(\mathcal{K})$ et $h^1(\mathcal{K})[-1]$ sont aussi des k-faisceaux pervers coniques sur Y, d'où la proposition. □

(2.1.4) Exemple. Avec les notations de §1, si l'on pose $S = \partial X_0$, l'espace $\overline{X}_0$ est homéomorphe au cône Y sur S, et le faisceaux pervers usuels sur $\overline{X}_0$ stratifiés par rapport à $\{0\}$ et $\overline{X}_0 - \{0\}$ s'identifient avec les $\mathbf{C}$-faisceaux pervers coniques sur Y, dont les fibres ont pour cohomologie des espaces vectoriels de dimension finie.

2.2 Les catégories et les foncteurs en jeu

Dorénavant, $\{H_i\}_{i\in I}$ désignera une famille finie de groupes et $\mathcal{A}, \mathcal{B}$ des catégories additives. Nous allons décrire des catégories et des foncteurs associés à ces objets et qui interviennent dans le formalisme des cycles évanescents (cf. [10, exp. XIII], [26, 2.1.2]).

(2.2.1) *Les catégories $\mathcal{C}_0$ et $\mathcal{C}$*

La catégorie $\mathcal{C}_0(\mathcal{A}; \{H_i\}_{i\in I})$ est par définition la catégorie dont

–) les objets sont les 4-uples $(U, \{V_i\}_{i\in I}, \{\alpha_i\}_{i\in I}, \{\rho_i\}_{i\in I})$ où U et les V_i sont des objets de $\mathcal{A}$, les $\alpha_i\colon U \to V_i$ sont des morphismes de $\mathcal{A}$ et les $\rho_i\colon H_i \to \mathrm{Aut}(V_i)$ sont des homomorphismes de groupes (réprésentations) tels que $\rho_i(h)\circ\alpha_i = \alpha_i$ pour tout $h \in H_i$ et $i \in I$.

–) les morphismes sont définis de façon naturelle.

La catégorie $\mathcal{C}(\mathcal{A}; \{H_i\}_{i\in I})$ est par définition la catégorie dont

–) les objets sont les 4-uples $(\{E_i\}_{i\in I}, F, \{u_i\}_{i\in I}, \{\{v_h^i\}_{h\in H_i}\}_{i\in I})$ où F et les E_i sont des objets de $\mathcal{A}$, et $u_i\colon E_i \to F$ et les $v_h^i\colon F \to E_i$ sont des morphismes de $\mathcal{A}$ tels que

- $v_{hh'}^i = v_h^i \circ u_i \circ v_{h'}^i + v_h^i + v_{h'}^i$ pour $h, h' \in H_i$ et $i \in I$
- $1_{E_i} + v_h^i \circ u_i$ et $1_F + u_i \circ v_h^i$ sont des isomorphismes, pour $h \in H_i$ et $i \in I$
- $v_h^i \circ u_j = 0$ pour $h \in H_i$ et $i \neq j$

–) les morphismes sont définis de façon naturelle.

Les catégories $\mathcal{C}_0(\mathcal{A}; \{H_i\}_{i\in I})$ et $\mathcal{C}(\mathcal{A}; \{H_i\}_{i\in I})$ sont aussi additives, et elles sont abéliennes si $\mathcal{A}$ est abélienne.

Tout foncteur additif $f\colon \mathcal{A} \to \mathcal{B}$ induit des foncteurs additifs $\mathcal{C}_0(f)$ et $\mathcal{C}(f)$ entre les catégories correspondantes. Ceci s'applique notamment au foncteur "translation" $\sqcup_{\mathcal{A}}$ et aux foncteurs de cohomologie $h_{\mathcal{A}}^n$, $n \in \mathbf{Z}$ sur les catégories des complexes d'objets d'une catégorie abélienne $\mathcal{A}$.

Remarquons que la catégorie des complexes d'objets de $\mathcal{C}_0(\mathcal{A}; \{H_i\}_{i\in I})$ (resp. de $\mathcal{C}(\mathcal{A}; \{H_i\}_{i\in I})$) s'identifie à la catégorie $\mathcal{C}_0(\mathbf{C}(\mathcal{A}); \{H_i\}_{i\in I})$ (resp. $\mathcal{C}(\mathbf{C}(\mathcal{A}); \{H_i\}_{i\in I})$).

(2.2.2) *Le foncteur Ω*

Si $\mathcal{A}$ est une catégorie abélienne, nous allons définir un foncteur additif

$$\Omega_{\mathcal{A}}\colon \mathbf{D}^*(\mathcal{C}_0(\mathcal{A}; \{H_i\}_{i\in I})) \longrightarrow \mathcal{C}(\mathbf{D}^*(\mathcal{A}); \{H_i\}_{i\in I}), \qquad * = \emptyset, +$$

suivant la construction de [10, exp. XIII, (1.4)], (cf. [26, (2.1.4)]).

Soit F un complexe d'objets de $\mathcal{C}_0(\mathcal{A}; \{H_i\}_{i\in I})$ et notons

$$F^n = (U^n, \{V_i^n\}_{i\in I}, \{\alpha_i^n\}_{i\in I}, \{\rho_i^n\}_{i\in I}), \quad n \in \mathbf{Z}$$
$$d_F^n = (d_U^n, \{d_{V_i}^n\}_{i\in I})\colon F^n \to F^{n+1}, \quad n \in \mathbf{Z}.$$

Les morphismes $\alpha_i\colon U \to V_i$ et $\rho_i\colon H_i \to \mathrm{Aut}(V_i)$ induisent des morphismes

$$\alpha\colon U \longrightarrow \oplus V_i, \quad \rho\colon \prod H_i \longrightarrow \mathrm{Aut}(\oplus V_i)$$

tels que $\rho(\underline{h})\circ\alpha = \alpha$ pour tout $\underline{h} \in \prod H_i$.

Rappelons que si $w\colon A \to B$ est un morphisme de complexes d'objets de $\mathcal{A}$, le cône de w, noté cône(w), est le complexe C défini par (cf. [13, I, §2])

$$C^n = B^n \oplus A^{n+1}, \qquad d_C^n = \begin{pmatrix} d_B^n & w^{n+1} \\ 0 & -d_A^{n+1} \end{pmatrix}.$$

Le complexe B s'injecte naturellement dans cône(w). On notera cône(A) := cône(1_A). Il s'agit d'un complexe homotopiquement trivial.

La correspondance $w \mapsto$ cône(w) est fonctorielle et additive dans un sens qu'on laisse au lecteur de préciser. En particulier, si $w_i : A_i \to B_i, i = 1, 2$, $a : A_1 \to A_2$, $b : B_1 \to B_2$ sont quatre morphismes de complexes tels que $w_2 \circ a = b \circ w_1$, ils induisent un morphisme

$$c(a, b)\colon \text{cône}(w_1) \longrightarrow \text{cône}(w_2).$$

Posons pour simplifier $V = \oplus V_i$ et $H = \prod H_i$, et soit $\widetilde{\alpha} : U \to V \oplus$ cône(U) le monomorphisme induit par α et l'inclusion $U \hookrightarrow$ cône(U). Notons $\widetilde{u}\colon V \oplus \text{cône}(U) \to Q$ le conoyau de $\widetilde{\alpha}$, et pour chaque $\underline{h} \in H$, soit $\widetilde{v}_{\underline{h}}\colon Q \to V \oplus \text{cône}(U)$ le seul morphisme tel que $1_{V\oplus\text{cône}(U)} + \widetilde{v}_{\underline{h}} \circ \widetilde{u} = \rho(\underline{h}) \oplus 1_{\text{cône}(U)}$. On a

$$\widetilde{v}_{\underline{h}\cdot\underline{h}'} = \widetilde{v}_{\underline{h}} \circ \widetilde{u} \circ \widetilde{v}_{\underline{h}'} + \widetilde{v}_{\underline{h}} + \widetilde{v}_{\underline{h}'}, \qquad \forall \underline{h}, \underline{h}' \in H.$$

En tenant compte des inclusions $H_j \hookrightarrow H$ et en composant avec les morphismes naturels $V_j \rightleftarrows V \oplus$ cône(U), on obtient des morphismes

$$u_j\colon V_j \longrightarrow Q \qquad \text{et} \qquad v_h^j\colon Q \longrightarrow V_j, \quad h \in H_j, j \in I.$$

On vérifie sans difficulté que le diagramme

$$\Omega_{\mathcal{A}}(F) := (\{V_i\}_{i\in I}, Q, \{[u_i]\}_{i\in I}, \{\{[v_h^i]\}_{h\in H_i}\}_{i\in I})$$

est un objet de $\mathcal{C}(\mathbf{K}(\mathcal{A}); \{H_i\}_{i\in I})$[3], et que la correspondance $F \mapsto \Omega_{\mathcal{A}}(F)$ s'étend naturellement en un foncteur additif

$$\Omega_{\mathcal{A}}\colon \mathbf{C}(\mathcal{C}_0(\mathcal{A}; \{H_i\}_{i\in I})) \longrightarrow \mathcal{C}(\mathbf{K}(\mathcal{A}); \{H_i\}_{i\in I}),$$

[3] Il est important ici de travailler dans la catégorie des complexes modulo homotopie. Ce procédé ne donne pas un objet de $\mathcal{C}(\mathbf{C}(\mathcal{A}); \{H_i\}_{i\in I})$.

Preuve : Il s'agit d'une conséquence directe de la définition du foncteur Ω et de l'identification canonique

$$\text{cône}(c(\boldsymbol{a}, \boldsymbol{a})) \equiv \text{cône}(\text{cône}(\boldsymbol{a}))$$

pour tout morphisme de complexes $\boldsymbol{a}$. Les détails sont laissés au lecteur. □

(2.2.5) Corollaire. Le foncteur

$$\mathcal{C}(h^0_{\mathcal{A}}) \circ \Omega_{\mathcal{A}} : \mathbf{K}(\mathcal{C}_0(\mathcal{A}; \{H_i\}_{i\in I})) \longrightarrow \mathcal{C}(\mathcal{A}; \{H_i\}_{i\in I})$$

est un foncteur cohomologique (voir [13, I, §1]).

(2.2.6) *Le foncteur* Λ

Gardons les notations de (2.2.2) et supposons que $\mathcal{A}$ est la catégorie des k-espaces vectoriels[4]. Etant donné un objet

$$\mathcal{O} = (\{E_i\}, F, \{u_i\}_{i\in I}, \{\{v^i_h\}_{h\in H_i}\}_{i\in I})$$

de $\mathcal{C}(\mathcal{A}; \{H_i\}_{i\in I})$, chaque E_i admet la structure de $k[H_i]$-module à gauche donnée par

$$h \cdot e := e + v^i_h(u_i(e)), \qquad h \in H_i, e \in E_i.$$

Rappelons d'abord quelques constructions élémentaires sur les répresentations de groupes. Soit H un groupe quelconque (par exemple, un des H_i) et E un $k[H]$-module à gauche. Le produit E^H est un $k[H]$-module à gauche, l'action de H étant

$$h \cdot \{\sigma \in H \mapsto e_\sigma\} = \{\sigma \mapsto e_{\sigma h}\}.$$

On dispose de

-) un monomorphisme $k[H]$-linéaire $\lambda : e \in E \mapsto \lambda(e) = \{\sigma \mapsto \sigma e\} \in E^H$

-) une répresentation $A : H \to \mathrm{Aut}_{k[H]}(E^H) : A(h)(\{\sigma \mapsto e_\sigma\}) = \{\sigma \mapsto he_{h^{-1}\sigma}\}$, avec $A(h) \circ \lambda = \lambda$ pour tout $h \in H$, d'où une autre répresentation $\bar{A} : H \to \mathrm{Aut}_{k[H]}(\mathrm{Coker}(\lambda))$ et un diagramme $\mathcal{E}$ de $k[H]$-modules

$$0 \longrightarrow E \xrightarrow{\ \lambda\ } E^H \xrightarrow{\ q\ } \mathrm{Coker}\,\lambda \longrightarrow 0, \qquad \{w_h\}_{h\in H} : \mathrm{Coker}\,\lambda \to E^H$$

[4] Plus généralement on pourrait prendre la catégorie des modules sur un faisceau d'anneaux défini sur un espace topologique.

qui passe tout d'abord au quotient "modulo homotopie" $\mathbf{K}(\mathcal{C}_0(\mathcal{A};\{H_i\}_{i\in I}))$, et puis induit un foncteur additif de $\mathbf{D}(\mathcal{C}_0(\mathcal{A};\{H_i\}_{i\in I}))$ dans $\mathcal{C}(\mathbf{D}(\mathcal{A});\{H_i\}_{i\in I})$, que l'on notera aussi $\Omega_{\mathcal{A}}$.

Remarquons que si

$$\boldsymbol{F} = (\boldsymbol{U}, \{\boldsymbol{V}_i\}_{i\in I}, \{\boldsymbol{\alpha}_i\}_{i\in I}, \{\boldsymbol{\rho}_i\}_{i\in I})$$

est un objet de $\mathbf{D}(\mathcal{C}_0(\mathcal{A};\{H_i\}_{i\in I}))$ et

$$\Omega_{\mathcal{A}}(\boldsymbol{F}) = (\{\boldsymbol{V}_i\}_{i\in I}, \boldsymbol{Q}, \{[\boldsymbol{u}_i]\}_{i\in I}, \{\{[\boldsymbol{v}_h^i]\}\}_{h\in H_i}\}_{i\in I}),$$

on a un triangle dans $\mathbf{D}(\mathcal{A})$

$$\boldsymbol{U} \xrightarrow{[\boldsymbol{\alpha}]} \oplus \boldsymbol{V}_i \xrightarrow{[\boldsymbol{u}]} \boldsymbol{Q} \longrightarrow \boldsymbol{U}[1], \tag{3}$$

fonctoriel par rapport à $\boldsymbol{F}$.

D'après les définitions, il est clair que

$$\Omega_{\mathcal{A}} \circ \sqcup_{\mathcal{C}_0(\mathcal{A};\{H_i\}_{i\in I})} \equiv \mathcal{C}(\sqcup_{\mathcal{A}}) \circ \Omega_{\mathcal{A}}.$$

Soit $F = (U, \{V_i\}_{i\in I}, \{\alpha_i\}_{i\in I}, \{\rho_i\}_{i\in I})$ un objet de $\mathcal{C}_0(\mathcal{A};\{H_i\}_{i\in I})$. Les morphismes $\alpha_i : U \to V_i$ et $\rho_i : H_i \to \mathrm{Aut}(V_i)$ induisent des morphismes $\alpha: U \longrightarrow \oplus V_i$, $\rho: H \longrightarrow \mathrm{Aut}(\oplus V_i)$ tels que $\rho(\underline{h}) \circ \alpha = \alpha$ pour tout $\underline{h} \in H$.

Posons pour simplifier $V = \oplus V_i$. Notons $\mu: V \to Q'$ le conoyau de α, et pour chaque $\underline{h} \in H$, soit $\nu_{\underline{h}}: Q' \to V$ le seul morphisme tel que $1_V + \nu_{\underline{h}} \circ \mu = \rho(\underline{h})$.

Ces morphismes induisent comme avant des morphismes

$$\mu_j: V_j \longrightarrow Q' \qquad \text{et} \qquad \nu_h^j: Q' \longrightarrow V_j, \quad h \in H_j, j \in I.$$

On vérifie que le diagramme

$$\Omega'_{\mathcal{A}}(F) := (\{V_i\}_{i\in I}, Q', \{\mu_i\}_{i\in I}, \{\{\nu_h^i\}_{h\in H_i}\}_{i\in I})$$

est un objet de $\mathcal{C}(\mathcal{A};\{H_i\}_{i\in I})$ et que la correspondance $F \mapsto \Omega'_{\mathcal{A}}(F)$ s'étend naturellement en un foncteur additif

$$\Omega'_{\mathcal{A}}: \mathcal{C}_0(\mathcal{A};\{H_i\}_{i\in I}) \longrightarrow \mathcal{C}(\mathcal{A};\{H_i\}_{i\in I}).$$

(2.2.3) Lemme. Soit $F = (U, \{V_i\}, \{\alpha_i\}, \{\rho_i\})$ un objet de $\mathcal{C}_0(\mathcal{A};\{H_i\})$ tel que $\alpha: U \to \oplus V_i$ est un monomorphisme. Alors

a) $\mathcal{C}(h^n)(\Omega_{\mathcal{A}}(F)) = 0$ si $n \neq 0$.

b) Il existe un isomorphisme naturel $\mathcal{C}(h^0)(\Omega_{\mathcal{A}}(F)) \simeq \Omega'_{\mathcal{A}}(F)$.

Preuve : Notons $\boldsymbol{F} = (\boldsymbol{U}, \{\boldsymbol{V}_i\}_{i\in I}, \{\boldsymbol{\alpha}_i\}_{i\in I}, \{\boldsymbol{\rho}_i\}_{i\in I})$ le complexe dont le seul terme non nul est $F^0 = F$ et gardons les notations précédentes. Posons

$$\begin{array}{rcl} \Omega_{\mathcal{A}}(\boldsymbol{F}) & = & (\{\boldsymbol{V}_i\}_{i\in I}, \boldsymbol{Q}, \{[\boldsymbol{u}_i]\}_{i\in I}, \{\{[\boldsymbol{v}_h^i]\}_{h\in H_i}\}_{i\in I}) \\ \Omega'_{\mathcal{A}}(\boldsymbol{F}) & = & (\{\boldsymbol{V}_i\}_{i\in I}, \boldsymbol{Q}', \{\boldsymbol{\mu}_i\}_{i\in I}, \{\{\boldsymbol{\nu}_h^i\}_{h\in H_i}\}_{i\in I}). \end{array}$$

Le complexe $\boldsymbol{Q}'$ est par définition le conoyau de $\boldsymbol{\alpha}$, et il est donc nul en dehors de 0. Plus précisément, le complexe $\Omega'_{\mathcal{A}}(\boldsymbol{F})$ ne rien d'autre que l'objet $\Omega'_{\mathcal{A}}(F)$ concentré en degré 0.

Comme les complexes $\boldsymbol{U}$ et $\boldsymbol{V}_i$ sont nuls en degrés $\neq 0$, le complexe $\boldsymbol{Q}$ est nul en degrés $\neq -1, 0$. De plus, le morphisme $h^0(\boldsymbol{\alpha}) = \alpha$ est injectif et d'après le triangle (3), le complexe $\boldsymbol{Q}$ est comologiquement nul en dehors de 0. Ceci démontre la partie a).

Comme $\boldsymbol{Q}$ est le conoyau de $\widetilde{\boldsymbol{\alpha}}$, il existe un seul morphisme $\boldsymbol{\xi}' : \boldsymbol{Q} \to \boldsymbol{Q}'$ tel que $\boldsymbol{\xi}' \circ \widetilde{\boldsymbol{u}} = \boldsymbol{\mu} \circ \boldsymbol{\pi}$, avec $\boldsymbol{\pi} : \boldsymbol{V} \oplus \text{cône}(\boldsymbol{U}) \to \boldsymbol{V}$ la première projection. Considérons le diagramme commutatif

$$\begin{array}{ccccccccc} U & \longrightarrow & \oplus V_i & \longrightarrow & h^0(\boldsymbol{Q}) & \longrightarrow & 0 & \longrightarrow & 0 \\ \downarrow{\scriptstyle 1} & & \downarrow{\scriptstyle 1} & & \downarrow{\scriptstyle h^0(\boldsymbol{\xi}')} & & \downarrow & & \downarrow \\ U & \longrightarrow & \oplus V_i & \longrightarrow & h^0(\boldsymbol{Q}') = Q'^0 & \longrightarrow & 0 & \longrightarrow & 0 \end{array}$$

où les lignes horizontales proviennent des suites exactes longues de cohomologie. D'après le lemme des cinq, $h^0(\boldsymbol{\xi}')$ est un isomorphisme, et

$$(\{1_{V_i}\}_{i\in I}, h^0(\boldsymbol{\xi}')) : \mathcal{C}(h^0)(\Omega_{\mathcal{A}}(F)) \longrightarrow \Omega'_{\mathcal{A}}(F)$$

est l'isomorphisme cherché. □

(2.2.4) Proposition. Soient $\boldsymbol{F}_1, \boldsymbol{F}_2$ deux complexes d'objets de $\mathcal{C}_0(\mathcal{A}; \{H_i\})$, et

$$\boldsymbol{\eta} = (\boldsymbol{g}, \{\boldsymbol{f}_i\}_{i\in I}) : \boldsymbol{F}_1 \longrightarrow \boldsymbol{F}_2$$

un morphisme. Posons

$$\Omega_{\mathcal{A}}(\boldsymbol{F}_k) = (\{\boldsymbol{V}_{k,i}\}_{i\in I}, \boldsymbol{Q}_k, \{[\boldsymbol{u}_{k,i}]\}_{i\in I}, \{\{[\boldsymbol{v}_h^{k,i}]\}\}_{h\in H_i}\}_{i\in I}), \quad k = 1, 2$$

et $\Omega_{\mathcal{A}}(\boldsymbol{\eta}) = (\{[\boldsymbol{f}_i]\}, [\boldsymbol{h}])$. On a un isomorphisme canonique entre $\Omega_{\mathcal{A}}(\text{cône}(\boldsymbol{\eta}))$ et

$$(\{\text{cône}(\boldsymbol{f}_i)\}_{i\in I}, \text{cône}(\boldsymbol{h}), \{[c(\boldsymbol{u}_{1,i}, \boldsymbol{u}_{2,i})]\}_{i\in I}, \{\{[c(\boldsymbol{v}_h^{1,i}, \boldsymbol{v}_h^{2,i})]\}\}_{h\in H_i}\}_{i\in I}).$$

avec q la projection naturelle et $1 + w_h \circ q = A(h)$, $1 + q \circ w_h = \overline{A}(h)$ pour tout $h \in H$.

-) un monomorphisme k-linéaire $c\colon e \in E \mapsto \{\sigma \mapsto e\} \in E^H$, qui établit un isomorphisme entre E et les invariants $(E^H)^{inv} = \{\boldsymbol{e} \in E^H \mid h \cdot \boldsymbol{e} = \boldsymbol{e}, \forall h \in H\}$, et satisfait la relation

$$c(h \cdot e) = A(h)(c(e)), \qquad e \in E, h \in H. \tag{4}$$

-) un k-isomorphisme $S : \{\sigma \mapsto e_\sigma\} \in E^H \mapsto \{\sigma \mapsto \sigma e_{\sigma^{-1}}\} \in E^H$, qui satisfait les relations $S^2 = 1_{E^H}$, $S \circ \lambda = c$, $S \circ c = \lambda$, $(S \circ A(h))(\boldsymbol{e}) = h \cdot S(\boldsymbol{e})$ et $S(h \cdot \boldsymbol{e}) = A(h)(S(\boldsymbol{e}))$ pour tout $h \in H$ et tout $\boldsymbol{e} \in E^H$.

-) une réprésentation $A'\colon H \to \mathrm{Aut}_k(\mathrm{Hom}_{k[H]}(I(H), E^H))$

$$A'(h)(\varphi)(x) = \varphi(xh), \quad \forall h \in H, \forall \varphi \in \mathrm{Hom}_{k[H]}(I(H), E^H), \forall x \in I(H)$$

où $I(H)$ est l'idéal d'augmentation de l'algèbre du groupe $k[H]$.

-) un diagramme $\mathcal{E}'$ de k-espaces vectoriels

$$0 \longrightarrow (E^H)^{inv} \xrightarrow{\text{inc.}} E^H \underset{\{w'_h\}_{h \in H}}{\overset{I^*}{\rightleftarrows}} \mathrm{Hom}_{k[H]}(I(H), E^H) \rightarrow 0$$

où $I^*(\boldsymbol{e})(x) = x \cdot \boldsymbol{e}$ pour tout $\boldsymbol{e} \in E^H$ et tout $x \in I(H)$, et $w'_h(\varphi) = \varphi(h-1)$ pour tout $h \in H$ et tout $\varphi \in \mathrm{Hom}_{k[H]}(I(H), E^H)$.

La correspondance qui à chaque $k[H]$-module E associe les diagrammes $\mathcal{E}$ et $\mathcal{E}'$, et les morphismes c et S, est évidemment fonctorielle.

(2.2.7) Lemme. Dans la situation précédente, la suite du diagramme $\mathcal{E}'$ est exacte, et les morphismes c, S induisent un isomorphisme naturel entre les diagrammes de k-espaces vectoriels $\mathcal{E}$ et $\mathcal{E}'$.

Preuve : La démonstration est facile et elle est laissée au lecteur. □

Revenons à notre objet de départ $\mathcal{O}$, et affectons de l'indice "i" les objets définis ci-dessus et provenant du $k[H_i]$-module E_i. Nous définissons un complexe $\boldsymbol{F}$ de $\mathbf{C}(\mathcal{C}_0(\mathcal{A}; \{H_i\}_{i \in I}))$, nul en degrés $\neq 0, 1$, de la façon suivante :

$$\begin{aligned} F^n &= (U^n, \{V_i^n\}_{i \in I}, \{\alpha_i^n\}_{i \in I}, \{\rho_i^n\}_{i \in I}), \quad n \in \mathbf{Z} \\ d_F^n &= (d_U^n, \{d_{V_i}^n\}_{i \in I}), \quad n \in \mathbf{Z} \end{aligned}$$

avec

- $U^0 = \oplus E_j$, $V_i^0 = E_i^{H_i}$, $\alpha_i^0 : U^0 \to V_i^0$ est la composante i-ième de $\oplus c_j$ et $\rho_i^0(h)(e) = h \cdot e$, $\forall h \in H_i, \forall e \in V_i^0$,
- $U^1 = F$, $V_i^1 = \mathrm{Coker}(\lambda_i)$, $\rho_i^1(h) \circ q_i = q_i \circ \rho_i^0(h)$, $\forall h \in H_i$, $\alpha_i^1 = q_i \circ \omega_i$ et $\omega_i : F \to E_i^{H_i}$ est définie par $\omega_i(f) = \{\sigma \in H_i \mapsto -v_\sigma^i(f) \in E_i\}$, $\forall f \in F$,
- $d_U^0 : \oplus E_j \to F$ est le morphisme induit par les u_i et $d_{V_i}^0 = q_i$.

De façon analogue on définit le complexe $\boldsymbol{F}'$ de $\mathbf{C}(\mathcal{C}_0(\mathcal{A}; \{H_i\}_{i\in I}))$, nul en degrés $\neq 0, 1$:

$$\begin{aligned} F'^n &= (U'^n, \{V_i'^n\}_{i\in I}, \{\alpha_i'^n\}_{i\in I}, \{\rho_i'^n\}_{i\in I}), \quad n \in \mathbf{Z} \\ d_{F'}^n &= (d_{U'}^n, \{d_{V_i'}^n\}_{i\in I}), \quad n \in \mathbf{Z} \end{aligned}$$

avec

- $U'^0 = U^0 = \oplus E_j$, $V_i'^0 = V_i^0 = E_i^{H_i}$, $\alpha_i'^0 : U^0 \to V_i^0$ est la composante i-ième de $\oplus \lambda_j$ et $\rho_i'^0 = A_i$,
- $U'^1 = U^1 = F$, $V_i'^1 = \mathrm{Hom}_{k[H_i]}(I(H_i), E_i^{H_i})$, $\rho_i^1(h) = (A_i(h))_*$, $\forall h \in H_i$, $\alpha_i'^1 = (I_i)^* \circ \omega_i'$ et $\omega_i' : F \to E_i^{H_i}$ est définie par $\omega_i'(f) = \{\sigma \in H_i \mapsto v_\sigma^i(f) \in E_i\}$, $\forall f \in F$,
- $d_{U'}^0 = d_U^0$, $d_{V_i'}^0 = (I_i)^*$.

Les correspondances $\mathcal{O} \mapsto \boldsymbol{F}$ et $\mathcal{O} \mapsto \boldsymbol{F}'$ s'étendent naturellement en des foncteurs additifs, notés

$$\Lambda, \Lambda' : \mathcal{C}(\mathcal{A}; \{H_i\}_{i\in I}) \longrightarrow \mathbf{C}(\mathcal{C}_0(\mathcal{A}; \{H_i\}_{i\in I}))$$

respectivement. D'après le lemme (2.2.7), les foncteurs Λ et Λ' sont naturellement isomorphes.

L'intérêt de ces foncteurs se trouve dans la proposition suivante (cf. [26, lemme (2.2.2.2)]).

(2.2.8) Proposition. Etant donné un objet $\mathcal{O}$ de $\mathcal{C}(\mathcal{A}; \{H_i\}_{i\in I})$, il existe un morphisme naturel $(\{[\gamma_i]\}_{i\in I}, [\beta]) : \mathcal{O} \to \Omega_{\mathcal{A}}(\Lambda(\mathcal{O}))$ dans $\mathcal{C}(\mathbf{K}(\mathcal{A}); \{H_i\}_{i\in I})$, qui devient un isomorphisme dans $\mathcal{C}(\mathbf{D}(\mathcal{A}); \{H_i\}_{i\in I})$. En particulier, on a des isomorphismes de foncteurs

$$(\mathrm{nat.}) \simeq \Omega_{\mathcal{A}} \circ \Lambda \simeq \Omega_{\mathcal{A}} \circ \Lambda',$$

où $(\mathrm{nat.}) : \mathcal{C}(\mathcal{A}; \{H_i\}_{i\in I}) \to \mathcal{C}(\mathbf{D}(\mathcal{A}); \{H_i\}_{i\in I})$ désigne le foncteur induit par l'immersion standard $\mathcal{A} \hookrightarrow \mathbf{D}(\mathcal{A})$.

Preuve : Posons, comme avant,

$$\mathcal{O} = (\{E_i\}, F, \{u_i\}_{i\in I}, \{\{v_h^i\}_{h\in H_i}\}_{i\in I}), \quad \boldsymbol{F} = \Lambda(\mathcal{O})$$

et gardons les notations de (2.2.2) et (2.2.6).

Soit $\gamma_i^0 = -\lambda_i : E_i \to V_i^0$ et $\gamma_i^n = 0\ \forall n \neq 0$. Il est clair que $d^0_{V_i} \circ \gamma_i^0 = -(q_i \circ \lambda_i) = 0$, d'où on obtient un morphisme de complexes $\boldsymbol{\gamma}_i : E_i \to \boldsymbol{V}_i$.

Soit $\beta^0 : F \to Q^0$ le morphisme défini par

$$\beta^0(f) = \widetilde{u^0}(\omega(f), 0, f), \qquad \forall f \in F$$

où $\omega : F \to \oplus E_i^{H_i}$ est le morphisme induit par les ω_i, et $\beta^n = 0\ \forall n \neq 0$. On a

$$(d_Q^0 \circ \beta^0)(f) = (d_Q^0 \circ \widetilde{u^0})(\omega(f), 0, f) = (\widetilde{u^1} \circ d^0_{V \oplus \text{cône}(U)})(\omega(f), 0, f) =$$
$$\widetilde{u^1}(d_V^0(\omega(f), f, 0) = \widetilde{u^1}(\alpha^0(f), f, 0) = \widetilde{u^1}(\widetilde{\alpha^0}(f)) = 0$$

d'où un morphisme de complexes $\boldsymbol{\beta} : F \to \boldsymbol{Q}$.

Voyons d'abord que $(\{[\gamma_i]\}_{i\in I}, [\boldsymbol{\beta}]) : \mathcal{O} \to \Omega_{\mathcal{A}}(\Lambda(\mathcal{O}))$ est un morphisme dans $\mathcal{C}(\mathbf{K}(\mathcal{A}); \{H_i\}_{i\in I})$. Notons $\boldsymbol{\pi}_i : \boldsymbol{V} \oplus \text{cône}(\boldsymbol{U}) \to \boldsymbol{V}_i$ la projection naturelle. Pour chaque $i \in I$ et chaque $h \in H_i$ on a

$$(v_h^i)^0(\beta^0(f)) = (v_h^i)^0(\widetilde{u^0}(\omega(f), 0, f)) = \pi_i(\widetilde{v_h^0}(\widetilde{u^0}(\omega(f), 0, f))) =$$
$$\pi_i(\rho_h(\omega(f)) - \omega(f), 0, 0) = h \cdot \omega_i(f) - \omega_i(f) =$$
$$= \cdots = -\lambda_i(v_h^i(f)) = \gamma_i^0(v_h^i(f))$$

$\forall f \in F$,d'où $\boldsymbol{v}_h^i \circ \boldsymbol{\beta} = \boldsymbol{\gamma} \circ v_h^i$.

Il nous reste à montrer que $[\boldsymbol{\beta}] \circ [u_i] = [\boldsymbol{u}_i] \circ [\boldsymbol{\gamma}_i], \forall i \in I$. Dans ce cas on a besoin de travailler dans la catégorie des complexes modulo homotopie, car $\beta^0 \circ u_i \neq u_i^0 \circ \gamma_i^0$. On doit donc trouver une homotopie $s_i : E_i \to Q^{-1}$ telle que

$$\beta^0 \circ u_i - u_i^0 \circ \gamma_i^0 = d_Q^{-1} \circ s_i. \tag{5}$$

Soit $t_i : E_i \hookrightarrow V^{-1} \oplus \text{cône}(\boldsymbol{U})^{-1} = V^{-1} \oplus U^{-1} \oplus U^0 = \oplus E_j$ l'inclusion naturelle et posons $s_i = -(\widetilde{u^{-1}} \circ t_i)$. On vérifie directement la relation (5).

Pour terminer la preuve de la proposition, il faut montrer que les morphismes $[\boldsymbol{\gamma}_i], i \in I$ et $[\boldsymbol{\beta}]$ sont quasi-isomorphismes. Pour les $[\boldsymbol{\gamma}_i]$ l'assertion est claire. A partir des définitions on voit facilement que $h^n(\boldsymbol{Q}) = 0$ pour $n \neq 0$. Plus précisement, la nullité de $h^{-1}(\boldsymbol{Q}) = \operatorname{Ker} d_Q^{-1}$ est une conséquence de l'injectivité des applications $c_i : E_i \to E_i^{H_i}$, et la nullité de $h^1(\boldsymbol{Q}) = \operatorname{Coker} d_Q^0$ est une conséquence de la surjectivité des différentielles $d^0_{\text{cône}(U)}$ et $d^0_{V_i}, i \in I$. Notons $\overline{\beta^0} : F \to h^0(\boldsymbol{Q})$ le morphisme induit par β^0. Soit $\overline{x} \in h^0(\boldsymbol{Q})$ avec $x \in \operatorname{Ker} d_Q^0$. Prenons un élément $y = (\{\boldsymbol{e}_i\}_{i\in I}, \{e_i\}_{i\in I}, f) \in V^0 \oplus U^0 \oplus U^1$

tel que $x = \widetilde{u^0}(y)$, et posons $f' = \sum u_i(e_i) + f$. On vérifie que $x = \beta^0(f' - \sum u_i(e_{i,1})) + d_Q^{-1}(\widetilde{u^{-1}}(0, 0, \{e_i + e_{i,1}\}_{i \in I}))$, où $e_{i,1} \in E_i$ désigne la composante relative à $1 \in H_i$ de $e_i \in E_i^{H_i}$. On en déduit la surjectivité de $\overline{\beta^0}$. L'injectivité de $\overline{\beta^0}$ est laissée au lecteur. □

2.3 Description des faisceaux pervers coniques

(2.3.1) *Rappels et compléments sur les systèmes locaux et le recollement des faisceaux*

A) Soit U un "bon" espace topologique connexe, $z_0 \in U$ un point de base et $H = \pi_1(U, z_0)$. Soit $p\colon (\widetilde{U}, \widetilde{z}_0) \to (U, z_0)$ le revêtement universel. Le groupe H s'identifie au groupe des automorphismes de $p : \widetilde{U} \to U$ et il opère donc à gauche sur le foncteur $p_* \circ p^{-1}$. En particulier, si $\mathfrak{L}$ est un système local sur U, le k-espace vectoriel $\Gamma(U, p_*p^{-1}\mathfrak{L})$ est un $k[H]$-module à gauche. La correspondance

$$\mathfrak{L} \mapsto \Gamma(U, p_*p^{-1}\mathfrak{L}) \tag{6}$$

est clairement fonctorielle, et établit une équivalence de catégories abéliennes entre la catégorie des systèmes locaux sur U et $\mathrm{Mod}(k[H])$. Pour les détails le lecteur pourra consulter [14, IV.9], [8].

Soit $\mathfrak{L}$ un système local sur U et notons E le $k[H]$-module qui lui est associé par (6). Le morphisme d'adjonction $\widetilde{\lambda}\colon \mathfrak{L} \hookrightarrow p_*p^{-1}\mathfrak{L}$ identifie $\mathfrak{L}$ avec le sous-faisceau des invariants de $p_*p^{-1}\mathfrak{L}$ par l'action de H, et les sections globales de $\mathfrak{L}$ s'dentifient avec E^{inv}. Considérons le diagramme $\widetilde{\mathcal{E}}$ de systèmes locaux sur U

$$0 \longrightarrow \mathfrak{L} \xrightarrow{\ \widetilde{\lambda}\ } p_*p^{-1}\mathfrak{L} \xrightarrow{\ \widetilde{q}\ } Coker\widetilde{\lambda} \longrightarrow 0, \qquad \{\widetilde{w}_h\}_{h \in H}\colon Coker\widetilde{\lambda} \to p_*p^{-1}\mathfrak{L}$$

où $\widetilde{q}$ est la projection naturelle et $\widetilde{w}_h \circ \widetilde{q}$ est égale a l'action de h sur $p_*p^{-1}\mathfrak{L}$ moins l'identité. Le diagramme $\mathcal{E}$ de (2.2.6) est naturellement isomorphe au diagramme de $k[H]$-modules obtenu par application de (6) à $\widetilde{\mathcal{E}}$. De plus, l'isomorphisme $E = \Gamma(U, p_*p^{-1}\mathfrak{L}) \simeq (E^H)^{inv}$ s'interprète comme l'application $c\colon e \in E \mapsto \{\sigma \mapsto e\} \in E^H$.

Si $\mathfrak{L}$ et $\mathfrak{L}'$ sont deux systèmes locaux sur U, $\varphi\colon \mathfrak{L} \to \mathfrak{L}'$ est un morphisme et $(p_*p^{-1}\varphi) \circ \lambda = 0$, alors $\varphi = 0$.

Si l'on note

$$\widetilde{\psi}_U : \mathrm{Mod}(k_U) \longrightarrow \mathcal{C}_0(\mathrm{Mod}(k_U); H)$$

le foncteur défini par $\widetilde{\psi}_U(\mathcal{F}) := (\mathcal{F}, p_*p^{-1}\mathcal{F}, \widetilde{\lambda}, \widetilde{\rho})$, où $\widetilde{\rho}$ réprésente l'action de H sur $p_*p^{-1}\mathcal{F}$, on a

$$\Omega'_{\mathrm{Mod}(k_U)}(\widetilde{\psi}_U)(\mathcal{L}) = (p_*p^{-1}\mathcal{L}, Coker\, \widetilde{\lambda}, \widetilde{q}, \{\widetilde{w}_h\}_{h\in H}). \tag{7}$$

B) Soit Z un espace topologique, $U \subset Z$ un ouvert non vide et $j : U \hookrightarrow Z$, $i : Z-U \hookrightarrow Z$ les inclusions. La donnée d'un faisceau de k-espaces vectoriels $\mathcal{F}$ sur Z est équivalente à la donnée de ses restrictions $j^{-1}\mathcal{F}, i^{-1}\mathcal{F}$ et du morphisme d'adjonction $\widetilde{\alpha}_{\mathcal{F}} : i^{-1}\mathcal{F} \to i^{-1}j_*j^{-1}\mathcal{F}$, car le diagramme suivant est cartésien

$$\begin{array}{ccc} \mathcal{F} & \longrightarrow & i_*i^{-1}\mathcal{F} \\ \downarrow & & \downarrow{\scriptstyle i_*\alpha_{\mathcal{F}}} \\ j_*j^{-1}\mathcal{F} & \longrightarrow & i_*i^{-1}j_*j^{-1}\mathcal{F}. \end{array}$$

En fait, les données du type $(\mathcal{F}_U, \mathcal{F}_{Z-U}, \widetilde{\alpha})$, où $\mathcal{F}_U$ est un faisceau de k-espaces vectoriels sur U, $\mathcal{F}_{Z-U}$ est un faisceau de k-espaces vectoriels sur $Z-U$ et $\widetilde{\alpha} : \mathcal{F}_{Z-U} \to i^{-1}j_*\mathcal{F}_U$ est un morphisme de faisceaux, forment une catégorie abélienne, notée $\mathcal{R}_0$, et la correspondance

$$\mathcal{F} \mapsto (j^{-1}\mathcal{F}, i^{-1}\mathcal{F}, \widetilde{\alpha}_{\mathcal{F}}) \tag{8}$$

est fonctorielle et établit une équivalence de catégories entre $\mathrm{Mod}(k_Z)$ et $\mathcal{R}_0$.

C) Soit T un "bon" espace topologique connexe et Z le cône sur T de sommet 0. Posons $U = Z - \{0\} \simeq]0,1] \times T$, $z_0 \in U$, $H = \pi_1(U, z_0)$ et gardons les notations précédentes. Si $\mathcal{L}$ est un système local sur U, le morphisme de restriction $\Gamma(U, \mathcal{L}) \to i^{-1}j_*\mathcal{L}$ est un isomorphisme, et en combinant (6) et (8) on obtient une équivalence de catégories abéliennes $\}_Z$ entre la catégorie des faisceaux de k-espaces vectoriels sur Z, constructibles par rapport à $\{0\}$ et U, et la catégorie dont les objets sont les 3-uples (E, F, α), où E est un $k[H]$-module, F est un k-espace vectoriel et $\alpha : F \to E^{inv}$ est une application linéaire, et les morphismes sont définis de façon naturelle. Si $\mathcal{F}$ est un faisceau constructible sur Z, et $\}_Z(\mathcal{F}) = (E, F, \alpha)$, E est le $k[H]$-module associé à la restriction $\mathcal{F}|_U$, $F = \mathcal{F}_0$ et α est le morphisme induit par l'adjonction.

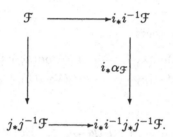

Si $\mathcal{L}$ est un système local sur U et E le $k[H]$-module qui lui est associé par (6), on a un isomorphisme naturel

$$\}_Z(j_*\mathcal{L}) \simeq (E, E^{inv}, 1).$$

Les résultats précédents se généralisent sans difficulté au cas où les espaces U et T ont plusieures composantes connexes.

Revenons à la situation décrite dans 2.1 et notons

$$\begin{array}{rcll} p_i : (\widetilde{Y_i^*}, \tilde{x}_i) & \longrightarrow & (Y_i^*, x_i), & i \in I \\ p = \coprod p_i : \coprod \widetilde{Y_i^*} & \longrightarrow & Y^* = \coprod Y_i^* & \end{array}$$

les revêtements universels, et $j : Y^* \hookrightarrow Y, j_i : Y_i^* \hookrightarrow Y, \kappa_i : Y_i^* \hookrightarrow Y^*$, $i \in I$, les inclusions.

Considérons les foncteurs

$$\begin{array}{rcl} \widetilde{\psi}_Y : \mathrm{Mod}(k_Y) & \longrightarrow & \mathcal{C}_0(\mathrm{Mod}(k_Y); \{H_i\}_{i\in I}) \\ \psi_Y : \mathrm{Mod}(k_Y) & \longrightarrow & \mathcal{C}_0(\mathrm{Vect}(k); \{H_i\}_{i\in I}) \end{array}$$

définis par

$$\begin{array}{rcl} \widetilde{\psi}_Y & = & (1_{\mathrm{Mod}(k_Y)}, \{(j_i)_*(p_i)_*p_i^{-1}j_i^{-1}\}_{i\in I}, \{\alpha_i\}_{i\in I}, \{\rho_i\}_{i\in I}) \\ \psi_Y & = & (\widetilde{\psi}_Y)_0 \end{array}$$

où $\alpha_i : 1_{\mathrm{Mod}(k_Y)} \to (j_i)_*(p_i)_*p_i^{-1}j_i^{-1}$ est le morphisme d'adjonction, et

$$\rho_i : H_i \to \mathrm{Aut}((j_i)_*(p_i)_*p_i^{-1}j_i^{-1})$$

est l'action induite par $H_i \equiv \mathrm{Aut}(\widetilde{Y_i^*}/Y_i^*)$ (cf. (2.3.1), A)).

On définit finalement les foncteurs additifs

$$\begin{array}{rcl} \widetilde{\Psi}_Y & := & \Omega_{\mathrm{Mod}(k_Y)} \circ \mathbf{R}\widetilde{\psi}_Y : \mathbf{D}^+(k_Y) \longrightarrow \mathcal{C}(\mathbf{D}^+(k_Y); \{H_i\}_{i\in I}) \\ \Psi_Y & := & \Omega_{\mathrm{Vect}(k)} \circ \mathbf{R}\psi_Y \equiv (\widetilde{\Psi}_Y)_0 : \mathbf{D}^+(k_Y) \longrightarrow \mathcal{C}(\mathbf{D}^+(k); \{H_i\}_{i\in I}). \end{array}$$

D'après (2.2.2), (3), si $\mathcal{K}$ est un objet de $\mathbf{D}^+_\Sigma(k_Y)$, et

$$\widetilde{\Psi}_Y(\mathcal{K}) = (\{\mathcal{E}_i\}_{i\in I}, \mathcal{F}, \{u_i\}_{i\in I}, \{\{v_h^i\}_{h\in H_i}\}_{i\in I}),$$

on a un triangle naturel par rapport à $\mathcal{K}$

$$T_{\mathcal{K}} := \mathcal{K} \xrightarrow{a} \oplus\mathcal{E}_i \xrightarrow{u} \mathcal{F} \longrightarrow \mathcal{K}[1]. \qquad (9)$$

(2.3.2) Proposition. Soit $\mathcal{K}$ un complexe de $\mathbf{D}^+_\Sigma(k_Y)$, et posons $\mathcal{L}_i^n = j_i^{-1}h^n(\mathcal{K})$, $\widetilde{\Psi}_Y(\mathcal{K}) = (\{\mathcal{E}_i\}_{i\in I}, \mathcal{F}, \{u_i\}_{i\in I}, \{\{v_h^i\}_{h\in H_i}\}_{i\in I})$. Il existent des isomorphismes naturels

a) $\mathbf{R}^n\widetilde{\psi}_Y(\mathcal{K}) \simeq \widetilde{\psi}_Y(h^n(\mathcal{K}))$.

b) $h^n(\mathcal{E}_i) \simeq (j_i)_*(p_i)_*p_i^{-1}\mathcal{L}_i^n$.

Preuve : D'après la suite spéctrale $\mathbf{R}^k\widetilde{\psi}_Y(h^l(\mathcal{K})) \Rightarrow \mathbf{R}^{k+l}\widetilde{\psi}_Y(\mathcal{K})$, pour la partie a) il suffit de montrer que $\mathbf{R}^k\widetilde{\psi}_Y(h^l(\mathcal{K})) = 0$ pour tout $l \in \mathbf{Z}$ et pour tout $k > 0$. Or, si $\mathcal{F}$ est un faisceau de k-espaces vectoriels sur Y, constructible par rapport à Σ, $\mathbf{R}^n(p_i)_*p_i^{-1}j_i^{-1}\mathcal{F} = 0$ pour tout $n > 0$ et

$$(\mathbf{R}^k(j_i)_*\mathbf{R}(p_i)_*p_i^{-1}j_i^{-1}\mathcal{F})_0 \simeq \mathbf{R}^k\Gamma(\widetilde{Y_i^*}, p_i^{-1}j_i^{-1}\mathcal{F}) = 0, \quad \forall k > 0$$

car $\widetilde{Y_i^*}$ est contractile. Par conséquent les faisceaux constructibles sont acycliques pour $\widetilde{\psi}_Y$. Ceci démontre a).

La partie b) est une conséquence de a) et de la définition du foncteur Ω (voir (2.2.2)). □

On a le résultat suivant, qui généralise celui de [9].

(2.3.3) Proposition. Soit $\mathcal{K}$ est un complexe de $\mathbf{D}^+_\Sigma(k_Y)$. Les propriétés suivantes sont équivalentes :

a) $\mathcal{K}$ est un faisceau pervers conique.

b) $\widetilde{\Psi}_Y(\mathcal{K})$ est concentré en degré 0, i.e. $\mathcal{C}(h^i)(\widetilde{\Psi}_Y(\mathcal{K})) = 0$ si $i \neq 0$.

c) $\Psi_Y(\mathcal{K})$ est concentré en degré 0.

Preuve : Posons $\widetilde{\Psi}_Y(\mathcal{K}) = (\{\mathcal{E}_i\}_{i\in I}, \mathcal{F}, \{u_i\}_{i\in I}, \{\{v^i_h\}_{h\in H_i}\}_{i\in I})$.
D'après la proposition (2.3.2), on a des isomorphismes naturels

$$j_*p_*p^{-1}j^{-1}h^n(\mathcal{K}) \simeq \oplus_i h^n(\mathcal{E}_i), \quad \forall n \in \mathbf{Z}. \tag{10}$$

Considérons le diagramme à lignes exactes suivant

$$
\begin{array}{ccccccc}
 & & & \vdots & & & \\
 & & & \downarrow & & & \\
 & & & \oplus h^{-1}(\mathcal{E}_i) & & & \\
 & & & \downarrow & & & \\
 & & & h^{-1}(\mathcal{F}) & & & \\
 & & & \downarrow & & & \\
\mathbf{R}^{-1}j_*j^{-1}\mathcal{K} & \to & \mathbf{R}^0\Gamma_0\mathcal{K} & \longrightarrow h^0(\mathcal{K}) & \to & \mathbf{R}^0j_*j^{-1}\mathcal{K} & \\
 & & & \downarrow & \swarrow_{\beta} & & \\
 & & & \oplus h^0(\mathcal{E}_i) & & & \\
 & & & \downarrow & & & \\
 & & & h^0(\mathcal{F}) & & & \\
 & & & \downarrow & & & \\
 & & & h^1(\mathcal{K}) & & & \\
 & & & \downarrow & & & \\
 & & & \vdots & & &
\end{array}
$$

où la ligne verticale (resp. horizontale) provient du triangle $T_{\mathcal{K}}$ 2.3, (9) (resp. du triangle de cohomologie locale à support) et $\beta\colon \mathbf{R}^0j_*j^{-1}\mathcal{K} \to \oplus h^0(\mathcal{E}_i) \simeq j_*p_*p^{-1}j^{-1}h^0(\mathcal{K})$ est le morphisme induit par adjonction. Ce diagramme est commutatif car le morphisme $h^0(\mathcal{K}) \to \oplus h^0(\mathcal{E}_i) \simeq j_*p_*p^{-1}j^{-1}h^0(\mathcal{K})$ est aussi le morphisme d'adjonction.

Si $\mathcal{K}$ est pervers conique, alors $\mathcal{K}$ est concentré en degrés 0 et 1, $j^{-1}\mathcal{K}$ est concentré en degré 0 et $\mathbf{R}^0\Gamma_0\mathcal{K} = 0$. D'après (10), les $\mathcal{E}_i$ sont aussi concentrés en degré 0, et d'après (2.1.2), $\mathcal{F}$ est concentré en degrés -1 et 0. Or, dans ce cas le morphisme β est injectif, et joint à la nullité de $\mathbf{R}^0\Gamma_0\mathcal{K}$ dans le diagramme ci-dessus on déduit que $h^{-1}(\mathcal{F}) = 0$, d'où l'implication a) $\Rightarrow$ b).

L'implication b) $\Rightarrow$ c) est triviale. Voyons l'implication c) $\Rightarrow$ a). Si $\Psi_Y(\mathcal{K}) = \widetilde{\Psi}_Y(\mathcal{K})_0$ est concentré en degré 0, on a

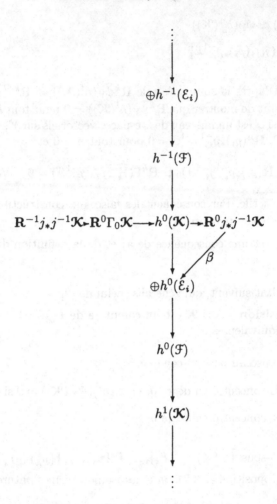

(i) $h^n(\mathcal{E}_i)_0 = 0, \quad n \neq 0, i \in I$

(ii) $h^n(\mathcal{F})_0 = 0, \quad n \neq 0.$

D'après (10), (i) et le fait que les $j^{-1}h^n(\mathcal{K})$ sont localement constants, on déduit que $j^{-1}\mathcal{K}$ est concentré en degré 0. D'après (i) et (ii) on déduit d'abord que $i^{-1}\mathcal{K}$ est concentré en degrés 0 et 1, puis la nullité de $\mathbf{R}^0\Gamma_0\mathcal{K}$. Par conséquent $\mathcal{K}$ est un faisceau pervers conique. □

Pour simplifier, notons par $\widetilde{\Psi}_Y$ et Ψ_Y les foncteurs

$$\begin{aligned} \mathcal{C}(h^0) \circ \widetilde{\Psi}_Y : \mathrm{Perv}_{con}(Y,k) &\longrightarrow \mathcal{C}(\mathrm{Mod}(k_Y); \{H_i\}_{i\in I}) \\ \mathcal{C}(h^0) \circ \Psi_Y : \mathrm{Perv}_{con}(Y,k) &\longrightarrow \mathcal{C}(\mathrm{Vect}(k); \{H_i\}_{i\in I}) \end{aligned}$$

respectivement. D'après le corollaire (2.2.5) et le théorème (2.3.3), les deux foncteurs précédents sont des foncteurs exacts entre catégories abéliennes.

On notera aussi $\Omega_Y = \Omega_{\mathrm{Mod}(k_Y)}, \Omega_U = \Omega_{\mathrm{Mod}(k_U)}, \Omega'_U = \Omega'_{\mathrm{Mod}(k_U)}, \Omega = \Omega_{\mathrm{Vect}(k)}$.

Le théorème suivant nous donne la description algébrique cherchée au début de ce paragraphe.

(2.3.4) Théorème. Avec les notations précédentes, le foncteur

$$\Psi_Y : \mathrm{Perv}_{con}(Y,k) \longrightarrow \mathcal{C}(\mathrm{Vect}(k); \{H_i\}_{i\in I})$$

est une équivalence de catégories abéliennes.

Preuve : Démontrons premièrement que Ψ_Y est essentiellement surjectif. Soit

$$\mathcal{O} = (\{E_i\}_{i\in I}, F, \{u_i\}, \{\{v_h^i\}_{h\in H_i}\}_{i\in I})$$

un objet de $\mathcal{C}(\mathrm{Vect}(k); \{H_i\}_{i\in I})$, et gardons le notations de (2.2.6), (2.3.1). Pour chaque $i \in I$, soit $\mathcal{L}_i$ le système local sur Y_i^* associé fonctoriellement au $k[H_i]$-module E_i (cf. (2.3.1), A)), et soit $\mathcal{L}$ le système local sur Y^* dont la restriction à chaque Y_i^* coïncide avec $\mathcal{L}_i$. Considérons la suite exacte de systèmes locaux sur Y^*

$$0 \to \mathcal{L} \xrightarrow{\widetilde{\lambda}} p_*p^{-1}\mathcal{L} \xrightarrow{\widetilde{q}} Coker\,\widetilde{\lambda} \to 0$$

et notons $\mathcal{G}$ le complexe de faisceaux sur Y^*

$$\cdots \longrightarrow 0 \longrightarrow \overset{\bullet}{p_*p^{-1}\mathcal{L}} \xrightarrow{\widetilde{q}} Coker\,\widetilde{\lambda} \longrightarrow 0 \longrightarrow \cdots.$$

Le complexe de k-espaces vectoriels $(j_*\mathcal{G})_0$ s'identifie naturellement au complexe (cf. (2.3.1), B))

$$\cdots \longrightarrow 0 \longrightarrow \overset{\bullet}{\oplus(E_i^{H_i})^{inv}} \xrightarrow{\oplus q_i} \oplus(\mathrm{Coker}\,\lambda_i)^{inv} \longrightarrow 0 \longrightarrow \cdots.$$

Soit G le complexe de k-espaces vectoriels

$$\cdots \longrightarrow 0 \longrightarrow \overset{\bullet}{\oplus E_i} \xrightarrow{u} F \longrightarrow 0 \longrightarrow \cdots,$$

avec $u = \sum u_i$, et soit $r\colon G \to (j_*\mathcal{G})_0$ le morphisme induit par

$$\begin{array}{ccc} \oplus E_i & \xrightarrow{\oplus c_i} & \oplus(E_i^{H_i})^{inv} \\ \Big\downarrow u & & \Big\downarrow \oplus q_i \\ F & \xrightarrow[V]{} & \oplus(\operatorname{Coker}\lambda_i)^{inv} \end{array}$$

où $V(f) = \{q_i(\{\sigma \in H_i \mapsto -v^i_\sigma(f)\})\}_{i\in I}$, $f \in F$.

Notons $\mathcal{K}$ le complexe de k_Y-modules obtenu en recollant $\mathcal{G}$ et G au moyen de r (cf. (2.3.1), B)). Il est clair que $\mathcal{K}$ est un complexe de faisceaux constructibles par rapport à Σ et donc ψ_Y-acyclique. On vérifie sans difficulté que l'on a un isomorphisme naturel $\psi_Y(\mathcal{K}) = \mathbf{R}\psi_Y(\mathcal{K}) \simeq \Lambda(\mathcal{O})$ (cf. (2.3.1), A)), et d'après la proposition (2.2.8)

$$\mathcal{O} \simeq \Omega(\Lambda(\mathcal{O})) \simeq \Psi_Y(\mathcal{K}). \tag{11}$$

De là on déduit que $\mathcal{K}$ est un faisceau pervers conique sur Y (prop. (2.3.3)) et Ψ_Y est essentiellement surjectif.

Il nous reste à montrer que Ψ_Y est pleinement fidèle. Pour cela, voyons d'abord comment $\widetilde{\Psi}_Y(\mathcal{K})$ peut se reconstruire à partir $\Psi_Y(\mathcal{K})$, pour $\mathcal{K}$ faisceau pervers conique. Posons

$$\begin{aligned} h^0(\mathcal{K})|_{Y_i^*} &= \mathcal{L}_i, \quad h^0(\mathcal{K})|_{Y^*} = \mathcal{L} = \oplus\mathcal{L}_i \\ \widetilde{\Psi}_Y(\mathcal{K}) &= (\{\mathcal{E}_i\}, \mathcal{F}, \{\tilde{u}_i\}, \{\{\tilde{v}^i_h\}_{h\in H_i}\}) \\ \Psi_Y(\mathcal{K}) &= (\{E_i\}, F, \{u_i\}, \{\{v^i_h\}_{h\in H_i}\}) = (\widetilde{\Psi}_Y(\mathcal{K}))_0 \equiv \Gamma(Y, \widetilde{\Psi}_Y(\mathcal{K})). \end{aligned}$$

D'après la proposition (2.3.2) , $\mathcal{E}_i \simeq (j_i)_*(p_i)_*p_i^{-1}\mathcal{L}_i$. En plus, on a des isomorphismes naturels

$$\begin{aligned} \mathcal{C}(j^{-1})\widetilde{\Psi}_Y(\mathcal{K}) &\overset{\text{(not.)}}{\equiv} \mathcal{C}(j^{-1})\mathcal{C}(h^0)\widetilde{\Psi}_Y(\mathcal{K}) \overset{\text{(def.)}}{=} \mathcal{C}(j^{-1})\mathcal{C}(h^0)\Omega_Y\mathbf{R}\widetilde{\psi}_Y(\mathcal{K}) \simeq \\ &\simeq \mathcal{C}(h^0)\Omega_U\mathbf{R}\widetilde{\psi}_U(j^{-1}\mathcal{K}) \overset{\star}{\simeq} \mathcal{C}(h^0)\Omega_U\widetilde{\psi}_U(\mathcal{L}) \overset{\star\star}{\simeq} \Omega'_U(\mathcal{L}), \end{aligned}$$

où l'isomorphisme $\star$ (resp. $\star\star$) provient de l'acyclicité des systèmes locaux pour $\widetilde{\psi}_U$ (resp. du lemme (2.2.3)). Autrement dit, la restriction à Y^* de $\widetilde{\Psi}_Y(\mathcal{K})$

est naturellement isomorphe au diagramme de systèmes locaux sur Y^*

$$(\{(\kappa_i)_*(p_i)_*p_i^{-1}\mathcal{L}_i\}, \oplus(\kappa_j)_*(Coker\,\widetilde{\lambda}_j), \{(\kappa_i)_*(\widetilde{q}_i)\}, \{\{(\kappa_i)_*(\widetilde{w}_h^i)\}_{h\in H_i}\}) \quad (12)$$

d'où $E_i = (\mathcal{E}_i)_0 \equiv \Gamma(Y, \mathcal{E}_i)$ muni de l'action de H_i définie dans (2.2.6), est naturellement isomorphe au $k[H_i]$-module associé à $\mathcal{L}_i$ (cf. (2.3.1), A)). Soit $\}_Y$ l'équivalence de catégories décrite dans (2.3.1), C). On en déduit des isomorphismes naturels

$$\}_Y(\mathcal{E}_i) \simeq (\{S_j^i\}_{j\in I}, E_i, c_i), \quad \}_Y(\mathcal{F}) \simeq (\{\mathrm{Coker}\,\lambda_j\}_{j\in I}, F, W) \quad (13)$$

où $S_j^i = 0$ si $i \neq j$, $S_i^i = E_i^{H_i}$ et $W: F \to \oplus(\mathrm{Coker}\,\lambda_j)^{inv}$ est un morphisme à déterminer.

Ces isomorphismes permettent d'identifier pour tout $i \in I$ et pour tout $h \in H_i$

$$\begin{array}{ccc} \}_Y(\widetilde{u}_i) & \longleftrightarrow & (\{\delta_{ij}\lambda_i\}_{j\in I}, u_i) \\ \}_Y(\widetilde{v_h^i}) & \longleftrightarrow & (\{\delta_{ij}w_h^i\}_{j\in I}, v_h^i) \end{array}$$

avec δ_{ij} le symbole de Kronecker.

Soit $f \in F$ et posons $W(f) = \{q_j(\boldsymbol{e}_j)\}_{j\in I}$. On peut supposer que $e_{j,1} = 0$ pour tout $j \in I$. On a pour tout $i \in I$ et tout $h \in H_i$

$$\{\sigma \in H_i \mapsto v_h^i(f)\} = c_i(v_h^i(f)) = w_h^i(q_i(\boldsymbol{e}_i)) = A_i(h)(\boldsymbol{e}_i) - \boldsymbol{e}_i,$$

d'où $v_h^i(f) = he_{i,h^{-1}\sigma} - e_{i,\sigma}$ pour tout $\sigma \in H_i$, ce qui donne pour $\sigma = h$ l'égalité $e_{i,h} = -v_h^i(f)$. On en déduit

$$W(f) = \{q_i(\{h \in H_i \mapsto -v_h^i(f)\}\}_{i\in I} = V(f).$$

Ceci nous dit que le diagramme de faisceaux constructibles $\widetilde{\Psi}_Y(\mathcal{K})$ est déterminé fonctoriellement par $\Psi_Y(\mathcal{K}) = (\widetilde{\Psi}_Y(\mathcal{K}))_0$.

Prenons un autre faisceau pervers conique $\mathcal{K}'$ et posons

$$\begin{aligned} h^0(\mathcal{K}')|_{Y_i^*} &= \mathcal{L}_i', \quad h^0(\mathcal{K}')|_{Y^*} = \mathcal{L}' = \oplus\mathcal{L}_i' \\ \widetilde{\Psi}_Y(\mathcal{K}') &= (\{\mathcal{E}_i'\}, \mathcal{F}', \{\widetilde{u'}_i\}, \{\{\widetilde{v'}_h^i\}_{h\in H_i}\}) \\ \Psi_Y(\mathcal{K}') &= (\{E'_i\}, F', \{u_i'\}, \{\{v'^i_h\}_{h\in H_i}\}) = (\widetilde{\Psi}_Y(\mathcal{K}'))_0 \equiv \Gamma(Y, \widetilde{\Psi}_Y(\mathcal{K}')). \end{aligned}$$

Affectons de " ′ " les objets qui s'en déduisent.

Soit $\boldsymbol{f}: \mathcal{K} \to \mathcal{K}'$ un morphisme. Posons $\varphi_i = h^0(\boldsymbol{f})|_{Y_i^*}: \mathcal{L}_i \to \mathcal{L}_i'$, $\varphi = h^0(\boldsymbol{f})|_{Y^*}: \mathcal{L} \to \mathcal{L}'$ et

$$\widetilde{\Psi}_Y(\boldsymbol{f}) = (\{\widetilde{g}_i\}_{i\in I}, \widetilde{h}), \qquad \Psi_Y(\boldsymbol{f}) = (\{g_i\}_{i\in I}, h).$$

Par les mêmes arguments que dans (12), on trouve que le morphisme $\widetilde{g}_i \colon \mathcal{E}_i \to \mathcal{E}'_i$ s'identifie à $(j_i)_*(p_i)_*p_i^{-1}\varphi_i$ et que la restriction à Y^* de $\widetilde{\Psi}_Y(\boldsymbol{f})$ s'identifie à

$$(\{(\kappa_i)_*(p_i)_*p_i^{-1}\varphi_i\}_{i\in I}, \oplus(\kappa_j)_*(\overline{(p_j)_*p_j^{-1}\varphi_j})),$$

où $\overline{(p_j)_*p_j^{-1}\varphi_j} \colon \operatorname{Coker}\widetilde{\lambda}_j \to \operatorname{Coker}\widetilde{\lambda}'_j$ est le morphisme induit par passage au quotient.

On en déduit que le morphisme $\varphi_i \colon \mathcal{L}_i \to \mathcal{L}'_i$ est le morphisme correspondant au morphismes de $k[H_i]$-modules $g_i \colon E_i \to E'_i$ (cf. (2.3.1), A)).

Supposons que $\Psi_Y(\boldsymbol{f}) = (\{g_i\}_{i\in I}, h) = 0$. Dans ce cas les morphismes φ_i doivent être nuls, ce qui entraîne la nullité des $\widetilde{g}_i$ et de $j^{-1}\widetilde{h}$. Or, comme $(\widetilde{h})_0 = h$ est aussi nul, $\widetilde{h} = 0$ et $\widetilde{\Psi}_Y(\boldsymbol{f}) = 0$. La fidelité de Ψ_Y est donc une conséquence de la proposition suivante

(2.3.5) Proposition. Le foncteur

$$\widetilde{\Psi}_Y \colon \operatorname{Perv}_{con}(Y,k) \longrightarrow \mathcal{C}(\operatorname{Mod}(k_Y); \{H_i\}_{i\in I})$$

est fidèle.

Preuve : Soient $\mathcal{K}$ et $\mathcal{K}'$ deux faisceaux pervers coniques sur Y et $\boldsymbol{f} \colon \mathcal{K} \to \mathcal{K}'$ un morphisme. Supposons que $\widetilde{\Psi}_Y(\boldsymbol{f}) = (\{\widetilde{g}_i\}_{i\in I}, \widetilde{h}) = 0$.

Notons (voir 2.3, (9))

$$\begin{aligned} T_{\mathcal{K}} &= \mathcal{K} \xrightarrow{a} \mathcal{E} \xrightarrow{u} \mathcal{F} \xrightarrow{b} \mathcal{K}[1] \\ T_{\mathcal{K}'} &= \mathcal{K}' \xrightarrow{a'} \mathcal{E}' \xrightarrow{u'} \mathcal{F}' \xrightarrow{b'} \mathcal{K}'[1]. \end{aligned}$$

Comme $\widetilde{g} = \oplus\widetilde{g}_i = 0$ et $\widetilde{h} = 0$ on a $a' \circ \boldsymbol{f} = 0$, $\boldsymbol{f}[1] \circ b = 0$. Considérons la suite exacte de faisceaux pervers de la proposition (2.1.3)

$$0 \to \mathcal{K}^1 \overset{\alpha_1}{\to} \mathcal{K} \overset{\alpha_2}{\to} \mathcal{K}^2 \to 0.$$

Par l'exactitude du foncteur $\widetilde{\Psi}_Y$, on obtient un diagramme commutatif à lignes exactes

$$\begin{array}{ccccccccc}
 & & \mathcal{K}^1[1] & \xrightarrow{\alpha_1[1]} & \mathcal{K}[1] & \xrightarrow{\alpha_2[1]} & \mathcal{K}^2[1] & \xrightarrow{+1} & \\
 & & {\scriptstyle b_1}\uparrow & & {\scriptstyle b}\uparrow & & {\scriptstyle b_2}\uparrow & & \\
0 & \longrightarrow & \mathcal{F}^1 & \xrightarrow{\gamma_1} & \mathcal{F} & \xrightarrow{\gamma_2} & \mathcal{F}^2 & \longrightarrow & 0 \\
 & & {\scriptstyle u_1}\uparrow & & {\scriptstyle u}\uparrow & & {\scriptstyle u_2}\uparrow & & \\
0 & \longrightarrow & \mathcal{E}^1 & \xrightarrow{\beta_1} & \mathcal{E} & \xrightarrow{\beta_2} & \mathcal{E}^2 & \longrightarrow & 0 \\
 & & {\scriptstyle a_1}\uparrow & & {\scriptstyle a}\uparrow & & {\scriptstyle a_2}\uparrow & & \\
 & & \mathcal{K}^1 & \xrightarrow{\alpha_1} & \mathcal{K} & \xrightarrow{\alpha_2} & \mathcal{K}^2 & \xrightarrow{+1} &
\end{array}$$

où la colonne à gauche (resp. à droite) est le triangle $T_{\mathcal{K}^1}$ (resp. $T_{\mathcal{K}^2}$). On a $a' \circ (f \circ \alpha_1) = 0, (f \circ \alpha_1)[1] \circ b_1 = 0$. Comme $\mathcal{F}'$ est concentré en degré 0, $h^0(a')$ est injective et $h^0(f \circ \alpha_1) = 0$. Or, comme $\mathcal{K}^1$ (resp. $\mathcal{K}$) est concentré en degré 0 (resp. degrés positifs), le morphisme $\boldsymbol{f} \circ \alpha_1$ est déterminé par son h^0, d'où la nullité de $\boldsymbol{f} \circ \alpha_1$. Par conséquent, le morphisme $\boldsymbol{f}$ passe au quotient et nous donne $\boldsymbol{f}' : \mathcal{K}^2 \to \mathcal{K}'$ tel que $\boldsymbol{f}' \circ \alpha_2 = \boldsymbol{f}$. Posons $\widetilde{\Psi}_Y(\boldsymbol{f}') = (\{\widetilde{g'}_i\}_{i \in I}, \widetilde{h'})$ et $\widetilde{g'} = \oplus \widetilde{g'}_i$. Par fonctorialité, on a $\widetilde{g'} \circ \beta_2 = \widetilde{g} = 0$, $\widetilde{h'} \circ \gamma_2 = \widetilde{h} = 0$, d'où $\widetilde{g'} = 0$, $\widetilde{h'} = 0$. Mais $\mathcal{K}^2$ étant supporté par $\{0\}$, $\mathcal{E}^2 = 0$ et le morphisme b_2 est un isomorphisme. On en déduit la nullité de $\boldsymbol{f}'$ et donc celle de $\boldsymbol{f}$. □

Pour terminer la preuve du théorème (2.3.4) il nous faut montrer que le foncteur Ψ_Y est surjectif au niveau des morphismes. Prenons deux faisceaux pervers coniques $\mathcal{K}$ et $\mathcal{K}'$ et gardons les notations précédentes.

Soit $(\{g_i\}_{i \in I}, h) : \Psi_Y(\mathcal{K}) \longrightarrow \Psi_Y(\mathcal{K}')$ un morphisme. Notons $\varphi_i : \mathcal{L}_i \to \mathcal{L}'_i$ le morphisme correspondant au morphisme $k[H_i]$-linéaire $g_i : E_i \to E'_i$ et $\widetilde{g}_i = (j_i)_*(p_i)_* p_i^{-1} \varphi_i : \mathcal{E}_i \to \mathcal{E}'_i$. Soit $\widetilde{h} : \mathcal{F} \to \mathcal{F}'$ le morphisme qu'on obtient en recollant h comme fibre en 0 et $\oplus (\kappa_j)_* (\overline{(p_j)_* p_j^{-1} \varphi_j})$ comme restriction à Y^*. On vérifie par construction que

$$(\{\widetilde{g}_i\}_{i \in I}, \widetilde{h}) : \widetilde{\Psi}_Y(\mathcal{K}) \longrightarrow \widetilde{\Psi}_Y(\mathcal{K}')$$

est un morphisme et $(\{\widetilde{g}_i\}_{i \in I}, \widetilde{h})_0 = (\{g_i\}_{i \in I}, h)$.

Posons $\widetilde{g} = \oplus \widetilde{g}_i$. D'après les axiomes des catégories triangulées, il existe un morphisme $\boldsymbol{f} : \mathcal{K} \to \mathcal{K}'$ tel que

$$(\boldsymbol{f}, \widetilde{g}, \widetilde{h}) : T_{\mathcal{K}} \longrightarrow T_{\mathcal{K}'}$$

est un morphisme de triangles. Il suffit de voir que $\widetilde{\Psi}_Y(\boldsymbol{f}) = (\{\widetilde{g}_i\}_{i \in I}, \widetilde{h})$.

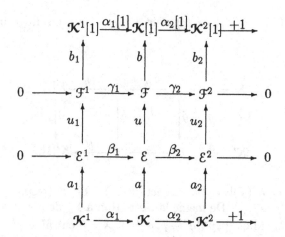

Posons $\widetilde{\Psi}_Y(\boldsymbol{f}) = (\{\widetilde{g'}_i\}_{i\in I}, \widetilde{h'})$ et $\widetilde{g'} = \oplus \widetilde{g'}_i$. On a un diagramme commutatif

$$\begin{array}{ccccccc}
\mathcal{K} & \xrightarrow{a} & \mathcal{E} & \xrightarrow{u} & \mathcal{F} & \xrightarrow{b} & \mathcal{K}[1] \\
\downarrow{\scriptstyle 0} & & \downarrow{\scriptstyle \widetilde{g'} - \widetilde{g}} & & \downarrow{\scriptstyle \widetilde{h'} - \widetilde{h}} & & \downarrow{\scriptstyle 0} \\
\mathcal{K}' & \xrightarrow{a'} & \mathcal{E}' & \xrightarrow{u'} & \mathcal{F}' & \xrightarrow{b'} & \mathcal{K}'[1].
\end{array}$$

Notons $\varphi' = h^0(\boldsymbol{f})|_{Y^*}$. La restriction à Y^* de $\widetilde{g'}$ (resp. $\widetilde{g}$) s'identifie à $p_*p^{-1}\varphi'$ (resp. $p_*p^{-1}\varphi$). De même la restriction à Y^* de a s'identifie au morphisme $\widetilde{\lambda} = \oplus \widetilde{\lambda}_i$. Mais $(p_*p^{-1}\varphi' - p_*p^{-1}\varphi) \circ \lambda = 0$ entraîne $\varphi' - \varphi = 0$ (voir (2.3.1), A)), d'où $\widetilde{g'} - \widetilde{g} \equiv j_*p_*p^{-1}(\varphi' - \varphi) = 0$. On en déduit $\widetilde{g'}_i = \widetilde{g}_i, \forall i \in I$ et $\widetilde{h'} = \widetilde{h}$. □

(2.3.6) Remarque. Dans la première partie de la preuve du théorème (2.3.4) on a construit, pour chaque objet $\mathcal{O}$ de $\mathcal{C}(\mathrm{Vect}(k); \{H_i\}_{i\in I})$, un faisceau pervers conique su Y, $\mathcal{K}$, tel que $\mathcal{O} \simeq \Psi_Y(\mathcal{K})$. En fait, la correspondance $\mathcal{O} \mapsto \mathcal{K}$ définit un foncteur, qui est un quasi-inverse à droite explicite de Ψ.

(2.3.7) Remarque. Dans la situation de 2.1, il est possible de considérer la t-structure sur $\mathbf{D}^+_\Sigma(k_Y)$ de perversité $p_d \colon \Sigma \to \mathbf{Z}$

$$p_d(\{0\}) = d, \quad p_d(Y_i^*) = 0, \ i \in I$$

pour $d \geq 1$, ou même pour $d \in \mathbf{Z}$. Notons $\mathrm{Perv}^{(d)}_{con}(Y, k)$ son coeur. Si $\mathcal{K}$ est un complexe de $\mathbf{D}^+_\Sigma(k_Y)$, et $\widetilde{\Psi}_Y(\mathcal{K}) = (\{\mathcal{E}_i\}_{i\in I}, \mathcal{F}, \boldsymbol{u}, \{\{v^i_h\}_{h\in H_i}\}_{i\in I})$, on peut démontrer que, pour $d \geq 2$, $\mathcal{K} \in \mathrm{Perv}^{(d)}_{con}(Y, k)$ si et seulement si les $\mathcal{E}_i$ sont concentrés en dégré 0 et $\mathcal{F} \in \mathrm{Perv}^{(d-1)}_{con}(Y, k)$, ce qui généralise la proposition (2.3.3). Ceci peut être utilisé pour donner une généralisation du théorème (2.3.4) au cas de $\mathrm{Perv}^{(d)}_{con}(Y, k)$, moyennant une récurrence par rapport à d. En fait il est possible de se débarrasser de l'hypothèse où les S_i ont un revêtement universel contractile, d'où on peut traiter par ces méthodes la catégorie des faisceaux pervers usuels sur un germe d'ensemble analytique complexe Y à singularité isolée, de dimension $d \geq 2$. Dans ce cas, cette catégorie coïncide avec $\mathrm{Perv}^{(d)}_{con}(Y, k)$. Ceci est l'objet d'un travail en préparation [27].

§3 Faisceaux pervers stratifiés par rapport au germe d'une courbe plane

3.1 Calcul du $\mathrm{R}\psi_f(\mathcal{L})$

Revenons aux invariants décrits dans (1.2.1) et (1.2.2) et gardons les notations introduites :

$$B = \pi_1(D^*, t_0) = \langle\delta\rangle, L = \pi_1(\overline{X}_{t_0}, y_0), \nu\colon B \to \mathrm{Aut}(L),$$
$$G = \pi_1(\overline{X}^*, y_0) \equiv L \times_\nu B, \Delta_i = [y_0, y_i]_{\overline{X}_{t_0}},$$
$$H_i = \pi_1(X^\partial_{0,i}, x_i) \hookrightarrow L_i = \mathrm{Aut}_L(\Delta_i)^{opp}, \rho_i\colon B \to \mathcal{S}(\Delta_i), i \in I.$$

Notons aussi $L' = \prod L_i$ et $\nu'\colon B \to \mathrm{Aut}(L')$ l'homomorphisme induit par les ν_i.

Posons $\Delta = \coprod \Delta_i$. Il s'agit d'un $(L; L')$-ensemble, qui est libre de rang $r = \#I$ en tant que L-ensemble à gauche. Soit $\rho\colon B \longrightarrow \mathcal{S}(\Delta)$ l'homomorphisme de groupes induit par les ρ_i, qui est compatible à gauche (resp. à droite) avec ν (resp. ν').

Pour un $(L; L')$-ensemble A, notons $k[A]$ le k-espace vectoriel de base A, qui a une structure naturelle de $(k[L]; k[L'])$-bimodule. On a évidemment $k[\Delta] = \oplus k[\Delta_i]$ et chaque $k[\Delta_i]$ est $k[L]$-libre à gauche de rang 1. Notons $\overline{\rho}\colon B \longrightarrow \mathrm{Aut}_k(k[\Delta])$ l'homomorphisme induit par ρ, et $I(\Delta)$ le noyau du morphisme d'augmentation $k[\Delta] \longrightarrow k$ qui envoit chaque $\alpha \in \Delta$ sur $1 \in k$. Il est clair que $\overline{\rho}(b)$ laisse invariant $I(\Delta)$, pour tout $b \in B$. Soit $\mathcal{U}\colon I(\Delta) \to k[\Delta]$ l'inclusion, et $\mathcal{U}_i\colon I(\Delta) \to k[\Delta_i]$, $i \in I$ les morphismes induits.

(3.1.1) Lemme. Le $k[L]$-module à gauche $I(\Delta)$ est libre de rang $\mu + r - 1$.

Preuve : Comme chaque Δ_i est un L-torseur à gauche, on a un isomorphisme $k[\Delta] \simeq k[L]^r$ et le morphisme d'augmentation $k[\Delta] \to k$ s'identifie au morphisme

$$p_r\colon (a_1, \ldots, a_r) \in k[L]^r \mapsto \sum p(a_i) \in k,$$

où $p\colon k[L] \to k$ est le morphisme d'augmentation usuel. Par conséquent $I(\Delta)$ est isomorphe au noyau de p_r, qui à son tour est isomorphe à $k[L]^{r-1} \oplus \mathrm{Ker}(p)$ par

$$(a_1, \ldots, a_r) \in \mathrm{Ker}(p_r) \mapsto (a_1, \ldots, a_{r-1}, \sum_{i=1}^{r} a_i) \in k[L]^{r-1} \oplus \mathrm{Ker}(p).$$

Or, L est un groupe libre de rang μ et $\mathrm{Ker}(p)$ est un $k[L]$-module à gauche (et à droite) libre de rang μ. □

(3.1.2) Les inclusions $H_i \hookrightarrow L_i$, $i \in I$, nous fournissent une action à droite du groupe $H = \prod H_i$ sur Δ. Notons pour chaque $\underline{h} \in H$

$$\mathcal{V}_{\underline{h}} \colon k[\Delta] \longrightarrow I(\Delta)$$

le morphisme $k[L]$-linéaire défini par

$$\mathcal{V}_{\underline{h}}(\alpha) = \alpha \cdot \underline{h} - \alpha, \quad \alpha \in \Delta,$$

et pour $h \in H_i \hookrightarrow H, i \in I$

$$\mathcal{V}_h^i \colon k[\Delta_i] \longrightarrow I(\Delta) \tag{14}$$

le morphisme induit par $\mathcal{V}_h$. On a les relations suivantes :

- $\mathcal{V}_{hh'}^i = \mathcal{V}_{h'}^i \circ \mathcal{U}_i \circ \mathcal{V}_h^i + \mathcal{V}_h^i + \mathcal{V}_{h'}^i$ pour $h, h' \in H_i$ et $i \in I$
- Les morphismes $1_{I(\Delta)} + \mathcal{V}_h^i \circ \mathcal{U}_i$ et $1_{k[\Delta_i]} + \mathcal{U}_i \circ \mathcal{V}_h^i$ coïncident avec l'action à droite de h sur $I(\Delta)$ et $k[\Delta_i]$ respectivement, et sont donc des isomorphismes, pour $h \in H_i$ et $i \in I$
- $\mathcal{V}_h^i \circ \mathcal{U}_j = 0$ pour $h \in H_i$ et $i \neq j$.

Considérons le diagramme de $k[L]$-modules à gauche libres suivant

$$(\{k[\Delta_i]\}_{i\in I}, I(\Delta), \{\mathcal{U}_i\}_{i\in I}, \{\{\mathcal{V}_h^i\}_{h\in H_i}\}_{i\in I}),$$

que l'on notera $\mathcal{P}_f$. Si E est un $k[L]$-module à gauche, notons $\mathrm{Hom}_{k[L]}(\mathcal{P}_f, E)$ le diagramme

$$(\{\mathrm{Hom}_{k[L]}(k[\Delta_i], E)\}_{i\in I}, \mathrm{Hom}_{k[L]}(I(\Delta), E), \{(\mathcal{U}_i)^*\}_{i\in I}, \{\{(\mathcal{V}_h^i)^*\}_{h\in H_i}\}_{i\in I}).$$

Il est clair que $\mathrm{Hom}_{k[L]}(\mathcal{P}_f, E)$ est un objet de la catégorie $\mathcal{C}(\mathrm{Vect}(k); \{H_i\}_{i\in I})$ (voir 2.2). Par conséquent, on a un foncteur exact

$$\mathrm{Hom}_{k[L]}(\mathcal{P}_f, -) \colon \mathrm{Mod}_g(k[L]) \longrightarrow \mathcal{C}(\mathrm{Vect}(k); \{H_i\}_{i\in I}).$$

Étant donné un $k[G]$-module E, notons $\mathcal{L}_E$ le système local sur $\overline{X}^*$ qui lui correspond fonctoriellement.

Notons pour simplifier $\Psi = \Psi_{\overline{X}_0}$ (voir. théorème (2.3.4)).

Le résultat principal de ce travail est le théorème suivant, qui généralise le théorème (3.2.1) de [26] au cas des courbes planes réductibles

(3.1.3) Théorème. Pour un $k[G]$-module E, il existe un isomorphisme

$$\Psi(\mathbf{R}\psi_f(\mathcal{L}_E)) \simeq \mathrm{Hom}_{k[L]}(\mathcal{P}_f, E)$$

naturel par rapport à E. De plus, cet isomorphisme fait correspondre à l'automorphisme de monodromie $\mathbf{T}_{\mathcal{L}_E} : \mathbf{R}\psi_f(\mathcal{L}_E) \xrightarrow{\sim} \mathbf{R}\psi_f(\mathcal{L}_E)$, l'automorphisme $(\{t_1^i(E)\}, t_2(E))$ de $\mathrm{Hom}_{k[L]}(\mathcal{P}_f, E)$ suivant :

$$\begin{aligned} t_1^i(E)(h)(\alpha) &= \delta^{-1}h(\rho_i(\delta)(\alpha)), \quad \alpha \in \Delta_i, h \in \mathrm{Hom}_{k[L]}(k[\Delta_i], E), i \in I \\ t_2(E)(h)(x) &= \delta^{-1}h(\overline{\rho}(\delta)(x)), \quad x \in I(\Delta), h \in \mathrm{Hom}_{k[L]}(I(\Delta), E). \end{aligned}$$

On donnera la démonstration de ce théorème dans 3.3.

(3.1.4) Il est possible d'énoncer une variante plus compacte du théorème précédent, exclusivement en termes de $k[G]$-modules. Pour cela, nous allons rappeler quelques compatibilités entre le produit semi-direct de $(L; L')$-ensembles et le produit croisé d'un bimodule par un groupe. Pour les détails, le lecteur pourra consulter [23, 1.5].

Notons

-) $G_i = L_i \times_{\nu_i} B, G' = L' \times_{\nu'} B \hookrightarrow \prod G_i$,

-) $\widetilde{\Delta_i} = \Delta_i \times_{\rho_i} B, \widetilde{\Delta} = \Delta \times_\rho B \equiv \coprod \widetilde{\Delta_i}$,

-) $\overline{\nu}: B \to \mathrm{Aut}(k[L]), \overline{\nu_i}: B \to \mathrm{Aut}(k[L_i])$, $i \in I$ les actions induites par ν et ν_i respectivement,

-) $\overline{\nu}' : B \longrightarrow \mathrm{Aut}(k[L'])$ l'action induite par les $\overline{\nu}_i$.

On a des isomorphismes canoniques d'anneaux (cf. [23, 1.5.7])

$$k[L]\#_{\overline{\nu}}B \simeq k[G], \quad k[L_i]\#_{\overline{\nu}_i}B \simeq k[G_i], \quad k[L']\#_{\overline{\nu}'}B \simeq k[G'].$$

Si M est un $(k[L]; k[L'])$-bimodule et $\overline{\mu} : B \to \mathrm{Aut}_k(M)$ est une action compatible à gauche (resp. à droite) avec $\overline{\nu}$ (resp. $\overline{\nu}'$), on peut définir le $(k[L]\#_{\overline{\nu}}B; k[L']\#_{\overline{\nu}'}B)$-bimodule $M\#_{\overline{\mu}}B$ de façon analogue à (1.2.3). On a un isomorphisme canonique de $(k[G]; k[G'])$-bimodules (resp. $k[G]$-modules à gauche, $k[G']$-modules à droite)

$$k[\widetilde{\Delta}] \simeq k[\Delta]\#_{\overline{\rho}}B, \quad (\text{resp. } k[\widetilde{\Delta}] \simeq k[G] \otimes_{k[L]} k[\Delta], \quad k[\widetilde{\Delta}] \simeq k[\Delta] \otimes_{k[L']} k[G']).$$

Notons $I(\widetilde{\Delta}/B)$ le noyau du morphisme d'augmentation $k[\widetilde{\Delta}] \longrightarrow k[B]$ et $\widetilde{\mathcal{U}}$: $I(\widetilde{\Delta}/B) \to k[\widetilde{\Delta}]$ l'inclusion. On définit aussi des morphismes de $(k[G]; k[B])$-bimodules (voir 1.2, (2)) :

$$\widetilde{\mathcal{V}}_h^i : k[\widetilde{\Delta_i}] \longrightarrow I(\widetilde{\Delta}/B), \quad h \in H_i, i \in I,$$

comme dans (14), et considérons le diagramme de $(k[G]; k[B])$-bimodules suivant

$$(\{k[\widetilde{\Delta}_i]\}_{i\in I}, I(\widetilde{\Delta}/B), \{\widetilde{\mathcal{U}}_i\}_{i\in I}, \{\{\widetilde{\mathcal{V}}^i_h\}_{h\in H_i}\}_{i\in I}),$$

que l'on notera $\widetilde{\mathcal{P}}_f$. Comme $k[G]$ est un $k[L]$-module libre à gauche et à droite, on a un isomorphisme canonique de $k[G]$-modules à gauche $k[G]\otimes_{k[L]}\mathcal{P}_f \simeq \widetilde{\mathcal{P}}_f$, et $\widetilde{\mathcal{P}}_f$ est un diagramme de $k[G]$-modules <u>libres</u> à gauche. On a donc un foncteur exact

$$\mathrm{Hom}_{k[G]}(\widetilde{\mathcal{P}}_f, -)\colon \mathrm{Mod}_g(k[G]) \longrightarrow \mathcal{C}(\mathrm{Vect}(k); \{H_i\}_{i\in I}).$$

Soit $\widetilde{t}^1_i$ l'automorphisme de $k[\widetilde{\Delta_i}]$ induit par l'action à droite de $\delta^{-1} \in B \hookrightarrow G_i$, $i \in I$, et $\widetilde{t}^2$ l'automorphisme de $I(\widetilde{\Delta}/B)$ induit par $\oplus \widetilde{t}^1_i$. D'après 1.2, (2), le couple $\widetilde{t} := (\{\widetilde{t}^1_i\}_{i\in I}, \widetilde{t}^2)$ est un automorphisme de $\widetilde{\mathcal{P}}_f$. En fait, $\widetilde{t}$ n'est rien d'autre que l'action de δ^{-1} pour la structure à droite sur $\widetilde{\mathcal{P}}_f$.

L'énoncé suivant est un corollaire du théorème (3.1.3).

(3.1.5) Théorème. Pour un $k[G]$-module E, il existe un isomorphisme

$$\Psi(\mathbf{R}\psi_f(\mathcal{L}_E), \mathbf{T}_{\mathcal{L}_E}) \simeq (\mathrm{Hom}_{k[G]}(\widetilde{\mathcal{P}}_f, E), \mathrm{Hom}_{k[G]}(\widetilde{t}, E))$$

naturel par rapport à E.

3.2 Applications aux faisceaux pervers

Nous allons donner quelques conséquences de 3.1, qui généralisent au cas des courbes planes réductibles ceux de [26, (3.2)]. Le théorème (3.1.5), joint aux résultats de [28], nous permet de décrire, en termes d'Algèbre Linéaire, les extensions d'un système local sur $\overline{X}^*$ à un faisceau pervers sur $\overline{X}$. En fait on dispose d'un résultat plus précis, qui s'énonce comme une équivalence de catégories.

Soit $\mathcal{C}_f$ la catégorie abélienne dont

-) les objets sont les 4-uples (E, C, U, V) où E est un $k[G]$-module à gauche de dimension finie sur k, C est un objet de $\mathcal{C}(\mathrm{Vect}(k); \{H_i\}_{i\in I})$ et U : $\mathrm{Hom}_{k[G]}(\widetilde{\mathcal{P}}_f, E) \to C$, $V : C \to \mathrm{Hom}_{k[G]}(\widetilde{\mathcal{P}}_f, E)$ sont des morphismes tels que $1 + V \circ U = \mathrm{Hom}_{k[G]}(\widetilde{t}, E)$.

-) les morphismes sont définis de façon naturelle.

(3.2.1) Théorème. Il existe une équivalence $\mathcal{R}_f$ entre la catégorie des faisceaux pervers sur $\overline{X}$ stratifiés par rapport aux strates de la courbe $\overline{X}_0$ et la catégorie $\mathcal{C}_f$.

La démonstration de ce théorème se fait comme celle du théorème (3.2.2) de [26]. Rappelons brièvement la version tu théorème de recollement des faisceaux pervers de Verdier [28]. Soit $\mathcal{R}c(\overline{X}; f)$ la catégorie (abélienne) dont les objets sont les 4-uples $(\mathfrak{F}, \boldsymbol{\Phi}, \boldsymbol{c}, \boldsymbol{v})$, où $\mathfrak{F}$ est un faisceau pervers sur $\overline{X}^*$, $\boldsymbol{\Phi}$ est un faisceau pervers sur $\overline{X}_0$ et $\boldsymbol{c}: \mathbf{R}\psi_f(\mathfrak{F}) \to \boldsymbol{\Phi}, \boldsymbol{v}: \boldsymbol{\Phi} \to \mathbf{R}\psi_f(\mathfrak{F})$ sont des morphismes tels que $1 + \boldsymbol{v} \circ \boldsymbol{c} = \mathbf{T}_{\mathfrak{F}}$, et les morphismes sont définis de la façon habituelle. Alors, le foncteur qui associe à chaque faisceau pervers $\mathcal{K}$ sur $\overline{X}$ l'objet $(j^{-1}\mathcal{K}, \Phi_f(\mathcal{K}), \mathrm{can}, \mathrm{var})$ de $\mathcal{R}c(\overline{X}; f)$, est une équivalence de catégories.

Ce théorème joint au théorème (3.1.5) nous donne le théorème (3.2.1).

(3.2.2) Proposition. Soit $\mathcal{K}$ un faisceau pervers sur $\overline{X}$, stratifié par rapport aux strates de la courbe $\overline{X}_0$, et

$$\mathcal{R}_f(\mathcal{K}) = (E, (\{C_1^i\}_{i\in I}, C_2, \{u_i\}, \{\{v_h^i\}_{h\in H_i}\}_{i\in I}), U, V).$$

Alors le cycle caractéristique de $\mathcal{K}$ est égal à

$$m_0 T^*_{\overline{X}}(\overline{X}) + \sum_{i\in I} m_1^i \overline{T^*_{\overline{X}_{0,i}-\{0\}}(\overline{X})} + m_2 T^*_0(\overline{X})$$

avec $m_0 = \dim(E), m_1^i = \dim(C_1^i)$ et $m_2 = \dim(C_2) + \sum_{i\in I}(e_i - 1)\dim(C_1^i) - \mu \dim(E)$, où e_i désigne la multiplicité à l'origine de la courbe C_i.

La proposition précédente est une conséquence de la formule de l'indice de Kashiwara [15] (voir [26, prop. (3.2.3)]). Pour cela considérons le triangle

$$\mathcal{K} \longrightarrow \mathbf{R}\psi_f(\mathcal{K}) \longrightarrow \Phi_f(\mathcal{K}) \xrightarrow{+1} .$$

Si x est un point lisse de la courbe C_i, on a $\mathbf{R}\psi_f(\mathcal{K})_x \simeq E$, $\Phi_f(\mathcal{K})_x \simeq C_1^i$, d'où

$$\chi(\mathcal{K}_x) = \dim(E) - \dim(C_1^i).$$

Prenons maintenant les fibres à l'origine. Le complexe $\mathbf{R}\psi_f(\mathcal{K})_0$ est quasi-isomorphe au complexe

$$\overset{\bullet}{\mathrm{Hom}_{k[G]}(k[\widetilde{\Delta}], E)} \longrightarrow \mathrm{Hom}_{k[G]}(I(\widetilde{\Delta}/B), E).$$

De même, le complexe $\Phi_f(\mathcal{K})_0$ est quasi-isomorphe au complexe $\overset{\bullet}{\oplus C_1^i} \longrightarrow C_2$. On en déduit l'égalité

$$\chi(\mathcal{K}_0) = [r \dim(E) - (\mu + r - 1)\dim(E)] - [\sum_i \dim(C_1^i) - \dim(C_2)].$$

Pour conclure il suffit d'appliquer la formule d'indice (*loc. cit.*).

(3.2.3) Corollaire. Soit E un $k[G]$-module à gauche, de dimension finie sur k. Le cycle caractéristique du complexe d'intersection $IC(\mathcal{L}_E)$ est égal à

$$m_0 T^*_{\overline{X}}(\overline{X}) + \sum_{i \in I} m_1^i \overline{T^*_{\overline{X}_{0,i}-\{0\}}(\overline{X})} + m_2 T^*_0(\overline{X})$$

avec

$$m_0 = \dim(E), \quad m_1^i = \mathrm{rg}(\mathrm{Hom}_{k[G]}(\tilde{t_i^1} - 1, E))$$

$$m_2 = \mathrm{rg}(\mathrm{Hom}_{k[G]}(\tilde{t^2} - 1, E)) + \sum_{i \in I} (e_i - 1)\, \mathrm{rg}(\mathrm{Hom}_{k[G]}(\tilde{t_i^1} - 1, E)) - \mu \dim(E).$$

La démonstration de ce résultat (cf. [26, cor. (3.2.6)]) est une conséquence de la proposition précédente et du fait que l'équivalence de catégories de Verdier rappelée ci-dessus, associe à $IC(\mathcal{L}_E)$ l'objet $(\mathcal{L}_E, \mathrm{Im}(\mathbf{T}_{\mathcal{L}_E} - 1), \mathbf{T}_{\mathcal{L}_E} - 1, \mathrm{incl.})$.

3.3 Démonstration du théorème (3.1.2)

Gardons les notations de §1, et posons pour simplifier

$Y = \overline{X}_0$, $Z = \overline{X}_{t_0}$, $Y_i^* = X^\partial_{0,i}$, $Z_i^* = X^\partial_{t_0,i}$, $i \in I$

$\kappa_i : Y_i^* \hookrightarrow Y$, $j_i : Z_i^* \hookrightarrow Z$, les inclusions

$\overline{p_i} : (\widetilde{Y_i^*}, \widetilde{x_i}) \to (Y_i^*, x_i)$, $p_i : (\widetilde{Z_i^*}, \widetilde{y_i}) \to (Z_i^*, y_i)$, $p : (\widetilde{Z}, \widetilde{y_0}) \to (Z, y_0)$, $i \in I$, les revêtements universels.

Plaçons nous dans la situation de l'énoncé du théorème (3.1.3). Vu que $f : \overline{X}^* \to D^*$ est une fibration C^∞ localement triviale de fibre au-dessus de t_0 l'espace Z et ayant $T : Z \xrightarrow{\cong} Z$ comme difféomorphisme caractéristique, le système local $\mathcal{L}_E$ se dévisse en $\mathcal{L}_1 = \mathcal{L}_E|_Z$ et un isomorphisme $\tau : \mathcal{L}_1 \xrightarrow{\cong} T_* \mathcal{L}_1$ (cf. [26, 2.1.6.6]). Ce dévissage correspond au dévissage $(E_L, \lambda : E_L \xrightarrow{\cong} E'_L)$ de E, où E_L est le $k[L]$-module obtenu à partir de E par restriction des scalaires ($\mathcal{L}_1$ est le système local sur Z associé au $k[L]$-module E_L), E'_L est le $k[L]$-module dont le k-espace vectoriel sous-jacent coïncide avec E_L et l'action de L est donnée par $l \cdot e = \nu(\delta^{-1})(l)e$, $l \in L$, $e \in E'_L$, ($T_* \mathcal{L}_1$ est le système local sur Z associé au $k[L]$-module E'_L), et $\lambda : e \in E_L \mapsto \delta^{-1} e \in E'_L$ (cf. [26, 3.3.2]).

Pour chaque $i \in I$ notons E_i le $k[L_i]$-module associé à $\mathcal{L}_1$. Les E_i s'identifient naturellement aux $\mathrm{Hom}_{k[L]}(\Delta_i, E_L)$. D'après [10, exp. XIV] et [26, prop. 2.1.6.7], il existe un isomorphisme naturel $\mathbf{R}sp_*(\mathcal{L}_1) \simeq \mathbf{R}\psi_f(\mathcal{L}_E)$. Nous allons donc calculer $\Psi(\mathbf{R}sp_*(\mathcal{L}_1))$. Soit $\mathcal{L}_1 \to \mathfrak{J}$ une résolution injective. Le complexe

$\mathbf{R}sp_*(\mathcal{L}_1)$ s'identifie au complexe $\mathfrak{J} = sp_*(\mathfrak{I})$, et $\Psi(\mathbf{R}sp_*(\mathcal{L}_1)) \simeq \Omega(\psi_Y(\mathfrak{J})) = \Omega(\widetilde{\psi}_Y(\mathfrak{J})_0)$. Or, par des raisons de constructibilité les complexes $\widetilde{\psi}_Y(\mathfrak{J})_0$ et $\Gamma(Y, \widetilde{\psi}_Y(\mathfrak{J}))$ sont quasi-isomorphes. Mais ce dernier complexe s'identifie au complexe de $\mathcal{C}_0(k; \{H_i\}_{i\in I})$:

$$(\Gamma(Z,\mathfrak{I}), \{\Gamma(Z_i^*, (p_i)_*(p_i)^{-1}(j_i)^{-1}\mathfrak{I})\}, \{\alpha_i(\mathfrak{I})\}, \{\rho_i(\mathfrak{I})\}) \tag{15}$$

provenant des diagrammes

$$\Gamma(Z,\mathfrak{I}) \xrightarrow{\text{restric.}} \Gamma(Z_i^*, (j_i)^{-1}\mathfrak{I}) \xrightarrow{\text{adjonc.}} \Gamma(Z_i^*, (p_i)_*\overset{\curvearrowleft H_i}{(p_i)^{-1}}(j_i)^{-1}\mathfrak{I}), \tag{16}$$

$i \in I$.

Notons $\boldsymbol{P}$ (resp. $\boldsymbol{Q}_i$, $i \in I$) la résolution $k[L]$-libre (resp. $k[H_i]$-libre) à gauche de k suivante :

$$\cdots \longrightarrow 0 \longrightarrow I(\Delta) \xrightarrow{u} k[\overset{\bullet}{\Delta}] \longrightarrow 0 \longrightarrow \cdots$$

$$(\text{resp.} \qquad \cdots \longrightarrow 0 \longrightarrow I(H_i) \longrightarrow k[\overset{\bullet}{H_i}] \longrightarrow 0 \longrightarrow \cdots)$$

(voir lemme (3.1.1)). Les inclusions $k[\Delta_i] \hookrightarrow k[\Delta]$ se prolongent en des morphismes $k[L]$-linéaires à gauche $\boldsymbol{d}_i : k[\Delta_i] \otimes_{k[H_i]} \boldsymbol{Q}_i \to \boldsymbol{P}$.

Notons $\mathfrak{I}_1 = p_*p^{-1}\mathfrak{I}$ (resp. $\mathfrak{I}_{2,i} = (p_i)_*(p_i)^{-1}(j_i)^{-1}\mathfrak{I}$, $\mathfrak{I}_{3,i} = (p_i)_*(p_i)^{-1}\mathfrak{I}_{2,i}$, $i \in I$). Il s'agit d'un complexe de faisceaux injectifs de $k[L]$-modules (resp. $k[H_i]$-modules) (cf. [26, lemme 3.3.1.4]). Le groupe $H_i = \mathrm{Aut}(p_i)$ agit d'abord sur le complexe de faisceaux de k-espaces vectoriels $\mathfrak{I}_{2,i}$, puis sur le complexe de faisceaux de $k[H_i]$-modules $\mathfrak{I}_{2,i}$ par application du foncteur $(p_i)_*(p_i)^{-1}$. Le diagramme (16) s'identifie à (cf. (2.3.1), A))

$$\mathrm{Hom}_{k[L]}(k,\Gamma(Z,\mathfrak{I}_1)) \xrightarrow{\text{restric.}} \mathrm{Hom}_{k[H_i]}(k,\Gamma(Z_i^*,\mathfrak{I}_{2,i})) \xrightarrow{\text{adjonc.}} \mathrm{Hom}_{k[H_i]}\overset{\curvearrowleft H_i}{(k,\Gamma(Z_i^*,\mathfrak{I}_{3,i}))}. \tag{17}$$

Or, $\Gamma(Z,\mathfrak{I}_1)$ (resp. $\Gamma(Z_i^*,\mathfrak{I}_{2,i})$, $\Gamma(Z_i^*,\mathfrak{I}_{3,i})$) est une résolution $k[L]$-injective (resp. $k[H_i]$-injective) de E_L (resp. de E_i, $E_i^{H_i}$). De plus, l'action de H_i sur $\Gamma(Z_i^*,\mathfrak{I}_{3,i})$ relève l'action A_i de H_i sur $E_i^{H_i}$ définie dans (2.2.6) (cf. *loc. cit.*). Le diagramme (17) est quasi-isomorphe à

$$\mathrm{Hom}_{k[L]}(\boldsymbol{P}, E) \xrightarrow{(\boldsymbol{d}_i)^*} \mathrm{Hom}_{k[H_i]}(\boldsymbol{Q}_i, E_i) \xrightarrow{(\lambda_i)_*} \overset{\curvearrowleft (A_i)_*}{\mathrm{Hom}_{k[H_i]}(\boldsymbol{Q}_i, E_i^{H_i})}\,. \tag{18}$$

Par conséquent le complexe (15) est quasi-isomorphe à (cf. [13, I, 6.2])

$$(\mathrm{Hom}_{k[L]}(\boldsymbol{P}, E), \{\mathrm{Hom}_{k[H_i]}(\boldsymbol{Q}_i, E_i^{H_i})\}, \{(\lambda_i)_* \circ (\boldsymbol{d}_i)^*\}, \{(A_i)_*\})$$

qui n'est rien d'autre que $\Lambda'(\mathrm{Hom}_{k[L]}(\mathcal{P}_f, E))$ et d'après la proposition (2.2.8), on trouve :

$$\Psi(\mathbf{R}\psi_f(\mathcal{L}_E)) \simeq \Psi(\mathbf{R}sp_*(\mathcal{L}_1)) \simeq \Omega(\psi_Y(\mathfrak{I})) = \Omega(\widetilde{\psi}_Y(\mathfrak{I})_0) \simeq$$
$$\Omega(\Gamma(Y, \widetilde{\psi}_Y(\mathfrak{I}))) \simeq \cdots \simeq \Omega(\Lambda'(\mathrm{Hom}_{k[L]}(\mathcal{P}_f, E))) \simeq \mathrm{Hom}_{k[L]}(\mathcal{P}_f, E).$$

On vérifie également que les isomorphismes précédents sont naturels par rapport à E, et la première partie du théorème (3.1.3) est démontrée.

Une fois qu'on a trouvé des isomorphismes naturels

$$\Psi(\mathbf{R}\psi_f(\mathcal{L}_E)) \simeq \Psi(\mathbf{R}sp_*(\mathcal{L}_1)) \simeq \mathrm{Hom}_{k[L]}(\mathcal{P}_f, E) \tag{19}$$

il nous reste à "lire"sur le terme de droite l'automorphisme de monodromie $\Psi(\mathbf{T}_{\mathcal{L}_E})$ agissant sur le terme de gauche. Or, d'après [26, prop. (2.1.6.7)] on a un diagramme commutatif

$$\begin{array}{ccccc}
\mathbf{R}sp_*(\mathcal{L}_1) & \xrightarrow{\mathbf{R}sp_*(\tau)} & \mathbf{R}sp_*(T_*\mathcal{L}_1) & \xrightarrow{\chi} & \mathbf{R}sp_*(\mathcal{L}_1) \\
\downarrow \cong & & & & \downarrow \cong \\
\mathbf{R}\psi_f(\mathcal{L}_E) & & \xrightarrow{\mathbf{T}_{\mathcal{L}_E}} & & \mathbf{R}\psi_f(\mathcal{L}_E)
\end{array}$$

où χ est l'identification canonique provenant de la relation $sp \circ T = sp$. L'isomorphisme que l'on cherche est celui qui correspond par (19) à la composition de $\Psi(\mathbf{R}sp_*(\tau))$ avec $\Psi(\chi)$. Par fonctorialité, celui qui correspond au premier coïncide avec $\mathrm{Hom}_{k[L]}(\mathcal{P}_f, \lambda)$.

Le difféomorphisme caractéristique $T\colon Z \xrightarrow{\cong} Z$ induit les $\rho_i(\delta)\colon \Delta_i \to \Delta_i$, $i \in I$, et $\overline{\rho}\colon I(\Delta) \to I(\Delta)$. De là on peut déduire comme dans [26, 3.3.2] que l'isomorphisme

$$\kappa = (\{\kappa_1^i\}, \kappa_2)\colon \mathrm{Hom}_{k[L]}(\mathcal{P}_f, E') \xrightarrow{\sim} \mathrm{Hom}_{k[L]}(\mathcal{P}_f, E)$$

correspondant à $\Psi(\chi)$ est donné par

$$\begin{aligned}
\kappa_1^i(h)(\alpha) &= h(\rho_i(\delta)(\alpha)), \quad \alpha \in \Delta_i, h \in \mathrm{Hom}_{k[L]}(k[\Delta_i], E'), i \in I \\
\kappa_2(h)(x) &= h(\overline{\rho}(\delta)(x)), \quad x \in I(\Delta), h \in \mathrm{Hom}_{k[L]}(I(\Delta), E').
\end{aligned}$$

Or, $\kappa \circ \mathrm{Hom}_{k[L]}(\mathcal{P}_f, \lambda) = (\{t_1^i(E)\}, t_2(E))$, ce qui termine la preuve du théorème (3.1.3). □

(3.3.1) Remarque. Le théorème (3.2.1) et le corollaire (3.2.3) nous donnent des résultats effectifs chaque fois qu'on connait explicitement les invariants décrits dans (1.2.1), i.e.

1) une base du groupe libre L

2) l'action sur cette base de $\nu(\delta)$

3) un élément $\alpha_i \in \Delta_i$

4) l'action sur α_i de γ_i et de $\rho_i(\delta)$

A partir de la résolution plongée de $(C, 0)$, et en utilisant des idées de A'Campo ([3, 1, 5, 24]) il est possible de trouver 1)–4). Ceci sera analysé dans un prochain travail.

Références

[1] N. A'Campo. La fonction zêta d'une monodromie. *Comment. Math. Helv.*, 50 :233–248, 1975.

[2] N. A'Campo. Le groupe de monodromie du déploiement des singularités isolées de courbes planes I. *Math. Ann.*, 213 :1–32, 1975.

[3] N. A'Campo. Le nombre de Lefschetz d'une monodromie. *Indag. Math.*, 35 :113–118, 1973.

[4] A.A. Beilinson, J. Bernstein et P. Deligne. *Faisceaux pervers. Astérisque*, 100, S.M.F., Paris, 1983.

[5] E. Brieskorn and H. Knörrer. *Plane Algebraic Curves.* Birkhäuser, 1986.

[6] J.L. Brylinski. Transformations canoniques, dualité projective, théorie de Lefschetz, transformation de Fourier et sommes trigonométriques. *Astérisque*, 140–141 :3–134, 1986.

[7] D. Cheniot et Lê Dũng Tráng. Remarques sur les deux exposés précedents. *Astérisque*, 7–8 :253–261, 1973.

[8] P. Deligne. *Equations Différentielles à Points Singuliers Réguliers. Lect. Notes in Math.*, 163, Springer-Verlag, Berlin-Heidelberg, 1970.

[9] P. Deligne. Lettre à R. MacPherson. 1981.

[10] P. Deligne et N. Katz. *Groupes de monodromie en Géométrie Algébrique (SGA 7 II). Lect. Notes in Math.*, 340, Springer-Verlag, Berlin-Heidelberg, 1973.

[11] A. Galligo, M. Granger et Ph. Maisonobe. $\mathcal{D}$-Modules et faisceaux pervers dont le support singulier est croisement normal. *Ann. Inst. Fourier*, 35 :1–48, 1985.

[12] M Goresky and R. MacPherson. Morse theory and intersection homology. *Astérisque*, 101 :135–192, 1983.

[13] R. Hartshorne. *Residues and Duality. Lect. Notes in Math.*, 20, Springer Verlag, Berlin-Heidelberg, 1966.

[14] B. Iversen. *Cohomology of sheaves. Universitext*, Springer Verlag, Berlin-Heidelberg, 1986.

[15] M. Kashiwara. *Systems of microdifferential equations. Progress in Math.*, 34, Birkhäuser, Boston, 1983.

[16] Lê Dũng Tráng. Sur les noeuds algébriques. *Comp. Math.*, 25(3) :281–321, 1972.

[17] Lê Dũng Tráng. *Three lectures on local monodromy.* Lect. Notes Series 43, Aarhus Universitet, 1974.

[18] Lê Dũng Tráng et Z. Mebkhout. Variétés caractéristiques et variétés polaires. *C. R. Acad. Sci. Paris*, 296 :129–132, 1983.

[19] E.J.N. Looijenga. *Isolated singular points on complete intersections. London Mathem. Soc. Lect. Notes Series*, 77, Cambridge Univ. Press, Cambridge, 1984.

[20] R. MacPherson and K. Vilonen. Elementary construction of perverse sheaves. *Invent. Math.*, 84 :403–436, 1986.

[21] Ph. Maisonobe. Faisceaux pervers dont le support singulier est une courbe plane. *Comp. Math.*, 62 :215–261, 1987.

[22] Ph. Maisonobe. Faisceaux pervers sur **C** relativement à $\{0\}$ et couple $E \rightleftarrows F$. Dans *Introduction à la théorie algébrique des systèmes différentiels, éd. par Lê Dũng Tráng*, pages 135–146, Hermann, Paris, 1988. (Travaux en cours, vol. 34).

[23] J.C. McConnell and J.C. Robson. *Noncommutative Noetherian Rings. Pure and applied Mathematics*, John Wiley & Sons, Chichester, 1987.

[24] F. Michel et C. Weber. Topologie des germes de courbes planes à plusieures branches. 1985. Prépub. de l'Univ. de Génève.

[25] J. Milnor. *Singular points of complex hypersurfaces. Ann. of Math. Studies*, 61, Princeton Univ. Press, Princeton, N.Y., 1968.

[26] L. Narváez-Macarro. Cycles évanescents et faisceaux pervers : cas des courbes planes irréductibles. *Comp. Math.*, 65 :321–347, 1988.

[27] L. Narváez-Macarro. t-structures coniques. (en préparation).

[28] J.L. Verdier. Extension of a perverse sheaf over a closed subspace. *Astérisque*, 130 :210–217, 1985.

L. Narváez-Macarro
Departamento de Algebra y Geometría
Fac. de Matemáticas, Univ. de Sevilla
Ap. 1160
41080 Sevilla (SPAIN)
E-mail : narvaez@algebra.us.es

A Desingularization Theorem for Systems of Microdifferential Equations

ORLANDO NETO

Centro de Matemática e Aplicações Fundamentais and Departamento de Matemática da Universidade de Lisboa

Abstract We prove that the blow up of a regular holonomic system of microdifferential equations is regular holonomic and calculate its support. We prove desingularization theorems for Lagrangian curves and for regular holonomic systems with support on a Lagrangian curve.

1: INTRODUCTION

In [N] we introduced a notion of blow up of a holonomic $\mathcal{E}_X$-module. In Section 4 we prove that, under reasonable assumptions, the blow up of a regular holonomic system is regular holonomic and calculate its support. These results motivate a desingularization game for Lagrangian subvarieties of a contact manifold. In Section 5 we show how to win the game when the contact manifold has dimension 3. As a consequence we get a desingularization theorem for regular holonomic $\mathcal{E}_X$-modules when the dimension of X equals 2. In Sections 2 and 3 we recall the main results of [N].

The author would like to thank M. Kashiwara for useful discussions.

2: LOGARITHMIC CONTACT MANIFOLDS

Let X be a complex manifold. A subset Y of X is called a *normal crossings divisor* if for any $x^o \in Y$ there is an open neighbourhood U of x^o, a system of local coordinates $(x_1, \dots x_n)$ defined on U and an integer ν such that

$$Y \cap U = \{x_1 \cdots x_\nu = 0\}. \tag{1}$$

Given a normal crossings divisor Y and an integer ν such that (1) holds, we say that the system of local coordinates $(x_1, \dots x_n)$ is subordinated to Y.

Let Θ_X be the sheaf of vector fields of X. Let I_Y be the defining Ideal of Y, that is, the Ideal of the local sections of $\mathcal{O}_X$ that vanish along Y. We say that a vector field u of X is *tangent* to Y if $uI_Y \subset I_Y$. Let $\Theta_{X\langle Y\rangle}$ be the sheaf of the vector fields on X that are tangent to Y.

Let Ω^1_X be the sheaf of differential forms on X. We say that a meromorphic differential form α on X has logarithmic poles along Y if $< \alpha, u >$ is an holomorphic function for any vector field u tangent to Y. Let $\Omega^1_{X\langle Y\rangle}$ denote the sheaf of differential forms on X with poles along Y.

The $\mathcal{O}_X$-modules $\Omega^1_{X\langle Y\rangle}$ and $\Theta_{X\langle Y\rangle}$ are locally free and dual of each other. Given an open set U of X and a system of local coordinates $(x_1, \ldots x_n)$ defined on U and subordinated to the normal crossings divisor Y, we have

$$\Theta_{X\langle Y\rangle} = \sum_{i=1}^{\nu} \mathcal{O}_X x_i \partial_{x_i} + \sum_{i=\nu+1}^{n} \mathcal{O}_X \partial_{x_i},$$

$$\Omega^1_{X\langle Y\rangle} = \sum_{i=1}^{\nu} \mathcal{O}_X \frac{dx_i}{x_i} + \sum_{i=\nu+1}^{n} \mathcal{O}_X dx_i.$$

We notice that if Y is the empty set then $\Theta_{X\langle Y\rangle}$ equals Θ_X and $\Omega^1_{X\langle Y\rangle}$ equals Ω^1_X. We put $\Omega^k_{X\langle Y\rangle} = \wedge^k \Omega^1_{X\langle Y\rangle}$, $\Omega^*_{X\langle Y\rangle} = \oplus_k \Omega^k_{X\langle Y\rangle}$. There is a differential d in $\Omega^*_{X\langle Y\rangle}$ that extends the differential of Ω^*_X (cf. [D]). We call $\Omega^*_{X\langle Y\rangle}$ the *Logarithmic de Rham complex of X with poles along Y.*

Let X be a complex manifold. The functor $\mathcal{S}$ that associates to a holomorphic vector bundle E its sheaf of sections $\mathcal{S}$(E) is an equivalence of categories between the category of holomorphic vector bundles on X and the category of locally free $\mathcal{O}_X$-modules. Let Y be a normal crossings divisor of X. We will denote by $T^*\langle X/Y\rangle$ the vector bundle with sheaf of sections $\Omega^1_{X\langle Y\rangle}$. We call this vector bundle the *Logarithmic cotangent bundle of X with poles along Y.*

Let X be a complex manifold. A group action $\alpha : \mathbf{C}^* \times X \to X$ is called a free group action of $\mathbf{C}^*$ on X if for each $x \in X$ the isotropy subgroup $\{t \in \mathbf{C}^* : \alpha(t, x) = x\}$ equals $\{1\}$. A manifold X with a free group action α of $\mathbf{C}^*$ is called a conic manifold. We associate to each free group action α of $\mathbf{C}^*$ on X a vector field ρ, the *Euler vector field of* α. We put

$$\mathcal{O}_{X(\lambda)} = \{f \in \mathcal{O}_X : \rho f = \lambda f\}$$

for any $\lambda \in \mathbf{C}$ and

$$\mathcal{O}^h_X = \oplus_{k\in\mathbf{Z}} \mathcal{O}_{X(k)}.$$

A section f of $\mathcal{O}_{X(\lambda)}$ is called a homogeneous function of degree λ. Given two conic complex manifolds (X_1, α_1) and (X_2, α_2), a holomorphic map $\varphi : X_1 \to X_2$ is called *homogeneous* if it commutes with the actions α_1, α_2.

2.1. Definition Let X be a complex manifold of dimension $2n$ and Y a normal crossings divisor of X. We say that $\sigma \in \Omega^2_{X\langle Y\rangle}$ is a *logarithmic symplectic form* if σ is locally exact and σ^n is a generator of the invertible $\mathcal{O}_X$-module $\Omega^{2n}_{X\langle Y\rangle}$. If X is a conic manifold, then we say that a logarithmic symplectic form σ is a *homogeneous logarithmic symplectic form* if $\alpha_t^* \sigma = t\sigma$, for any $t \in \mathbf{C}^*$. Here α_t equals $\alpha(t, *)$.

If X is a [conic] manifold and σ is a [homogeneous] logarithmic symplectic form, we call the pair (X, σ) a [homogeneous] logarithmic symplectic manifold. If (X_1, σ_1), (X_2, σ_2) are [homogeneous] logarithmic symplectic manifolds, then a [homogeneous] holomorphic map $\varphi : X_1 \to X_2$ is called a *morphism of [homogeneous] logarithmic symplectic manifolds* if $\varphi^*\sigma_2 = \sigma_1$.

We notice that if the normal crossings divisor Y equals the empty set, then we get the usual definition of [homogeneous] symplectic manifold. We remark that a morphism of logarithmic symplectic manifolds is not necessarily a local homeomorphism, as it happens in the category of symplectic manifolds.

If (X, σ) is a homogeneous symplectic manifold with poles along Y and ρ is its Euler vector field, then the interior product $\theta = i(\rho)\sigma$ is a logarithmic differential form with poles along Y. Moreover, $d\theta$ equals σ. We call θ the *canonical* 1*-form* of the homogeneous logarithmic symplectic manifold (X, σ).

2.2. Theorem *Let σ be a logarithmic symplectic form on a complex manifold X with poles along a divisor with normal crossings Y. Given $x^o \in X$, let ν be the number of irreducible components of Y at x^o. Then there is a system of local coordinates $(x_1, \dots, x_n, \xi_1, \dots, \xi_n)$ on U such that $Y \cap U = \{x_1 \cdots x_\nu = 0\}$ and $\sigma|_U$ equals*

$$\sum\nolimits_{i=1}^{\nu} d\xi_i \frac{dx_i}{x_i} + \sum\nolimits_{i=\nu+1}^{n} d\xi_i dx_i. \tag{2}$$

Morover, if σ is a homogeneous logarithmic symplectic form then we can suppose that the functions x_i are homogeneous of degree 0 and the functions ξ_i are homogeneous of degree 1.

This theorem generalizes the classical Darboux Theorem for [homogeneous] symplectic manifolds. See [N] for its proof.

2.3. Example Let X be a complex manifold. Let ρ be the Euler vector field of T^*X. We can define in a canonical way a 1-differential form of degree θ such that $d\theta$ is a symplectic form and $i(\rho)d\theta = \theta$. This differential form θ is called the canonical 1-form of T^*X.

Let $\pi_X : T^*X \to X$ be the canonical projection. Given a system of local coordinates $(x_1, ..., x_n)$ on an open set U of X there is one and only one family of holomorphic functions ξ_i, $1 \le i \le n$, defined on $\pi^{-1}(U)$ such that $\theta|_{\pi^{-1}(U)}$ equals

$$\sum_{i=1}^{n} \xi_i dx_i. \tag{3}$$

Moreover, the functions $(x, \xi) = (x_1, ..., x_n, \xi_1, ..., \xi_n)$ define a system of local coordinates on $\pi_X^{-1}(U)$. Here we still denote by x_i the function $x_i\pi_X$ defined on $\pi_X^{-1}(U)$.

Let ρ' be the Euler vector field of $T^*\langle X/Y\rangle$. Given a normal crossings divisor Y on X, we can define on $T^*\langle X/Y\rangle$ in a similar way, a logarithmic differential form θ' of degree 1, with poles along Y, such that $d\theta'$ is a logarithmic symplectic form with poles along Y and $i(\rho')d\theta' = \theta'$. The differential form θ' is called the canonical 1-form of $T^*\langle X/Y\rangle$. Let $\pi' : T^*\langle X/Y\rangle \to X$ be the canonical projection. Given a system of local coordinates $(x_1, ..., x_n)$ on an open set U of X subordinated to Y, there is one and only one family of holomorphic functions ξ_i', $1 \le i \le n$ defined on $\pi^{-1}(U)$ such that $\theta'|_U$ equals

$$\sum_{i=1}^{\nu} \xi_i' \frac{dx_i}{x_i} + \sum_{i=\nu+1}^{n} \xi_i' dx_i. \tag{4}$$

Moreover, the functions $(x, \xi') = (x_1, ..., x_n, \xi_1', ..., \xi_n')$ also define a system of local coordinates on $\pi'^{-1}(U)$. The system of local coordinates (x, ξ') is called *the canonical system of local coordinates on $T^*\langle X/Y\rangle$ associated to* (x). We notice that

$$\rho'|_U = \sum_i \xi_i' \frac{\partial}{\partial \xi_i'}.$$

Since $\Omega^1_{X\langle Y\rangle}|_{X\setminus Y} = \Omega^1_X|_{X\setminus Y}$, then $T^*|_{X\setminus Y} = T^*\langle X/Y\rangle|_{X\setminus Y}$. If we put $V = \pi_X^{-1}(U \setminus Y)$ then

$$\xi_i'|_V = \begin{cases} x_i \xi_i|_V, & \text{if } \; 1 \le i \le \nu, \\ \xi_i|_V, & \text{if } \; \nu + 1 \le i \le n. \end{cases}$$

Given a vector bundle E over a complex manifold X, we will identify X with the zero section of E. Put $\mathring{E} = E \setminus X$. The manifold $\mathring{E}$ has a natural structure of conic manifold, coming from the action of $\mathbf{C}^*$ on the fibers of E. If θ is the canonical 1-form of $T^*\langle X/Y\rangle$ then $\theta|_{\mathring{T}^*\langle X/Y\rangle}$ is a homogenous logarithmic symplectic form.

2.4. Definition Given a complex manifold X we say that a $\mathbf{C}$-bilinear morphism

$$\{\star, \star\} : \mathcal{O}_X \times \mathcal{O}_X \to \mathcal{O}_X$$

is a *Poisson bracket* if it verifies the following conditions:

(i) $\{f, g\} = -\{g, f\}$

(ii) $\{fg, h\} = f\{g, h\} + g\{f, h\}$

(iii) $\{\{f, g\}, h\} + \{\{g, h\}, f\} + \{\{h, f\}, g\} = 0$

We call $\{f, g\}$ the Poisson bracket of f and g. We call a complex manifold X endowed with a Poisson bracket a *Poisson manifold.* An analytical subset V of X is called *involutive* if $\{I_V, I_V\} \subset I_V$. If $(X_1, \{\star, \star\}_1)$, $(X_2, \{\star, \star\}_2)$ are Poisson manifolds and $\varphi : X_1 \to X_2$ is a holomorphic map such that $\{\varphi^* f, \varphi^* g\}_1 = \varphi^*\{f, g\}_2$, for any holomorphic functions f, g defined in an open set of X_2, we call φ a morphism of Poisson manifolds.

If V is an involutive subvariety of X_2 and $\varphi : X_1 \to X_2$ is a morphism of Poisson manifolds then $\varphi^{-1}(V)$ is an involutive subvariety of X_1.

2.5. Example A logarithmic symplectic manifold has a canonical structure of Poisson manifold. Given $\alpha \in \Omega^1_X(Y)$ let $H(\alpha)$ denote the only vector field $u \in \Theta_X(Y)$ such that

$$< u, \alpha >= \sigma(u, H(\alpha)).$$

The bilinear form $(f,g) \mapsto < dg, H(df) >$ is a Poisson bracket on $\mathcal{O}_X$. If (x,ξ) is a system of local coordinates on an open set U of X such that $\sigma|_U$ equals (2) then $\{f,g\}$ equals

$$\sum_{i=1}^{\nu} x_i \left(\frac{\partial f}{\partial \xi_i}\frac{\partial g}{\partial x_i} - \frac{\partial f}{\partial x_i}\frac{\partial g}{\partial \xi_i}\right) + \sum_{i=\nu+1}^{n} \left(\frac{\partial f}{\partial \xi_i}\frac{\partial g}{\partial x_i} - \frac{\partial f}{\partial x_i}\frac{\partial g}{\partial \xi_i}\right). \tag{6}$$

In particular

$$\{\xi_i, x_j\} = \begin{cases} \delta_{ij} x_j, & \text{if } 1 \leq j \leq \nu; \\ \delta_{ij}, & \text{if } \nu+1 \leq j \leq n. \end{cases}$$

Let X be a homogeneous logarithmic symplectic manifold. Suppose that its set of poles Y is smooth. We can define in a canonical way a morphism of sheaves

$$\mathrm{Res}_Y : \Omega^1_X(Y)|_Y \to \mathcal{O}_Y,$$

the *Poincaré residue along* Y (cf. [D]). Let us fix a system of local coordinates $(x_1, ..., x_n)$ in an open set U of X such that $Y \cap U = \{x_1 = 0\}$. Then

$$\mathrm{Res}_Y(a_1 \frac{dx_1}{x_1} + \sum_{i=2}^{n} a_i dx_i) = a_1 \qquad (\mathrm{mod}\ I_Y).$$

Let X be a homogeneous logarithmic symplectic manifold. Suppose that its set of poles Y is smooth. We call *residual submanifold* of X the set of points of Y where the Poincaré residue of the canonical 1-form θ vanishes. We call *residual function* to a holomorphic function ξ defined on an open set U of X such that $\xi|_Y$ equals the residue of θ.

The set of poles and the residual submanifold of a homogeneous logarithmic symplectic manifold X are involutive submanifolds of X.

2.6. Definition. Let X be a complex manifold of dimension $2n-1$ and Y a normal crossings divisor of X. We say that a logarithmic differential form $\omega \in \Omega^1_X(Y)$ is a *logarithmic contact form* if $\omega(d\omega)^{n-1}$ is a local generator of $\Omega^{2n-1}_X(Y)$. We say that an invertible sub$\mathcal{O}_X$-module $\mathcal{L}$ of $\Omega^1_X(Y)$ is a *logarithmic contact structure* with poles along Y if $\mathcal{L}$ is locally generated by logaritmic contact forms with poles along Y.

We call *logarithmic contact manifold* a pair $(X, \mathcal{L})$, where X is a complex manifold and $\mathcal{L}$ is a logarithmic contact structure on X. Let $(X_1, \mathcal{L}_1)$, $(X_2, \mathcal{L}_2)$ be two logarithmic contact manifolds. Let $\varphi : X_1 \to X_2$ be a holomorphic map. We say that φ is a contact transformation if given an open set U of X_2 and a generator ω of $\mathcal{L}_2|_U$ then $\varphi^*\omega$ is a generator of $\mathcal{L}_1|_{\varphi^{-1}(U)}$.

If Y is the empty set, then we get the usual definition of contact manifold.

2.7. Remark (i) Given a logarithmic contact form ω and a nowhere vanishing holomorphic function φ, we have that $\varphi\omega$ is a logarithmic contact form.
(ii) We say that two logarithmic contact forms ω_1 and ω_2, are equivalent if there is a nowhere vanishing holomorphic function φ such that $\omega_2 = \varphi\omega_1$.
(iii) We notice that it is equivalent to give a structure of logarithmic contact manifold along Y and to give the following data:
- an open covering $(U_i)_{i\in I}$ of X;
- logarithmic contact forms $\omega_i \in \Gamma(U_i, \Omega^1_{X}\langle Y\rangle)$, $i \in I$, with poles along $U_i \cap Y$;
verifying the condition "ω_i is equivalent to ω_j on $U_i \cap U_j$".

We say that the logarithmic contact forms (ω_i) generate the logarithmic contact structure $\mathcal{L}$.

The following theorem generalizes the classical Darboux Theorem for contact manifolds.

2.8. Theorem *Let X be a complex manifold of dimension $2n+1$. Let Y be a smooth hypersurface of X. Let $\mathcal{L}$ be a logarithmic contact structure on X with poles along Y. Let us fix $x^o \in X$. Let Z be the residual submanifold of X.*
(i) If $x^o \notin Y$, then there is an open neigbourhood U of x^0 and a system of local coordinates $(x_1, ..., x_n)$ on U such that

$$dx_{n+1} - \sum_{i=1}^{n} p_i dx_i \tag{7}$$

is a generator of $\mathcal{L}|_U$.
(ii) If $x^o \in Z$, then there is an open neigbourhood U of x^o and a system of local coordinates $(x_1, ..., x_n)$ on U such that

$$dx_{n+1} - p_1\frac{dx_1}{x_1} - \sum_{i=2}^{n} p_i dx_i \tag{8}$$

is a generator of $\mathcal{L}|_U$.
(iii) If $x^o \notin Z$, then there is an open neigbourhood U of x^o and a system of local coordinates $(x_1, ..., x_n)$ on U such that

$$\frac{dx_{n+1}}{x_{n+1}} - \sum_{i=1}^{n} p_i dx_i \tag{9}$$

is a generator of $\mathcal{L}|_U$.

2.9. Example Let X be a complex manifold and Y a normal crossings divisor of X. We call *Projective cotangent bundle of X with poles along*

Y the projective bundle $\pi : \mathbf{P}^*\langle X/Y\rangle \to X$ associated to the vector bundle $\pi : T^*\langle X/Y\rangle \to X$. The projective bundle $\mathbf{P}^*\langle X/Y\rangle$ has a canonical structure of logarithmic contact manifold with poles along $\pi^{-1}(Y)$.

Let θ be the canonical 1-form of $T^*\langle X/Y\rangle$. Near each point of $\dot{T}^*\langle X/Y\rangle$ there is a local section η of $\mathcal{O}_{T^*\langle X/Y\rangle}(1)$ that does not vanish at the point. Then the logarithmic differential form $\eta^{-1}\theta$ induces a local section of $\Omega^1_{\mathbf{P}^*\langle X/Y\rangle}\langle\pi^{-1}(Y)\rangle$. The forms obtained in this way generate a logarithmic contact structure on $\mathbf{P}^*\langle X/Y\rangle$ with poles along $\pi^{-1}(Y)$.

Suppose that X is a copy of $\mathbf{C}^2$ with coordinates (x, y) and that Y equals $\{x = 0\}$. Hence $T^*\langle X/Y\rangle$ is a copy of $\mathbf{C}^4$ with coordinates (x, y, ξ, η) and θ equals $\xi dx/x + \eta dy$. Put $V = \{\xi \neq 0\}$, $W = \{\eta \neq 0\}$. Then $\dot{T}^*\langle X/Y\rangle$ equals $V \cup W$. The sets V and W are invariant by the$\mathbf{C}^*$-action. Put $p = -\xi/\eta$, $q = -\eta/\xi$. We can identify $V/\mathbf{C}^*$ $[W/\mathbf{C}^*]$ with a copy of $\mathbf{C}^3$ with coordinates (x, y, p) $[(x, y, q)]$. Moreover,

$$\frac{1}{\eta}\theta = dy - p\frac{dx}{x}, \quad \frac{1}{\xi}\theta = \frac{dx}{x} - qdy.$$

The forms $\frac{1}{\eta}\theta$, $\frac{1}{\xi}\theta$ generate the canonical logarithmic contact structure of $\mathbf{P}^*\langle X/Y\rangle$.

2.10. Theorem *There is an equivalence of categories between the category of homogeneous logarithmic symplectic manifolds and the category of logarithmic contact manifolds.*

Given a homogeneous logarithmic symplectic manifold X, we endow the manifold $\mathbf{X}$ of the orbits of its $\mathbf{C}^*$-action with a logarithmic contact structure, just like in Example 2.9. Given a logarithmic contact manifold $(X, \mathcal{L})$ we associate to it an homogeneous logarithmic symplectic manifold in the following way. Let $\gamma_o : L \to X$ be the vector bundle associated to $\mathcal{L}$. Put $\hat{X} = L \setminus X$, $\gamma = \gamma_o|_{\hat{X}}$. Let U be an open set of X. Let ω be a generator of $\mathcal{L}|_U$. The section ω determines a trivialization

$$\gamma^{-1}(U) \to U \times \mathbf{C}^*.$$

Hence ω determines a holomorphic function $\eta : \gamma^{-1}(U) \to \mathbf{C}^*$. Suppose that the dimension of X equals $2n - 1$. Since

$$(d(\eta\gamma^*\omega))^n = Const \cdot d\eta \wedge \gamma^*\omega \wedge (d\gamma^*\omega)^{n-1}$$

then $\eta\gamma^*\omega$ is an homogeneous canonical 1-form on $\gamma^{-1}(U)$. Let ω' be another generator of $\mathcal{L}|_U$. There is an invertible holomorphic function φ on U such that $\omega' = \varphi\omega$. If η' is the holomorphic function determined by ω' then

$$\eta'\gamma^*\omega' = (\gamma^*\varphi)^{-1}\eta \cdot \gamma^*\varphi\gamma^*\omega = \eta\gamma^*\omega.$$

Therefore the logarithmic contact structure $\mathcal{L}$ induces in a canonical way a homogeneous logarithmic symplectic form on $\hat{X}$.

3: LOGARITHMIC MICRODIFFERENTIAL OPERATORS.

Given a fiber bundle $\pi : E \to X$, we will denote by $\mathcal{O}_{[E]}$ the subsheaf of $\mathcal{O}_E$ of sections wich are polynomial in the fibers of π.

Let X be a complex manifold and Y a normal crossings divisor of X. We will denote by $\mathcal{D}_{X\langle Y\rangle}$ the sub$\mathcal{O}_X$-algebra of $\mathcal{H}om_{\mathbf{C}_X}(\mathcal{O}_X, \mathcal{O}_X)$ generated by $\Theta_{X\langle Y\rangle}$. We call the elements of $\mathcal{D}_{X\langle Y\rangle}$ differential operators on X tangent to Y. If Y equals the empty set then $\mathcal{D}_{X\langle Y\rangle}$ equals the ring $\mathcal{D}_X$ of the differential operators on X.

3.1. Definition Let U be an open set of X and let $(x_1, \dots, x_n)$ be a system of local coordinates subordinated to the normal crossings divisor Y. Let (x, ξ') be the associated system of canonical coordinates on $\pi'^{-1}(U) \subset T^*\langle X/Y\rangle$. Given a section P of $\mathcal{D}_{X\langle Y\rangle}$ we define the *total symbol* of P as the element (P_j) of $\mathcal{O}_{[T^*\langle X/Y\rangle]}$

$$e^{-\langle x,\xi'\rangle_\nu} P e^{\langle x,\xi'\rangle_\nu},$$

where $\langle x, \xi'\rangle_\nu = \sum_{i=1}^{\nu} \xi_i' log x_i + \sum_{i=\nu+1}^{n} x_i \xi_i'$ and each P_j is a homogeneous polynomial of degree j relative to the action of $\mathbf{C}^*$ on the fibers of $T^*\langle X/Y\rangle$ and coefficients in $\pi'^{-1}\mathcal{O}_X$.

The following proposition tells us how to calculate the total symbol of the sum and product of two differential operators tangent to Y. Moreover, the following proposition tells us how to compare the total symbols of a differential operator tangent to Y relatively to two different systems of local coordinates.

3.2. Proposition (i) *Given two sections P, Q of $\mathcal{D}_{X\langle Y\rangle}$,*

$$(P+Q)_l = P_l + Q_l \tag{1}$$

$$(PQ)_l = \sum_{\substack{l=j+k-|\alpha| \\ \alpha\in\mathbf{N}^n}} \frac{1}{\alpha!} \left(\partial_\xi^\alpha P_j\right)\left(\delta_x^\alpha Q_k\right). \tag{2}$$

(ii) *If $(\tilde{x}_1, \dots, \tilde{x}_n)$ is another system of local coordinates of X_n on U subordinated to Y then the associated systems of canonical coordinates are related by*

$$\xi_k = \frac{\delta\langle \tilde{x}, \tilde{\xi}\rangle_\nu}{\delta x_k}, \quad 1 \le k \le n.$$

Moreover, for any $l \in \mathbf{Z}$,

$$\tilde{P}_l(\tilde{x}, \tilde{\xi}) = \sum_{\substack{\sigma \\ \alpha_1,\dots,\alpha_\sigma}} \frac{1}{\sigma!\alpha_1!\cdots\alpha_\sigma!} \langle \tilde{\xi}, \delta_x^{\alpha_1}\tilde{x}\rangle_\nu \cdots \langle \tilde{\xi}, \delta_x^{\alpha_\sigma}\tilde{x}\rangle_\nu \partial_\xi^{\alpha_1+\dots+\alpha_\sigma} P_k(x, \xi). \tag{3}$$

Here the indexes run over $k \in \mathbf{Z}$, $\sigma \in \mathbf{N}$, $\alpha_1, \ldots, \alpha_\sigma \in \mathbf{N}^n$, such that $|\alpha_1|, \ldots, |\alpha_\sigma| \geq 2$ and $l = k + \sigma - \sum_{i=1}^{\sigma} |\alpha_i|$. Given $\beta \in \mathbf{N}^n$, $\langle \tilde{\xi}, \delta_x^\beta \tilde{x} \rangle_\nu$ denotes

$$\sum_{j=1}^{\nu} \tilde{\xi}_j \delta_x^\beta \log \tilde{x}_j + \sum_{j=\nu+1}^{n} \tilde{\xi}_j \delta_x^\beta \tilde{x}_j.$$

Moreover, $\delta_{x_i} = \frac{\delta}{\delta x_i} = x_i \frac{\partial}{\partial x_i}$, if $1 \leq i \leq \nu$, $\delta_{x_i} = \frac{\delta}{\delta x_i} = \frac{\partial}{\partial x_i}$, if $\nu + 1 \leq i \leq n$.

We call *principal symbol of P relative to Y* to the homogeneous part $\sigma_Y(P)$ of higher degree of the total symbol relative to Y and to some system of local coordinates.

The ring $\mathcal{D}_{X\langle Y \rangle}$ is endowed with a natural filtration $F.\mathcal{D}_{X\langle Y \rangle}$ by the order of the operators. Moreover, $\mathcal{D}_{X\langle Y \rangle}$ is the enveloping algebra of $\Theta_{X\langle Y \rangle}$. The filtration by the order and the filtration of the enveloping algebra coincide. Therefore the graded ring associated to the filtered ring $\mathcal{D}_{X\langle Y \rangle}$ is the Symmetric algebra of $\Theta_{X\langle Y \rangle}$. Hence the analytic spectrum of gr$\mathcal{D}_{X\langle Y \rangle}$ equals $T^*\langle X/Y \rangle$. Therefore π^{-1}gr$\mathcal{D}_{X\langle Y \rangle}$ equals $\mathcal{O}_{[T^*\langle X/Y \rangle]}$. Hence we have canonical morphisms of sheaves

$$\sigma_{Y,n} : \pi^{-1} F_n \mathcal{D}_{X\langle Y \rangle} \to \mathcal{O}_{[T^*\langle X/Y \rangle](n)}.$$

The morphism $\sigma_{Y,n}$ associates to a differential operator of order n tangent to Y its principal symbol. Put $\sigma_n = \sigma_{\phi,n}$. Let X_n be a copy of $\mathbf{C}^n$ with coordinates $(x_1, ..., x_n)$. Put $Y = \{x_1 = 0\}$. Let $\theta = \sum_{i=1}^{n} \xi_i dx_i$ be the canonical 1-form of T^*X. Let $\theta' = \xi_1' dx_1 / x_1 \sum_{i=2}^{n} \xi_i' dx_i$ be the canonical 1-form of $T^*\langle X/Y \rangle$. The following equalities hold:

$$\sigma(x_1 \partial_{x_1}) = x_1 \xi_1, \quad \sigma_Y(x_1 \partial_{x_1}) = \xi_1', \quad \sigma(\partial_{x_1}) = \xi_1.$$

3.3. Definition Let n be a positive integer. Let ν be a nonnegative integer smaller or equal then n. Let X_n be a copy of $\mathbf{C}^n$ with coordinates $(x_1, \ldots, x_n)$. Let Y_ν be the normal crossings divisor $\{x_1 \cdots x_\nu = 0\}$ of X_n. Let U be an open set of $T^*\langle X_n/Y_\nu \rangle$. Let m be an integer. Let $\mathcal{E}_{(x_1,\ldots,x_n,\nu)(m)}(U)$ be the space of formal series $\sum_{j \leq m} P_j$ where P_j is a section of $\mathcal{O}_{T^*\langle X_n/Y_\nu \rangle(j)}$ on U, such that for any compact set K contained in U there is a real number C such that

$$\sup_K |P_{-j}| \leq C^j j!, \qquad \text{for each } j \geq 0.$$

Given $P, Q \in \mathcal{E}_{(x_1,\ldots,x_n,\nu)(m)}(U)$, we define their sum and product respectively by (1) and (2). The ring $\mathcal{E}_{(x_1,\ldots,x_n,\nu)}$ is endowed in this way with a structure of filtered $\mathbf{C}$-algebra. Given two systems of local coordinates $(x_1, \ldots, x_n)$ and $(\tilde{x}_1, \ldots, \tilde{x}_n)$ on X_n verifying the conditions of Proposition 3.2 (ii), we glue the rings $\mathcal{E}_{(x_1,\ldots,x_n,\nu)}$ and $\mathcal{E}_{(\tilde{x}_1,\ldots,\tilde{x}_n,\nu)}$ by (3).

Given a complex manifold X and a normal crossings divisor Y of X, we can associate in this way a canonical sheaf $\mathcal{E}_{\langle X/Y \rangle}$ on $T^*\langle X/Y \rangle$ that equals $\mathcal{E}_X$

on $T^*(X \setminus Y)$. We call this sheaf the sheaf of *logarithmic microdifferential operators on X with poles along Y* (cf. [N]).

Given a section P of $\mathcal{E}_{\langle X/Y\rangle}$, we can consider a total symbol $(P_j) \in \mathcal{E}_{(x_1,\dots,x_n,\nu)}$ representing P as the "power expansion" of P relative to the system of local coordinates $(x_1, ..., x_n)$. From now on we will identify the sheaves $\mathcal{E}_{(x_1,\dots,x_n,\nu)}$ and $\mathcal{E}_{\langle X_n/Y_\nu\rangle}$.

If Y equals the empty set, then $\mathcal{E}_{\langle X/Y\rangle}$ equals the sheaf of logarithmic microdifferential operators $\mathcal{E}_X$. The sheaf $\mathcal{E}_{\langle X/Y\rangle}$ is a microlocalization of the sheaf $\mathcal{D}_{X\langle Y\rangle}$ in the same sence that the sheaf $\mathcal{E}_X$ (cf. [SKK]) is a microlocalization of the sheaf $\mathcal{D}_X$. The sheaf $\mathcal{E}_{\langle X/Y\rangle}$ has the following properties.

3.4. Theorem (i) *The sheaf $\mathcal{E}_{\langle X/Y\rangle}$ is a (left and right) noetherian Ring with zariskian fibers.*
(ii) *The following equality holds.*

$$\mathcal{E}_{\langle X/Y\rangle} \mid_{\pi^{-1}(X\setminus Y)} = \mathcal{E}_X \mid_{\pi^{-1}(X\setminus Y)}$$

(iii) *There is a canonical flat imersion*

$$\pi^{-1}\mathcal{D}_{X\langle Y\rangle} \hookrightarrow \mathcal{E}_{\langle X/Y\rangle}.$$

Moreover, $\mathcal{E}_{\langle X/Y\rangle} \mid_X = \mathcal{D}_{X\langle Y\rangle}$.
(iv) *We have an isomorphism*

$$gr\mathcal{E}_{\langle X/Y\rangle} \xrightarrow{\sim} \mathcal{O}^h_{T^*\langle X/Y\rangle}.$$

Moreover, $\sigma_{m+n-1}([P,Q]) = \{\sigma_m(P), \sigma_n(Q)\}$, *for any* P, $Q \in \mathcal{E}_{\langle X/Y\rangle}$.
(v) *If the symbol of* $P \in \mathcal{E}_{\langle X/Y\rangle,p}$ *does not vanish, then* P *is invertible.*

Richard Melrose introduced independently a notion of "Fourier integral operators with corners". These operators seem to be the equivalent in the C^∞ category of the operators defined above.

3.5. Definition Let $(X, \mathcal{A})$ be a ringed space. We call *adjoint morphism* of $(X, \mathcal{A})$ to an anti-isomorphism $(a,*) : (X, \mathcal{A}) \to (X, a^{-1}\mathcal{A})$ such that $a^{-1}\mathcal{A}$ is isomorphic to $\mathcal{A}$ and $(a,*)^2 = \mathrm{id}_{\mathcal{A}}$.

We will write $\mathcal{A}^a$ and $\mathcal{A}^*$ instead of $a^{-1}\mathcal{A}$ and $*(\mathcal{A})$.

We call a ringed space with an adjoint morphism a *self dual* ringed space. We say that a subsheaf $\mathcal{B}$ of a self dual ringed space is *self dual* if $\mathcal{B}^* = \mathcal{B}$. We say that a morphism of ringed spaces between two self dual ringed spaces is self dual if it commutes with the adjoint morphisms.

Fixed a system of local coordinates $(x_1, \dots, x_n)$ in X, we can define a notion of *adjoint* of a differential operator on $\mathcal{D}_X$. We put $\partial^*_{x_i} = -\partial_{x_i}$, $f^* = f$, for any holomorphic function f. Hence $(x_i\partial_{x_i})^* = (-\partial_{x_i})x_i = -(x_i\partial_{x_i}+1)$. In order to define globally this adjoint morphism we have to twist $\mathcal{D}_X$. Actually,

$$(dx_1\cdots dx_n)^{-\frac{1}{2}} \otimes P^* \otimes (dx_1\cdots dx_n)^{\frac{1}{2}}$$

does not depend on $(x_1,\ldots,x_n)$.

Put $\omega_X = \Omega^{dim X}$. Let $\mathcal{L}_1$, $\mathcal{L}_2$ be invertible $\mathcal{O}_X$-modules defined in an open connected subset of X. If $\varphi_1,\varphi_2 : \mathcal{L}_1 \to \mathcal{L}_2$ are isomorphisms such that $\varphi_i(s)^{\otimes 2} = s^{\otimes 2}$, for any local section s of $\mathcal{L}_1$, then $\varphi_1 = \pm\varphi_2$. Locally there is always an isomorphism $\varphi : \mathcal{L}_1 \to \mathcal{L}_2$. Moreover, locally there is always an invertible $\mathcal{O}_X$-module $\mathcal{L}_U$ such that $\mathcal{L}_U^{\otimes 2} \xrightarrow{\sim} \omega_X|_U$. The sheaf

$$\mathcal{L}_U \otimes \mathcal{D}_X|_U \otimes \mathcal{L}_U^{\otimes(-1)}, \tag{4}$$

does not depend on the choice of $\mathcal{L}_U$. We can glue the sheaves (4) into a sheaf on X wich we will denote as

$$\omega_X^{\otimes\frac{1}{2}} \otimes \mathcal{D}_X \otimes \omega_X^{\otimes -\frac{1}{2}} \quad \text{or as} \quad \tilde{\mathcal{D}}_X.$$

In a similar way, we define

$$\tilde{\mathcal{E}}_X = \pi_X^{-1}\omega_X^{\otimes\frac{1}{2}} \otimes \mathcal{E}_X \otimes \pi_X^{-1}\omega_X^{\otimes-\frac{1}{2}},$$

$$\tilde{\mathcal{E}}_{\langle X/Y\rangle} = \pi_{X/Y}^{-1}\omega_X^{\otimes\frac{1}{2}} \otimes \mathcal{E}_{\langle X/Y\rangle} \otimes \pi_{X/Y}^{-1}\omega_X^{\otimes-\frac{1}{2}}.$$

Here π_X, $\pi_{X/Y}$ denote respectively the canonical projections of T^*X and $T^*\langle X/Y\rangle$.

3.6. Example (i) *Let X_n be a copy of $\mathbf{C}^n$ with coordinates $(x_1,\ldots,x_n)$. Given an open set U of T^*X_n and a total symbol $P = \sum_k P_k \in \mathcal{E}_{X_n}(U)$ we denote by P^* the total symbol $\sum_k Q_k \in \mathcal{E}^a_{X_n}(U)$, where*

$$Q_l(x,-\xi) = \sum_{\substack{l=k-|\alpha| \\ \alpha\in\mathbf{N}^n}} \frac{(-)^{|\alpha|}}{\alpha!}\partial_\xi^\alpha\partial_x^\alpha P_k.$$

We will denote by * the adjoint morphism of $\mathcal{E}_{X_n}$. The pair $(a,*)$ is an adjoint morphism.
(ii) *There is one and only one anti-isomorphism of filtered $\mathbf{C}$-Algebras*

$$^* : \mathcal{E}_{\langle X_n/Y_\nu\rangle} \to \mathcal{E}^a_{\langle X_n/Y_\nu\rangle}$$

such that the following diagram commutes.

$$\begin{array}{ccc} \mathcal{E}_{\langle X_n/Y_\nu\rangle}|_{X_n\setminus Y_\nu} & \xrightarrow{\sim} & \mathcal{E}_{X_n}|_{X_n\setminus Y_\nu} \\ {*}\downarrow & & \downarrow{*} \\ \mathcal{E}^a_{\langle X_n/Y_\nu\rangle}|_{X_n\setminus Y_\nu} & \xrightarrow{\sim} & \mathcal{E}^a_{X_n}|_{X_n\setminus Y_\nu} \end{array}$$

This anti-isomorphism is an adjoint morphism. Let U be an open set of $T^*\langle X_n/Y_\nu\rangle$. If $P = \sum_j P_j \in \mathcal{E}_{\langle X_n/Y_\nu\rangle}$ and $P^* = \sum_j Q_j$, then

$$Q_l(x, -\xi) = \sum_{\substack{l=k-|\alpha| \\ \alpha\in\mathbf{N}^n}} \frac{(-)^{|\alpha|}}{\alpha!}\partial_\xi^\alpha \Delta_x^\alpha P_k.$$

Here $\Delta_x^\alpha = \Delta_{x_1}^{\alpha_1}\cdots\Delta_{x_n}^{\alpha_n}$, $\Delta_{x_i}^{\alpha_i} = \vartheta_i(\vartheta_i - 1)\cdots(\vartheta_i - \alpha_i + 1)$, if $1 \le i \le \nu$, $\Delta_{x_i}^{\alpha_i} = \partial_{x_i}^{\alpha_i}$, if $\nu + 1 \le i \le n$, $\vartheta_i = x_i\partial_{x_i} + \xi_i'\partial_{\xi_i'}$.

For the proof of (i) see [SKK]. For the proof of (ii) see [N].

3.7. Definition. Let $(X, \mathcal{E})$ be a quantized contact manifold. Let P be a section of $\mathcal{E}$ of order smaller or equal than n. We define

$$\sigma_m'(P) = \tfrac{1}{2}\sigma_{m-1}(P - (-)^m P^*).$$

We refer to $\sigma_m'(P)$ as the *subprincipal symbol* of P of order m. We define $\sigma'(P)$ as $\sigma_m'(P)$ if P has order $m \in \mathbf{Z}$ and zero if P equals zero. We refer to $\sigma'(P)$ as the *subprincipal symbol of P*.

3.8. Proposition. (i) *If $P \in \mathcal{E}(m-1)$, then $\sigma_m'(P)$ equals $\sigma_{m-1}(P)$.*
(ii) *If $P \in \mathcal{E}(m)$, $Q \in \mathcal{E}(n)$, then*

$$\sigma_{m+n}'(PQ) = \sigma_m(P)\sigma_n'(Q) + \sigma_m'(P)\sigma_n(Q) + \frac{1}{2}\{\sigma_m(P), \sigma_n(Q)\}.$$

(iii) *Let $(x_1, \ldots, x_n)$ be a system of coordinates on a open set U of a complex manifold X and let (x, ξ) be the associated system of canonical coordinates of T^*X. Let V be an open set of $\pi^{-1}(U)$ and P a section of $\mathcal{E}_X(m)(V)$. Then $\sigma_m'(dx^{\otimes\frac{1}{2}} \otimes P \otimes dx^{\otimes(-\frac{1}{2})})$ equals*

$$dx^{\otimes\frac{1}{2}} \otimes \left(P_{m-1} - \sum_{i=1}^{n} \frac{\partial^2 P_m}{\partial\xi_i\partial x_i}\right) \otimes dx^{\otimes(-\frac{1}{2})}.$$

Let $(X, \mathcal{A})$ be a ringed space. We say that a subsheaf $\mathcal{S}$ of $\mathcal{A}$ is *proper* along a subset Y of X if $\mathcal{S}_x \neq \mathcal{A}_x$, for any $x \in Y$. Let $\mathcal{I}$ be a both side Ideal of $\mathcal{A}$. We say that $\mathcal{I}$ is *prime* [*maximal*] along a subset Y of X if $\mathcal{I}_x$ is a prime [maximal] ideal of $\mathcal{A}_x$, for any $x \in Y$ and if $\mathcal{I}_x$ equals $\mathcal{A}_x$, for any $x \notin Y$.

3.9. Proposition. *Let X be a complex manifold and Y a closed submanifold of X. Let Z be the residual submanifold of $T^*\langle X/Y\rangle$.*
(i) *There is one and only one both side Ideal $\mathcal{I}_Y$ of $\mathcal{E}_{\langle X/Y\rangle}$ that is prime along the set of poles of X. This Ideal is generated by $I_{\pi^{-1}(Y)}$.*
(ii) *For each $\lambda \in \mathbf{C}$, there is one and only one both side Ideal $\mathcal{I}_\lambda$ that is maximal along Z and it is contained in the sheaf of the local sections P of $\mathcal{E}$ such that*

$$\xi\sigma'(P) \equiv \lambda\sigma(P) \qquad (\text{mod } I_Y + I_Z^2).$$

Here ξ denotes an arbitrary residual function. Given a system of local coordinates $(x_1, \ldots, x_n)$ such that Y equals $\{x_1 = 0\}$ then $\mathcal{I}_\lambda$ is locally generated by $\delta_{x_1} + \lambda + \frac{1}{2}$. The ideal $\mathcal{I}_\lambda$ is self dual iff $\lambda = 0$. The both side Ideal $\mathcal{I}_\lambda$ is generated has a left side Ideal and has a right side Ideal by $\delta_{x_1} + \lambda + \frac{1}{2}$ and x_1.

We call $\mathcal{I}_Y$ the *ideal of the set of poles of $T^*\langle X/Y\rangle$*. We call $\mathcal{I}_{-\frac{1}{2}}$ the *blow up residual ideal.*

Let γ be the canonical projection from $\mathring{T}^*\langle X/Y\rangle$ into $\mathbf{P}^*\langle X/Y\rangle$ associated to the $\mathbf{C}^*$-action of $\mathring{T}^*\langle X/Y\rangle$. We will still denote by $\mathcal{E}_{\langle X/Y\rangle}$ the sheaf $\gamma_*\mathcal{E}_{\langle X/Y\rangle}$. We notice that the sections of $\mathcal{E}_{\langle X/Y\rangle}$ are locally constant along the $\mathbf{C}^*$-orbits. Hence $\gamma^{-1}\gamma_*\mathcal{E}_{\langle X/Y\rangle} \tilde{\to} \mathcal{E}_{\langle X/Y\rangle}$.

3.10. Definition Let X be a homogeneous logarithmic symplectic manifold. We call a quantization [self dual quantization] of X to a sheaf of filtered $\mathbf{C}$-algebras $\mathcal{E}$ [endowed with an adjoint morphism $*$] on X such that given an open set U of X and an homogeneous logarithmic symplectic transformation $\varphi : U \to T^*\langle X/Y\rangle$ there is an isomorphism of filtered $\mathbf{C}$-algebras [self dual filtered $\mathbf{C}$-algebras] $\Phi : \mathcal{E}|_U \to \varphi^{-1}\mathcal{E}_{\langle X/Y\rangle}$. We call the pair $(X, \mathcal{E})$ a *quantized logarithmic symplectic manifold.* We can define in a similar way *quantized logarithmic contact manifold* and *quantized contact manifold.*

4: BLOW UP OF A REGULAR HOLONOMIC SYSTEM

Let $(X, \mathcal{L})$ be a contact manifold. Let Λ be a closed Lagrangian submanifold of X and I_Λ the defining ideal of Λ. Let $\tau_0 : X_0 \to X$ be the *blow up* of X along Λ. Let E_0 be the exceptional divisor of τ_0. Let $j : X_0 \setminus E_0 \hookrightarrow \tilde{X}$ be the inclusion map. Let $\mathcal{O}_{(E_0)}$ be the subsheaf of $j_*\mathcal{O}_{X_0\setminus E}$ of sections f such that fg is holomorphic for every function g in I_{E_0}.

4.1. Proposition *The $\mathcal{O}_{X_0}$-module $\tilde{\mathcal{L}} = \mathcal{O}_{(E_0)}\tau_0^*\mathcal{L}$ is a structure of logarithmic contact manifold with poles along E_0. Moreover, $\tau_0|_{X_0\setminus E_0} : X_0 \setminus E_0 \to X$ is a contact transformation.*

The main tool in the proof of Proposition 4.1 is the following classical Darboux type Theorem: "*Let $(X, \mathcal{L})$ be a contact manifold of dimension $2n+1$. Let Λ be a Lagrangian submanifold of X. Let us fix $p \in \Lambda$. Then there is an open neighbourhood U of p and a system of local coordinates $(x_1, \ldots, x_{n+1}, p_1, \ldots, p_n)$ on U such that the logarithmic differential form $dx_{n+1} - \sum_{i=1}^n p_i dx_i$ generates $\mathcal{L}|_U$ and $\Lambda \cap U = \{x_1 = \cdots = x_{n+1} = 0\}$.*"

Suppose that X is a copy of $\mathbf{C}^3$ with coordinates (x, y, p) and $\mathcal{L}$ is generated by $\omega = dy - pdx$. Put $z = y - xp$. Suppose that $\Lambda = \{x = y = 0\} = \{x = z = 0\}$. Let $\tau_0 : X_0 \to X$ be the blow up of X along Λ. Let E_0 be the exceptional divisor of τ_0. The manifold X_0 is the patching of two copies X', X'' of $\mathbf{C}^3$

with coordinates $(x, \frac{y}{x}, \frac{z}{z})$ and $(x, \frac{z}{x}, p)$ respectivelly. The patching is given by

$$x = x, \quad z = x\frac{z}{x}, \quad p = \frac{y}{z} - (\frac{z}{x})^{-1}.$$

Moreover, $E_0 \cap X_0' = \{x = 0\}$, $E_0 \cap X_0'' = \{z = 0\}$,

$$\omega_0' = \frac{1}{x}(\tau_0^* \omega|_{X_0'}), \quad \omega_0'' = \frac{1}{z}(\tau_0^* \omega|_{X_0''}).$$

The logarithmic differential forms ω_0', ω_0'' generate $\mathcal{L}_0$. Now

$$\omega_0' = d\frac{y}{x} + \frac{z}{x}\frac{dx}{x}, \quad \omega_0'' = \frac{dz}{z} + \frac{x}{z}dp,$$

and so ω_0', ω_0'' are logarithmic contact forms with poles along E_0. Hence $\mathcal{L}_0$ is a logarithmic contact structure with poles along E_0.

4.2. Definition We call the pair $(\tau_0 : X_0 \to X, \tilde{\mathcal{L}})$ the *blow up* of the contact manifold $(X, \mathcal{L})$ along its Lagrangian submanifold Λ.

Let $(X, \mathcal{L})$ be a logarithmic contact manifold with poles along a smooth hypersurface Y. Let Z be the residual submanifold of X. Let $\tau_1 : X_1 \to X$ be the blow up of X along Z. Let Y_1 the proper inverse image of Y.

4.3. Proposition (i) *The $\mathcal{O}_{X_1}$-module $\mathcal{L}_1 = \tau_1^* \mathcal{L}$ is a structure of logarithmic contact manifold with poles along Y_1. Moreover, $\tau_1|_{X_1 \setminus Y_1} : X_1 \setminus Y_1 \to X$ is a contact transformation.*
(ii) *Let $(\hat{X}, \theta)$ $[(\hat{X}_1, \theta_1)]$ be the homogeneous logarithmic symplectic manifold associated with $(X, \mathcal{L})$ $[(X_1, \mathcal{L}_1)]$. Let $\hat{Z}$ be the inverse image of Z by $\hat{X} \to X$. Let $\hat{\tau}_1 : \hat{X}_1 \to \hat{X}$ be the blow up of $\hat{X}$ along $\hat{Z}$. Then $\widetilde{\hat{X}} = \hat{X}_1$ and the map $\hat{\tau}_1$ is a morphism of homogeneous Poisson manifolds. Moreover, the diagram bellow commutes.*

$$\begin{array}{ccc} \hat{X}_1 & \longrightarrow & \hat{X} \\ \downarrow & & \downarrow \\ X_1 & \longrightarrow & X \end{array} \qquad (1)$$

The proof of this Proposition is quite similar to the proof of Proposition 4.1.

4.4. Definition (i) We call the pair $(\tau_1 : X_1 \to X, \mathcal{L}_1)$ the *blow up of the contact manifold $(X, \mathcal{L})$ along its residual submanifold.*
(ii) We call the pair $(\hat{\tau}_1 : \hat{X}_1 \to \hat{X}, \theta_1)$ the *blow up of the homogeneous symplectic manifold (X, θ) along its residual submanifold.*
(iii) *Let $(X, \mathcal{L})$ be a contact manifold. Let Λ be a closed Lagrangian submanifold of X. Let $(\tau_0 : X_0 \to X, \tilde{\mathcal{L}})$ be the blow up of the contact manifold $(X, \mathcal{L})$ along Λ. Let $(\tau_1 : X_1 \to X_0, \mathcal{L}_1)$ be the blow up of the contact manifold $(X, \mathcal{L})$ along its residual submanifold. Let Y_1 be the set of poles of X_1.*

Put $\tilde{X} = X_1 \setminus Y_1$, $\tilde{\mathcal{L}} = \mathcal{L}_1|_{\tilde{X}}$, $\tau = \tau_0\tau_1|_{\tilde{X}}$. *We call the pair* $(\tau : \tilde{X} \to X, \tilde{\mathcal{L}})$ *the contact blow up of* $(X, \mathcal{L})$ *along* Λ.
(iv) Let Γ be a Lagrangian subvariety of X. Let Γ_0 be the proper inverse image of Γ by τ_0. Let Γ_1 be the inverse image of Γ_0 by τ_1. Put $\tilde{\Gamma} = \Gamma_1 \cap \tilde{X}$. We call $\tilde{\Gamma}$ the *semiproper inverse image of* Γ *by* τ.

4.5. Proposition *Let* M *be a complex manifold. Let* Z *be a discrete subset of* X. *Let* $\sigma : \tilde{M} \to M$ *be the blow up of* $\tilde{M}$ *along* Z. *Let* E *be the excepcional divisor of* σ. *Let* $\tilde{\sigma} : \mathbf{P}^*\tilde{M} \to M$ *be the bimeromorphic map associated with* σ. *Put* $\Lambda = \pi_M^{-1}(Z) \subset \mathbf{P}^*M$.
(i) *The blow up of* $\mathbf{P}^*M$ *along* Λ *equals the logarithmic contact manifold* $\mathbf{P}^*\langle \tilde{M}/E\rangle$.
(ii) *Let* $\tau : \widetilde{\mathbf{P}^*M} \to \mathbf{P}^*M$ *be the contact blow up of* $\mathbf{P}^*M$ *along* Λ. *Then* $\widetilde{\mathbf{P}^*M}$ *equals* $\mathbf{P}^*M \setminus \mathbf{P}^*_E M$, *the domain of* $\tilde{\sigma}$ *equals* $\mathbf{P}^*M \setminus \mathbf{P}^*_E M$ *and* $\tau = \tilde{\sigma}$. *The following diagram commutes.*

$$\begin{array}{ccccc} \mathbf{P}^*\tilde{X} \setminus \mathbf{P}^*_E\tilde{X} & \longrightarrow & \mathbf{P}^*\langle \tilde{M}/E\rangle & \longrightarrow & \mathbf{P}^*M \\ & \searrow & \downarrow & & \downarrow \\ & & \tilde{M} & \longrightarrow & M \end{array} \qquad (2)$$

(iii) *Let* $\hat{X}_1$ *be the blow up of* $T^*\langle \tilde{M}/E\rangle$ *along its residual submanifold. Let* $\hat{Y}_1$ *be the set of poles of* $\hat{X}_1$. *Then* $\hat{X}_1 \setminus \hat{Y}_1$ *equals* $T^*\tilde{M} \setminus T^*_E\tilde{M}$. *Moreover, the diagram bellow commutes.*

$$\begin{array}{ccc} T^*\tilde{M} \setminus T^*_E\tilde{M} & \longrightarrow & T^*\langle \tilde{M}/E\rangle \\ \downarrow & & \downarrow \\ \mathbf{P}^*\tilde{M} \setminus \mathbf{P}^*_E\tilde{M} & \longrightarrow & \mathbf{P}^*\langle \tilde{M}/E\rangle \end{array} \qquad (3)$$

All morphisms considered in statements (i) and (ii) are extensions of one of the maps $id_{M\setminus Z}$, $id_{\mathbf{P}^*(M\setminus Z)}$ and $\pi : \mathbf{P}^*(M \setminus Z) \to M \setminus Z$. All the manifolds considered contain $M \setminus Z$ or $\mathbf{P}^*(M \setminus Z)$ as Zariski open sets. Therefore if any of the horizontal morphisms in diagram (2) exist it is unique. Hence it is suficient to prove its existence locally. This can be done once we choose convenient systems of local coordinates. The proof of (iii) is similar.

Let $(X, \mathcal{E})$ be a self dual quantized contact manifold. Let $\mathcal{I}_\Lambda$ be the $\mathcal{E}(0)$-submodule of $\mathcal{E}(1)$ consisting of the microdifferential operators $P \in \mathcal{E}(1)$ such that $\sigma_1(P) \in I_\Lambda$. Following Kashiwara-Oshima [KO] we define

$$\mathcal{E}_\Lambda = \sum_{k\geq 1} \mathcal{I}^k_\Lambda.$$

The $\mathbf{C}$-algebra $\mathcal{E}_\Lambda$ is noetherian and has zariskian fibers. Moreover it is a self dual subalgebra of $\mathcal{E}$.

Let $\mathcal{A}$ be the $\mathcal{E}_{(0)}$ subalgebra of $\mathcal{E}$ locally generated by $\mathcal{E}_\Lambda$ and $\mathcal{E}_{(1)}\vartheta$, where ϑ is a microdifferential operator verifying the following conditions:

(i) $\vartheta \in \mathcal{I}_\Lambda$.

(ii) $d\sigma(\vartheta) \equiv \theta \pmod{I_\Lambda \Omega^1_X}$. (4)

(iii) $\sigma'(\vartheta) + \frac{1}{2} \in I_\Lambda$.

The C-Algebra $\mathcal{A}$ is noetherian, with zariskian fibers and self dual (cf. [N]).

4.6. Theorem. (i) *Let $(X, \mathcal{E})$ be a self dual quantized contact manifold. Let Λ be a Lagrangian submanifold of X. Let $\tau_0 : X_0 \to X$ be the blow up of X along Λ. Let E_0 be the exceptional divisor of τ_0. Then there is a self dual quantization $\mathcal{E}_0$ of the logarithmic contact manifold X_0 and a morphism of self dual filtered* **C***-algebras*

$$\Phi_0 : \tau_0^{-1}\mathcal{E}_\Lambda \to \mathcal{E}_0 \tag{5}$$

such that $\Phi_0 \,|_{X_0\setminus E_0}: \tau_0^{-1}\mathcal{E}_\Lambda \,|_{X_0\setminus E_0} \to \mathcal{E}_0 \,|_{X_0\setminus E_0}$ is an isomorphism. The morphism Φ_0 is flat.

Given another self dual quantization $\mathcal{E}'_0$ of X_0 and another morphism of self dual filtered **C***-algebras $\Phi'_0 : \tau_0^{-1}\mathcal{E}_\Lambda \to \mathcal{E}'_0$, there is one and only one isomorphism of filtered* **C***-algebras $\Psi : \mathcal{E}_0 \to \mathcal{E}'_0$ such that $\Psi\Phi_0 = \Phi'_0$.*

We refer to (5) as the *blow up of the quantized contact manifold $(X, \mathcal{E})$ along Λ.*

(ii) *Let $(X, \mathcal{E})$ be a self dual quantized logarithmic contact manifold. Let Y be the set of poles of X. Let $\tau_1 : X_1 \to X$ be the blow up of X along its residual submanifold. Let Y_1 be the proper inverse image of Y. Let E_1 be the exceptional divisor of τ_1. There is a self dual quantization $\mathcal{E}_1$ of the logarithmic contact manifold X_1 and a morphism of self dual filtered* **C***-algebras*

$$\Phi_1 : \tau_1^{-1}\mathcal{E} \to \mathcal{E}_1 \tag{6}$$

such that $\Phi_1 \,|_{X_1\setminus Y_1}: \tau_1^{-1}\mathcal{E}_\Lambda \,|_{X_1\setminus Y_1} \to \mathcal{E}_1 \,|_{X_1\setminus Y_1}$ is an isomorphism. In general, this morphism is not flat.

Given another self dual quantization $\mathcal{E}'_1$ of X_1 and another morphism of self dual filtered **C***-algebras $\Phi'_1 : \tau_1^{-1}\mathcal{E} \to \mathcal{E}'_1$ there is one and only one isomorphism of self dual filtered* **C***-algebras $\Psi : \mathcal{E}_1 \to \mathcal{E}'_1$ such that $\Psi\Phi_1 = \Phi'_1$.*

We refer to (6) as the blow up of the quantized logarithmic contact manifold $(X, \mathcal{E})$ along Λ.

(iii) *Let $(X, \mathcal{E})$ be a self dual quantized contact manifold and Λ a closed Lagrangian submanifold of X. Let $\tilde{X}$ be the contact blow up of X along Λ. Put $\tilde{\mathcal{E}} = \mathcal{E}_1|_{\tilde{X}}$. Put*

$$\Phi = \Phi_0\Phi_1|_{\tilde{X}} : \tau^{-1}\mathcal{E}_\Lambda \to \tilde{\mathcal{E}}. \tag{7}$$

There is one and only one morphism from $\tau^{-1}\mathcal{A}$ into $\tilde{\mathcal{E}}$ that extends Φ. This morphism is flat. We will still denote it by Φ.

We refer to $\Phi : \tau^{-1}\mathcal{A} \to \tilde{\mathcal{E}}$ as the *contact blow up of the quantized contact manifold $(X, \mathcal{E})$ along Λ.*

The proof of Theorem 4.6 has similarities with the proofs of Propositions 4.1 and 4.5. It is enough to prove the existence of the morphisms (5), (6) and (7) locally. In order to prove the local existence we can choose local models. We sketch the proof of the existence of (7) in the following example.

Let M be a complex manifold. Let Z be a discrete subset of M. Let $\sigma : \tilde{M} \to M$ be the blow up of M along Z. Let E be the exceptional divisor of σ. Put $X = \mathbf{P}^*M$. Put $\Lambda = \pi_M^{-1}(Z) \subset X$. With the notations of the theorem above, if $\mathcal{E} = \mathcal{E}_M$, then $\tilde{\mathcal{E}} = \mathcal{E}_{\tilde{M}}|_{\mathbf{P}^*\tilde{M}\setminus\mathbf{P}^*_E\tilde{M}}$.

Let M be a copy of $\mathbf{C}^2$ with coordinates (x, y). Put $Z = \{x = y = 0\}$. The manifold $\tilde{M}$ is the patching of two copies M_1, M_2 of $\mathbf{C}^2$ with coordinates respectively $(x, \frac{y}{x})$ and $(y, \frac{x}{y})$. Put $x_0 = x$, $y_0 = \frac{y}{x}$. This change of coordinates induces a morphism Φ from $\mathcal{E}_{(x,y)}|_{\mathbf{P}^*(M\setminus Z)}$ onto $\tilde{\mathcal{E}}_{(x_0,y_0)}|_{\mathbf{P}^*(M_1\setminus E)}$ given by

$$\Phi(x) = x_0, \quad \Phi(y) = x_0 y_0, \quad \Phi(\partial_y) = \frac{1}{x_0}\partial_{y_0},$$

$$\Phi(\partial_x) = \frac{1}{x_0}(x_0\partial_{x_0} - y_0\partial_{y_0} - \tfrac{1}{2}).$$

Hence

$$\Phi(x\partial_y) = \partial_{y_0}, \quad \Phi(y\partial_y) = y_0\partial_{y_0}, \quad \Phi(\partial_y^{-1}) = x_0\partial_{y_0}^{-1},$$

$$\Phi(\partial_y(x\partial_x + y\partial_y + \tfrac{1}{2})) = \partial_{x_0}\partial_{y_0}. \tag{8}$$

We notice that the algebra $\mathcal{A}$ associated with the Lagrangian variety Λ is the sub$\mathcal{E}(0)$-algebra of $\mathcal{E}_{(x,y)}$ generated by $x\partial_y$, $y\partial_y$ and $\partial_y(x\partial_x + y\partial_y + \frac{1}{2})$. We can use (8) to show that Φ can be extended to a morphism from $\tau^{-1}\mathcal{A}$ into $\tilde{\mathcal{E}}_{\tilde{M}}$.

4.7. Definition. Let $(X, \mathcal{E})$ be a quantized contact manifold. $\mathcal{M}$ be a coherent $\mathcal{E}$-module. The module $\mathcal{M}$ is called *holonomic* if its support is a Lagrangian variety. Let Λ be a Lagrangian submanifold of X. Let $\Phi : \tau^{-1}\mathcal{E} \to \tilde{\mathcal{E}}$ be the blow up of the quantized contact manifold $(X, \mathcal{E})$ along Λ. We call the $\tilde{\mathcal{E}}$-module $\tilde{\mathcal{M}} = \tilde{\mathcal{E}} \otimes_{\tau^{-1}\mathcal{A}} \tau^{-1}\mathcal{M}$ the *contact blow up of $\mathcal{M}$ along* Λ.

4.8. Theorem. *Let $(X, \mathcal{E})$ be a quantized contact manifold of dimenson $2n - 1$. Let Λ be a Lagrangian submanifold of X. Let $\Phi : \tau^{-1}\mathcal{E} \to \tilde{\mathcal{E}}$ be the contact blow up of the quantized contact manifold $(X, \mathcal{E})$ along Λ.*
(i) *The semiproper inverse image of a Lagrangian variety is Lagrangian.*
(ii) *If $\mathcal{M}$ is a regular holonomic $\mathcal{E}$-module, then the support of $\tilde{\mathcal{M}}$ is the semiproper inverse image of the support of $\mathcal{M}$.*

Proof of (i). The variety Γ_0 is the closure of an involutive variety of codimension n. Therefore it is involutive and has codimension n. The variety $\Gamma_0 \cap E_0$ is also involutive. It follows from Proposition 4.3(ii) that Γ_1 is involutive. The variety Γ_1 is the union of $\tau_1^{-1}(\Gamma_0 \cap E_0)$ and the proper inverse image of

Γ_0 by τ_1. Since the codimension of $\Gamma_0 \cap E_0$ equals $n+1$, then the codimension of $\tau_1^{-1}(\Gamma_0 \cap E_0)$ equals n. Hence $\tilde{\Gamma}$ is Lagrangian.

Proof of (ii). We have the following Lemma (cf. [N]).

4.9. Lemma. *Let $\mathcal{M}$ be a holonomic $\mathcal{E}$-module and $\mathcal{N}$ a coherent $\mathcal{E}_\Lambda$-submodule of $\mathcal{M}$ such that there is a microdifferential operator ϑ verifying the conditions (i), (ii) and (iii) of (4) and a polynomial b verifying*
(iv) $b(k) \neq 0$, $k = 0, 1, 2, \ldots$
(v) $b(\vartheta)\mathcal{N} \subset \mathcal{N}(-1)$.
Then $\mathcal{A} \otimes_{\mathcal{E}_\Lambda} \mathcal{N}$ is isomorphic to $\mathcal{M}$.

As a consequence of this Lemma we have

$$\begin{aligned}\tilde{\mathcal{E}} \otimes_{\mathcal{A}} \mathcal{M} &= \tilde{\mathcal{E}} \otimes_{\mathcal{A}} \mathcal{A} \otimes_{\mathcal{E}_\Lambda} \mathcal{N} \\ &= \tilde{\mathcal{E}} \otimes_{\mathcal{E}_o} \mathcal{E}_o \otimes_{\mathcal{E}_\Lambda} \mathcal{N}.\end{aligned}$$

Hence supp $\tilde{\mathcal{E}} \otimes_{\mathcal{A}} \mathcal{M}$ equals the intersection of $\tilde{X}$ and $\tau_1^{-1}(\mathrm{supp}\mathcal{E}_0 \otimes_{\mathcal{E}_\Lambda} \mathcal{N})$. Therefore it is enough to prove the following Theorem.

4.10. Theorem *Let $\mathcal{M}$ be a regular holonomic $\mathcal{E}$-module and $\mathcal{N}$ a $\mathcal{E}_\Lambda$-submodule of $\mathcal{M}$ wich is $\mathcal{E}_{(0)}$-coherent and generates $\mathcal{M}$ has a $\mathcal{E}$-module. Then the analytic set supp $\mathcal{E}_0 \otimes_{\mathcal{E}_\Lambda} \mathcal{N}$ is the proper inverse image of supp $\mathcal{N}$.*

In order to prove Theorem 4.10 we will recall the notions of normal cone, normal deformation, microcharacteristic variety and 1-microcharacteristic variety.

4.11. Definition. Let X be a complex manifold and Y a submanifold of X. Let $\mathcal{A}_Y$ be the subring $\oplus_{k\in\mathbf{Z}} I_Y^k c^{-k}$ of $\mathcal{O}_X[c, c^{-1}]$. We call the analytic space $\check{X}$ =Specan$\mathcal{A}_Y$ the *normal deformation of X along Y*. Let χ denote the canonical projection $\check{X} \to X$. Given a subset S of X, we put

$$C_Y(S) = \overline{\chi^{-1}(S) \setminus c^{-1}(0)} \cap c^{-1}(0).$$

We call $C_Y(S)$ the *normal cone* of S along Y. Put $C_p(S) = C_{\{p\}}(S)$.

Let X be a complex manifold and Y a closed submanifold of X. Let $\tau_0 : X_0 \to X$ be the blow up of X along Y. Put $\check{X}_0 = \check{X} \setminus C_Y(Y)$. We have a canonical morphism $\rho : \check{X}_0 \to X_0$ such that the diagram bellow commutes.

$$\begin{array}{ccc} X_0 & \leftarrow & \check{X}_0 \\ \downarrow & \swarrow & \\ X & & \end{array}$$

Suppose that X is a copy of $\mathbf{C}^2$ with coordinates (x, y). Put $Y = \{x = y = 0\}$. Then $\check{X}$ is a copy of $\mathbf{C}^3$ with coordinates $(c, \tilde{x}, \tilde{y})$. Moreover, $C_Y(Y) = \{\tilde{x} =$

$\tilde{y} = 0\}$. The manifold X_0 is the obvious patching of the two copies of $\mathbf{C}^2$ with coordinates $(x, \frac{y}{x})$, $(\frac{x}{y}, y)$. The map χ is given by

$$x = c\tilde{x}, \qquad y = c\tilde{y}.$$

The map ρ is given by

$$\begin{cases} x = c\tilde{x}, \\ \frac{y}{x} = \tilde{y}\tilde{x}^{-1}, \end{cases} \qquad \begin{cases} y = c\tilde{y}, \\ \frac{x}{y} = \tilde{x}\tilde{y}^{-1}. \end{cases}$$

There is a canonical $\mathbf{C}^*$-action on $\check{X}_0$ such that $\check{X}_0/\mathbf{C}^* \tilde{\to} X_0$. In the particular case we are considering, the action is given by

$$t \cdot (c, \tilde{x}, \tilde{y}) = (tc, t^{-1}\tilde{x}, t^{-1}\tilde{y}).$$

Let $(X, \mathcal{E})$ be a quantized contact manifold. Let $\mathcal{M}$ be a coherent $\mathcal{E}$-module. We call *microcharacteristic variety of $\mathcal{M}$ along* Λ (cf. [KS]) the analytic set

$$C_\Lambda(\mathcal{M}) = C_\Lambda(\mathrm{supp}\mathcal{M}).$$

4.12. Definition Let $\mathcal{N}$ be a coherent $\mathcal{E}_\Lambda$-module. We say that an increasing filtration $(\mathcal{N}_k)$ of $\mathcal{N}$ is a good filtration if

(i) $\mathcal{N} = \cup_k \mathcal{N}_k$.

(ii) $\mathcal{E}_{\Lambda(m)}\mathcal{N}_k \subset \mathcal{N}_{k+m}$, $\qquad m, k \in \mathbf{Z}$.

(iii) $\mathcal{E}_{\Lambda(m)}\mathcal{N}_k = \mathcal{N}_{k+m}$, $\qquad m \geq 0,\ k >> 0$.

(iv) $\mathcal{E}_{\Lambda(m)}\mathcal{N}_k = \mathcal{N}_{k+m}$, $\qquad m \leq 0,\ k << 0$.

Given a coherent $\mathcal{E}_\Lambda$-module $\mathcal{N}$ with a good filtration we put

$$Ch_\Lambda(\mathcal{N}) = \mathrm{supp}(\mathcal{O}_{X_0} \otimes_{\tau_0^{-1}\mathrm{gr}\mathcal{E}_\Lambda} \tau_0^{-1}\mathrm{gr}\mathcal{N}).$$

Following [F] we put

$$\check{C}h_\Lambda(\mathcal{N}) = \mathrm{supp}(\mathcal{O}_{\check{X}} \otimes_{\chi^{-1}\mathrm{gr}\mathcal{E}_\Lambda} \chi^{-1}\mathrm{gr}\mathcal{N}).$$

We notice that $Ch_\Lambda(\mathcal{N})$ and $\check{C}h_\Lambda(\mathcal{N})$ do not depend of the choice of the good filtration (cf. [F]).

Let $\mathcal{M}$ be a coherent $\mathcal{E}$-module. If $\mathcal{N}$ is a coherent $\mathcal{E}_\Lambda$-submodule of $\mathcal{M}$, then $Ch_\Lambda(\mathcal{N})$ $[\check{C}h_\Lambda(\mathcal{N})]$ does not depend of the choice of $\mathcal{N}$ and we can define $Ch_\Lambda(\mathcal{M}) = Ch_\Lambda(\mathcal{N})$ $[\check{C}h_\Lambda(\mathcal{M}) = \check{C}h_\Lambda(\mathcal{N})]$. Following [F] we put

$$C^1_\Lambda(\mathcal{M}) = Ch_\Lambda(\mathcal{M}) \cap c^{-1}(0).$$

The analytic set $C^1_\Lambda(\mathcal{M})$ is called the *1-microcharacteristic variety of $\mathcal{M}$ along* Λ.

The following theorem was proved by Kashiwara [K] for $\mathcal{D}$-modules and was generalized by Kashiwara [K] and Laurent [L] to $\mathcal{E}$-modules. It shows that the microcharacteristic variety of a regular holonomic module only depends on its support.

4.13. Theorem If $\mathcal{M}$ is regular holonomic, then

$$C^1_\Lambda(\mathcal{M}) = C_\Lambda(\mathcal{M}).$$

It follows from the flateness of the morphism $\tau_0^{-1}\mathcal{E}_\Lambda \to \mathcal{E}_0$ that, if $\mathcal{N}$ is a coherent $\mathcal{E}_\Lambda$-module with a good filtration, then

$$\operatorname{supp}(\mathcal{E}_0 \otimes_{\mathcal{E}_\Lambda} \mathcal{N}) = \mathcal{C}h_\Lambda(\mathcal{N}). \tag{8}$$

Since the morphism $\rho^{-1}\mathcal{O}_X \to \mathcal{O}_{X_o}$ is faithfully flat

$$\rho^{-1}\mathcal{C}h_\Lambda(\mathcal{M}) = \check{\mathcal{C}}h_\Lambda(\mathcal{M}) \cap \check{X}_0.$$

Therefore

$$\begin{aligned}\rho^{-1}(\operatorname{supp}(\mathcal{E}_o \otimes_{\mathcal{E}_\Lambda} \mathcal{N}) \cap E_o) &= C^1_\Lambda(\mathcal{M}) \cap \hat{X}_o\\ &= C_\Lambda(\mathcal{M}) \cap \hat{X}_o\\ &= \rho^{-1}(\Gamma_o \cap E_o).\end{aligned}$$

This ends the proof of Theorem 4.8.

5: A DESINGULARIZATION THEOREM

We say that two Lagrangian submanifolds Λ_1, Λ_2 of a contact manifold $(X, \mathcal{L})$ have *clean intersection* at $p \in X$ if $T_p(\Lambda_1 \cap \Lambda_2) = T_p\Lambda_1 \cap T_p\Lambda_2$. We say that Λ_1 and Λ_2 have a *clean intersection* if they have a clean intersection at p, for any $p \in \Lambda_1 \cap \Lambda_2$.

Let $(X, \mathcal{L})$ be a contact manifold of dimension 3. If Λ_1 and Λ_2 are Lagrangian submanifolds of X with clean intersections at p^o, then there is an open neighbourhood U of p^o and a system of local coordinates (x, y, p) on U such that $dy - pdx$ is a generator of $\mathcal{L}|_U$ and $\Lambda_1 = \{x = y = 0\}$, $\Lambda_2 = \{y = p = 0\}$.

5.1. Theorem *Let X be a contact manifold of dimension 3 and Γ a Lagrangian subvariety of X. Then there are an open neighbourhood X' of Γ, a contact manifold $\tilde{X}$, a holomorphic map $\tau : \tilde{X} \to X$, a Lagrangian subvariety $\tilde{\Gamma}$ of $\tilde{X}$, a normal crossings divisor H of $\tilde{X}$ and a closed Lagrangian submanifold Λ of X' such that:*

(i) *The variety $\tilde{\Gamma}$ is a union of closed Lagrangian submanifolds with clean intersections. Moreover, $\tau(\tilde{\Gamma})$ equals Γ.*

(ii) *The map $\tau|_{\tilde{X}\setminus H} : \tilde{X} \setminus H \to X' \setminus \Lambda$ is a biholomorphic contact transformation. Moreover,*

$$\tau(\tilde{\Gamma} \setminus H) = \Gamma \setminus \Lambda, \quad \Gamma \cap \Lambda = \Gamma_{Sing}.$$

Let $\mathcal{E}$ be a self dual quantization of X and $\mathcal{M}$ a regular holonomic $\mathcal{E}$-module with support Γ. Then there is one and only one self dual quantization $\tilde{\mathcal{E}}$ of

$\tilde{X}$ and a regular holonomic $\tilde{\mathcal{E}}$-module $\tilde{\mathcal{M}}$ such that:
(iii) supp $\tilde{\mathcal{M}} = \tilde{\Gamma}$,
(iv) $\tilde{\mathcal{E}}|_{\tilde{X}\setminus H}$ is isomorphic to $\mathcal{E}|_{X'\setminus\Lambda}$ and $\tilde{\mathcal{M}}|_{\tilde{X}\setminus H}$ equals $\mathcal{M}|_{X'\setminus\Lambda}$.

We will first prove a local version of the Theorem.

5.2. Lemma *Let M be a complex manifold of dimension 2. Let U be an open set of $\mathbf{P}^*M$ such that $\pi_M(U) = M$. Let Γ be a closed Lagrangian subvariety of U. Suppose that Γ is in generic position. Then there is a complex manifold $\tilde{M}$, a proper map $\tilde{\sigma} : \tilde{M} \to M$, an open set $\tilde{U}$ of $\mathbf{P}^*\tilde{M}$ an holomorphic map $\tau : \tilde{U} \to U$, a Lagrangian subvariety $\tilde{\Gamma}$ of $\tilde{U}$ and a normal crossings divisor H of $\tilde{U}$ such that:*
(i) The variety $\tilde{\Gamma}$ is a union of Lagrangian submanifolds with clean intersections. Moreover, $\tau(\tilde{\Gamma}) = \Gamma$.
(ii) The map $\tau|_{\tilde{U}\setminus H} : \tilde{U}\setminus H \to U\setminus\Lambda$ is a biholomorphic contact transformation. Moreover,

$$\tau(\tilde{\Gamma} \setminus H) = \Gamma \setminus \Lambda, \quad \Gamma \cap \Lambda = \Gamma_{Sing}.$$

Let $\mathcal{M}$ be a regular holonomic $\mathcal{E}_M|_U$-module with support Γ. Then there is a regular holonomic $\mathcal{E}_{\tilde{M}}|_{\tilde{U}}$-module with support $\tilde{\Gamma}$ such that $\tilde{\mathcal{M}}|_{\tilde{U}\setminus H}$ is isomorphic to $\mathcal{M}|_{U\setminus\Lambda}$.

Proof. Put $M_0 = M$, $N_0 = \pi_M(\Gamma)$. Let Z_0 be the singular support of N_0. Let $\sigma_k : M_{k+1} \to M_k$ be the blow up of M_k along Z_k. Let N_{k+1} be the proper inverse image of N_k by σ_k. Let Z_{k+1} be the singular set of N_{k+1}. There is an integer n such that N_n is nonsingular. Put $\tilde{M} = M_n$. Let $\sigma : \tilde{M} \to M$ be the composition of the morphisms σ_k, $1 \leq k \leq n$. Put $U_0 = U$, $\Gamma_0 = \Gamma$, $\Lambda_0 = \Lambda$. Let $\imath : U_0 \hookrightarrow \mathbf{P}^*M_0$ be the open inclusion. Let U_k be a contact manifold. Let $\imath_k : U_k \to \mathbf{P}^*M_k$ be a contact transformation. Put $p_k = \pi_{M_k}\imath_k$. Put $\Lambda_k = p_k^{-1}(Z_k)$. Let $\tau_{k+1} : \tilde{U}_k \to U_k$ be the blow up of U_k along Λ_k. Put $U_{k+1} = \tilde{U}_k$. Let Γ_k be a submanifold of U_k such that $p_k(\Gamma_k) = N_k$. Let Γ_{k+1} be the semiproper inverse image of Γ_k by τ_k. It follows from Propositon 4.5(iii) that there is an open inclusion $\imath_{k+1} : U_{k+1} \to \mathbf{P}^*M_{k+1}$. It follows from the commutativity of the diagram (2) of Section 4 that $p_{k+1}(\Gamma_{k+1}) = N_{k+1}$.

Put $\tilde{U} = U_n$, $\tilde{\Gamma} = \Gamma_n$, $\tilde{N} = N_n$. Let τ be the compositon of the morphisms τ_k. Since $\tilde{N}$ is nonsingular, then $\tilde{\Gamma}$ is the union of $\mathbf{P}^*_{\tilde{N}}\tilde{M} \cap \tilde{U}$ and $\mathbf{P}^*_A\tilde{M} \cap \tilde{U}$, where A is a subvariety of $\tilde{N}$ of dimension smaller than the dimension of $\tilde{N}$, therefore a discrete set. Hence $\tilde{\Gamma}$ is a union of smooth Lagrangian submanifolds with clean intersections. Since Γ_{k+1} is the semiproper inverse image of Γ_k, then we have $\tau_{k+1}(\Gamma_{k+1}) \subset \Gamma_k$. Hence $\tau(\tilde{\Gamma}) \subset \Gamma$. Put $\Gamma'_k = \mathbf{P}^*_{N_k}M_k \cap U_k$. The Lagrangian variety Γ'_k is contained in Γ_k, for all k. The map $p_k|_{\Gamma'_k} : \Gamma'_k \to N_k$ is a bijection, for all k. Therefore $\tau(\tilde{\Gamma}) = \Gamma'_0$. Since Γ_0 is in generical position, $\Gamma'_0 = \Gamma_0$. We have now finished the proof of (i).

Put $E_0 = Z_0$, $E_{k+1} = \sigma_{k+1}^{-1}(E_k)$. The variety E_k is a normal crossings divisor of M_k. Put $H_k = p_k^{-1}(E_k)$, $E = E_n$, $H = H_n$. Hence $H_0 = \Lambda$. Statement (ii) follows from the following facts:

• $\tau_{k+1}|_{U_{k+1}\setminus H_{k+1}} : U_{k+1} \setminus H_{k+1} \to U_k \setminus H_k$ is a biholomorphic contact transformation,

• $\tau_{k+1}(\Gamma_{k+1} \setminus H_{k+1}) = \Gamma_k \setminus H_k$,

• $\tau_{k+1}(\Lambda_{k+1}) \subset \Lambda_k$.

Put $\mathcal{M}_0 = \mathcal{M}$. Let $\mathcal{M}_{k+1}$ be the contact blow up of $\mathcal{M}_k$ along Λ_k. Put $\tilde{\mathcal{M}} = \mathcal{M}_n$. Since $\mathcal{M}_0$ is in generic position, then the module $\mathcal{M}_0$ comes locally from a $\mathcal{D}$-module. Theorem 5.3 shows that if $\mathcal{M}_k$ is regular holonomic and comes locally from a $\mathcal{D}$-module, then $\mathcal{M}_{k+1}$ is regular holonomic and comes locally from a $\mathcal{D}$-module. It follows from Theorem 4.8 that if $\mathcal{M}_k$ is regular holonomic and supp $\mathcal{M}_k$ equals Γ_k, then $\mathcal{M}_{k+1}$ is regular holonomic and supp $\mathcal{M}_{k+1}$ equals Γ_{k+1}.

This ends the proof of Lemma 5.2. Let us now prove Theorem 5.1.

Let Γ be a Lagrangian variety. Let p be a singular point of Γ. We can choose open sets V, V' contained in X and a contact transformation φ from V into $\mathbf{P}^*\mathbf{C}^2$ verifying the following properties:
(i) The open set V is a relatively compact neigbourhood of p with real smooth boundary ∂U.
(ii) The open set V'contains the closure of V.
(iii) The point p is the only singular point of $\Gamma \cap V$ and $\varphi(\Gamma \cap V')$ is in generic position.
(iv) The point p is contained in the fiber $\{x = y = 0\}$ of the origin of $\mathbf{C}^2$.
(v) There is a regular holonomic $\tilde{\mathcal{D}}_{\mathbf{C}^2}$-module $\mathcal{N}$defined in a neigbourhood of the origin such that $\mathcal{M}|_{\varphi(V)}$ is isomorphic to $\tilde{\mathcal{E}}_{\mathbf{C}^2} \otimes_{\tilde{\mathcal{D}}_{\mathbf{C}^2}} \mathcal{M}|_{\varphi(V)}$.

Put $\Lambda = \varphi^{-1}\{x = y = 0\} \cap V$. Put X'$= X \setminus (\varphi^{-1}\{x = y = 0\} \cap \partial V)$. The Lagrangian manifold Λ is closed in X'. Hence we can apply the lemma above to the open set U of X and desingularize Γ near p. We can repeat the procedure near each singularity of Γ. If $\mathcal{M}$ is a regular holonomic $\mathcal{E}$-module with characteristic variety Γ the procedure described above desingularizes Γ and produces a coherent $\tilde{\mathcal{E}}$-module $\tilde{\mathcal{M}}$ with support $\tilde{\Gamma}$.

This ends the proof of Theorem 5.1.

Let X and Y be complex manifolds. Let $f : X \to Y$ be a holomorphic map. Put $\mathcal{D}_{X\to Y} = \mathcal{O}_X \otimes_{f^{-1}\mathcal{O}_Y} \mathcal{D}_Y$. Let 1_X, 1_Y denote the identities of the rings $\mathcal{O}_X(X)$ and $\mathcal{D}_Y(Y)$, respectivelly. We will denote by $1_{X\to Y}$ the section $1_X \otimes 1_Y$

of $\mathcal{D}_{X\to Y}$ as well as any of its restrictions.

5.3. Theorem *Let M be a complex manifold. Let Z be a discrete submanifold of M. Let $\sigma : \tilde{M} \to M$ be the blow up of M along Z. Let E be the excepcional divisor of σ. Put $\Lambda = \pi_M^{-1}(Z)$. Let $\tau : \mathbf{P}^*M \to \tilde{X} = \mathbf{P}^*\tilde{M} \setminus \mathbf{P}^*_E\tilde{M}$ be the contact blow up of M along Λ. Then*

$$\tilde{\mathcal{E}} \otimes_{\mathcal{A}} \mathcal{E} \otimes_{\tilde{\mathcal{D}}_M} \mathcal{N} \tilde{\to} \tilde{\mathcal{E}} \otimes_{\tilde{\mathcal{D}}_{\tilde{M}\to M}} \otimes^{\mathbf{L}}_{\tilde{\mathcal{D}}_M} \mathcal{N}.$$

This Theorem is a consequence of the following Lemma:

5.4. Lemma *There is an isomorphism of $(\tilde{\mathcal{E}}, \tilde{\mathcal{D}}_M)$-bimodules*

$$\tilde{\mathcal{E}} \otimes_{\tilde{\mathcal{D}}_{\tilde{M}}} \tilde{\mathcal{D}}_{\tilde{M}\to M} \tilde{\to} \tilde{\mathcal{E}} \otimes_{\mathcal{A}} \mathcal{E}. \tag{1}$$

It follows from this lemma that, if $\mathcal{N}$ is a regular holonomic $\tilde{\mathcal{D}}_M$-module, then the complex

$$\tilde{\mathcal{D}}_{\tilde{M}\to M} \otimes^{\mathbf{L}}_{\tilde{\mathcal{D}}_M} \mathcal{N} \tag{2}$$

has regular holonomic cohomology groups. By (1), $\tilde{\mathcal{E}} \otimes_{\tilde{\mathcal{D}}_{\tilde{M}}} \tilde{\mathcal{D}}_{\tilde{M}\to M}$ is a flat $\pi_{\tilde{M}}^{-1}\tilde{\mathcal{D}}_M \mid_{\tilde{X}}$-module. Therefore the higher torsion groups of (2) have characteristic variety contained in $\mathbf{P}^*_E\tilde{M}$. Hence, if $\mathcal{M}$ equals $\mathcal{E} \otimes_{\tilde{\mathcal{D}}_m} \mathcal{N}$, then

$$\tilde{\mathcal{E}} \otimes_{\mathcal{A}} \mathcal{M} = \tilde{\mathcal{E}} \otimes_{\tilde{\mathcal{D}}_{\tilde{M}}} \tilde{\mathcal{D}}_{\tilde{M}\to M} \otimes_{\tilde{\mathcal{D}}_M} \mathcal{N}$$

is regular holonomic.

We will now prove the Lemma.

The sections of $\tilde{\mathcal{E}} \otimes_{\mathcal{D}_{\tilde{M}}} \mathcal{D}_{\tilde{M}\to M}$ are locally sums of sections of the type

$$P \otimes 1_{\tilde{M}\to M} Q, \tag{3}$$

with $P \in \tilde{\mathcal{E}}$, $Q \in \mathcal{D}_M$. The morphism (1) maps (3) into $P \otimes Q \in \tilde{\mathcal{E}} \otimes_{\mathcal{A}} \mathcal{E}$. This morphism is a morphism of $(\tilde{\mathcal{E}}, \tilde{\mathcal{D}}_M)$-bimodules and is an isomorphism outside $\pi_{\tilde{M}}^{-1}(E)$. Let p be a point of $\pi_{\tilde{X}}^{-1}(E) \setminus \mathbf{P}^*_E\tilde{X}$ and put $q = \pi(p)$. We can fix systems of local coordinates $(x_1, \dots, x_n, \xi_1, \dots, \xi_n)$ in a neighbourhood of p and $(y_1, \dots, y_n, \eta_1, \dots, \eta_n)$ in a neighbourhood of q, such that

$$\{p\} = \{y = \eta' = 0\}, \quad \eta' = (\eta_1, \dots, \eta_{n-1}),$$

$$\{q\} = \{x = \xi' = 0\}, \quad \xi' = (\xi_1, \dots, \xi_{n-1}), \qquad \text{and}$$

$$x_1 = y_1, \quad x_k = x_1 y_k, \quad 1 \le k \le n-1,$$

$$\xi_1 = \eta_1 - \frac{1}{y_1}\sum_{i=2}^{n} x_i \eta_i, \quad \xi_i = \frac{1}{y_1}\eta_i, \quad 1 \le k \le n-1.$$

Put $E = \tilde{\mathcal{E}}_p$, $A = \mathcal{A}_q$. Let E be the filtered A-module generated by u_k, $k \geq 0$, with relations

$$\partial_{y_n}^{-1} u_{k+1} = u_k, \qquad k \geq 0.$$

Let D be the filtered $\mathcal{D}_{\tilde{X},p}$-module generated by v_k, $k \geq 0$, with relations

$$\partial_{y_n} v_k = y_1 v_{k+1}, \qquad k \geq 0.$$

If we prove that E is isomorphic to $\mathcal{E}_q$ and D is isomorphic to $\mathcal{D}_{\tilde{M}\to M,p}$, then the morphism (1) takes u_k into v_k and therefore it is an isomorphism.

Let us call φ the epimorphism from E onto $\mathcal{E}_q$ that maps u_k into $\partial_{y_n}^k$. If $u \in E$, then u is a finite sum of elements of the type $Q_{kj}\partial_{y_n}^k u_j$, $k, j \geq 0$, such that $\mathrm{ord} Q_{kj} = 0$ or $Q_{kj} = 0$ or $k = j = 0$ and in this case $\mathrm{ord} Q_{oo} \leq 0$. Put $l = \max\{k + j : \sigma(Q_{kj}) \neq 0\}$. If $l \geq 0$, then

$$\sigma_l(\varphi(u)) = \sum_{k+j=0} \sigma_o(Q_{kj})\xi^k \eta^j \neq 0.$$

Therefore if $\varphi(u)$ equals 0 then $Q_{kj} = 0$ for $(k, j) \neq (0, 0)$.

Hence $\varphi(u) = Q_{oo}$ and therefore u equals 0. The other proof is similar.

Bibliography

[D] Deligne P. - "Equations differentielles a points singuliers reguliers", Springer Lect. Notes 163 (1970)

[F] Fernandes T. M. - "Problème de Cauchy pour les systèmes microdifférentiels", Astérisque 140-141, 135-220 (1986)

[EGA] Grothendieck, A. et Dieudonné J. - "Eléments de Géométrie Algébrique I" Springer Verlag, II Publ. Math. IHES 4 (1960)

[G] Grothendieck A. - "Techniques de construction en Géometrie analytic", Séminaire Henri Cartan, 13ᵉ année, 1960/61 W.A. Benjamin (1967)

[Hi] Hironaka H. - "Resolution of singularities of an algebraic variety over a field of characteristic 0", Ann. of Math. 79 (1964)

[Hz] Houzel C. - "Géometrie analytic locale" Seminaire Henri Cartan 13ᵉ année 1960/61 W.A. Benjamin (1967)

[K] Kashiwara M. Unpublished

[KK1] Kashiwara M. and Kawai T. - "On Holonomic systems of Microdifferential Equations. III - Systems with Regular Singularities" Publ. RIMS,

Kyoto Univ. 17, 813-979 (1981)

[KO] Kashiwara M. and Oshima T. - "Systems of differential equations with regular singularities and their boundary value problems", Ann. of Maths., 106, 145-200 (1977)

[KS] Kashiwara M. and Schapira P. - Problème de Cauchy pour les systèmes microdifférentielles dans le domaine complexe, Inventiones Math. 46, 17-38 (1978)

[L] Laurent Y. Oral comunication

[N] Neto O. - "Blow up for a holonomic system" Publ. RIMS, Kyoto Univ. 29, 167-233 (1993)

[S] Schapira P. - "Microdifferential systems in the Complex Domain" Springer-Verlag (1985)

[SKK] Sato M., Kawai T. and Kashiwara M. - "Microfunctions and Pseudo-differential Equations", in Springer Lect. Notes 287, 265-529 (1973)

Centro de Matemática e Aplicações Fundamentais
Universidade de Lisboa
Av. Prof. Gama Pinto, 2
1699 Lisboa Codex
Portugal

TOPOLOGICAL STABILITY

A.A. DU PLESSIS AND C.T.C. WALL

Aarhus University (AAdP)
Liverpool University (CTCW)

This article is an introductory account of work by the authors which is to appear in a book entitled "The geometry of topological stability".

A C^∞ map $f : N \to P$ between smooth manifolds is called C^r-stable if it has a neighbourhood W such that for any $g \in W$ there exist C^r-diffeomorphisms ρ of N, λ of P with $g = \lambda \circ f \circ \rho$. If λ and ρ depend continuously on g, we call f strongly stable.

We will concentrate on the case when N is compact. The results can be extended to the general case, but matters become much more complicated, partly because we have a choice of topologies on the spaces $C^\infty(N, P)$ (of C^∞-mappings) etc - on the whole, the Whitney topologies are the most appropriate - leading to many variants of the definition of stability, which are demonstrably not all equivalent, so that it is harder to match up necessary with sufficient conditions for stability. To hint at the kind of considerations involved we draw attention to a generalisation of properness which plays an important rôle. We say f is *quasi-proper* if the discriminant $\Delta(f)$ has a neighbourhood U in P such that the restriction of f to $f^{-1}U$ is proper. Then the condition of being quasi-proper is *necessary* for versions of C^0-stability requiring some kind of uniformity in the Whitney topology though not for C^0-stability *per se*.

The theory of C^∞-stability was developed by Mather in 1968-1970 in an important series of papers $[M1 - M6]$, in which he introduced many new ideas and techniques. We summarise some of his main conclusions.

For each $k \in \mathbf{N}$, the map f induces a jet section $j^k f : N \to J^k(N, P)$. Now $J^k(N, P)$ is fibred over $N \times P$; the generic fibre is the space $J^k(n, p)$ of k-jets of germs $h : (\mathbf{R}^n, 0) \to (\mathbf{R}^p, 0)$. This space admits equivalence relations induced by the groups of local diffeomorphisms of source and of target ; important is also the relation of contact- (or $\mathcal{K}-$) equivalence. Roughly speaking, germs h_i $(i = 0, 1)$ are $\mathcal{K}$-equivalent if and only if the local algebras $Q(h_i) = \mathcal{O}_n/(h_i^* \mathbf{m}_p.\mathcal{O}_n)$ are isomorphic : if $n > p$, in the complex case, this means that the fibres have isomorphic germs.

Now if $S \subset J^k(n,p)$ is invariant under diffeomorphisms of source and target, it induces a subbundle $S(N,P)$ of $J^k(N,P)$. In fact we partition $J^k(n,p)$ by orbits under the cruder $\mathcal{K}$-equivalence.

Theorem 1 *For N compact, the map f is C^∞-stable if and only if for some (hence all) $k \geq p+1, J^k f$ is multitransverse to the partition of $J^k(N,P)$ by $\mathcal{K}$-orbits.*

Multitransversality can be interpreted either by taking preimages $(j^k f)^{-1}S$ of orbits, pushing forward $f((j^k f)^{-1}S)$ to P and seeking transversality of such manifolds in P to each other and (at self-intersections) to themselves ; or equivalently by considering the r-fold product $(j^k f)^r : N^r \to (J^k(N,P))^r$, restricting to the open subset $N^{(r)}$ of r-tuples of distinct points, and demanding transversality to the intersections of submanifolds $T_1 \times \ldots \times T_r$ with each T_i a $\mathcal{K}$-orbit as above (for any $r \geq p+1$) with the preimage of the diagonal in P^r.

If (n,p) is such that, for large enough k, there is a subset X of $J^k(n,p)$, of codimension greater than n, whose complement is a finite union of $\mathcal{K}$-orbits, then for f generic, $j^k f$ will avoid $X(N,P)$ and multitransversality will hold. Thus C^∞-stable maps are dense (and open) in $C^\infty(N,P)$. Such dimensions (n,p) Mather called "nice", and determined them. Here are the conditions.

$p-n$:	≤ -3	-2	-1	$\in [0,3]$	≥ 4
"nice":	$p<7$	$p<6$	$p<8$	$7n<6p+9$	$7n<6p+8$

Otherwise, the argument fails and in fact, as Mather showed, C^∞-stable maps are not dense. One thus seeks an alternative theory. Now outside the nice dimensions, C^1-stable maps are not dense either, as we have shown elsewhere [du Plessis & Wall, 1989] : in fact C^1-stability seems to be much closer to C^∞-stability than one would at first guess. We thus turn to C^0-stability.

Again, such a theory was developed by Mather around 1970 (see [Mather, 1970, 1973, 1976], [GWPL]), following earlier work and ideas of Thom [Thom, 1955, 1964]. This main conclusion here can be formulated as follows.

Theorem 2 *There exists a "canonical" Whitney regular stratification $\mathcal{A}^k(n,p)$ of an open subset $J^k(n,p) - W^k(n,p)$: codim $W^k(n,p) \to \infty$ with k ; and if f is quasi-proper, $j^k f$ avoids $W^k(N,P)$ and is multitransverse to $\mathcal{A}^k(N,P)$, then f is C^0-stable.*

This implies that C^0-stable maps are always dense and open in the space

$C^\infty_{pr}(N,P)$ of proper maps $N \to P$, which is enough for many applications. Maps satisfying this sufficient condition for stability we will refer to as *MT-stable*. Here MT may be interpreted as standing for Mather-Thom or, if you prefer, MultiTransverse.

We return to the case N, compact. Observe that any C^0- stable map f is then C^0-equivalent to an MT-stable one. In particular, as the stratification $\mathcal{A}^k(n,p)$ is semialgebraic, hence finite, there are only finitely many topological types available for germs of C^0-stable maps. We call a map f (or a germ $f^\wedge$) of *MT-type* if it is topologically equivalent to an MT-stable map (or germ).

Among the key ideas in the proof of this result are use of maps of "finite singularity type" (a globalisation of finite $\mathcal{K}$-determinacy ; the complement of these has infinite codimension) and the theorem that such a map has a proper, C^∞-stable unfolding

$$\begin{array}{ccc} N & \xrightarrow{f} & P \\ \downarrow i & & \downarrow j \\ N' & \xrightarrow{F} & P' \end{array}$$

Mather showed that the map F can be canonically stratified, and the diagram is a pullback, so that one can perturb j and induce a perturbation of f. Moreover, f is multitransverse to $\mathcal{A}^k(n,p)$ if and only if j is transverse to the stratification of P'. One can now deform j to satisfy this ; the proof of C^0-stability uses the isotopy lemmas of Thom applied to this stratification.

The above theory has, for us, two principal defects. One is that it seems to be virtually impossible to determine explicitly the stratification $\mathcal{A}^k(n,p)$: indeed, except in "trivial" cases (where no moduli are involved in the $\mathcal{K}$-classification) only one stratum is known explicitly [Wall, 1980]. Secondly, the above sufficient condition for stability is known not to be necessary : this follows from an example due to [Looijenga, 1977], where A-regularity was shown to fail by [Bruce, 1980].

Thus problems of the following type remain open : are C^0-stability and C^∞-stability equivalent in the nice dimensions ? Does global stability follow from local stability in the case when N is compact ? (That this is true for C^∞-stability follows from Theorem 1, at least if local stability is required at all finite subsets S of N with a common target (and at most $p+1$ points)).

Thus rather than seek to determine $\mathcal{A}^k(n,p)$ explicitly, it seems more pertinent to seek necessary and sufficient conditions for stability. We would like to prove the

Conjecture 3 *There exist algebraic subsets $X^k(n,p)$ of $J^k(n,p)$, with codim $X^k(n,p) \to \infty$ with k, and a stratification $\mathcal{B}^k(n,p)$ of $J^k(n,p) - X^k(n,p)$*

such that if N is compact, f is C^0-stable if and only if, for k such that codim $X^k(n,p) > n$, $j^k f$ avoids $X^k(N,P)$ and is multitransverse to $\mathcal{B}^k(N,P)$.

This would resolve all the above questions. Unfortunately, we are far from being able to prove it. What we can prove is

Theorem 4 *There exist algebraic subsets $Y^k(n,p)$ of $J^k(n,p)$, with $W^k(n,p) \subseteq J^k(n,p)$, and a stratification $\mathcal{B}^k(n,p)$ of $J^k(n,p) - Y^k(n,p)$ such that if N is compact and $j^k f$ avoids $Y^k(N,P)$ then f is C^0-stable if and only if $j^k f$ is multitransverse to $\mathcal{B}^k(N,P)$.*

For k large enough codim $Y^k(n,p) = n - p + \tau(p-n)$, where τ is as follows :

$p-n$:	≤ -7	$[-7,-3]$	-2	-1	0	1	2	≥ 3
$\tau(p-n)$:	11	$n-p+4$	11	10	13	21	26	$8(p-n)+7$

If N is compact and $p < \tau(p-n)$, then the condition that $j^k f$ avoids $Y^k(N,P)$ and is multitransverse to $\mathcal{B}^k(N,P)$ is necessary and sufficient for C^0-stability and also for strong C^0-stability.

In fact, it it not difficult to say what $\mathcal{B}$ has to be in general. We first recall from Mather's theory that a k-jet z is $\mathcal{K}$-sufficient if any two representative germs of z are $\mathcal{K}$-equivalent. A $\mathcal{K}$-sufficient jet z thus has a representative f unique up to $\mathcal{K}$-equivalence and in turn this has a stable unfolding F unique up to $\mathcal{A}$- (i.e. right-left) equivalence, provided the dimension of its source is specified. Moreover, two k-jets z_i $(i = 0,1)$ are $\mathcal{K}$-equivalent if and only if corresponding ($C^\infty-$) stable germs F_i (with the same source dimension) are $\mathcal{A}$-equivalent. Now we define the z_i to be ST-equivalent (ST stands for stably topologically) if F_0 and F_1 are C^0-equivalent germs. The strata of $\mathcal{B}^k(n,p)$ are the ST-equivalence classes in $J^k(n,p) - W^k(n,p)$.

This definition becomes ambigous if it is possible to have F_0 and F_1 C^0-inequivalent while their trivial unfoldings $F_0 \times \mathbf{R}$ and $F_1 \times \mathbf{R}$ are C^0-equivalent. Our methods are not sufficiently delicate to handle such a case : we conjecture that it does not arise.

A large part of the proof of Theorem 4 consists in an explicit determination of these classes ; eventually, $Y^k(n,p)$ is the complement of the subset where we have achieved a complete classification. Notice that we show *inter alia* that the strata of $\mathcal{B}^k(n,p)$ so far discovered are smooth submanifolds. As the above remark about suspension illustrates, it is not even clear in general that we have topological submanifolds.

The theorem implies that C^0-stabililty and C^∞-stability are indeed equivalent in the nice dimensions, and characterises C^0-stability in the "semi-nice" dimensions where $\mathcal{K}$-modality ≤ 1 is generic. It also implies that in the domain of its applicability, C^0-stability is indeed a local condition (for N compact). It follows from the example mentioned above that $\mathcal{B}$ cannot be Whitney regular, though we would expect $\mathcal{B}$ to be A-regular, and indeed the stratification constructed in Theorem 4 is C-regular in the sense of [Bekka, 1988].

We now outline a few of the ideas involved in the proof of the above result. We begin by discussing necessary conditions for C^0-stability. In fact, a first step towards such results was taken by May [May, 1973, 1974] and Damon, in a series of papers leading up to [Damon, 1979]. Their work was addressed to the problem of proving that C^0-stability was equivalent to C^∞-stability in the nice dimensions. Such a theorem was in fact obtained, but only after replacing the hypothesis of C^0-stability by a much stronger one. However, many ideas were introduced in these papers which form a basis for our work.

A crucial point is to obtain some local necessary condition for global stability. We have looked at half a dozen conditions, but it seems that the key to success is counting local contributions. Suppose, for example, Δ is a class of germs, saturated for topological equivalence, which is presented at isolated points, so that for any map f,

$$\Delta(f) = \{x \in N \mid \text{the germ } f_x^\wedge \text{ of } f \text{ at } x \text{ is in } \Delta\}$$

is discrete. Then for N compact, the number of points in $\Delta(f)$ is finite, and a perturbation that adds or loses a point of $\Delta(f)$ locally will change the gobal topological type. If $f^\wedge$ is a germ admitting two deformations, in one of which Δ does not appear while in the other it does, then a map f having $f^\wedge$ as a germ has two deformations which are not topologically equivalent ; thus f cannot be C^0-stable.

This basic idea needs to be generalised in several directions. First, it is best for all the theory to work in the target rather than in the source, so for $y \in P$ we consider the germ of f at $\Sigma(f, y) = f^{-1}(y) \cap \Sigma(f)$. Secondly, we have to contemplate sets Δ of germs more general than those above, and so count compact components (rather than points) of $\Delta(f)$. Making sense of the creation or destruction of such "locally" in the target of a smooth map is the key to the generalisation. The details are rather technical, but give a working definition of "disruptive" germ and a theorem that a C^0-stable map with N compact has no discuptive germ.

A germ $f^\wedge$ not of MT-type is easily shown to be disruptive, so if $f^\wedge$ is not disruptive, $\Sigma(f, y)$ has a neighbourhood U such that the restriction of f to $U \cap \Sigma_{Top}(f)$ is proper. Improving this to $U \cap \Sigma(f)$ is a key step in the proof.

Now let $\Delta_1, \ldots, \Delta_r$ be pairwise distinct submanifolds of $J^k(n,p)$, each saturated under ST-equivalence. Then their occurence, or simultaneous occurrence (in the same $\Sigma(f,y)$) as transverse germ is a C^0-invariant property. We can now show that, provided that each cod $(\Delta_i) \leq n$, any germ not multi-transverse to the Δ_i is disruptive. Note that for a Δ_i-transverse germ $h^\wedge$ at a generic point we can show that a stable unfolding of $h^\wedge$ is C^0-equivalent to the product of $h^\wedge$ with the identity map of $\mathbf{R}^a$, so our hypothesis about stable topological type comes into play. However if cod $(\Delta_i) > n$ there are no Δ_i-transverse germs and a different approach is needed. We deal with this by an inductive technique based on the relation of the given class Δ to others in its neighbourhood. This is not so inefficient as might appear at first sight since the methods used to classify C^∞-stable germs up to topological equivalence are also inductive, and we can ensure that the discussions of the two problems march together.

Defining submanifolds of jet space as *fine* ST-invariants when their ST-invariance is proved by the particular methods mentioned above, we obtain

Theorem 5 *If a smooth map $f: N \to P$ (with N compact) is C^0-stable, then it is multi-transverse to all fine ST-invariants.*

We now turn to sufficent conditions for C^0-stability. The first explicit result giving such sufficient conditions in cases where moduli are involved appears in [Looijenga, 1977]. His technique for proving topological triviality was used in [Wirthmüller, 1979] to obtain a more general result, and was given a further considerable extension in [Damon, 1980].

The basic local question to be resolved is a follows. Let $g^\wedge$ be a k-modal (for $\mathcal{K}$-equivalence) germ, with versal unfolding $F^\wedge$, so the "base" B of $F^\wedge$ is a k-dimensional subset. Then we seek to prove C^0-triviality "along the base", *i.e.* of $F^\wedge$, considered as k-parameter family of map-germs

$$F : N' = N \times B \to P' = P \times B$$

Such a trivialisation yields a retraction (R,S) of $F^\wedge$ to its restriction $f^\wedge$: $N \to P$. For stability, more is needed. Consider the diagram

$$\begin{array}{ccc} N & \xrightarrow{f} & P \\ \downarrow i_0 & & \downarrow j_0 \\ N' & \xrightarrow{F} & P' \end{array}$$

Any j close enough to j_0 is transverse to B, and the pullback f_j of F by j also unfolds to F ; we need to show all these f_j C^0-equivalent. For the retraction (R,S) to yield this, we need that $S \circ j$ is a homeomorphism (germ) of P. A retraction with this property is called *tame*.

A map f is called *tamely* $P-C^0$*-stable* if for some (hence any) stable unfolding F of f as above there exist a neighbourhood V of $j(P)$ and a retraction $(R,S) : F \to f$ with S tame. This implies strong C^0 stability. We make a corresponding definition for germs. Then we can prove globalisation.

Theorem 6 *If f is quasi-proper, smooth of finite singularity type and locally tamely $P-C^0$-stable then f is strongly C^0-stable.*

The proof extends the local trivialisations first to a neighbourhood of the critical set in the fibre (this is just a reinterpretation of the definition), then to a neighbourhood of a compact set in a fibre, and then globally. The crucial step is the second: to obtain the desired continuous maps of function spaces we make use of the Kirby-Edwards theorem on spaces of embeddings. The final globalisation is a routine piecing together using a suitable covering of the target.

We are now again reduced to a question on germs. In general we think of F as a k-parameter family of maps $f_t : N \to P$ and define the instability locus

$$V(F) = \{(y,b) \in P \times B \mid \text{ the germ of } f_b \text{ at } \Sigma(f_b, y) \text{ is not } C^\infty\text{-stable}\}$$

We assume F weighted homogeneous, with weights > 0 on N and P and ≤ 0 on B (eventually some other cases have to be included). First suppose $V(F) = 0 \times B$. The Looijenga-Damon method shows that the constant vector fields on B lift to F-trivialising vector fields outside $F^{-1}(0 \times B)$ (source) and $0 \times B$ (target); piecing together allows us to extend over the complement of $0 \times B$ (source), and show that the result is integrable, and yields topological trivialisations. Unfortunately if dim $B > 1$, the resulting retraction is not obviously tame.

Instead, we choose a weighted homogeneous norm function ρ on P and consider the restriction to a sphere $S\,(\rho = 1)$. There are no unstable points on $S \times B$, and we can show that the restriction of F to $F^{-1}(S \times B)$ has no unstable germs. This restriction can then be smoothly trivialised. The corresponding retraction is extended using the $\mathbf{R}^*$-action, and we can prove that this extension is tame.

Indeed, even this is not enough. For we aim to prove that transversality of a map-germ to the $\mathcal{K}$-invariant submanifold of jet-space represented by the jets of the f_t at 0 implies tame $P-C^0$-stability of the germ, and this requires (via a now standard principle of transfer of transversality in jet space to transversality in a versal unfolding) that for *any* submanifold of $P \times B$, transverse to B and of dimension *dim* P, the model retraction above should restrict (near 0) to a homeomorphism. This condition we call *V-tame* ("very tame"), and we do obtain it in the situation described.

We call submanifolds of jet space represented by families (f_t) as above *civilised* when such a V-tame model retraction exists. Then since transversality to such submanifolds implies local tame $P - C^0$-stability, Theorem 6 implies.

Theorem 7 *A quasi-proper smooth map $f : N \to P$ everywhere multitransverse to civilised submanifolds is strongly C^0-stable.*

We can thus take as strata of $\mathcal{B}$ submanifolds of jet space that are both fine ST-invariant and civilised.

The discussion of tame and V-tame retractions had as original motivation a globalisation of the Looijenga-Damon theory. But it is also the key to progress beyond the (rather small) collection of strata to which that theory applies. We do not need to restrict to the situation $V(F) = 0 \times B$, for we are able to prove that any tame retraction of $S \times B$ extends to a V-tame retraction of $P \times B$ near $0 \times B$. On $S \times B - V(F)$ we have local C^∞-stability, so a smooth (hence V-tame) retraction. If multi-transversality to civilised submanifolds holds everywhere on $S \times B \cap V(F)$ we have local tame retraction there. Glueing all these gives a tame retraction on $S \times B$, hence civilisation. (There is a technical difficulty here : we have not (yet) a completely general technique for glueing local tame retractions to obtain a global one. Our methods are, however, adequate for all examples considered so far).

The stage is now set for a construction of civilised submanifolds by induction on codimension. We need to compute $V(F)$. Here it is easiest to complexify, in order to use methods of complex algebraic geometry.

According to Mather, C^∞-stable germs f are characterised by the condition $\theta(f) = tf(\theta_n) + \omega f(\theta_p)$. Now consider an unstable map-germ $f : \mathbf{R}^n \to \mathbf{R}^p$ with C^∞-stable unfolding $F : \mathbf{R}^{n+a} \to \mathbf{R}^{p+a}$. Then the restrictions ϕ_i to $\mathbf{R}^n(u = 0)$ of $tF(\partial/\partial u_i) - \omega F(\partial/\partial u_i)$ $(1 \le i \le a)$ project to an $\mathcal{O}_p$ -basis of the "instability module" $I(f) = \Theta(f)/\{tf(\Theta_n) + \omega f(\Theta_p)\}$ of f. $I(f)$ thus admits an $\mathcal{O}_p$ -presentation

$$0 \longrightarrow \mathcal{O}_p^b \xrightarrow{E(f)} \mathcal{O}_p^a \longrightarrow I(f) \longrightarrow 0,$$

(indeed, if $n \ge p$ we may take $b = p + a$); and the support of $I(f)$ is the instability locus $V(f)$ for f. (Thus if, for example, $a = 1$, the entries in the row matrix $E(f)$ generate an ideal defining $V(f)$.) Considering the module rather than the support *per se* has the usual advantages ; in particular, the required transversality to civilised submanifolds can often (perhaps always) be deduced algebraically.

We have established an efficient algorithm for computing instability modules in the quasi-homogeneous situation considered geometrically above. In some situations, considerations of commutative algebra and the structure of the algorithm above are such that only minimal geometric input is required to determine the instability locus. In other situations, we have used a computer. Fortunately, while *finding* an instability module may require much computer calculation, the output can be so arranged that *checking* that the module has indeed been found is very much simpler, and can even sometimes be done by hand.

To conclude, we indicate the results of these calculations in a few not untypical cases. The "unstable deformations" listed below constitute a parametrisation of the instability locus. In each case, we tabulate the (multi-)strata of $\mathcal{B}$ presented by points on the locus. The condition that multi-transversality to civilised submanifolds holds everywhere on $S \times B \cap V(F)$ translates in these examples to the codimensions of these strata being strictly less than that of the stratum under consideration : when this fails, no tame retraction can exist, and the corresponding f_t are $ST-$ distinct from the rest.

E_{16} ("the Pham example") $\mu = 16$ weights (3,1;9)
Normal form $x^3 + pxy^6 + qy^9$ $(4p^3 + 27q^2 \neq 0)$.
Unfold by $\{y^i \mid 1 \leq i \leq 7\}$, $\{xy^i \mid 0 \leq i \leq 7\}$ $(q \neq 0)$ or $\{y^i \mid 1 \leq i \leq 10\}, \{xy^i \mid 0 \leq i \leq 4\}$ $(p \neq 0)$; seek to trivialise over xy^6, xy^7 resp y^9, y^{10}, *so* $n = 15, p = 14, k = 2$.
Unstable deformations are :
(1) $x^3 + px(y+2t)^2(y-t)^4 + q(y+2t)^3(y-t)^6$
with D_4 at $(0, -2t)$, E_{10} at $(0,t)$; codimensions sum to 13 ;
(2) $x^3 + (y + (9-i)t)^i (y - it)^{9-i} (1 \leq i \leq 4)$ $(if \quad p = 0)$
to E_{14}, $E_{12}.A_2$, $E_{10}.D_4$, $E_8.E_6$ for $i = 1,2,3,4$; with codimensions 13, 13, 13, 14 ;
(3) $x^3 + x(y + (6-i)t)^i (y - it)^{6-i} (1 \leq i \leq 3)$ (if $q = 0$)
to $E_{13}.A_1$, $E_{10}.D_4$, $E_7.E_7$ for $i = 1,2,3$; with codimensions 13, 13, 14.
The two cases with codimension 14 obstruct tame retractions ; the cases $p = 0$, $q = 0$ are ST-distinct from the rest.

$J^2_{(1)}$ $\mu = 15$ weights 13, 8, 3, 16, 21
Normal form $< xz + y^2, xy + z^7 >$.
Unfold by $\varepsilon_2(z, z^2, y, z^3, yz, z^4, x, yz^2, z^5, yz^3, yz^4, yz^5)$;
seek to trivialise over $\varepsilon_2 yz^5$, so $n = 15, p = 14, k = 1$.
The unstable deformations is

$$< xz + y^2 + \tfrac{3}{4}t^{16}, \quad x(y - \tfrac{3}{2}t^8) + z^{-1}\{(z+t^3)^8 - t^{24}\} >,$$

with an E_{14} singularity at $(t^{13}, \frac{1}{2}t^8, -t^3)$. This has codimension 13, so we have a tame retraction.

HD_{14} $\mu = 14$ weights 9, 6, 5; 15, 18
Normal form $< xy + z^3, x^2 + y^3 >$.
Unfold by $\varepsilon_1(z)$, $\varepsilon_2(z, y, x, z^2, yz, y^2, xz, z^3, yz^2, y^2z, z^4)$;
seek to trivialise over $\varepsilon_2 z^4$, so $n = 14, p = 13, k = 1$.
There are no unstable deformations, so there is a tame retraction.

$J(3,4)$ $\delta = 14$ $\mathcal{K}$-codimension $= 16$ weights $3,1;6,7$
Normal form $< x^2 \pm y^6, xy^4 + vy^7 >$.
Unfold by $\varepsilon_1\{y^i \mid 1 \leq i \leq 4\}$ $\varepsilon_2\{y^i \mid 1 \leq i \leq 7\}$, $\varepsilon_2\{xy^i \mid 0 \leq i \leq 3\}$; seek to trivialise over $\varepsilon_2 y^7$, so $n = p = 16, k = 1$.
The unstable deformations are
(1) $< x^2 \pm (y+2t)^2(y-t)^4, x(y+2t)(y-t)^3 + v(y+2t)^2(y-t)^5 >$,
with a $B_{2,2}$ at $(0,-2t)$ and $D_{3,5}$ at $(0,t)$;
(2) $< x^2 \pm (y^2-t^2)^3, x(y^2-t^2)^2 >$ (if $v = 0$),
with C_2 singularities at $(0, \pm t)$;
(3) $< x^2 \pm (y+t)^5(y-5t), \quad x(y+t)^4 >$ (if $v = 0$),
with an $I_{2,4,8}$ at $(0,-t)$ and $(0,5t)$ also in $f^{-1}(0)$.
Here we have a civilised stratum $I(2,4)$ consisting of the $\mathcal{K}$-classes $I_{2,4,8}$ of $< x^2 + y^5, xy^4 >$ and $I_{2,4,7}$ of $< x^2 + y^5, xy^4 + y^7 >$.
The respective codimensions of these strata are $4 + 11 = 15$, $8 + 8 = 16$, 15; the second of these thus obstructs tame retractions.

FC_0 $\delta = 9$ cod $= 10$ weights $1,1;3,3$
Normal form $< x^3 - 3xy^2, -3x^2y + y^3 >$.
Unfold by $\varepsilon_1(x, y, xy, y^2)$, $\varepsilon_2(x, y, x^2, xy, x^2y, xy^2)$; seek to trivialise over $\varepsilon_2(x^2y, xy^2)$, so $n = p = 12, k = 2$. There are no real unstable deformations, so there is a tame retraction.

This last case is of interest because it is equivalent over $\mathbf{C}$ to FC_1, with normal form $< x^3, y^3 >$. This germ is ST-distinct from germs with normal form

$$f_\rho = < x^3 + \rho xy^2, \quad \rho x^2y + y^3 > \quad \text{for} \quad \rho \neq 0, 3.$$

Thus FC_1 defines a stratum od $\mathcal{B}$, while FC_0 belongs to the stratum containing all f_ρ with $\rho < -1$. So the stratification $\mathcal{B}$ is not the real part of a complex algebraic stratification of jet space.

References

Bekka, K., 1988, *Sur les propriétés topologiques et métriques des espaces stratifiés*, doctoral thesis, Orsay.

Bruce, J.W., 1980, A stratification of the space of cubic surfaces, Math. Proc. Camb. Phil. Soc. 87, 427-441.

Damon, J., 1979, Topological stability in the nice dimensions, Topology 18 129-142.

Damon, J., 1980, Finite determinacy and toplogical triviality I, Invent. Math. 62 299-324.

Du Plessis, A.A., and C.T.C. Wall, 1989, On C^1-stability and $\mathcal{A}^{(1)}$-determinacy, Publ. Math. IHES 70, 5-46.

GWPL C.G. Gibson, K. Wirthmüller, A.A. Du Plessis and E.J.N. Looijenga, *"Topological stability of smooth mappings"*, Springer Lecture Notes in Math. 552 (1976).

Looijenga, E.J.N., 1977, On the semi-universal deformation of a simple-elliptic singularity I : unimodularity, Topology 16 257-262.

Mather, J.N. Stability of C^∞-mappings.

M1 The division theorem, Ann. of Math. 87 (1968) 89-104.

M2 Infinitesimal stability implies stability, Ann. of Math. 89 (1969) 254-291.

M3 Finitely determined map-germs, Publ. Math. IHES 35 (1969) 127-156.

M4 Classification of stable germs by **R**-algebras, Publ. Math. IHES 37 (1970) 223-248.

M5 Transversality, Advances in Math. 4 (1970) 301-335.

M6 The nice dimensions, Springer lecture notes in math. 192 (1971) 207-253.

Mather, J.N., 1970, Notes on topological stability, preprint, Harvard.

Mather, J.N., 1973, Stratifications and mappings, pp. 195-232 in *"Proc. conference on dynamical systems"* (ed. M.M. Peixoto) Academic Press.

Mather, J.N., 1976, How to stratify manifolds and jet spaces, Springer lecture notes in math. 535 128-176.

May, R.D., 1973, *Transversality properties of topologically stable mappings* Ph. D. thesis, Harvard.

May, R.D., 1974, Stability and transversality, Bull. Amer. Math. Soc. 80 85-89.

Thom, R., 1955, Les singularités des applications différentiables, Ann. Inst. Fourier 6 (1955-1956) 43-87. (see also Sém. Bourbaki 134).

Thom, R., 1964, Local topological properties of differentiable mappings, pp. 191-202 in *"Differential Analysis"*, Oxford University Press.

Wall, C.T.C., 1980, The first canonical stratum, Jour. London Math. Soc. 21 419-433.

Wirthmüller, K., 1979, *Universell topologisch trivialen deformationen*, Ph. D. thesis, Regensburg.

Boundary Fronts and Caustics and their Metamorphoses

INNA SCHERBACK

Program Systems Institute of Russian Academy of Sciences

TABLE OF CONTENT

Abstract In this paper the subsequent wavefronts and caustics generated by an initial wavefront which is a surface with boundary are studied. It appears that such boundary fronts and caustics are connected with boundary singularities (i.e. singularities of functions on a manifold with boundary) in the same way as usual fronts and caustics are connected with the usual singularities of functions.
Boundary fronts and caustics are described in terms of generating families of corresponding objects in contact and symplectic spaces. These objects are pairs of Legendrian and Lagrangian submanifolds. In two- and three-dimensional spaces, the stable singularities of boundary fronts and caustics are classified and metamorphoses of simple singularities of boundary fronts and caustics are described.

INTRODUCTION

In this paper we study the subsequent wavefronts and caustics generated by an initial wavefront which is a surface with a boundary.

Wavefronts and caustics created by radiating surfaces without boundary have been studied in many papers (for references see [4]). In particular Arnold established a connection between their singularities and singularities of functions in [2]. Singularities of wavefronts and caustics in spaces of dimension < 10 were classified in [9]. The papers [1], [10] are devoted to an investigation of bifurcations (or metamorphoses) in families of wavefronts and caustics.

If the radiating surface has boundary, new singularities arise. The resulting wavefronts and caustics are then called *boundary fronts* and *boundary caustics.* The boundary front has two components which correspond to the surface and its boundary respectively. These components are tangent to each other. The singular points of the moving boundary front sweep out the boundary caustic which has three components. Two of these are the caustics of the surface and the boundary respectively. The third component consists of their points of tangency.

It follows from [5] that singularities of boundary fronts and caustics are connected with *boundary singularities.* Recall that a boundary singularity is the germ of function with critical value 0 at the critical point $O \in R^n \subset R^{n+1}$ which is considered up to diffeomorphisms preserving the boundary R^n. We illustrate this connection by the following simple example.

Example. Let us consider an ellipse E in the plane as an initial front (fig.1). The boundary is the single point $\partial E \in E$. We are interested in the part of E in a neighborhood of ∂E. Let the wave propagate inside E with unit velocity. The boundary front at the time t consists of two components: the front $\Phi(t)$ of E and the front $\Phi_\partial(t)$ of the boundary, ∂E.

In accordance with Huygens principle, $\Phi(t)$ is the envelope of the circles of radius t centered at the points of E, while $\Phi_\partial(t)$ is the circle of radius t centered at ∂E. It is well known that the only stable singularities of the usual (non-boundary) fronts in the plane are cusps (singularities of type A_2) and transversal self-intersection points. In the boundary case a new singularity is added. It is just the point of tangency of $\Phi(t)$ and $\Phi_\partial(t)$. Such points lie in the normal B_2 to E at ∂E. For almost all t, curve $\Phi(t)$ is smooth at the point of line B_2. In a neighborhood of this point the boundary front is two smooth curves which are tangent to each other.

To see the connection with boundary singulartities theory, consider the squared distance $F(x, \lambda_1, \lambda_2)$ between a point $x \in E$ and a point (λ_1, λ_2) in the plane. F is a family of functions on manifold E with boundary ∂E, and λ_1, λ_2 are the parameters of this family. For generic $(\lambda_1^0, \lambda_2^0) \in B_2$ the corresponding squared distance function has non-degenerate critical point at ∂E. So this function has a boundary singularity of type B_2 at ∂E (i.e. is equivalent to germ z^2 at $z = 0$). This just means that corresponding $\Phi(t)$ is smooth at $(\lambda_1^0, \lambda_2^0)$ (and is the reason of our notation). Family F is a versal deformation of this function. The boundary front in a neighborhood of $(\lambda_1^0, \lambda_2^0)$ is the bifurcation set of family $F(x, \lambda_1, \lambda_2) - t_0^2$, where t_0 is the distance between ∂E and $(\lambda_1^0, \lambda_2^0)$. It is diffeomorphic to the "standard" bifurcation set $\Sigma(B_2)$ of the boundary singularity B_2 (fig.2). It appears that it is the only new stable singularity of boundary fronts in the plane.

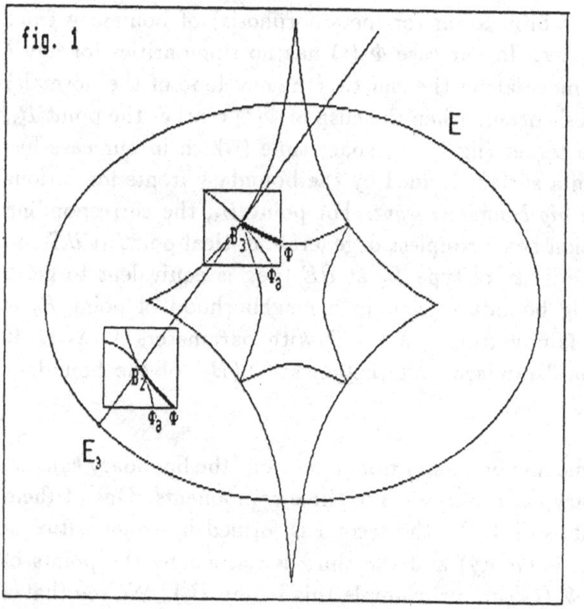

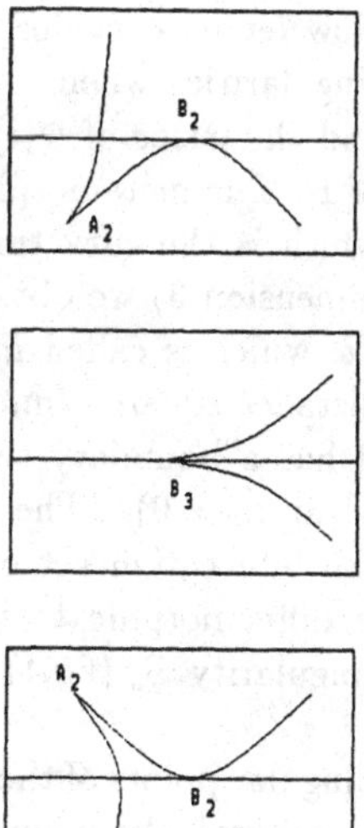

fig.3

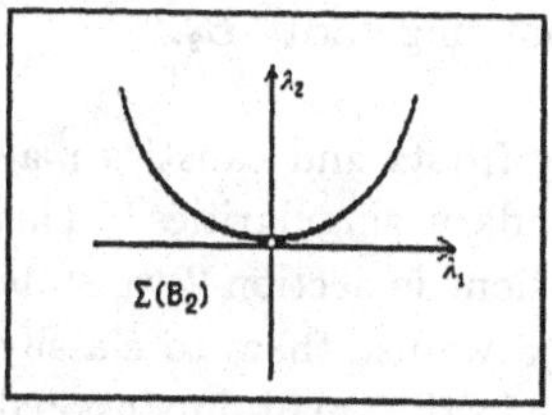

fig. 2

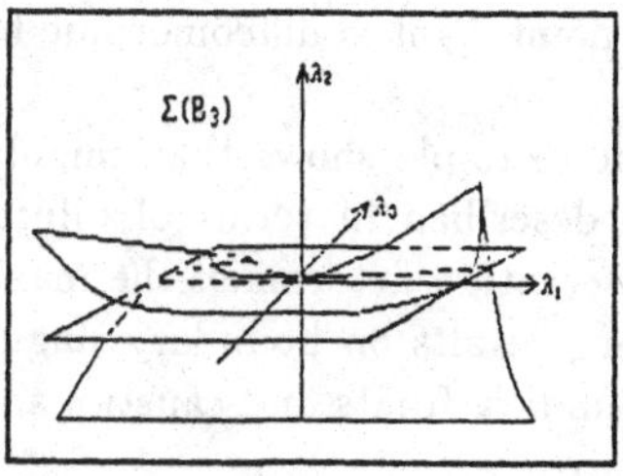

fig. 4

Now let us consider the bifurcation (or metamorphosis) of boundary front singularities when t varies. In our case $\Phi_\partial(t)$ has no singularities for any t and the cusps of $\Phi(t)$ move along the caustic (the envelope of the normals) of E. The metamorphosis occurs when the cusp of $\Phi(t)$ reaches the point B_3, which is the curvature center (fig.3). In space-time (which in our case has dimension 3) we obtain a surface formed by the boundary fronts for various t's, which is called *the big boundary front.* For point B_3, the corresponding distance squared function has a simplest degenerate critical point at ∂E , so it has a boundary singularity of type B_3 at ∂E (i.e. is equivalent to germ z^3 at $z = 0$). The big boundary front in a neighborhood of point B_3 is the bifurcation set of family $F(x, \lambda_1, \lambda_2) - t^2$ with parameters λ_1, λ_2, t. It is diffeomorphic to the "standard" bifurcation set $\Sigma(B_3)$ of the boundary singularity B_3 (fig.4).

Singular points of the moving boundary front sweep out the boundary caustic. In general, the boundary caustic consists of three components. One of them is formed by singularities of $\Phi(t)$, the second is formed by singularities of $\Phi_\partial(t)$ (in our case this is empty) and the third is formed by the points of tangency of $\Phi(t)$ and $\Phi_\partial(t)$ (in our example this is line B_2). We see that a new stable singularity appears for boundary caustics too. In a neighborhood of point B_3 it is diffeomorphic to the boundary front singularity B_2.

The example shows that singularities of boundary fronts and caustics may be described in terms of bifurcation sets of boundary singularities. This observation determines the main line of our exposition: in section 2 we state some results on boundary singularities, in section 3 we use them to classify boundary fronts and caustics and their bifurcations in the small-dimensional spaces.

2: BOUNDARY SINGULARITIES

Consider germs of functions in a neighborhood of the origin $O \in R^n \subset R^{n+1}$ (or $O \in C^n \subset C^{n+1}$). *The boundary singularity* is the class of stably equivalent germs (in other words the germ up to diffeomorphisms in R^{n+1} (or C^n) preserving the boundary R^n (or C^n), and adding nondegenerate quadratic forms in new variables).

In [3], [6] simple, unimodal and bimodal boundary singularities were classified.

There is another approach to the classification of boundary singularities based on the idea of *decomposition.* This approach is convenient for applications to boundary fronts and caustics, so we will consider it in more detail.

For a given boundary singularity X one can obtain two ordinary singularities X_1 and X_2. The singularity X_1 is the original germ considered up to diffeomorphisms of R^{n+1} (or C^{n+1}) preserving the origin O. On the other hand

X_2 is the singularity of the restriction of the original germ to the boundary considered up to diffeomorphisms of the boundary. The pair (X_1, X_2) is called *the decomposition* of X. The decomposition is called *simple* if both X_1 and X_2 are simple singularities (i.e. A_k, D_k or E_k). For instance, the simple boundary singularities have the following (simple) decompositions:

$$B_k - (A_{k-1}, A_1), \quad C_k - (A_1, A_{k-1}), \quad F_4 - (A_2, A_2).$$

In [8] all boundary singularities with simple decomposition are classified.

Let $(x, y) = (x, y_1, \cdots, y_n)$ be the coordinates on R^{n+1} (or C^{n+1}) with boundary $x = 0$. Let $f(x, y)$ be a function with critical value 0 in the critical point O. *Lagrangian transformation* of f is the function $f^*(z, x, y) = zx + f(x, y)$ on R^{n+2} (or C^{n+2}) with coordinates (z, x, y) and the boundary $z = 0$. It is easy to prove that f^{**} and f are stably equivalent as functions on the manifold with boundary [7]. Moreover the Lagrangian transformation induces an involution on the set of boundary singularities: $X \to X^*, X^{**} = X$. If (X_1, X_2) is the decomposition of X then (X_2, X_1) is the decomposition of X^*. We say that X and X^* are *dual*. In particular, the simple boundary singularities B_k and C_k are dual, and F_4 is self-dual (as well as the root systems of the same types).

It is easy to see that the bifurcation set of X consists of two irreducible components which are the bifurcation sets of X_1 and X_2, and the multiplicity of X is the sum of those of X_1 and X_2. The bifurcation sets and the multiplicities of dual boundary singularities coincide.

3: BOUNDARY FRONTS AND CAUSTICS

The machinery for the investigation of fronts and caustics is that of contact and symplectic geometries (see [4 ch.3]).

The boundary fronts (caustics) are connected with objects of contact (symplectic) space which we call *Lagrangian (Legendrian) pairs.*

3.1. Generating families of Lagrangian and Legendrian pairs.

Definition 1. An ordered pair (L, L_0) of Lagrangian (Legendrian) submanifolds L and L_0 in the symplectic (contact) space is called *a Lagrangian (Legendrian) boundary surface* if the intersection of L and L_0 is a smooth hypersurface in each of them and these submanifolds are not tangent at intersection points.

Definition 2. *A Lagrangian (Legendrian) pair* is the composite

$$(L, L_0) \xrightarrow{i} M^{2n} \xrightarrow{\pi} B^n$$

$$((L, L_0) \xrightarrow{i} M^{2n+1} \xrightarrow{\pi} B^{n+1})$$

where i is embedding of (L, L_0) in the Lagrangian (Legendrian) fibration

$$M^{2n} \xrightarrow{\pi} B^n \quad (M^{2n+1} \xrightarrow{\pi} B^{n+1}).$$

It is well known that any Lagrangian (Legendrian) map

$$L \xrightarrow{i} M^{2n} \xrightarrow{\pi} B^n \qquad (L \xrightarrow{i} M^{2n+1} \xrightarrow{\pi} B^{n+1})$$

may be given locally by the germ at O of a generating family $F(y, \lambda)$ of functions of $y = (y_1, \cdots, y_k)$ with parameters $\lambda = (\lambda_1, \cdots, \lambda_n)$ (resp. a family of surfaces $V(\lambda_0, \lambda) = \{y \mid F(y, \lambda) = \lambda_0\}$) in such a way that $L = \{(\lambda, \kappa) \mid \exists y : F_y = 0, \kappa = F_\lambda\}$ (resp. $L = \{(\lambda_0, \lambda, \kappa) \mid \exists y : \lambda_0 = F, F_y = 0, \kappa = F_\lambda\}$) is the embedding of Lagrangian (Legendrian) submanifold of symplectic (contact) space with canonical coordinates $\lambda, \kappa = (\kappa_1, \cdots, \kappa_n)$ (resp. $(\lambda_0, \lambda, \kappa)$) in the Lagrangian (Legendrian) fibration $\pi(\lambda, \kappa) = \lambda$ (resp. $\pi(\lambda_0, \lambda, \kappa) = (\lambda_0, \lambda)$) [4].

It follows from the results of [5] that any Lagrangian (reesp. Legendrian) pair may be given locally by a generating family $F(x, y, \lambda)$ (resp. $V(\lambda_0, \lambda) = \{(x, y) \mid F(x, y, \lambda) = \lambda_0\}$) in such a way that $F(x, y, \lambda)$ (resp. $V(\lambda_0, \lambda) = \{(x, y) \mid F(x, y, \lambda) = \lambda_0\}$) is the generating family for L and $F(0, y, \lambda)$ (resp. $V(\lambda_0, \lambda) = \{y \mid F(0, y, \lambda) = \lambda_0\}$) is the generating family for L_0. Moreover Lagrangian (Legendrian) equivalent pairs are in correspondence with R^+-stable (resp. fibre stable) equivalent generating families, and the stable Lagrangian (Legendrian) pairs are in correspondence with ((R^+-) versal unfoldings of boundary singularities. One can find the definitions of all these equivalences in [4].

Thus, classification of boundary singularities leads to a classification of stable Lagrangian and Legendrian pairs up to corresponding equivalences. We say that the Lagrangian (Legendrian) pair has a simple decomposition if the Lagrangian (Legendrian) maps forming the pair correspond to simple singularities of functions. The classification of boundary singularities with a simple decomposition [8] leads to the classification of Lagrangian (Legendrian) pairs with simple decomposition. It appears that all stable germs of Lagrangian (Legendrian) pairs in the space of dimension < 11 (resp. < 12) have simple decompositions.

There is a natural involution on the set of Lagrangian (Legendrian) pairs which transposes the Lagrangian (Legendrian) maps forming the pair. Obviously the dual pairs (L, L_0) and (L_0, L) are stable or unstable simultaneously. The generating families of stable dual pairs are versal deformations of dual boundary singularities. It was in this way that the duality of boundary singularities was discovered initially.

3.2. Description of boundary fronts and caustics in terms of generating families.

The boundary fronts are images of the Legendrian pairs in the same way as the usual fronts are images of the Legendrian maps. Similarly, *the boundary caustic* is the set of critical values of the Lagrangian pair. The boundary front consists of two components which are the fronts of the Legendrian maps forming the pair. In general, the boundary caustic consists of three components. Two of these are the caustics of the Lagrangian maps which form the pair. They may be empty. The third one - the image of intersection - is never empty.

The bifurcation set of the family of hypersurfaces

$$V(\lambda_0, \lambda) = \{(x, y) \mid F(x, y, \lambda) = \lambda_0\}$$

consists of those (λ_0, λ) such that $V(\lambda_0, \lambda)$ is non-smooth or non-transversal to $x = 0$.

Theorem 1. *The boundary front is the bifurcation set of the generating family of its Legendrian pair.*

If a boundary singularity has decomposition (X_1, X_2), then the bifurcation sets of X_1 and X_2 are the two components of the corresponding boundary front. From the classification of boundary singularities with a simple decomposition it follows that in any space of dimersion < 12 each component of a stable boundary front is the bifurcation set of a simple singularity.

A function on a manifold with boundary is called *a Morse function* if all critical points of the function and of its restriction to the boundary are non-degenerate, and the function has no critical points on the boundary. Consider the family of functions on a manifold with boundary vanishing at the point O of boundary. *The caustic* of this family is formed by those parameter values which determine the non-Morse functions of the family.

Theorem 2. *The boundary caustic is the caustic of the generating family of its Lagrangian pair.*

3.3. Classification of stable singularities of boundary fronts and caustics in two - and three - dimensional spaces.

Obviously, the lists of stable singularities of boundary fronts and caustics contain the lists of all stable singularities of non-boundary fronts and caustics (which can be found, for example, in [4]). Here we consider only those points of boundary fronts and caustics which are images of points of intersection $L \cap L_0$. Such points we call *boundary points.* According to theorems 1 and 2 in a neighborhood of a boundary point the boundary front (caustic) is defined by a generating family which is the unfolding of a boundary singularity. We say that *the boundary front (caustic) has singularity* X at the boundary point if the corresponding generating family is a versal unfolding of a boundary singularity X.

Below we give the classification in the spaces R^2 and R^3. To obtain C-classification one must replace all $\pm$'s by $+$'s in the formulas for the generating families.

Everywhere the boundary is $x = 0$. Coordinates in the plane are (λ_1, λ_2) and those in space are $(\lambda_1, \lambda_2, \lambda_3)$.

Proposition 1. *The stable singularities of boundary fronts at the boundary points are as follows:*
$B_2 = C_2$ *(the generating family is* $\pm x^2 \pm y^2 + \lambda_1 x + \lambda_2$ *in* R^2*;*
$B_2 = C_2$, $B_3(x^3 \pm y^2 + \lambda_1 x^2 + \lambda_2 x + \lambda_3)$, $C_3(xy + y^3 + \lambda_1 y^2 + \lambda_2 y + \lambda_3)$ *in* R^3*;*
$B_k, C_k, F_4 (k = 2, 3, 4)$ *in* R^4.

Remark. We will use the classification in R^4 to describe the bifurcations of boundary fronts in R^3.

Proposition 2. *The stable singularities of boundary caustics at boundary points are as follows:*
$B_2 = C_2$ *(the generating family is* $\pm x^2 \pm y^2 + \lambda_1 x$*),*
$B_3(x^3 \pm y^2 + \lambda_1 x^2 + \lambda_2 x)$, $C_3(xy + y^3 + \lambda_1 y^2 + \lambda_2 y)$ *in* R^2*;*
$B_2 = C_2, B_3, C_3$, $B_4(\pm x^4 \pm y^2 + \lambda_1 x^3 + \lambda_2 x^2 + \lambda_3 x)$,
$C_4(xy \pm y^4 + \lambda_1 y^3 + \lambda_2 y^2 + \lambda_3 y)$, $F_4(\pm x^2 + y^3 + \lambda_1 xy + \lambda_2 x + \lambda_3 y$ *in* R^3.

3.4. Metamorphoses of boundary fronts and caustics.

Let us consider a general one-parameter family of boundary fronts (caustics) in n-dimensional space. As in the non-boundary case [1], the union of all boundary fronts (caustics) of the family is a boundary front (caustic) in $(n+1)$-dimensional space, which is called *the big boundary front (caustic)*. The investigation of metamorphoses in general one-parameter families of boundary fronts (caustics) is reduced to the problem of transforming the *time function* t (i.e. a function without critical points) into normal form in a neighborhood of a point of the big boundary front (caustic) by means of diffeomorphisms preserving the big boundary front (caustic). This problem was solved by Arnol'd [1] when the big front corresponds to a simple boundary or non-boundary singularity. In spaces of dimension < 5 a general boundary front only has such singularities. It enables one to describe the bifurcations in general one-parameter families of boundary fronts in the plane and in three-dimensional space. Then the big boundary fronts lie in spaces of dimensions 3 and 4 respectively. To obtain the bifurcation one must intersect the big boundary front (caustic) with hypersurfaces $t = \text{const} < 0, t = 0, t = \text{const} > 0$. So we describe the bifurcations of boundary fronts by the generating families of big boundary fronts and normal forms for the function t. The bifurcations corresponding to non-boundary points of big boundary fronts may be found, for example, in [10].

Proposition 3. *The list of all bifurcations in general one-parameter families of boundary fronts at boundary points of the big boundary fronts is as folllows:*
$B_2 = C_2(\pm x^2 \pm y^2 + \lambda_1 x + \lambda_2 = 0, t = \lambda_1 \pm \tau^2)$,
$B_3(x^3 \pm y^2 + \lambda_1 x^2 + \lambda_2 x + \lambda_3 = 0, t = \lambda_1)$,
$C_3(xy + y^3 + \lambda_1 y^2 + \lambda_2 y + \lambda_3 = 0, t = \lambda_1)$ *in* R^2;
$B_2(t = \lambda_1 \pm \tau_1^2 \pm \tau_2^2)$, $B_3(t = \lambda_1 \pm \tau^2)$, $C_3(t = \lambda_1 \pm \tau^2)$,
$B_4(\pm x^4 \pm y^2 + \lambda_1 x^3 + \lambda_2 x^2 + \lambda_3 x + \lambda_4 = 0, t = \lambda_1)$,
$C_4(xy \pm y^4 + \lambda_1 y^3 + \lambda_2 y^2 + \lambda_3 y + \lambda_4 = 0, t = \lambda_1)$,
$F_4(\pm x^2 + y^3 + \lambda_1 xy + \lambda_2 x + \lambda_3 y + \lambda_4 = 0, t = \lambda_1)$ *in* R^3
(everywhere the boundary is $x = 0$*, the* τ*'s are coordinates on which the generating family does not depend).*

Remark. Some metamorphoses of one-parameter families of boundary fronts can't be realized as bifurcations of a moving boundary front. For example, the metamorphosis B_2 changes the topology of the Legendrian pair and so does not correspond to any moving boundary front.

The general boundary caustic in four-dimensional space has non-simple singularities. Corresponding metamorphoses are unknown. Therefore we are forced to restrict ourselves to the metamorphoses of boundary caustics related to simple boundary singularities. They are described by the following two propositions.

Proposition BC. *The boundary front* B_k *(* C_k *) is diffeomorphic to the boundary caustic* B_{k+1} *(* C_{k+1} *).*

A simple calculation shows the boundary caustic F_4 is given by the equation

$$(\lambda_1^4 \pm 24\lambda_1\lambda_2 - 48\lambda_3)(\lambda_1^2\lambda_3 + 3\lambda_2^2)\lambda_3 = 0$$

in the space with coordinates $\lambda_1, \lambda_2, \lambda_3$.

Proposition F. *There exists a diffeomorphism of* $(\lambda_1, \lambda_2, \lambda_3)$*-space preserving the boundary caustic* F_4 *and transforming the time function* t *into the form*

$$t = \lambda_1 + a\lambda_2 + b\lambda_3, \quad a \neq 0, b \neq 0$$

The metamorphoses F_4 are presented in fig. 5a and 5b.

PERESTROIKAS of CAUSTIC F_4

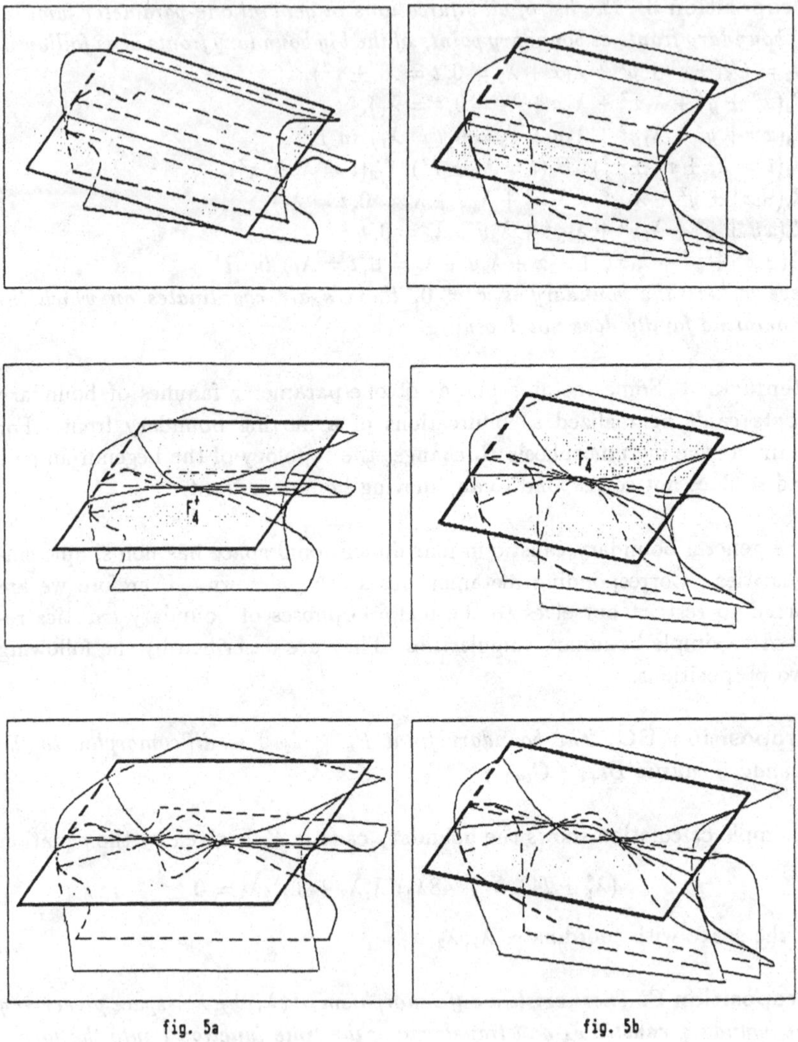

fig. 5a

fig. 5b

References

[1] Arnol'd V.I. Wave Fronts Evolution and Equivariant Morse Lemma. *Commentarii Pure Appl. Math.*29 (1976) 557 - 582.

[2] Arnol'd V.I. Normal forms of functions near degenerate critical points, Weyl groups A_k, D_k, E_k and Lagrangian singularities. *Functional Anal. Appl.* 6:4 (1972), p.3 - 25.

[3] Arnol'd V.I. Critical points of functions on the manifold with boundary, simple Lie groups B_k, C_k, F_4 and singularities of evolutes. *Russian Math. Syrveys* 33:5 (1978), p.91 -105.

[4] Arnol'd V.I., S.M. Gusein-Zade and A.N. Varchenko. *Singularities of differentiable maps*, I (in Russian). Nauka, Moscow, 1982. English translation: Birkhauser, Boston, 1985.

[5] Nguyen huu Duc, Nguyen tien Dai. Stabilite de l'interaction geometrique entre deux composantes holonomes simples. *C. R. Acad. Sci. Paris,* Ser. A 291 (1980), p.113 - 116.

[6] Matov V.I. Unimodal and bimodal germs of functions on the manifold with boundary. *Trudy Semin. im. Petrovskogo* 7 (1981), p.174 - 189.

[7] Scherbak I.G. Duality of boundary singularities. *Russian Math. Syrveys* 39:2 (1984), p.207 - 208.

[8] Scherbak I.G. Boundary singularities with simple decomposition. *Trudy Semin. im. I. G. Petrovskogo* 15 (1990).

[9] Zakalukin V.M. On Lagrangian and Legendre singularities. *Functional Anal. Appl.* 10:1 (1976), p.26 - 36.

[10] Zakalukin V.M. Metamorphoses of fronts and caustics depending on parameter, versality of maps. In *Itogi Nauki Tecn., Ser. Sovrem. Probl. Mat., VINITI, Moscow,* 22 (1983), p.56 - 93.

Inna Scherback
Program Systems Institute
of Russian Academy of Sciences
152140 Pereslavl-Zalessky, Russia

Quid des stratifications canoniques

RENÉ THOM

IHES Paris

Soit donné un compact K de $\mathbf{R}^n$ qui est un stratifié de Whitney, dont les strates (en nombre fini) sont des variétés C^∞. Il est probablement démontrable (1)*, dans l'état actuel de la littérature, que pour presque toute projection linéaire $\hat{q} : \mathbf{R}^n \to \mathbf{R}^q$, $q < n$, l'image $B = \hat{q}(K)$ est elle-même un stratifié de Whitney dont les strates sont C^∞. De plus, l'application $p : K \to B$, restriction de $\hat{q}$ à K, est elle-même stratifiée : il existe une substratification (S) de (K) et une stratification (S') de B telle que l'application $\hat{q}$ restreinte à toute strate $X \in (S)$ de K est une submersion sur la strate-image $Y = p(X) \in (S')$ de B. Le corang $crg(X) = \dim\ker(p)$ sur une strate (X) est constant sur X, et pour toute relation d'incidence $X < Z$ dans (S), on peut obtenir $crg(X) \leq crg(Z)$ (c'est la condition de *non-éclatement*, vérifiée dès que la propriété $A(p)$ *est satisfaite pour tout couple de strates* $X < Z$ *de* (S)). Si on peut s'assurer que les directions $(\hat{q})$ qui satisfont à ces conditions forment un ouvert dense dans la grassmannienne, alors (en vertu du second lemme d'isotopie), le type topologique d'un morphisme stratifié $p : K \to B$ est stable par rapport aux petites variations de toute "bonne" projection p. Une telle condition est satisfaite si (K) est semi-analytique (et probablement sous-analytique ?). Mais dans le cas où K est semi-algébrique, le théorème classique de Tarski-Seidenberg donne plus : quel que soit A semi-algébrique dans $\mathbf{R}^n$, même non compact, l'image $P(A)$ est semi-algébrique pour *toute* projection p. Le principe de la démonstration reste le même : stratifier la source par le corang de p sur toutes les strates de A (faire appel à tous les symboles de Boardman et leurs généralisations nécessaires), prendre leurs images dans le but (ici Tarski-Seidenberg évite d'avoir à utiliser le théorème de Varchencko pour stabiliser les strates images), substratifier pour satisfaire à Whitney partout, et appliquer le même traitement au saturé $p^{-1}(p(\Sigma))$, en considérant toute strate définie dans l'ensemble critique Σ. Comme tous les ensembles à considérer sur la source sont semi-algébriques, il en va de même pour les images dans le but, ce qui assure en particulier le caractère localement fini des strates potentielles dans le but.

Les fibrés obtenus via le théorème conjectural énoncé plus haut sont ceux que

* NDE Pour un contre-exemple, voir : Michał Kwieciński et Laurent Noirel *"Sur une question de René Thom à propos des projections d'une stratification de Whitney"*, à paraître aux CRAS Paris

dans mon article du B.A.M.S. (1969), j'appelais les morphismes faiblement stratifiés (weakly stratified). Ils peuvent ne pas remplir la condition $A(p)$ et présenter de l'éclatement. Mais dans toute statification d'un tel morphisme de projection, $p : K \to B$, une propriété subsiste : L'image réciproque $p^{-1}(Y)$ d'une strate du but est un fibré localement trivial.

Esquissons ici une démonstration de ce fait : Soit Y une strate de la stratification S' du but B, y un point (arbitraire) de Y. La fibre $p^{-1}(y)$ dans K est un espace stratifié compact, section transversale de la stratification (S) de K par la variété lisse (ouverte) $\hat{q}^{-1}(y)$, où $\hat{q}$ est la projection linéaire initiale $\hat{q} : R^n \to R^q$. En un point z quelconque de Y, on considère une petite boule $b(z)$ de centre z, ayant pour bord une petite sphère δW de dimension $d-1$, où d est la dimension de Y. On peut considérer cette boule W comme fibrée en diamètres par les trajectoires d'un champ radial local. Chacune de ces trajectoires radiales t, paramétrée par s, définit une isotopie de la contre-image $p_t^{-1}(m(s))$. Or le 2e lemme d'isotopie marche aussi avec paramètres (ici l'espace des paramètres est la sphère δW, qui est compacte), et ceci définit une trivialisation de l'application $p \mid p^{-1}(Y) \to Y$ sur un voisinage du point z de Y.

Ceci ne suffit pas à assurer que dans un projection à la Tarski-Seidenberg d'un semi-algébrique A sur B, 1) la projection p puisse être stratifiée, 2) La contre-image d'une strate du but est elle-même un fibré localement trivial sur l'image. Ils n'en sont pas moins vrais.

Le point 1) résulte du fait que la non-compacité d'un semi-algébrique peut être contrôlée via une compactification à l'infini, où on remplace la projection linéaire par un éclatement convenable de la "variété à l'infini". Quant au point 2), il sort immédiatement du théorème précédent, l'éclatement de la variété à l'infini donnant précisément "radialement" la projection fibrée, et "sphériquement" une compactification adéquate de la fibre.

Il résulte de tout ceci qu'une projection de Tarski-Seidenberg peut être stratifiée, en sorte que la contre-image de toute strate du but est un fibré (stratifié) localement trivial. Une telle sratification étant acquise, on s'efforcera de la simplifier en éliminant du but toute strate qui peut être "effacée" sans compromettre les propriétés de stratification. S'il ne subsiste plus aucune strate effaable dans le but, ni dans la stratification associée de la source, on dira que la stratification est *minimale*. Je ne me hasarderai pas à affirmer qu'il n'existe qu'une seule stratification minimale (à l'équivalence près) quoiqu'il soit difficile de fabriquer des contre-exemples (en théorie C^∞, l'effacement d'une strate soulève des problèmes liés aux structures différentiables exotiques de la sphère...). Mais, de toute manière, la procédure de stratification du morphisme projection peut être décrite selon un "protocole" canonique (1), en sorte qu'on peut parler de la stratification canonique d'une projection de Tarski-Seidenberg. Cette stratification est alors minimale, parce que toutes les strates successivement introduites sont nécessaires et ne peuvent être effacées, si l'on sait qu'au départ les stratifications données de la source et du but sont elles-mêmes minimales.

Ces considérations peuvent s'étendre à tout morphisme stratifié propre p d'un ensemble stratifié A sur un stratifié B (A, B stratifiés C^∞ de Whitney). Définissons sur B une certaine propriété (E) de sous-variétés C^∞ connexes de dimension $k \leq \dim B$: Deux points quelconques P, Q de E peuvent être reliés par un chemin (C), tel que la contre-image de C dans A définisse une famille $E(s)$ de stratifiés dont la variation lorsque s parcourt c peut être plongé dans une isotopie (stratifiée) dans A. Dans nos fibrés *faiblement stratifiés* cette propriété est vraie de toute strate de B. Si aucune strate ne peut être effacée de B (ni de A), on dira que la stratification est minimale. Il peut arriver – pour des morphismes suffisamment explicites géométriquement – , qu'on puisse démontrer que deux stratifications minimales sont nécessairement isomorphes selon des isomorphies globales de A et B commutant avec p. On dira en ce cas que la stratification du morphisme $p : A \to B$ est *canonique*.

Application : Une stratification canonique de l'espace des jets ?

Soit G une groupe algébrique sur le corps $\mathbf{R}$ des réels. Par définition, c'est une variété algébrique (sur $\mathbf{R}$), munie d'une loi de groupe définie par un morphisme M (multiplication) $\mathbf{R}$-algébrique de $G \times G$ sur G. Considérons le cas où G agit algébriquement dans une variété $\mathbf{R}$-algébrique A. Alors (A) est feuilletée (avec singularités) par les orbites de l'action multiplication à gauche de G dans A. Une orbite est la classe d'équivalence ainsi définie : deux points x, y de A sont G-équivalents, s'il existe un élément g de G tel que $y = g \cdot x$. On peut alors former *le graphe* de cette relation d'équivalence, c'est-à-dire l'espace K constitué des couples (x, y) de points de A qui appartiennent à une même orbite. Formons le produit $G \times A_1 \times A_2$ avec deux copies de A, et soit Γ l'espace défini dans ce produit par le système algébrique d'équations exprimant la relation $b = g \cdot a$, où $g \in G$, $a \in A_1$, $b_2 \in A$. Cet espace K n'est autre que l'image de Γ dans la projection canonique : $G \times A_1 \times A_2 \to A_1 \times A_2$. Il en résulte que K, projection d'un ensemble algébrique, est lui-même semi-algébrique dans $A_1 \times A_2$, est la projection sur un facteur du carré $A_1 \times A_2$ définit un morphisme de Tarski-Seidenberg de K *sur* A (tout élément de A appartient à une orbite ...) On appellera *stratification canonique* de A (en tant que G-espace) la stratification canonique de ce morphisme $K \to A$. (Le choix du facteur A_i, $i = 1$ ou 2, choisi est, par symétrie de la relation d'équivalence $b = g \cdot a$, indifférent).

Si au lieu d'opérer sur un espace $\mathbf{R}$-algébrique avec un groupe $\mathbf{R}$- algébrique, on avait pris un espace A' $\mathbf{C}$-algébrique où opère un groupe G' $\mathbf{C}$-algébrique, toutes ces constructions fonctionnent, à l'exception du fait qu'on doit remplacer Tarski-Seidenberg par son **C**-équivalent, à savoir qu'une projection linéaire d'un ensemble **C**- algébrique est un ensemble **C**-constructible. Dans la définition locale des **C**-constructibles, on doit remplacer l'inégalité réelle par l'admission en nombre fini d'ensembles différences $X - Y$, où Y est un **C**-sous-ensemble du **C**-ensemble X. Les **C**-constructibles sont de ce fait des ensembles qui se déduisent d'ensembles stratifiés réels par l'omission de certaines strates ; observer que tout **C**-ensemble, tout **C**-morphisme est aussi un **R**-ensemble, resp. un **R**-morphisme, quand on identifie **C** à $\mathbf{R}^2$. (Les strates omises sont attachées aux strates "collapsées" par la projection, comme le montre l'exemple standard de l'axe $0y$ dans la projection parallèle à $0z$ du paraboloide hyperbolique d'équation $y = xz$ dans $0xyz$). On peut se demander dans quelle mesure ces strates omises ne proviennent pas de l'existence d'une loi de groupe, comme l'homothétie sur $0z : z \to xz$ exige $x \neq 0$.

On peut faire agir un groupe G dans lui-même, soit par multiplication à gauche $x \to g \cdot x$, soit par la représentation adjointe $y \to x^{-1}yx$. La construction précédente du graphe $K(a)$ de l'action (a) conduira à la stratification d'un morphisme canonique $K(a) \to G$. Bien que la stratification d'une projection de Tarski-Seidenberg soit en principe arbitraire (on peut partir de stratifications semi-algébriques arbitraires de la source et du but), dans le cas considéré, on doit s'imposer en priorité de n'utiliser en principe que des strates G-invariantes. On observera que la translation à gauche dans G donne un résultat trivial, $K(a)$ isomorphe à G, mais il n'en va plus de même pour la représentation adjointe. Il est à espérer que les strates ainsi canoniquement définies permettront de retrouver les objets classiques de la théorie des groupes de Lie – peut-être de manière plus intelligible que par la voie algébrique qui fut celle de l'Histoire.

Nous allons appliquer les constructions précédentes à la stratification des ensembles algébriques, ainsi qu'aux espaces de jets $\mathbf{J}^k(n,p)$ de $\mathbf{R}^n$ dans $\mathbf{R}^p$.

Stratification des ensembles algébriques

Soit A un **R**-ensemble algébrique affine, A^c sa **C**-complexification. Disons que A possède une stratification cohérente, $A = \cup_i X_i$, où tous les X_i sont des **R**-ensembles, si le complexifié A^c admet une C-stratification formée des complexifications X_i^c des X_i réelles (les relations d'incidence respectives étant les complexifiées de celles liant les strates réelles correspondantes). Par exemple, un ensemble réunion de variétés linéaires possède une stratification cohérente (celle définie par les variétés et leurs intersections). Les grassmanniennes possèdent des stratifications cohérentes, celles définies par leurs cycles de Schubert. Mais je ne crois pas qu'une caractérisation des variétés cohérentes soit connue. Enonçons à ce sujet une conjecture "sauvage" : toute intersection transverse de variétés cohérentes est cohérente.

Application aux espaces de jets

On appliquera la construction décrite plus haut à l'action standard introduite par Ehresmann de $L^r(n) \times L^r(p)$ dans $J^r(n,p)$ pris sur $\mathbf{R}$ ou sur $\mathbf{C}$. On obtiendra ainsi les stratifications canoniques de $J^r(n,p)$, ainsi que des groupes d'automorphismes locaux $L^r(n)$, $L^r(p)$. Un problème majeur est de comprendre le comportement de ces stratifications par rapport à la projection canonique d'"oubli" $g(r+1,r) : J^{r+1} \to J^r$. Si on se restreint aux singularités transversalement définies en codimension finie, il n'y a pas de problème à s'assurer que les strates correspondantes s'envoient par les projections $g(k+r,r)$ strate sur strate, justifiant ainsi la hiérarchie observée dans les symboles de Boardman. Ceci est une conséquence de la théorie du déploiement alias théorie de la déformation plate universelle d'un germe d'ensemble analytique, lorsque la singularité du germe est isolée. Par contre, si la singularité n'est pas isolée, il peut y avoir des déformations plus simples ayant le même défaut, et on est alors amené à déplier ces singularités secondaires en familles, ce qui conduit à une structure peu comprise actuellement (la théorie de Siersma des singularités non isolées). Observer que même dans le cas isolé,le déploiement est de dimension finie, mais peut comporter des singularités. En utilisant les notations précédentes, on voit que le diagramme :

$$\begin{array}{ccc} K(r+1) & \longrightarrow & J^{r+1} \\ \downarrow g & & \downarrow g \\ K(r) & \longrightarrow & J^r \end{array}$$

commute. Observer que si l'on veut un groupe $L^j(n) \times L^j(p)$ agissant dans $J^r(n,p)$ quel que soit r, il faut prendre les groupes $L(n) \times L(p)$ pour $r = \infty$. Nous appellerons ce groupe produit Λ pour simplifier.

Toute strate (X) de la stratification canonique de $J^{(r \text{ ou } r+1)}$ noté $\mathbf{J}$ par la suite, est caractérisée par le fait que si on associe à un jet $x \in X$ l'ensemble $k(x)$ des jets dans $\mathbf{J}$ qui lui sont équivalents sous l'action de Λ (i.e. l'ensemble des orbites issues x), alors cet ensemble varie isotopiquement dans $\mathbf{J}$ lorsque x parcourt la strate X. La strate de x peut alors être définie comme le plus grand ouvert d'une variété invariante W selon Λ contenant x pour laquelle cette propriété d'isotopie des classes $k(y)$ reste valable quel que soit le chemin décrit par y dans W. Ceci entraîne que là projection d'une strate par toute projection g est dans une strate.

Mais, comme on suppose que les jets sont définis par des germes d'applications C^∞, toute strate d'ordre r est projection par g d'une strate d'ordre $r+k$ supérieur. Une telle strate n'a aucune raison d'être unique, il suffit de remplacer un représentant z du jet dans J^{r+1} par un autre représentant de la forme $z+u$, où $u \in \ker g$ pour obtenir une autre strate. On voit qu'ainsi ces stratification se comportent de manière naturelle par rapport aux projections

(g) – les stratifications se ramifient lorsqu'on remonte les flèches (g) du bas vers le haut. Si l'on opère avec le groupe Λ (les automorphismes locaux de la source et du but sont alors C^∞), on obtiendra au-dessus d'un jet $z \in J^r$, dans J^{r+k} pour k assez grand, un ouvert partout dense dans $g^{-1}(z)$ constitué de jets topologiquement stables, où toutes les strates de la source et du but dont définies transversalement (ceci est une reformulation de la stabilité des singularités C^∞ "génériques"). Après tout, le but essentiel de la théorie des stratifications est d'extraire de ces constructions plutôt laides des résultats sur la topologie des espaces et morphismes stratifiés... Il y a une morale à tirer de cette tentative de stratifier les espaces de jets. C'est qu'au fond les stratifications intéressantes sont celles qui portent sur les processus de stabilisation, c'est-à-dire sur les espaces de jets d'ordre m borné inférieurement et supérieurement $r < m < r + s$. C'est en faisant tendre s vers l'infini qu'on a, à la limite, des stratifications qui se stabilisent. Tout se passe comme si la bonne stratification – la limite projective pour $s = \infty$ – était définie sur un espace fonctionnel de dimension infinie, où elle resterait de codimension finie. Les autres strates, définies en $J^r(n,p)$ ne jouent qu'un rôle préparatoire à l'obtention des strates "stables".

Dans l'article-bilan paru récemment [Thom 89], j'ai évoqué la possibilité théorique de stratifier les espaces de jets en partant d'une stratification du diagramme de morphismes $\mathbf{R}^n \times \mathbf{R}^p \times J^r(n,p) \to \mathbf{R}^p \times J^r(n,p) \to J^r(n,p)$. Je pense actuellement que cette tentative est illusoire, en raison du fait que la projection g – en ordre fini – se comporte mal vis-à-vis de la L-équivalence.

Exemple simple : dans $J^4(2,1)$, considérons les deux fonctions $(x^2, x^2+2xy^2+y^4)$; elles sont L-équivalentes puisque la seconde est $(x.+y^2)^2$. L'image de ces deux fonctions par g_3^4 dans $J^3(2,1)$ est (x^2, x^2+2xy^2).

Or, ces deux fonctions ne peuvent être L-équivalentes si l'on s'interdit d'écrire des puissances de degré > 3 ; en effet, les courbes définies par ces expressions dans $\mathbf{R}^2(x,y)$ ne sont pas – à l'origine – topologiquement équivalences.

Si au lieu d'espaces de jets, on parlait de système de p polynômes de degré $\leq r$, la considération des stratifications naturelles des espaces de coefficients aurait peut-être quelque intérêt, en vue de détecter précisément les singularités hautement non génériques, comme celles comportant des composantes multiples ou des composantes de dimension excédentaire. Il y aurait là comme une voie opposée – duale – à celle de la transversalité : après tout, comme l'a suggéré Siersma, dans le cas des fonctions $(p = 1)$, les singularités non isolées servent de guide générateur aux familles de singularités stables de codimension finie.

Notes

(1) Cette affirmation – telle qu'elle est ici énoncée – me semble pécher par un optimisme excessif. Très probablement, les conditions (a, b) de Whitney ne suffisent pas – seules – à assurer que "presque toute projection" de K est stratifiée. Il semble indispensable de renforcer les propriétés de Whitney par une condition du type suivant :

Soient $X < Y$ un couple de strates de K, X incidente à Y, $\dim X = \xi < \dim Y = \eta$. Désignons par $N(Y)$ l'éclatement de Nash de Y dont la fibre est la Grassmannienne G des η-plans, par $Q(X)$ l'éclatement "grassmannien" de X (au sens de Kuo) dont la fibre est le cycle de Schubert de G formé des η-plans qui passent par un ξ-plan donné (tangent à X en un point $p \in X$). Soit Z l'intersection de $Q(X)$ par l'adhérence de $N(Y)$. Il faut sans doute postuler que Z est lui-même un ensemble stratifié C^∞ et que la projection canonique de Z sur X est elle-même un morphisme stratifié C^∞ dont les valeurs régulières sont partout denses. Sans doute même faut-il généraliser l'assertion pour toute chaîne de strates de K. Ce type de condition est visiblement satisfait pour les semi-analytiques.

(2) L'Auteur voudrait exprimer sa reconnaissance pour une communication écrite du Dr. Iséo Nakai dont j'adopte ici les conclusions : à savoir l'impossibilité de définir dans un ordre r fini *à ne pas dépasser* une stratification raisonnable des espaces de jets $J^r(n,p)$ (i.e. une stratification qui commute aux projections $J^k \rightarrow J^{k-i}$ pour tous les $i < k$). Par contre, il n'est pas impossible que, dans une stratification à k fini, *certains* blocs de strates commutent à *certaines* de ces flèches, constituant un système infini dont la limite pourrait constituer la stratification associée à une singularité de codimension infinie (non isolée dans le cas des fonctions).

Références

R. Thom, *Ensembles et Morphismes stratifiés*, Bull. Amer. Math. Soc. 75 (1969), pp. 240-284.

R. Thom, *Problèmes rencontrés dans mon parcours mathématique : un bilan*, Publ. Math. I.H.E.S. 70 (1989), pp. 199-214.

Irrégularité des revêtements cycliques

MICHEL VAQUIÉ

Ecole Normale Supérieure, Paris

INTRODUCTION

Dans un article de 1931 ([Za.3]), Zariski calcule l'irrégularité d'un revêtement cyclique du plan projectif complexe $\mathbf{P}^2_{\mathbf{C}}$ ramifié au dessus d'une courbe C ayant comme seules singularités des points doubles ordinaires et des points cuspidaux. Il peut en déduire des résultats d'annulation pour certains groupes de cohomologie de faisceaux associés à la courbe C.

Récemment, en utilisant une construction différente plusieurs auteurs ont généralisé l'étude des revêtements cycliques (en particulier H. Esnault et E. Viehweg [Es.], [Es.-Vi.1], et F. Loeser et l'auteur [Lo.-Va.], [Va.]). Mais si l'utilisation de théorèmes puissants déduits essentiellement de la théorie de Hodge permet d'obtenir des résultats plus généraux, la beauté de la démonstration originale a disparu.

Dans cet article je poursuis d'une part l'étude des revêtements cycliques et surtout je montre comment il est possible d'utiliser les méthodes originales de Zariski pour démontrer ces résultats.

Rappelons la situation étudiée par Zariski.

Soit C une courbe dans le plan affine complexe $\mathbf{A}^2_{\mathbf{C}} = \operatorname{Spec} \mathbf{C}[x,y]$, définie par un polynôme $f(x,y)$, et soit X^o le revêtement cyclique de $\mathbf{A}^2_{\mathbf{C}}$ de degré n ramifié au dessus de la courbe C, c'est à dire la surface définie dans l'espace affine $\mathbf{A}^3_{\mathbf{C}} = \operatorname{Spec} \mathbf{C}[x,y,z]$ par le polynôme $z^n - f(x,y)$.
L'irrégularité q du revêtement cyclique X^o est l'irrégularité d'une résolution des singularités $\tilde{X}$ d'une compactification X de X^o, où l'irrégularité de $\tilde{X}$ est par définition la dimension du groupe de cohomologie $\mathrm{H}^1(\tilde{X}, \mathcal{O}_{\tilde{X}})$, c'est à dire est égale à la moitié du premier nombre de Betti b_1 de la surface $\tilde{X}$.
L'irrégularité est un invariant birationnel de la surface $\tilde{X}$, par conséquent le nombre q ne dépend ni de la compactification X choisie, ni de la résolution des singularités $\tilde{X}$.

Nous pouvons choisir pour la compactification X de X^o un revêtement cyclique du plan projectif $\mathbf{P}^2_{\mathbf{C}}$ qui étend le revêtement du plan affine $\mathbf{A}^2_{\mathbf{C}}$, mais alors le morphisme $\pi: X \to \mathbf{P}^2_{\mathbf{C}}$ est ramifié le long de la courbe C et de la droite à l'infini L.

Si la courbe C est définie dans $\mathbf{P}^2_{\mathbf{C}} = \text{Proj } \mathbf{C}[u,x,y]$ par le polynôme $f(u,x,y)$ homogène de degré m et si la droite L est définie par u, la surface X est alors définie dans l'espace projectif $\mathbf{P}^3_{\mathbf{C}} = \text{Proj } \mathbf{C}[u,x,y,z]$ par le polynôme homogène g, avec $g(u,x,y,z) = z^n - u^{n-m} f(u,x,y)$ si le degré n du revêtement est supérieur ou égal au degré m de la courbe C et avec $g(u,x,y,z) = u^{m-n}z^n - f(u,x,y)$ sinon.
Le morphisme $\pi: X \to \mathbf{P}^2_{\mathbf{C}}$ est le morphisme induit par la projection de $\mathbf{P}^3_{\mathbf{C}} \setminus \{\infty\}$ dans $\mathbf{P}^2_{\mathbf{C}}$, où ∞ est le point de coordonnées homogènes $(0;0;0;1)$.

L'étude de l'irrégularité des revêtements cycliques est motivée par la question posée par Zariski dans un article de 1929 ([Za.1]) : soit C une courbe du plan affine complexe $\mathbf{A}^2_{\mathbf{C}} = \text{Spec } \mathbf{C}[x,y]$ définie par un polynôme $f(x,y)$, alors quelle condition doit vérifier une surface X de $\mathbf{A}^3_{\mathbf{C}} = \text{Spec } \mathbf{C}[x,y,z]$ pour que la courbe C soit le discriminant de l'application π de X dans $\mathbf{A}^2_{\mathbf{C}}$, restriction de la projection de $\mathbf{A}^3_{\mathbf{C}}$ dans $\mathbf{A}^2_{\mathbf{C}}$?
Si la surface X est définie dans $\mathbf{A}^3_{\mathbf{C}}$ par le polynôme $\varphi(x,y,z)$, le lieu critique C_π du morphisme π est la courbe définie dans X par l'idéal de Fitting $F_0(\Omega^1_{X/\mathbf{A}^2_{\mathbf{C}}})$, c'est à dire la courbe définie par $(\partial\varphi/\partial z)$.

Par définition le discriminant D_π est l'image réduite du lieu critique C_π dans $\mathbf{A}^2_{\mathbf{C}}$, par conséquent l'équation de la courbe D_π dans le plan est obtenue en éliminant z entre les équations:

$$\begin{cases} \varphi(x,y,z) = 0 \\ \dfrac{\partial\varphi}{\partial z}(x,y,z) = 0. \end{cases}$$

Pour pouvoir répondre à cette question il suffit d'après un résultat de Enriques ([En.]) de connaître le groupe fondamental $\pi_1(\mathbf{A}^2_{\mathbf{C}} \setminus C)$ du complémentaire de la courbe C.

Zariski fait remarquer que pour étudier le groupe fondamental $\pi_1(\mathbf{A}^2_{\mathbf{C}} \setminus C)$ il est très utile de connaître l'irrégularité des revêtements cycliques du plan $\mathbf{A}^2_{\mathbf{C}}$ ramifiés au dessus de la courbe C, c'est à dire l'irrégularité des revêtenents X^o définis par $\varphi(x,y,z) = z^n - f(x,y)$.
Dans le cas où la courbe C a pour seuls points singuliers des points doubles ordinaires et des points cuspidaux Zariski calcule l'irrégularité q du revêtement cyclique en fonction de la surabondance des systèmes linéaires pré-adjoints à la courbe C de degré inférieur à $m-3$, c'est à dire en fonction des dimensions des groupes de cohomologie $\mathrm{H}^1(\mathbf{P}^2, \mathcal{A}^o_C(\mu))$ pour $0 \leq \mu \leq m-3$, où $\mathcal{A}^o_C$ est le faisceau des fonctions g de $\mathcal{O}_{\mathbf{P}^2}$ s'annulant aux points cuspidaux de $\mathbf{P}^2$ avec multiplicité au moins un.

Zariski peut alors déduire de ce résultat et de la nullité de l'irrégularité quand le degré du revêtement est une puissance d'un nombre premier ([Za.2]) des théorèmes d'annulation pour les groupes $\mathrm{H}^1(\mathbf{P}^2, \mathcal{A}_C^o(\mu))$.

Dans un article précédent j'ai généralisé ces résultats au cas des revêtements cycliques d'une surface projective complexe non singulière S ramifiés au dessus d'une courbe C non nécessairement irréductible et ayant des singularités quelconques ([Va.]).
Dans cet article je donnais l'irrégularité du revêtement en fonction de la dimension de groupes de cohomologie $\mathrm{H}^1(S, \mathcal{A}_\alpha \otimes_{\mathcal{O}_S} \omega_S(\mu))$, où les faisceaux $\mathcal{A}_\alpha$ sont des faisceaux d'idéaux de $\mathcal{O}_S$ définis à partir des points singuliers de la courbe C.
Je déduisais de ce résultat et de théorèmes d'annulation obtenus grâce à la théorie de Hodge (cf. [Es.-Vi.2]), des conditions d'annulations pour les espaces $\mathrm{H}^1(S, \mathcal{A}_\alpha \otimes_{\mathcal{O}_S} \omega_S(\mu))$.

Dans le présent article je vais généraliser les résultats de l'article [Va.].

Pour tout revêtement cyclique X d'une variété projective complexe non singulière S de dimension d, $d \geq 2$, je calcule la dimension des groupes de cohomologie $\mathrm{H}^r(\tilde{X}, \mathcal{O}_{\tilde{X}})$, où comme précédemment $\tilde{X}$ est une résolution des singularités du revêtement X.
J'obtiens cette dimension en fonction de groupes de cohomologie de certains faisceaux $\mathcal{A}_\alpha \otimes_{\mathcal{O}_S} \omega_S(\mu)$, où les $\mathcal{A}_\alpha$ sont encore des faisceaux d'idéaux de $\mathcal{O}_S$ définis à partir des singularités du lieu de ramification D du revêtement cyclique X.
Puis je généralise à ce cas les théorèmes d'annulation de [Va.].
Je montre ensuite comment il est possible de généraliser la démonstration de Zariski.
Plus précisément je décris le revêtement cyclique X de S comme diviseur de l'espace total d'un fibré projectif $Z = \mathbf{P}(\mathcal{E})$ associé à un faisceau localement libre de rang deux sur S. Pour trouver la dimension des groupes de cohomologie $\mathrm{H}^r(\tilde{X}, \mathcal{O}_{\tilde{X}})$, où $\tilde{X}$ est une résolution des singularités de X, je calcule l'idéal d'adjonction $\mathcal{A}_X$ du diviseur X dans Z.

Ensuite, suivant la démonstration originale de Zariski, je peux déduire du théorème d'annulation de l'irrégularité de tout revêtement cyclique du plan projectif complexe $\mathbf{P}^2_{\mathbf{C}}$ de degré n égal à une puissance d'un nombre premier([Za.2]), une généralisation des résultats d'annulation de ([Za.3]) au cas d'une courbe C du plan projectif complexe ayant des singularités quelconques.
Je rappelle aussi comment ces résultats sont liés à l'étude du groupe fondamental du complémentaire d'une courbe plane C.

Il est en effet très difficile de calculer le groupe $G = \pi_1(\mathbf{P}^2_{\mathbf{C}} \setminus C)$, mais il est possible de déterminer un polynôme $\Delta_C(t)$, appelé le polynôme d'Alexander de la courbe C, qui donne le groupe quotient $G'\backslash G''$, où G' et G'' sont les deux premiers groupes dérivés du groupe G.

Ce polynôme $\Delta_C(t)$ a été étudié par plusieurs auteurs, en particulier par A. Libgober ([Li.1], [Li.2]), et plus récemment par F. Loeser et l'auteur ([Lo.-Va.]) qui donnent sa valeur dans le cas d'une courbe quelconque de $\mathbf{P}^2$ en fonction de la dimension des groupes de cohomologie $\mathrm{H}^1(S, \mathcal{A}_\alpha \otimes_{\mathcal{O}_S} \omega_S(\mu))$, c'est à dire exactement les groupes apparaissant dans le calcul de l'irrégularité d'un revêtement cyclique.

Nous pouvons ainsi déduire de ce résultat la nullité de l'irrégularité de tout revêtement cyclique du plan projectif quand le groupe fondamental du complémentaire de la courbe est abélien.

Notation

Pour toute variété algébrique Y nous notons ω_Y le faisceau dualisant sur Y. Si la variété Y est non singulière et si D est un diviseur sur Y dont le support est à croisements normaux nous notons respectivement Ω^p_Y et $\Omega^p_Y(\log D)$ le faisceau des p-formes différentielles sur Y et le faisceau des p-formes différentielles sur Y à pôles logarithmiques le long de D.

Pour tout diviseur D sur Y, la formule d'adjonction donne un isomorphisme canonique entre le faisceau dualisant ω_D et le faisceau $\omega_Y \otimes_{\mathcal{O}_Y} \mathcal{O}_D(D)$, et toute résolution des singularités de D, $\pi: \tilde{D} \to D$, donne une application canonique injective de l'image directe $\pi_*(\omega_{\tilde{D}})$ dans le faisceau ω_D.

Alors, si Y est une variété non singulière et si D est un diviseur effectif réduit sur Y, l'idéal d'adjonction de D dans Y est le faisceau d'idéaux $\mathcal{A}_D$ de $\mathcal{O}_Y$ qui rend commutatif le diagramme suivant:

$$(0.1) \qquad \begin{array}{ccccccccc}
 & & 0 & & 0 & & & & \\
 & & \downarrow & & \downarrow & & & & \\
 & & \omega_Y & \longrightarrow & \omega_Y & & & & \\
 & & \downarrow & & \downarrow & & & & \\
0 & \longrightarrow & \mathcal{A}_D \otimes_{\mathcal{O}_Y} \omega_Y(D) & \longrightarrow & \omega_Y(D) & \longrightarrow & \mathcal{B} & \longrightarrow & 0 \\
 & & \downarrow & & \downarrow & & \downarrow & & \\
0 & \longrightarrow & \pi_*(\omega_{\tilde{D}}) & \longrightarrow & \omega_D & \longrightarrow & \mathcal{B} & \longrightarrow & 0 \\
 & & \downarrow & & \downarrow & & & & \\
 & & 0 & & 0 & & & &
\end{array}$$

En particulier le cosupport de l'idéal d'adjonction $\mathcal{A}_D$ de D dans Y, c'est à dire le support du fermé qu'il définit dans Y, est inclus dans le lieu singulier du diviseur D.

Pour tout nombre réel y nous notons $[y]$ la partie entière de y et $\lceil y \rceil$ le plus petit entier supérieur ou égal à y, c'est à dire $-[-y]$.

Si nous avons une famille de diviseurs irréductibles $(E_j)_{j\in J}$ sur une variété S, si nous nous donnons des nombres réels y_j, $j \in J$, et si nous posons $D = \sum_{j\in J} y_j E_j$, nous pouvons définir les diviseurs $[D]$ et $\lceil D \rceil$ par:

$$[D] = \sum_{j\in J} [y_j] E_j \quad \text{et} \quad \lceil D \rceil = \sum_{j\in J} \lceil y_j \rceil E_j.$$

Nous utiliserons cette notation dans le cas particulier suivant :
Soit $D = \sum_{j\in J} m_j E_j$ un diviseur sur S et soit γ un nombre réel, alors nous posons:

$$(0.2) \qquad [\gamma D] = \sum_{j\in J} [\gamma m_j] E_j \quad \text{et} \quad \lceil \gamma D \rceil = \sum_{j\in J} \lceil \gamma m_j \rceil E_j.$$

Dans cet article nous considérons une variété projective non singulière S de dimension d, $d \geq 2$, définie sur le corps des nombres complexes $\mathbf{C}$, et nous nous donnons un faisceau très ample $\mathcal{L}$ sur S.

I IRRÉGULARITÉ D'UN REVÊTEMENT CYCLIQUE

Nous allons rappeler la construction de H. Esnault :
Soit $\mathcal{L}$ un faisceau inversible sur une variété non singulière S de dimension d et soit D un diviseur sur S appartenant au système linéaire $|\mathcal{L}^{\otimes n}|$, c'est à dire défini par une section globale g du faisceau $\mathcal{L}^{\otimes n}$, où n est un entier strictement positif.

Cette section globale permet alors de définir une structure d'algèbre sur le faisceau $\bigoplus_{p=0}^{n-1} \mathcal{L}^{\otimes -p}$, et la variété $X = \mathrm{Spec}_S(\bigoplus_{p=0}^{n-1} \mathcal{L}^{\otimes -p})$ est le revêtement cyclique de S ramifié au dessus du diviseur D.

Par construction le morphisme ψ de X dans S est fini, non ramifié au dessus de l'ouvert $U = S \setminus D$, et le groupe de Galois du revêtement est le groupe cyclique $\mathbf{Z}/n\mathbf{Z}$.

Nous appelons $\sigma: \tilde{S} \to S$ une résolution plongée des singularités du diviseur réduit D_{red}, $\tilde{D} = \sigma^*(D)$ l'image inverse totale de D, et $\bar{X}$ la normalisée du produit fibré $\tilde{S} \times_S X$.

Le diviseur $\tilde{D}$ a pour support un diviseur à croisements normaux et l'ouvert $\tilde{S} \setminus \tilde{D}$ est isomorphe à $S \setminus D$. Le morphisme $\bar{\psi}: \bar{X} \to \tilde{S}$ est un revêtement cyclique de degré n non ramifié au dessus de l'ouvert $\tilde{S} \setminus \tilde{D}$, et le morphisme $\bar{\sigma}: \bar{X} \to X$ est un morphisme propre birationnel.

Nous considérons aussi une résolution des singularités de $\bar{X}$, $\bar{\pi}: \tilde{X} \to \bar{X}$, telle que l'image inverse Δ du diviseur $\tilde{D}$ par le morphisme composé de $\tilde{X}$ dans

$\tilde{S}$ ait pour support un diviseur à croisements normaux. Alors le morphisme composé $\pi: \tilde{X} \to X$ est une résolution des singularités de X.

Nous avons le diagramme suivant:

$$(1.1) \qquad \begin{array}{ccccc} & & \tilde{X} & & \\ & \overset{\bar\pi}{\swarrow} & & \overset{\pi}{\searrow} & \\ \bar{X} & & \xrightarrow{\bar\sigma} & & X \\ \downarrow{\scriptstyle \bar\psi} & & & & \downarrow{\scriptstyle \psi} \\ \tilde{S} & & \xrightarrow{\sigma} & & S \end{array}$$

Le diviseur $\tilde{D}$ sur la variété $\tilde{S}$ s'écrit sous la forme $\tilde{D} = \sum_{j\in J} m_j D_j$, où les D_j sont les composantes irréductibles réduites. Pour tout p, $0 \leq p \leq n-1$, nous définissons le diviseur effectif $\tilde{D}(p)$ et le faisceau inversible $\mathcal{L}^{[p]}$ sur $\tilde{S}$ par:

$$(1.2) \quad \tilde{D}(p) = \left[\frac{p}{n}\tilde{D}\right] = \sum_{j\in J}\left[\frac{m_j p}{n}\right] D_j \quad \text{et} \quad \mathcal{L}^{[p]} = \sigma^*(\mathcal{L}^{\otimes p}) \otimes_{\mathcal{O}_{\tilde{S}}} \mathcal{O}_{\tilde{S}}(-\tilde{D}(p)).$$

Nous pouvons maintenant rappeler le résultat de H. Esnault([Es.] lemmes 1 et 2, [Es.-Vi.1] lemme 1.8 ou [Vi.1] lemmes 1.3 et 1.4)

Lemme I.1 *La variété $\bar{X}$ a pour seules singularités des singularités rationnelles.*

Lemme I.2 *Le revêtement cyclique $\bar{X}$ de $\tilde{S}$ est égal à* $\mathrm{Spec}_{\tilde{S}}\left(\bigoplus_{p=0}^{n-1} \mathcal{L}^{[p]-1}\right)$.

Nous pouvons en déduire le résultat suivant.

Proposition I.1 *Pour tout entier r, $0 \leq r \leq d$, la dimension q de l'espace de cohomologie $\mathrm{H}^r(\tilde{X}, \mathcal{O}_{\tilde{X}})$ est égale à*

$$q = \dim \mathrm{H}^r(\tilde{X}, \mathcal{O}_{\tilde{X}}) = \sum_{p=0}^{n-1} \dim \mathrm{H}^{d-r}\Big(S, \sigma_*(\mathcal{L}^{[p]} \otimes_{\mathcal{O}_{\tilde{S}}} \omega_{\tilde{S}})\Big).$$

Démonstration : Nous déduisons des lemmes précédents un isomorphisme entre

$$\mathrm{H}^r(\tilde{X}, \mathcal{O}_{\tilde{X}}) \quad \text{et} \quad \bigoplus_{p=0}^{n-1} \mathrm{H}^r(\tilde{S}, \mathcal{L}^{[p]-1}),$$

d'où par dualité l'égalité $q = \sum_{p=0}^{n-1} \dim \mathrm{H}^{d-r}(\tilde{S}, \mathcal{L}^{[p]} \otimes_{\mathcal{O}_{\tilde{S}}} \omega_{\tilde{S}})$.

Pour avoir la proposition il suffit alors de montrer que les images directes supérieures $\mathrm{R}^q\sigma_*(\mathcal{L}^{[p]} \otimes_{\mathcal{O}_{\tilde{S}}} \omega_{\tilde{S}})$, $q > 0$, sont nulles, c'est à dire d'après la formule de projection que les faisceaux $\mathrm{R}^q\sigma_*(\omega_{\tilde{S}} \otimes_{\mathcal{O}_{\tilde{S}}} \mathcal{O}_{\tilde{S}}(-\tilde{D}(p)))$ sont nuls pour tout entier p, $0 \leq p \leq n-1$.

La nullité de ces faisceaux a été démontrée par E. Viehweg ([Vi.2] Proposition 2.3); ce résultat est une conséquence d'un théorème d'annulation dont la démonstration utilise la théorie de Hodge.

Si le diviseur D a uniquement des singularités isolées, en particulier il faut que le diviseur D soit réduit, nous pouvons exprimer les faisceaux $\sigma_*(\mathcal{L}^{[p]} \otimes_{\mathcal{O}_{\tilde{S}}} \omega_{\tilde{S}})$ qui apparaissent dans la proposition en fonction des exposants de Hodge des singularités de D.

Soit (D, x) une singularité isolée d'hypersurface de dimension $d-1$, c'est à dire nous avons $(D, x) \subset (S, x)$ où S est une variété non singulière de dimension d, et soit $\sigma: \tilde{S} \to S$ une résolution plongée de la singularité (D, x). Nous appelons $E = \bigcup_{j\in J} E_j$ le diviseur exceptionnel réduit de σ et nous appelons ν_j la valuation discrète associée à la composante irréductible E_j du diviseur.

L'image inverse de D dans $\tilde{S}$ s'écrit $\sigma^*(D) = \tilde{D} + \sum_{j\in J} m_j E_j$, où $\tilde{D}$ est la transformée stricte de D par σ, et le diviseur canonique $K_{\tilde{S}}$ sur $\tilde{S}$ s'écrit $K_{\tilde{S}} = \sigma^*(K_S) + \sum_{j\in J} k_j E_j$; c'est équivalent à dire que pour tout j l'entier m_j est égal à $\nu_j(f)$ où f est une équation de D dans S, et que l'entier k_j est égal à $\nu_j(\eta)$ où η est un générateur du faisceau dualisant ω_S.

Alors si φ est un élément de l'anneau local $\mathcal{O}_{S,x}$, c'est à dire un germe de fonction en x sur S, nous définissons le nombre rationnel $\beta_{D,x}(\varphi)$ de la manière suivante:

$$\beta_{D,x}(\varphi) = \inf_{j\in J}\Big(\frac{1 + k_j + \nu_j(\varphi)}{m_j} - 1\Big). \tag{1.3}$$

Le nombre $\beta_{D,x}(\varphi)$ ainsi défini est indépendant de la résolution plongée des singularités $\sigma: \tilde{S} \to S$ choisie.

Nous pouvons définir pour tout nombre α l'idéal $\mathcal{A}_\alpha$ de l'anneau local $\mathcal{O}_{S,x}$ comme l'idéal des éléments φ vérifiant $\beta_{D,x}(\varphi) > \alpha$.

Alors l'idéal $\mathcal{A}_\alpha$ peut être défini comme l'image directe par le morphisme σ du faisceau inversible $\mathcal{O}_{\tilde{S}}\big(-\sum_{j\in J}([(\alpha+1)m_j] - k_j)E_j\big)$ (cf [Lo.-Va.] démonstration de la proposition 4.5).

Si nous avons une situation globale, c'est à dire un diviseur D à singularités isolées sur une variété non singulière S, la construction précédente se globalise et nous pouvons définir les faisceaux d'idéaux $\mathcal{A}_\alpha$ de $\mathcal{O}_S$ à partir d'une résolution plongée des singularités de D. En particulier une section globale φ de $\mathcal{O}_S$ appartient à $\mathcal{A}_\alpha$ si pour tout point singulier x du diviseur D nous avons l'inégalité $\beta_{D,x}(\varphi) > \alpha$.

Nous nous intéressons aux faisceaux $\mathcal{A}_\alpha$ uniquement pour α compris entre

-1 et 0. Nous pouvons remarquer facilement que le faisceau $\mathcal{A}_{-1}$ est égal au faisceau $\mathcal{O}_S$ et que le faisceau $\mathcal{A}_0$ est l'idéal d'adjonction $\mathcal{A}_D$ de la variété D dans S ([Va.] Proposition 2.1).

Nous dirons qu'un nombre rationnel α, $-1 \leq \alpha \leq 0$, appartient au spectre de la singularité (D,x) s'il existe un élément φ de l'anneau local $\mathcal{O}_{S,x}$ vérifiant $\beta_{D,x}(\varphi) = \alpha$, et nous appellerons spectre du diviseur D de S la réunion des spectres de toutes les singularités (D,x) de D. Nous noterons $Sp(D)$ le spectre de D.

Pour tout nombre α et pour tout nombre λ strictement positif nous avons l'inclusion $\mathcal{A}_\alpha \subset \mathcal{A}_{\alpha-\lambda}$, et nous avons égalité entre les faisceaux $\mathcal{A}_\alpha$ et $\mathcal{A}_{\alpha-\lambda}$ pour λ suffisamment petit si et seulement si le nombre α n'appartient pas au spectre de D.

Dans le cas où la singularité (D,x) est définie sur le corps des nombres complexes $\mathbf{C}$ nous pouvons relier les nombres α définis précédemment à partir d'une résolution plongée des singularités à la structure de Hodge mixte sur la fibre de Milnor. Nous avons plus précisément le résultat suivant ([Vr.1], [Vr.2]).

A toute singularité isolée d'hypersurface (D,x) de dimension $d-1$ nous pouvons associer une suite de nombres rationnels, le spectre singulier de (D,x), définie à partir de la structure de Hodge mixte sur le $(d-1)$-ième groupe de cohomologie H de la fibre de Milnor de (D,x) de la manière suivante : Le nombre rationnel α apparait dans le spectre singulier de (D,x) avec la multiplicité k, où k est un entier strictement positif, si $d-p-2 < \alpha \leq d-1-p$, $\exp(2\pi i\alpha) = \lambda$, $\dim F^p H_\lambda / F^{p+1} H_\lambda = k$, $0 \leq p \leq d-1$, où $H = \oplus H_\lambda$ est la décomposition de H en sous-espaces propres généralisés pour l'action de la monodromie et où $F^\cdot$ est la filtration de Hodge sur H_λ

Les deux définitions du spectre que nous avons données coïncident, plus précisément un nombre rationnel négatif ou nul α apparait dans le spectre de la singularité (D,x) si et seulement si α appartient à $Sp(D,x)$.

Nous disons aussi que dans ce cas le nombre α est un exposant de Hodge de la singularité (D,x).

Revenons à la situation initiale: nous considérons un faisceau inversible $\mathcal{L}$ sur la variété non singulière S et un diviseur D sur S appartenant au système linéaire $|\mathcal{L}^{\otimes n}|$.

Nous supposons maintenant que le diviseur D est formé de la réunion de deux diviseurs D' et D'', avec D' réduit et ayant uniquement des singularités isolées et avec D'' non nécessairement réduit mais tel que son support D''_{red} soit à croisements normaux et coupe transversalement le diviseur D' en des points non singuliers de celui-ci.

Alors l'image inverse totale $\sigma^*(D)$ du diviseur D par une résolution plongée $\sigma\colon \tilde{S} \to S$ des singularités du diviseur D' a pour support un diviseur à croisements normaux et nous pouvons déduire des définitions précédentes le théorème suivant.

Théorème I.1 *Soit S une variété non singulière de dimension d définie sur le corps des nombres complexes* **C**, *et soit D un diviseur sur S vérifiant la propriété précédente.*
Soit X le revêtement cyclique de degré n de S défini par le diviseur D et soit $\tilde{X}$ une résolution des singularités de X, alors la dimension q de l'espace de cohomologie $\mathrm{H}^r(\tilde{X}, \mathcal{O}_{\tilde{X}})$, $0 \le r \le d$, est égale à:

$$\begin{aligned} q &= \dim \mathrm{H}^r(\tilde{X}, \mathcal{O}_{\tilde{X}}) \\ &= \sum_{p=0}^{n-1} \dim \mathrm{H}^{d-r}\Big(S, \mathcal{A}_\alpha \otimes_{\mathcal{O}_S} \omega_S \otimes_{\mathcal{O}_S} \mathcal{L}^{\otimes p} \otimes_{\mathcal{O}_S} \mathcal{O}_S\big(-\big[\tfrac{p}{n} D''\big]\big)\Big), \end{aligned}$$

avec $\alpha = \frac{p}{n} - 1$ et où le diviseur $\left[\frac{p}{n} D''\right]$ est défini à partir du diviseur D'' (cf (0.3)), où le faisceau $\mathcal{A}_\alpha$ est le faisceau des germes de fonctions φ sur S vérifiant l'inégalité $\beta_{D',x}(\varphi) > \alpha$ pour tout point singulier x du diviseur D'.

Démonstration : D'après la proposition I.1 il suffit de démontrer que les faisceaux $\mathcal{A}_\alpha \otimes_{\mathcal{O}_S} \omega_S \otimes_{\mathcal{O}_S} \mathcal{L}^{\otimes p} \otimes_{\mathcal{O}_S} \mathcal{O}_S\big(-\big[\frac{p}{n} D''\big]\big)$ et $\sigma_*(\mathcal{L}^{[p]} \otimes_{\mathcal{O}_{\tilde{S}}} \omega_{\tilde{S}})$ sont isomorphes.

Comme σ est un isomorphisme en dehors des points singuliers de D', les diviseurs $\tilde{D}(p) = \left[\frac{p}{n}\sigma^*(D)\right]$ et $\sum_{j\in J}\left[\frac{pm_j}{n}\right] E_j + \sigma^*\left(\left[\frac{p}{n} D''\right]\right)$ sont égaux et $\mathcal{L}^{[p]} \otimes_{\mathcal{O}_{\tilde{S}}} \omega_{\tilde{S}}$ est isomorphe au faisceau suivant:

$$\sigma^*\Big(\mathcal{L}^{\otimes p} \otimes_{\mathcal{O}_S} \mathcal{O}_S\big(-\big[\tfrac{p}{n} D''\big]\big) \otimes_{\mathcal{O}_S} \omega_S\Big) \otimes_{\mathcal{O}_{\tilde{S}}} \mathcal{O}_{\tilde{S}}\Big(-\sum_{j\in J}\Big[\frac{pm_j}{n}\Big] E_j + \sum_{j\in J} k_j E_j\Big).$$

Le théorème est alors une conséquence de la formule de projection et de la définition des faisceaux $\mathcal{A}_\alpha$ à partir de la résolution plongée des singularités du diviseur D'.

Si nous nous donnons un diviseur réduit D' sur la variété S dont toutes les singularités sont des singularités isolées et appartenant au système linéaire $|\mathcal{L}^{\otimes m}|$, avec $1 \le m \le n$, et si nous supposons qu'il existe un diviseur L sur S appartenant au système linéaire $|\mathcal{L}|$ avec L non singulier et coupant transversalement D' en des points non singuliers, un tel diviseur L existe toujours si nous supposons que le faisceau inversible $\mathcal{L}$ est très ample sur S, alors nous pouvons prendre pour D'' le diviseur $(n-m)L$.

Nous considérons alors le revêtement cyclique X de S ramifié au dessus de

$D' \bigcup L$ défini par le diviseur $D = D' + (n-m)L$, et nous avons le résultat suivant.

Corollaire I.1 *La dimension de l'espace de cohomologie* $\mathrm{H}^r(\tilde{X}, \mathcal{O}_{\tilde{X}})$ *est égale à:*

$$q = \dim \mathrm{H}^r(\tilde{X}, \mathcal{O}_{\tilde{X}}) = \sum_{p=0}^{n-1} \dim \mathrm{H}^{d-r}\left(S, \mathcal{A}_\alpha \otimes_{\mathcal{O}_S} \omega_S \otimes_{\mathcal{O}_S} \mathcal{L}^{\otimes \lceil \frac{pm}{n} \rceil}\right) \tag{1.4}$$

$$avec \quad \alpha = \frac{p}{n} - 1$$

Le résultat précédent donnant la dimension des groupes $\mathrm{H}^r(\tilde{X}, \mathcal{O}_{\tilde{X}})$ en fonction des groupes de cohomologie des faisceaux $\mathcal{A}_\alpha$ tordus par une puissance convenable de $\mathcal{L}$, est une généralisation d'une partie du théorème de Zariski ([Za.3] fin du §7, p.503).

Pour avoir une généralisation du théorème en entier, il faut démontrer des résultats d'annulation sur ces groupes, résultats dont la démonstration utilise de façon essentielle la théorie de Hodge et des théorèmes d'annulation de la cohomologie singulière des variétés complexes affines.

Les deux propositions suivantes sont des généralisations à une variété S de dimension quelconque des propositions 3.3 et 3.4 de [Va.].

Nous supposons dans la suite que la variété S est projective et que le faisceau inversible $\mathcal{L}$ est très ample. Nous considérons de nouveau un diviseur D' à singularités isolées sur la variété S, appartenant au système linéaire $|mL|$, et nous voulons étudier les groupes de cohomologie $\mathrm{H}^r(S, \mathcal{A}_\alpha \otimes_{\mathcal{O}_S} \omega_S \otimes_{\mathcal{O}_S} \mathcal{L}^{\otimes \lceil (\alpha+1)m \rceil})$ pour un nombre α de $\mathbf{Q} \bigcap]-1, 0[$ n'appartenant pas au spectre singulier de D'.

Proposition I.2 *Si le nombre* α *n'appartient pas au spectre singulier* $Sp(D')$ *nous avons la relation:*

$$\mathrm{H}^r\left(S, \mathcal{A}_\alpha \otimes_{\mathcal{O}_S} \omega_S \otimes_{\mathcal{O}_S} \mathcal{L}^{\otimes \lceil (\alpha+1)m \rceil}\right) = 0, \quad pour \quad r \geq 1.$$

Démonstration : Comme le nombre α n'appartient pas à $Sp(D')$, pour tout $\lambda > 0$ suffisamment petit les faisceaux $\mathcal{A}_\alpha$ et $\mathcal{A}_{\alpha-\lambda}$ sont égaux, et nous avons l'égalité $\lceil (\alpha+1)m \rceil = \lceil (\alpha - \lambda + 1)m \rceil$.

Soit $E = \bigcup_{j=1}^r E_j$ le diviseur exceptionnel de la résolution plongée $\sigma: \tilde{S} \to S$ des singularités de D', nous notons m_j le coefficient de la composante irréductible E_j dans l'image inverse $\sigma^*(D') = \tilde{D}' + \sum_{j=1}^r m_j E_j$.

Alors, quitte à remplacer α par $\alpha - \lambda$ pour un $\lambda > 0$ suffisamment petit, nous pouvons supposer que le nombre rationnel $\alpha = \frac{p}{n} - 1$, avec $(p, n) = 1$, vérifie:

$$(\alpha+1)m \notin \mathbf{Z} \quad \text{et} \quad n > m,$$
$$(\alpha+1)m_j \notin \mathbf{Z} \quad \forall j, \quad 1 \leq j \leq r.$$

Comme précédemment nous considérons le revêtement cyclique de degré n, $\psi: X \to S$ ramifié au dessus du diviseur $D' \bigcup L$ et le revêtement cyclique $\bar{\psi}: \bar{X} \to \tilde{S}$ défini grâce aux faisceaux inversibles

$$\mathcal{L}^{[p]} = \sigma^*(\mathcal{L}^{\otimes p}) \otimes_{\mathcal{O}_{\tilde{S}}} \mathcal{O}_{\tilde{S}}\big(-\big[\tfrac{p}{n}\tilde{D}\big]\big),$$

où $\tilde{D}$ est l'image inverse $\sigma^*(D' + (n-m)L) = \tilde{D}' + (n-m)\tilde{L} + \sum_{j=1}^r m_j E_j$.

Nous choisissons encore un nombre $\lambda > 0$ suffisamment petit tel que les faisceaux $\mathcal{A}_\alpha$ et $\mathcal{A}_{\alpha-\lambda}$ soient égaux et nous définissons un nouveau faisceau inversible $\mathcal{L}'^{[p]}$ sur $\tilde{S}$ par:

$$\mathcal{L}'^{[p]} = \sigma^*(\mathcal{L}^{\otimes p}) \otimes_{\mathcal{O}_{\tilde{S}}} \mathcal{O}_{\tilde{S}}\big(-\big[\big(\tfrac{p}{n} - \lambda\big)\tilde{D}\big]\big). \tag{1.5}$$

Comme les nombres $\frac{p}{n}m$ et $\frac{p}{n}m_j$ ne sont pas entiers, pour λ convenablement choisi nous avons les égalités: $\big[\frac{p}{n}(n-m)\big] = \big[\big(\frac{p}{n} - \lambda\big)(n-m)\big]$ et $\big[\frac{p}{n}m_j\big] = \big[\big(\frac{p}{n} - \lambda\big)m_j\big]$, $1 \leq j \leq r$. Nous en déduisons que les diviseurs $\big[\frac{p}{n}\tilde{D}\big]$ et $\big[\big(\frac{p}{n} - \lambda\big)\tilde{D}\big]$ sont égaux, d'où l'égalité entre les faisceaux $\mathcal{L}'^{[p]}$ et $\mathcal{L}^{[p]}$.

Alors les groupes $\mathrm{H}^r\big(S, \mathcal{A}_\alpha \otimes_{\mathcal{O}_S} \omega_S \otimes_{\mathcal{O}_S} \mathcal{L}^{\otimes\lceil(\alpha+1)m\rceil}\big)$ et $\mathrm{H}^r\big(\tilde{S}, \mathcal{L}'^{[p]} \otimes_{\mathcal{O}_{\tilde{S}}} \omega_{\tilde{S}}\big)$ sont égaux, et le faisceau $\mathcal{L}'^{[p]}$ vérifie l'égalité:

$$\mathcal{L}'^{[p]} \otimes_{\mathcal{O}_{\tilde{S}}} \mathcal{L}^{[n-p]} = \sigma^*(\mathcal{L}^{\otimes n}) \otimes_{\mathcal{O}_{\tilde{S}}} \mathcal{O}_{\tilde{S}}\big(-\big[\big(\tfrac{p}{n} - \lambda\big)\tilde{D}\big] - \big[\tfrac{n-p}{n}\tilde{D}\big]\big).$$

Nous pouvons choisir le nombre λ tel que pour tout entier l apparaissant comme coefficient d'une composante irréductible du diviseur $\tilde{D}$ nous ayons l'égalité $\big[\big(\frac{p}{n} - \lambda\big)l\big] + \big[\frac{n-p}{n}l\big] = l - 1$, d'où la relation: $\big[\big(\frac{p}{n} - \lambda\big)\tilde{D}\big] + \big[\frac{n-p}{n}\tilde{D}\big] = \tilde{D} - \tilde{D}_{\mathrm{red}}$.

Nous avons alors l'isomorphisme: $\mathcal{L}'^{[p]} \otimes_{\mathcal{O}_{\tilde{S}}} \mathcal{L}^{[n-p]} \simeq \mathcal{O}_{\tilde{S}}(\tilde{D}_{\mathrm{red}})$.

Nous en déduisons que les faisceaux $\mathcal{L}'^{[p]} \otimes_{\mathcal{O}_{\tilde{S}}} \omega_{\tilde{S}}$ et ${\mathcal{L}^{[n-p]}}^{-1} \otimes_{\mathcal{O}_{\tilde{S}}} \Omega^d_{\tilde{S}}(\log\tilde{D})$ sont isomorphes, où nous notons $\Omega^d_{\tilde{S}}(\log\tilde{D})$ le faisceau des d-formes différentielles sur $\tilde{S}$ à pôles logarithmiques le long du diviseur $\tilde{D}_{\mathrm{red}}$.

Alors le groupe $\mathrm{H}^r(\tilde{S}, \mathcal{L}'^{[p]} \otimes_{\mathcal{O}_{\tilde{S}}} \omega_{\tilde{S}})$ est isomorphe pour tout r à

$$\mathrm{H}^r(\tilde{S}, {\mathcal{L}^{[n-p]}}^{-1} \otimes_{\mathcal{O}_{\tilde{S}}} \Omega^d_{\tilde{S}}(\log\tilde{D})),$$

et d'après un résultat de H. Esnault et E. Viehweg ([Es.] Corollaire 4, [Es.-Vi.1] Lemme 1.2) ce groupe est un facteur direct du groupe $H^r(\tilde{X}, \Omega^d_{\tilde{X}}(\log\Delta))$.

D'après la théorie de Hodge pour le complémentaire d'un diviseur à croisements normaux ([De.] th. (3.2.5)) ce groupe est lui-même facteur direct du groupe $H^{d+r}(\tilde{U}, \mathbf{C})$, où $\tilde{U}$ est le complémentaire du diviseur Δ dans la résolution $\tilde{X}$ de X.

Comme l'ouvert U est le complémentaire d'un diviseur très ample de S, c'est un ouvert affine. L'ouvert $\tilde{U}$, qui est fini au dessus de l'ouvert U, est aussi un ouvert affine, par conséquent les groupes $H^i(\tilde{U}, \mathbf{C})$ sont nuls pour $i > d$, où d est la dimension de S.

Nous en déduisons que les groupes $H^r(\tilde{S}, \mathcal{L}'^{[p]} \otimes_{\mathcal{O}_{\tilde{S}}} \omega_{\tilde{S}})$ sont nuls pour $r > 0$.

Nous notons dans la suite de ce paragraphe β_1, β_2,..., β_l les éléments non entiers du spectre singulier du diviseur D', c'est à dire appartenant à $Sp(D') \bigcap]-1, 0[$, avec pour tout i, $1 \leq i \leq l$, $\beta_i = \dfrac{p_i}{n_i} - 1$ où p_i et n_i sont des entiers positifs premiers entre eux.

Nous fixons un entier r, $0 \leq r \leq d-1$, et nous appelons q et q_0 les dimensions respectives des groupes de cohomologie $H^r(\tilde{X}, \mathcal{O}_{\tilde{X}})$ et $H^r(S, \mathcal{O}_S)$.

Nous déduisons du théorème I.1 que la différence $q - q_0$ est toujours positive ou nulle, et est égale à:

$$(1.6) \qquad q - q_0 = \sum_{p=1}^{n-1} \dim H^{d-r}\Big(S, \mathcal{A}_\alpha \otimes_{\mathcal{O}_S} \omega_S \otimes_{\mathcal{O}_S} \mathcal{L}^{\otimes \lceil \frac{pm}{n} \rceil}\Big).$$

Nous avons alors le résultat suivant:

Proposition I.3 *Pour que la différence $q - q_0$ soit non nulle il faut qu'au moins un des entiers n_i, $1 \leq i \leq l$, divise à la fois le degré m du diviseur D' et le degré n du revêtement $\psi: X \to S$.*

Démonstration : D'après la relation (1.6), pour que la différence $q - q_0$ soit non nulle, il faut que l'un des groupes $H^{d-r}(S, \mathcal{A}_\alpha \otimes_{\mathcal{O}_S} \omega_S \otimes_{\mathcal{O}_S} \mathcal{L}^{\otimes \lceil \frac{pm}{n} \rceil})$ ne soit pas nul; nous déduisons de la proposition I.2 que le nombre rationnel $\alpha = \dfrac{p}{n} - 1$ appartient au spectre singulier de D', c'est à dire est égal à $\beta_i = \dfrac{p_i}{n_i} - 1$ pour un indice i, par conséquent l'entier n_i doit diviser le degré n du revêtement.

Nous devons montrer que cet entier n_i divise aussi aussi le degré m du diviseur D'.

Supposons que ce n'est pas le cas, alors le nombre $(\beta_i + 1)m = \frac{p_i m}{n_i}$ n'est pas entier, et pour tout nombre strictement positif λ suffisamment petit, les entiers $\lceil(\beta_i + 1)m\rceil$ et $\lceil(\beta + 1)m\rceil$, où $\beta = \beta_i + \lambda$, sont égaux. De plus pour λ suffisamment petit, les faisceaux $\mathcal{A}_\beta$ et $\mathcal{A}_{\beta_i}$ sont égaux.

Le groupe $\mathrm{H}^{d-r}(S, \mathcal{A}_{\beta_i} \otimes_{\mathcal{O}_S} \omega_S \otimes_{\mathcal{O}_S} \mathcal{L}^{\otimes\lceil\frac{pm}{n}\rceil})$ est alors égal au groupe

$$\mathrm{H}^{d-r}(S, \mathcal{A}_\beta \otimes_{\mathcal{O}_S} \omega_S \otimes_{\mathcal{O}_S} \mathcal{L}^{\otimes\lceil(\beta+1)m\rceil}),$$

avec β n'appartenant pas à $Sp(D')$, et ce groupe est nul d'après la proposition précédente.

II IDÉAL D'ADJONCTION

Pour calculer l'irrégularité d'un revêtement cyclique X de degré n du plan projectif $\mathbf{P}^2$ ramifié au dessus d'une courbe C, Zariski donne une construction explicite de ce revêtement comme fermé de degré n de l'espace projectif $\mathbf{P}^3$. Pour trouver le genre géométrique de cette surface X, il calcule son idéal d'adjonction $\mathcal{A}_X$ dans $\mathbf{P}^3$ en utilisant le fait que la courbe C a pour seuls points singuliers des points doubles ordinaires et des points cuspidaux ([Za.3]).

Dans ce paragraphe nous allons généraliser la démonstration de Zariski.

Nous allons d'abord montrer comment le revêtement cyclique X d'une variété S de dimension d, défini par un faisceau inversible $\mathcal{L}$ sur S et une section globale g du faisceau $\mathcal{L}^{\otimes n}$, peut être plongé dans l'espace total d'un fibré projectif $Z = \mathbf{P}(\mathcal{E})$, où $\mathcal{E}$ est le faisceau localement libre de rang deux sur S défini par $\mathcal{E} = \mathcal{O}_S \oplus \mathcal{L}$.

En particulier si la variété S est régulière, le revêtement X est un diviseur sur une variété régulière Z de dimension $d + 1$.

Nous appelons ρ le morphisme naturel de Z dans S et $\mathcal{O}_Z(1)$ le faisceau inversible tautologique sur Z, quotient inversible "universel" du faisceau $\rho^*(\mathcal{E})$. L'image directe $\rho_*(\mathcal{O}_Z(1))$ est isomorphe au faisceau $\mathcal{E}$, par conséquent les groupes $\mathrm{H}^0(Z, \mathcal{O}_Z(1))$ et $\mathrm{H}^0(S, \mathcal{E}) = \mathrm{H}^0(S, \mathcal{O}_S) \oplus \mathrm{H}^0(S, \mathcal{L})$ sont égaux et la section $1 \in \mathrm{H}^0(S, \mathcal{O}_S)$ définit une section globale h du faisceau $\mathcal{O}_Z(1)$, c'est à dire un diviseur H sur la variété Z.

La restriction de ρ à H induit un isomorphisme ι de H dans S. L'immersion fermée ι' de S dans Z ainsi définie peut être aussi construite à partir de la propriété universelle de $\rho: Z = \mathbf{P}(\mathcal{E}) \to S$: le morphisme $\iota': S \to Z$ est le morphisme correspondant au quotient inversible $\mathcal{L}$ de $\mathcal{E}$. En particulier le faisceau $\mathcal{L}$ sur S est isomorphe à $\iota'^*(\mathcal{O}_Z(1))$, c'est à dire à la restriction $\mathcal{O}_H(1)$ du faisceau tautologique au diviseur H.

Comme le faisceau $\mathcal{E}$ est de la forme $\mathcal{E} = \mathcal{O}_S \oplus \mathcal{L}$, nous avons les isomorphismes:

$$\rho_*(\mathcal{O}_Z(n)) \simeq \mathcal{S}^n\mathcal{E} = \bigoplus_{p=0}^{n} \mathcal{L}^{\otimes p} \quad \text{si } n \geq 0, \tag{2.1}$$
$$\rho_*(\mathcal{O}_Z(n)) = 0 \quad \text{si } n < 0;$$

(2.2)

$$\mathrm{R}^1\rho_*(\mathcal{O}_Z(n)) \simeq \underline{\mathcal{H}om}_{\mathcal{O}_S}(\mathcal{S}^{-n-2}\mathcal{E} \otimes_{\mathcal{O}_S} \det\mathcal{E}, \mathcal{O}_S) = \bigoplus_{p=n+1}^{-1} \mathcal{L}^{\otimes p} \quad \text{si } n \leq -2,$$

$$\mathrm{R}^1\rho_*(\mathcal{O}_Z(n)) = 0 \quad \text{si } n > -2.$$

Nous avons alors l'égalité:

$$\mathrm{H}^0(Z, \mathcal{O}_Z(n)) = \bigoplus_{p=0}^{n} \mathrm{H}^0(S, \mathcal{L}^{\otimes p}), \tag{2.3}$$

et nous appelons h^n la section globale de $\mathcal{O}_Z(n)$ correspondant à la section 1 de $\mathcal{O}_S$, nous trouvons en particulier $h^1 = h$ où h est la section définissant le diviseur H.

Nous nous donnons maintenant un diviseur C sur la variété S défini par une section globale f du faisceau inversible $\mathcal{L}^{\otimes m}$ et un diviseur L défini par une section globale u de $\mathcal{L}$, dans le cas où la variété S est régulière nous supposerons que le diviseur C est réduit, que le diviseur L est régulier et coupe C transversalement.

Grâce à l'égalité (2.3) nous pouvons considérer la section globale $g = u^{n-m}f$ de $\mathcal{L}^{\otimes n}$ comme une section du faisceau $\mathcal{O}_Z(n)$ sur Z. Nous définissons alors le diviseur X de Z associé à la section $h^n - u^{n-m}f$ de ce faisceau.

Proposition II.1 *Le morphisme $\psi: X \to S$ obtenu comme restriction au diviseur X du morphisme $\rho: Z \to S$ est le revêtement cyclique de degré n de S, ramifié au dessus du diviseur $C \bigcup L$, défini par la section globale $u^{n-m}f$ de $\mathcal{L}^{\otimes n}$.*

Démonstration : Nous pouvons vérifier localement sur S que le morphisme $\psi: X \to S$ est un morphisme fini. De plus nous avons par définition du diviseur X la suite exacte suivante de faisceaux sur Z:

$$0 \longrightarrow \mathcal{O}_Z(-n) \longrightarrow \mathcal{O}_Z \longrightarrow \mathcal{O}_X \longrightarrow 0$$

et en prenant l'image directe par le morphisme ρ nous trouvons grâce aux égalités (2.1) et (2.2) la suite exacte de faisceaux sur S:

$$0 \longrightarrow \mathcal{O}_S \longrightarrow \psi_*(\mathcal{O}_X) \longrightarrow \bigoplus_{p=1}^{n-1} \mathcal{L}^{\otimes -p} \longrightarrow 0,$$

ce qui nous donne bien un isomorphisme entre le faisceau $\psi_*(\mathcal{O}_X)$ et l'algèbre $\bigoplus_{p=0}^{n-1} \mathcal{L}^{\otimes -p}$.

La section globale u du faisceau $\mathcal{L}$ sur S peut aussi être considérée grâce à l'égalité (2.3) comme une section globale du faisceau $\mathcal{O}_Z(1)$ sur Z, et nous appelons V le diviseur sur Z défini par cette section u.

L'isomorphisme $\iota: H \to S$ permet d'identifier les diviseurs $X \cap H$ et $V \cap H$ de H avec les diviseurs $C + (n-m)L$ et L de S.

Dans le cas où la variété S est l'espace projectif $\mathbf{P}^d$ et le faisceau inversible $\mathcal{L}$ est le faisceau ample $\mathcal{O}_S(1)$, la variété $Z = \mathbf{P}(\mathcal{O}_S \oplus \mathcal{O}_S(1))$ est obtenue comme l'éclatement du point ∞ de l'espace projectif $\mathbf{P}^{d+1}$ et le morphisme $\rho: Z \to S$ est induit par la projection de $\mathbf{P}^{d+1} \setminus \{\infty\}$ dans $\mathbf{P}^d$.

Comme le diviseur X défini précédemment ne rencontre pas le diviseur exceptionnel de l'éclatement $Z \to \mathbf{P}^{d+1}$, nous pouvons le considérer comme un diviseur de $\mathbf{P}^{d+1}$ ne contenant pas le point ∞. Nous retrouvons ainsi la description faite par Zariski du revêtement cyclique de $\mathbf{P}^d$.

Nous supposons maintenant que la variété S est régulière, que le diviseur C est réduit et que le diviseur L est régulier, et nous voulons étudier les singularités de la variété X.

Soit o un point fermé de la variété S, nous notons $\mathcal{O}_{S,o}$ l'anneau local de S en o, c'est un anneau régulier de dimension d et nous notons $(x_1, \ldots, x_d)$ un système régulier de paramètres de cet anneau.

Alors pour tout point $\tilde{o}$ de la variété Z au dessus du point o, l'anneau $\mathcal{O}_{S,o}$ est inclus dans l'anneau local $\mathcal{O}_{Z,\tilde{o}}$ de Z en $\tilde{o}$, et nous pouvons trouver un système régulier de paramètres $(x_1, \ldots, x_d, z)$ de $\mathcal{O}_{Z,\tilde{o}}$ qui étend la système régulier de $\mathcal{O}_{S,o}$ et tel que l'anneau quotient $\mathcal{O}_{Z,\tilde{o}}/(z)$ soit isomorphe à $\mathcal{O}_{S,o}$.

Si le diviseur $C + (n-m)L$ est défini localement au voisinage du point o par un élément g de $\mathcal{O}_{S,o}$, alors le diviseur X est défini au voisinage du point $\tilde{o}$ par l'élément $z^n - g$.

Nous en déduisons que le lieu singulier $\mathrm{Sing}(X)$ du revêtement cyclique X de S est inclus dans l'image inverse du lieu singulier du diviseur $C + (n-m)L$, c'est à dire dans $\rho^{-1}(L) \bigcup \rho^{-1}(\mathrm{Sing}(C))$.

Nous en déduisons aussi que $\mathrm{Sing}(X)$, défini par l'idéal de Fitting $F_d(\Omega^1_X)$, est inclus dans le diviseur $(n-1)H$ de Z, en fait l'isomorphisme ι induit un isomorphisme entre $\mathrm{Sing}(X) \bigcap H$ et $\mathrm{Sing}(C + (n-m)L)$.

Nous voulons calculer l'idéal d'adjonction du diviseur X dans la variété régulière Z, c'est à dire le faisceau d'idéaux $\mathcal{A}_X$ de $\mathcal{O}_Z$ rendant commutatif le diagramme suivant:

$$
(2.4)\qquad
\begin{array}{ccccccccc}
0 & \longrightarrow & \omega_Z & \longrightarrow & \mathcal{A}_X \otimes \omega_Z(X) & \longrightarrow & \pi_*(\omega_{\tilde{X}}) & \longrightarrow & 0 \\
 & & \downarrow & & \downarrow & & \downarrow & & \\
0 & \longrightarrow & \omega_Z & \longrightarrow & \omega_Z(X) & \longrightarrow & \omega_X & \longrightarrow & 0
\end{array}
$$

où nous appelons $\pi: \tilde{X} \to X$ une résolution des singularités de X.

Si nous sommes au voisinage d'un point singulier $\tilde{o}$ de X situé au dessus d'un point o appartenant au diviseur L et n'appartenant pas au diviseur C, l'équation g de X dans Z est de la forme $z^n - u^{n-m}$, où u est un élément d'une suite régulière de $\mathcal{O}_{S,o}$ définissant le diviseur L au voisinage du point o, nous supposons dans ce cas $n \geq m+2$.

De même si nous sommes au voisinage d'un point $\tilde{o}$ de X situé au dessus d'un point o appartenant à $C \bigcap L$, l'équation de X est de la forme $z^n - u^{n-m}x$, où (u,x) est une suite régulière de $\mathcal{O}_{S,o}$ telle que u et x définissent respectivement les diviseurs L et C au voisinage de o, dans ce cas nous supposons $n \geq m+1$.

Dans ces deux cas, il est très facile de calculer directement l'idéal d'adjonction $\mathcal{A}_X$.

Si nous sommes au voisinage d'un point singulier $\tilde{o}$ de X qui se trouve au dessus d'un point singulier o du diviseur C, c'est à dire si l'équation g de X dans Z est de la forme $z^n - f$, où f est l'élément de $\mathcal{O}_{S,o}$ définissant la singularité (C,o) dans S, il est en général plus difficile de calculer l'idéal d'adjonction $\mathcal{A}_X$.

Remarque II.1 *Si (C,o) est une singularité isolée d'hypersurface non dégénérée par rapport à son polyèdre de Newton, il est encore possible de trouver une description simple de l'idéal $\mathcal{A}_X$. C'est en particulier le cas pour les singularités considérées par Zariski et plus généralement pour les singularités de courbes planes d'équation de la forme $x^p + y^q$.*

En effet si la singularité (C,o) définie dans S par $f = f(x_1,\ldots,x_d)$ est non dégénérée par rapport à son polyèdre de Newton, la singularité $(X,\tilde{o})$ définie par $z^n - f$ est aussi non dégénérée. Alors, en utilisant la résolution des singularités de Kouchnirenko, il est possible de donner une description de l'idéal d'adjonction $\mathcal{A}_X$ ([Me.-Te.] Théorème 2.1.1.). L'idéal $\mathcal{A}_X$ est l'idéal de $\mathcal{O}_{Z,\tilde{o}}$ engendré par les monômes $x_1^{i_1}\ldots x_d^{i_d} z^{i_{d+1}}$ tels que $(i_1,\ldots,i_d,i_{d+1})$ soit à l'intérieur du polyèdre de Newton de la singularité $(X,\tilde{o})$.

Pour calculer l'idéal d'adjonction $\mathcal{A}_X$ dans le cas général nous utilisons la description du revêtement cyclique X donnée au paragraphe I, c'est à dire: $X = \mathrm{Spec}_S\left(\bigoplus_{p=0}^{n-1} \mathcal{L}^{\otimes -p}\right)$.

Nous pouvons trouver un recouvrement du fibré projectif $Z = \mathbf{P}(\mathcal{E})$ par les deux ouverts Z^+ et Z^- définis par les égalités suivantes:

$$(2.5) \qquad Z^+ = \mathrm{Spec}_S\Big(\bigoplus_{p\geq 0} \mathcal{L}^{\otimes p}\Big) \qquad \text{et} \qquad Z^- = \mathrm{Spec}_S\Big(\bigoplus_{p\leq 0} \mathcal{L}^{\otimes p}\Big).$$

Ces ouverts, munis des morphismes $\rho^+ : Z^+ \to S$ et $\rho^- : Z^- \to S$ obtenus par restriction, sont les espaces fibrés vectoriels sur S associés aux faisceaux inversibles $\mathcal{L}$ et $\mathcal{L}^{-1}$.

L'intersection $Z^+ \cap Z^-$ est égale à $\mathrm{Spec}_S\big(\bigoplus_{p\in\mathbf{Z}} \mathcal{L}^{\otimes p}\big)$ et les complémentaires respectifs H^+ et H^- de Z^+ et Z^- dans Z sont des sous-variétés fermées de Z isomorphes par ρ à S (cf. [Dm.] Partie 2).

En particulier nous avons avec les notations précédentes $H = H^+$ et l'ouvert Z^-, affine au dessus de S, contient les deux sous-variétés fermées H et X de Z.

Dans la suite, comme nous nous intéressons à ce qui se passe au voisinage de X dans Z, nous nous plaçons sur l'ouvert Z^- de Z. De plus, comme le lieu singulier $\mathrm{Sing}(X)$ est inclus dans H, la restriction à l'ouvert Z^+ de l'idéal d'adjonction $\mathcal{A}_X$ de X dans Z est égale au faisceau $\mathcal{O}_Z$.

Nous allons définir une notion d'idéal "homogène" de l'anneau local $\mathcal{O}_{Z,\tilde{o}}$.

Nous nous plaçons au voisinage d'un point $\tilde{o}$ de H, nous appelons o son image par ρ et grâce à l'isomorphisme ι' entre H et S, l'anneau local $\mathcal{O}_{S,o}$ est isomorphe à $\mathcal{O}_{H,\tilde{o}}$.

Si nous appelons z l'élément de $\mathcal{O}_{Z,\tilde{o}}$ définissant le diviseur H, l'anneau $\mathcal{O}_{Z,\tilde{o}}$ est isomorphe au localisé en (z) de l'anneau $\mathcal{O}_{S,o}[z]$ et tout élément φ de l'anneau $\mathcal{O}_{Z,\tilde{o}}$ peut s'écrire sous la forme $\varphi = \sum_{n\geq 0} \varphi_n z^n$, où les φ_n sont des éléments de l'anneau $\mathcal{O}_{S,o}$.

Un idéal $\mathcal{A}$ de l'anneau $\mathcal{O}_{Z,\tilde{o}}$ est "homogène" si nous pouvons l'écrire sous la forme $\mathcal{A} = \bigoplus_{n\geq 0} \mathcal{I}^{(n)} z^n$ où les $\mathcal{I}^{(n)}$ sont des idéaux de $\mathcal{O}_{S,o}$ vérifiant $\mathcal{I}^{(n)} \subset \mathcal{I}^{(n+1)}$.

Si nous considérons un faisceau d'idéaux $\mathcal{A}$ de $\mathcal{O}_Z$, nous dirons de la même manière que la restriction de ce faisceau $\mathcal{A}$ à l'ouvert Z^- est "homogène", si nous pouvons écrire la restriction $\mathcal{A}_{|Z^-}$ sous la forme $\mathcal{A}_{|Z^-} = \bigoplus_{n\geq 0} \mathcal{I}^{(n)} \otimes_{\mathcal{O}_S} \mathcal{L}^{\otimes -n}$, où les $\mathcal{I}^{(n)}$ sont des faisceaux d'idéaux de $\mathcal{O}_S$ vérifiant $\mathcal{I}^{(n)} \subset \mathcal{I}^{(n+1)}$.

Nous appelons g l'élément de $\mathcal{O}_{S,o}$ qui définit le diviseur $D = C + (n-m)L$ au voisinage de o. Alors la variété X est définie dans Z au voisinage de $\tilde{o}$ par l'élément $z^n - g$ de l'anneau $\mathcal{O}_{Z,\tilde{o}}$. L'anneau local $\mathcal{O}_{X,\tilde{o}}$ de X au point $\tilde{o}$ est égal à l'anneau quotient $\mathcal{O}_{X,\bar{o}} = \mathcal{O}_{Z,\bar{o}}/(z^n - g)$ et nous pouvons l'écrire sous la forme suivante: $\mathcal{O}_{X,\bar{o}} = \bigoplus_{p=0}^{n-1} \mathcal{O}_{S,o} z^p$.

Si nous regardons globalement cette égalité, nous retrouvons l'égalité entre $\mathcal{O}_X$ et $\bigoplus_{p=0}^{n-1} \mathcal{L}^{\otimes -p}$.

Avec cette écriture l'image $\bar{\varphi}$ d'un élément $\varphi = \sum_{j\geq 0} \varphi_j z^j$ de $\mathcal{O}_{Z,\bar{o}}$ dans $\mathcal{O}_{X,\bar{o}}$ est égale à:

(2.6)
$$\bar{\varphi} = \sum_{p=0}^{n-1} \Big(\sum_{k\geq 0} g^k \varphi_{p+nk} \Big) z^p .$$

Nous considérons dans la suite un diviseur C sur S à singularités isolées appartenant au système linéaire $|\mathcal{L}^{\otimes m}|$, et un diviseur régulier L appartenant au système linéaire $|\mathcal{L}|$ et transverse à C. Comme au paragraphe I, pour tout nombre réel α, $-1 \leq \alpha < 0$, nous définissons le faisceau $\mathcal{A}_\alpha$ comme l'idéal de l'anneau $\mathcal{O}_S$ constitué des germes de fonctions $\bar{\varphi}$ vérifiant $\beta_{C,x}(\bar{\varphi}) > \alpha$ en tout point singulier x du diviseur C.

Alors pour tout entier $n \geq m$, nous pouvons définir les faisceaux d'idéaux $\mathcal{I}_F^{(p)}$ par les égalités suivantes:

(2.7)
$$\mathcal{I}_F^{(p)} = \mathcal{A}_\alpha \otimes_{\mathcal{O}_S} \mathcal{O}_S(-e(p)L)$$
$$\text{avec} \quad \alpha = -\frac{p+1}{n} \quad \text{et} \quad e(p) = \mathrm{Sup}\Big(\Big[\frac{(n-m)(n-1-p)}{n} \Big], 0 \Big).$$

Remarques II.2 *(i) Comme les faisceaux $\mathcal{A}_\alpha$ vérifient les inclusions $\mathcal{A}_\alpha \subset \mathcal{A}_\beta$ pour $\alpha \geq \beta$, les faisceaux $\mathcal{I}_F^{(p)}$ donnés par les égalités (2.7) vérifient bien $\mathcal{I}_F^{(p)} \subset \mathcal{I}_F^{(p+1)}$.*
(ii) Comme le faisceau $\mathcal{A}_{-1}$ est égal au faisceau $\mathcal{O}_S$, le faisceau $\mathcal{I}_F^{(p)}$ est isomorphe au faisceau $\mathcal{O}_S$ pour tout $p \geq n-1$.

Proposition II.2 *L'idéal d'adjonction $\mathcal{A}_X$ de X dans Z, où le diviseur X de Z est le revêtement cyclique de degré n de S défini par le diviseur $C+(n-m)L$, est l'idéal de $\mathcal{O}_Z$ égal à $\mathcal{O}_Z$ sur l'ouvert Z^+ et dont la restriction à l'ouvert Z^- est égale à:*

(2.8)
$$\mathcal{A}_X = \bigoplus_{p\geq 0} \mathcal{I}_F^{(p)} \otimes_{\mathcal{O}_S} \mathcal{L}^{\otimes -p}.$$

Démonstration : Comme le faisceau dualisant ω_X sur X est inversible, nous pouvons définir un idéal $\mathcal{I}_F$ de $\mathcal{O}_X$ par $\mathcal{I}_F = (\pi_*(\omega_{\tilde{X}})) \otimes_{\mathcal{O}_X} (\omega_X)^{-1}$ et nous appelons alors F le fermé de X défini par cet idéal. C'est un particulier le support du faisceau $\mathcal{B}$ défini en (0.2).

Par définition l'idéal d'adjonction $\mathcal{A}_X$ de X dans Z est l'idéal de $\mathcal{O}_Z$ qui définit F comme fermé de Z.

Comme la variété $\bar{X}$ définie en (1.1) à partir de la résolution plongée $\sigma: \tilde{S} \rightarrow S$ des singularités de C a pour seules singularités des singularités rationnelles, les faisceaux $\bar{\sigma}_*(\omega_{\bar{X}})$ et $\pi_*(\omega_{\tilde{X}})$ sont isomorphes. Nous pouvons alors utiliser les descriptions des variétés X et $\bar{X}$ comme revêtements cycliques respectivement de S et $\tilde{S}$ pour calculer l'idéal $\mathcal{I}_F$, nous en déduiron ensuite l'idéal d'adjonction $\mathcal{A}_X$.

Comme le morphisme $\psi: X \rightarrow S$ est fini et plat, le faisceau dualisant ω_X sur X se déduit du faisceau dualisant ω_S par la relation $\omega_X = \psi^{-1}\underline{\mathcal{H}om}_{\mathcal{O}_S}(\psi_*\mathcal{O}_X, \omega_S)$, c'est à dire:

$$\omega_X = \bigoplus_{p=0}^{n-1} \mathcal{L}^{\otimes p} \otimes_{\mathcal{O}_S} \omega_S = \mathcal{O}_X \otimes_{\mathcal{O}_S} (\mathcal{L}^{\otimes n-1} \otimes_{\mathcal{O}_S} \omega_S). \tag{2.9}$$

De la même manière, le faisceau dualisant $\omega_{\bar{X}}$ sur $\bar{X}$ se déduit du faisceau dualisant $\omega_{\tilde{S}}$ par la relation $\omega_{\bar{X}} = \tilde{\psi}^{-1}\underline{\mathcal{H}om}_{\mathcal{O}_{\tilde{S}}}(\tilde{\psi}_*\mathcal{O}_{\bar{X}}, \omega_{\tilde{S}})$, c'est à dire:

$$\omega_{\bar{X}} = \bigoplus_{p=0}^{n-1} \mathcal{L}^{[p]} \otimes_{\mathcal{O}_{\tilde{S}}} \omega_{\tilde{S}}. \tag{2.10}$$

L'application de faisceaux sur X: $\bar{\sigma}_*(\omega_{\bar{X}}) \longrightarrow \omega_X$ donne alors par image directe l'application de faisceaux sur S:

$$\bigoplus_{p=0}^{n-1} \sigma_*\left(\mathcal{L}^{[p]} \otimes_{\mathcal{O}_{\tilde{S}}} \omega_{\tilde{S}}\right) \longrightarrow \bigoplus_{p=0}^{n-1}\left(\mathcal{L}^{\otimes p} \otimes_{\mathcal{O}_S} \omega_S\right).$$

Cette application est "homogène", par conséquent l'idéal $\mathcal{I}_F$ qu'elle définit dans $\mathcal{O}_X = \bigoplus_{p=0}^{n-1} \mathcal{L}^{\otimes -p}$ est lui aussi "homogène", c'est à dire que $\mathcal{I}_F$ peut s'écrire:

$$\mathcal{I}_F = \bigoplus_{p=0}^{n-1} \mathcal{I}_F^{(p)} \otimes_{\mathcal{O}_S} \mathcal{L}^{\otimes -p}, \tag{2.11}$$

Localement au voisinage d'un point o, les idéaux $(\mathcal{I}_F^{(p)})_o$ de l'anneau $\mathcal{O}_{S,o}$ vérifient $(\mathcal{I}_F^{(p)})_o \subset (\mathcal{I}_F^{(p+1)})_o$, pour tout p, $0 \leq p < n-1$ et $g \in (\mathcal{I}_F^{(0)})_o$.
vPar définition de l'idéal d'adjonction $\mathcal{A}_X$ de X dans Z, un élément φ de l'anneau $\mathcal{O}_{Z,\bar{o}}$ appartient à $\mathcal{A}_X$ si et seulement si son image $\bar{\varphi}$ dans l'anneau $\mathcal{O}_{X,\bar{o}} = \mathcal{O}_{Z,\bar{o}}/(z^n - g)$ appartient à l'idéal $\mathcal{I}_F$.

Nous déduisons alors de la relation (2.6) et de la définition (2.11) des idéaux $\mathcal{I}_F^{(p)}$ que l'idéal $\mathcal{A}_X$ est homogène et peut s'écrire localement sous la forme $(\mathcal{A}_X)_o = \bigoplus_{p \geq 0}(\mathcal{I}^{(p)})_o z^p$ avec $(\mathcal{I}^{(p)})_o = (\mathcal{I}_F^{(p)})_o$ pour $0 \leq p \leq n-1$ et $(\mathcal{I}^{(p)})_o = \mathcal{O}_{S,o}$ pour $p \geq n$.

Le faisceau d'idéaux $\mathcal{I}_F$ de $\mathcal{O}_X$ est isomorphe à $\underline{\mathcal{H}om}_{\mathcal{O}_X}(\omega_X, \pi_*(\omega_{\tilde{X}}))$, et nous déduisons des égalités (2.9) et (2.10) l'isomorphisme de $\mathcal{I}_F$ avec :

$$\underline{\mathcal{H}om}_{\mathcal{O}_S}\Big(\mathcal{L}^{\otimes n-1}\otimes_{\mathcal{O}_S}\omega_S, \bigoplus_{p=0}^{n-1}\sigma_*(\mathcal{L}^{[p]}\otimes_{\mathcal{O}_{\tilde{S}}}\omega_{\tilde{S}})\Big) =$$
$$\bigoplus_{p=0}^{n-1}\sigma_*(\mathcal{L}^{[p]}\otimes_{\mathcal{O}_{\tilde{S}}}\omega_{\tilde{S}})\otimes_{\mathcal{O}_S}\left(\mathcal{L}^{\otimes n-1}\otimes_{\mathcal{O}_S}\omega_S\right)^{-1}.$$

Pour trouver les idéaux $\mathcal{I}_F^{(p)}$, $0 \leq p \leq n-1$, il suffit alors d'écrire que l'idéal $\mathcal{I}_F$ de $\mathcal{O}_X$ est égal à la somme directe $\bigoplus_{p=0}^{n-1}\mathcal{I}_F^{(p)}\otimes_{\mathcal{O}_S}\mathcal{L}^{\otimes -p}$, où les faisceaux $\mathcal{I}_F^{(p)}$ sont des idéaux de $\mathcal{O}_S$, et nous trouvons l'égalité:

$$\mathcal{I}_F^{(p)} = \sigma_*(\mathcal{L}^{[n-1-p]}\otimes_{\mathcal{O}_{\tilde{S}}}\omega_{\tilde{S}})\otimes_{\mathcal{O}_S}\left(\mathcal{L}^{\otimes n-1-p}\otimes_{\mathcal{O}_S}\omega_S\right)^{-1}. \tag{2.12}$$

Pour conclure il suffit alors d'utiliser la définition du faisceau $\mathcal{A}_\alpha$ de $\mathcal{O}_S$ comme faisceau image directe par le morphisme σ du faisceau $\mathcal{O}_{\tilde{S}}\Big(-\sum([(\alpha+1)m_j]-k_j)E_j\Big)$, c'est à dire l'isomorphisme:

$$\sigma_*(\mathcal{L}^{[n-1-p]}\otimes_{\mathcal{O}_{\tilde{S}}}\omega_{\tilde{S}}) \simeq \mathcal{A}_\alpha\otimes_{\mathcal{O}_S}\omega_S\otimes_{\mathcal{O}_S}\mathcal{L}^{\otimes\lceil\frac{(n-1-p)m}{n}\rceil}, \quad \text{avec} \quad \alpha = -\frac{p+1}{n}.$$

Remarque II.3 *Plaçons nous de nouveau au voisinage d'un point $\tilde{o}$ de $X\bigcap H$ et appelons o son image dans S. Si $(dx_1,\dots,x_d)$ est un système régulier de paramètres sur S au point o, nous pouvons noter $(dx_1\wedge\ldots\wedge dx_d)$ un générateur du faisceau dualisant ω_S au point o.*
Alors $(x_1,\dots,x_d,z)$ est un système régulier de paramètres sur Z au point $\tilde{o}$, et comme le diviseur X est défini dans Z par z^n-g, avec $g\in\mathcal{O}_{S,o}$, le faisceau dualisant ω_X sur X admet comme générateur au point $\tilde{o}$ l'élément $\left(\dfrac{dx_1\wedge\ldots\wedge dx_d}{z^{n-1}}\right)$.

Comme nous pouvons considérer que (z^{1-n}) est le générateur du faisceau $\mathcal{L}^{\otimes n-1}$, nous retrouvons bien l'écriture précédente $\omega_X = \mathcal{O}_X\otimes_{\mathcal{O}_S}(\mathcal{L}^{\otimes n-1}\otimes_{\mathcal{O}_S}\omega_S)$.

Pour tout entier k de $\mathbf{Z}$ et pour tout faisceau $\mathcal{F}$ sur Z, nous notons $\mathcal{F}(k)$ le nouveau faisceau sur Z égal à $\mathcal{F}\otimes_{\mathcal{O}_Z}\mathcal{O}(k)$.

Pour tout entier positif k, la section globale h^k du faisceau inversible $\mathcal{O}_Z(k)$ définit une application injective de $\mathcal{O}_Z$ dans $\mathcal{O}_Z(k)$ et nous voulons définir pour tout germe φ du faisceau $\mathcal{O}_Z$ un entier $\kappa=\kappa(\varphi)$ comme le plus petit entier $k\geq 0$ tel que le germe $h^k\varphi$ appartienne au sous-faisceau $\mathcal{A}_X(k)$, c'est à dire l'entier κ est défini par:

$$h^k\varphi\in\mathcal{A}_X(k) \iff k\geq\kappa. \tag{2.13}$$

L'existence de l'entier $\kappa(\varphi)$ est une conséquence du résultat suivant.

Lemme II.1 *Tout germe φ de $\mathcal{O}_Z$ vérifie les deux propriétés:*
(i) $h^k\varphi \in \mathcal{A}_X(k) \Longrightarrow h^{k+1}\varphi \in \mathcal{A}_X(k+1)$,
(ii) $\exists k \in \mathbf{N}$, *tel que* $h^k\varphi \in \mathcal{A}_X(k)$.

Démonstration : La première propriété est évidente car le faisceau $\mathcal{A}_X$ est un idéal de $\mathcal{O}_Z$.

De même il suffit de montrer la deuxième propriété pour $\varphi = 1$, c'est à dire il suffit de montrer que pour k suffisamment grand, h^k appartient à l'idéal d'adjonction $\mathcal{A}_X$.

C'est équivalent à montrer que pour k suffisamment grand, le fermé F de Z défini par $\mathcal{A}_X$ est inclus dans le diviseur kH. Or nous savons que le sous espace réduit F_{red} est inclus dans $\mathrm{Sing}(X)_{\mathrm{red}}$, qui est lui même inclus dans H.

Grâce à l'isomorphisme ι entre H et S nous pouvons définir une application injective de $\mathcal{O}_H$ dans $\mathcal{O}_Z$, section de la surjection naturelle de $\mathcal{O}_Z$ dans $\mathcal{O}_H$. Nous pouvons ainsi considérer tout germe $\bar{\varphi}$ de $\mathcal{O}_H$ comme un germe de $\mathcal{O}_Z$ et définir l'entier $\kappa(\bar{\varphi})$.

Nous allons maintenant étudier l'entier $\kappa(\bar{\varphi})$ localement au voisinage d'un point singulier o de C et montrer que $\kappa(\bar{\varphi})$ dépend essentiellement de la singularité (C, o).

Si $\tilde{o}$ est un point de Z appartenant à H, et si o est son image dans S, l'anneau local $\mathcal{O}_{S,o}$ est isomorphe à $\mathcal{O}_{H,\tilde{o}}$, c'est à dire à l'anneau quotient $\mathcal{O}_{Z,\tilde{o}}/(z)$, où z est l'élément régulier de $\mathcal{O}_{Z,\tilde{o}}$ définissant H dans Z.

Soit f un élément de $\mathcal{O}_{S,o}$ définissant le germe d'hypersurface (C, o) à singularité isolée, et pour tout entier n, $n \geq 2$, nous appelons $X(n)$ le germe d'hypersurface de Z en $\tilde{o}$ défini par $z^n - f$.

Pour tout élément $\bar{\varphi}$ de $\mathcal{O}_{S,o}$ et pour tout entier n, nous considérons l'entier $\kappa = \kappa(\bar{\varphi}, n)$ défini comme précédemment par le revêtement cyclique $X(n)$ de S, c'est à dire défini par:

$$z^k\bar{\varphi} \in \mathcal{A}_{X(n)} \Longleftrightarrow k \geq \kappa(\bar{\varphi}, n), \tag{2.14}$$

où $\mathcal{A}_{X(n)}$ est l'idéal d'adjonction de $X(n)$ dans Z au point $\tilde{o}$.

Proposition II.3 *Soit $\beta_{C,o}(\bar{\varphi})$ le nombre rationnel défini en (1.3) à partir d'une résolution plongée de la singularité (C, o), alors pour tout entier $n \geq 2$, nous avons l'égalité :*

$$\kappa(\bar{\varphi}, n) = [-n.\beta_{C,o}(\bar{\varphi})]. \tag{2.15}$$

Remarques II.4 *(i) Libgober démontre l'existence d'un entier β ne dépendant que de la singularité (C, o) vérifiant l'égalité $\kappa(\bar{\varphi}, n) = [n\beta]$ dans le cas*

où (C, o) est une singularité isolée non dégénérée par rapport à son polyhèdre de Newton (cf. [Li.2] Proposition 5.1).
Pour cela il utilise la description de l'idéal d'adjonction $\mathcal{A}_{X(n)}$ du revêtement cyclique $X(n)$ donnée dans ([Me.-Te.]), et en particulier le fait que $\mathcal{A}_{X(n)}$ est engendré par des monômes $x_1^{i_1} \ldots x_d^{i_d} z^{i_{d+1}}$.
(ii) Loeser et l'auteur ont donné une démonstration de cette proposition pour une singularité isolée (C, o) quelconque.
Dans cette démonstration ils utilisent la définition du nombre $\beta_{C,o}$ à partir de la structure de Hodge sur la fibre de Milnor de la singularité (cf. [Lo.-Va.] Définition-Proposition 3.2 et Proposition 3.3).

Démonstration de la proposition II.3 : Grâce à la description de l'idéal d'adjonction donnée à la proposition II.2, nous trouvons l'équivalence:

$$z^k \bar{\varphi} \in \mathcal{A}_{X(n)} \Longleftrightarrow \bar{\varphi} \in \mathcal{I}_F^{(k)},$$

où l'idéal $\mathcal{I}_F^{(k)}$ de $\mathcal{O}_{S,o}$ est égal à l'idéal $\mathcal{A}_\alpha$ avec $\alpha = -\dfrac{k+1}{n}$.

En effet, comme nous nous plaçons au voisinage du point o de C, le diviseur L n'apparait pas.

Par définition de l'idéal $\mathcal{A}_\alpha$, le germe $\bar{\varphi}$ appartient à cet idéal si et seulement si le nombre $\beta_{C,o}(\bar{\varphi})$ est strictement plus grand que α.

Nous avons donc l'équivalence:

$$k \geq \kappa(\bar{\varphi}, n) \Longleftrightarrow \beta_{C,o}(\bar{\varphi}) > -\frac{k+1}{n},$$

dont nous déduisons l'égalité voulue: $\kappa(\bar{\varphi}, n) = [-n.\beta_{C,o}(\bar{\varphi})]$.

Nous allons maintenant reprendre la démonstration de Zariski dans [Za.3]. Pour trouver l'irrégularité q du revêtement cyclique X de $\mathbf{P}^2$, il considère pour certaines valeurs de v dans $\mathbf{Z}$ les faisceaux $\mathcal{A}_X(v) \otimes_{\mathcal{O}_\mathbf{P}} \omega_{\mathbf{P}^3}$ sur $\mathbf{P}^3$, et il cherche à calculer les dimensions $\mathrm{h}^0(\mathbf{P}^3, \mathcal{A}_X(v) \otimes_{\mathcal{O}_\mathbf{P}} \omega_{\mathbf{P}^3})$ des espaces de leurs sections globales et leurs caractéristiques d'Euler-Poincaré $\chi(\mathbf{P}^3, \mathcal{A}_X(v) \otimes_{\mathcal{O}_\mathbf{P}} \omega_{\mathbf{P}^3})$.

Pour cela il a besoin d'introduire de nouveaux faisceaux $\mathcal{A}^{(d)}$ sur $\mathbf{P}^3$ définis à partir de l'idéal d'adjonction $\mathcal{A}_X$. Dans le cas plus général que nous regardons, nous allons définir de manière analogue des faisceaux $\mathcal{A}^{(d)}$ sur Z, et nous montrons comment nous pouvons les interpréter à partir des faisceaux $\mathcal{A}_\alpha$ définis par les exposants de Hodge.

Pour tout entier positif d, nous appelons $\mathcal{A}^{(d)}$ l'idéal de $\mathcal{O}_Z$ image réciproque du sous-faisceau $\mathcal{A}_X(d)$ de $\mathcal{O}_Z(d)$ par l'application injective de $\mathcal{O}_Z$ dans $\mathcal{O}_Z(d)$ définie par la section globale h^d. Nous pouvons noter: $\mathcal{A}^{(d)} = \{l \in \mathcal{O}_Z \ / \ h^d l \in \mathcal{A}_X(d)\}$.

En particulier, pour tout entier v, toute section globale s du faisceau $\mathcal{A}^{(d)}(v - d)$ définit un diviseur F sur Z appartenant au système linéaire $|(v-d)H|$ et tel que le diviseur $F + dH$ soit adjoint à la variété X.

Nous définissons le faisceau $\mathcal{B}^{(d)}$ comme la restriction au diviseur H du faisceau $\mathcal{A}^{(d)}$, c'est à dire le faisceau défini par la suite exacte suivante:

$$(2.16) \qquad 0 \longrightarrow \mathcal{A}^{(d+1)}(v-d-1) \xrightarrow{h} \mathcal{A}^{(d)}(v-d) \longrightarrow \mathcal{B}^{(d)}(v-d) \longrightarrow 0,$$

où nous notons h l'application définie grâce à la section globale h de $\mathcal{O}_Z(1)$.

Lemme II.2 *Le faisceau $\mathcal{B}^{(d)}$ est un faisceau d'idéaux du faisceau $\mathcal{O}_H$.*

Démonstration : Comme les faisceaux $\mathcal{A}^{(d+1)}$ et $\mathcal{A}^{(d)}$ sont des idéaux de $\mathcal{O}_Z$, il existe une application naturelle du faisceau $\mathcal{B}^{(d)}(v-d)$ dans $\mathcal{O}_H(v-d)$. Il faut montrer que cette application est injective.

Soit $\bar{\varphi}$ un germe de section du faisceau $\mathcal{B}^{(d)}(v-d)$ et soit φ un germe de section de $\mathcal{A}^{(d)}(v-d)$ dont l'image est égale à $\bar{\varphi}$. Il suffit de montrer que si l'image de $\bar{\varphi}$ dans $\mathcal{O}_H(v-d)$ est nulle, c'est à dire s'il existe un germe de section φ_1 de $\mathcal{O}_Z(v-d-1)$ tel que $\varphi = h\varphi_1$, alors $\bar{\varphi}$ est nul dans $\mathcal{B}^{(d)}(v-d)$, c'est à dire que le germe φ_1 appartient au sous-faisceau $\mathcal{A}^{(d+1)}(v-d-1)$.

Par définition des faisceaux d'idéaux $\mathcal{A}^{(d)}$ de $\mathcal{O}_Z$, nous avons:

$$\begin{aligned} \varphi \in \mathcal{A}^{(d)}(v-d) &\Longrightarrow h^d\varphi \in \mathcal{A}_X(v) \\ &\Longrightarrow h^{d+1}\varphi_1 \in \mathcal{A}_X(v) \Longrightarrow \varphi_1 \in \mathcal{A}^{(d+1)}(v-d-1), \end{aligned}$$

d'où le résultat.

Proposition II.4 *Pour tout entier $d \geq 0$ nous avons l'isomorphisme:*

$$\mathcal{B}^{(d)} \simeq \mathcal{I}_F^{(d)} := \mathcal{A}_\alpha \otimes_{\mathcal{O}_S} \mathcal{O}_S(-e(d)L)$$

$$\textit{avec} \quad \alpha = -\frac{d+1}{n} \quad \textit{et} \quad e(d) = \mathrm{Sup}\Big(0, \Big[\frac{(n-m)(n-d-1)}{n}\Big]\Big).$$

Démonstration : Plaçons nous sur l'ouvert $Z^- = \mathrm{Spec}_S \bigoplus_{p\geq 0} \mathcal{L}^{\otimes -p}$ de Z. Alors pour tout $k \in \mathbf{Z}$, la restriction à Z^- du faisceau inversible $\mathcal{O}_Z(k)$ est isomorphe au faisceau $\bigoplus_{p\geq 0} \mathcal{L}^{\otimes k-p}$.
vLa section globale h de $\mathcal{O}_Z(1)$ définissant le diviseur H sur Z correspond à l'inclusion naturelle de $\mathcal{O}_Z(-1) = \bigoplus_{p\geq 1} \mathcal{L}^{\otimes -p}$ dans $\mathcal{O}_Z = \bigoplus_{p\geq 0} \mathcal{L}^{\otimes -p}$, et nous retrouvons ainsi l'isomorphisme entre $\mathcal{O}_H$ et $\mathcal{O}_S$.

Grâce à la proposition II.2, qui donne la restriction à Z^- de l'idéal $\mathcal{A}_X$, nous vérifions que pour tout entier $d \geq 0$ la restriction de l'idéal $\mathcal{A}^{(d)}$ à Z^- est égale à:

$$(2.17) \qquad \mathcal{A}^{(d)} = \bigoplus_{p\geq 0} \mathcal{I}_F^{(p+d)} \otimes_{\mathcal{O}_S} \mathcal{L}^{\otimes -p}.$$

Le conoyau de l'application injective de $\mathcal{A}^{(d+1)}(-1)$ dans $\mathcal{A}^{(d)}$ définie par h est alors isomorphe à $\mathcal{I}_F^{(d)}$, et nous déduisons de la suite exacte (2.16) l'isomorphisme cherché entre les faisceaux $\mathcal{B}^{(d)}$ et $\mathcal{I}_F^{(d)}$.

Proposition II.5 *Pour tout entier $d \geq 0$ et pour tout entier v, nous avons des applications canoniques entre les espaces des sections globales:*

$$\zeta \colon \mathrm{H}^0(Z, \mathcal{A}^{(d)}(v-d)) \longrightarrow \mathrm{H}^0(S, \mathcal{I}_F^{(d)} \otimes_{\mathcal{O}_S} \mathcal{L}^{\otimes v-d}),$$
$$\xi \colon \mathrm{H}^0(Z, \mathcal{A}^{(d)}(v-d) \otimes_{\mathcal{O}_Z} \omega_Z) \longrightarrow \mathrm{H}^0(S, \mathcal{I}_F^{(d)} \otimes_{\mathcal{O}_S} \mathcal{L}^{\otimes v-d-1} \otimes_{\mathcal{O}_S} \omega_S).$$

Les applications ζ et ξ sont surjectives respectivement pour $v \geq d$ et pour $v \geq d+2$.

Démonstration : Nous allons donner la démonstration de l'existence et de la surjectivité de l'application ξ, la démonstration pour l'application ζ est à peu près identique, en plus simple.

De la proposition II.4 et de l'isomorphisme entre les faisceaux $\rho^*(\omega_S \otimes_{\mathcal{O}_S} \mathcal{L}) \otimes_{\mathcal{O}_Z} \mathcal{O}_Z(-2)$ et ω_Z nous déduisons l'isomorphisme :

$$\mathcal{B}^{(d)}(v-d) \otimes_{\mathcal{O}_Z} \omega_Z \simeq \mathcal{I}_F^{(d)} \otimes_{\mathcal{O}_S} \mathcal{L}^{\otimes v-d-1} \otimes_{\mathcal{O}_S} \omega_S. \tag{2.18}$$

L'application ξ cherchée est obtenue alors à partir de l'application naturelle du faisceau $\mathcal{A}^{(d)}$ dans le faisceau $\mathcal{B}^{(d)}$.

Nous avons le diagramme commutatif suivant:

$$\begin{array}{ccc}
\mathrm{H}^0(Z, \mathcal{A}^{(d)}(v-d) \otimes_{\mathcal{O}_Z} \omega_Z) & \hookrightarrow & \mathrm{H}^0(Z, \mathcal{O}_Z(v-d) \otimes_{\mathcal{O}_Z} \omega_Z) \\
\downarrow \xi & & \downarrow \xi' \\
\mathrm{H}^0(S, \mathcal{I}_F^{(d)} \otimes_{\mathcal{O}_S} \mathcal{L}^{\otimes v-d-1} \otimes_{\mathcal{O}_S} \omega_S) & \hookrightarrow & \mathrm{H}^0(S, \mathcal{L}^{\otimes v-d-1} \otimes_{\mathcal{O}_S} \omega_S)
\end{array}$$

L'image directe par le morphisme ρ du faisceau $\mathcal{O}_Z(v-d) \otimes_{\mathcal{O}_Z} \omega_Z$ est isomorphe pour $v-d-2 \geq 0$ au faisceau $\omega_S \otimes_{\mathcal{O}_S} \bigoplus_{p=0}^{v-d-2} \mathcal{L}^{\otimes p+1}$, d'où l'isomorphisme entre les espaces $\mathrm{H}^0(Z, \mathcal{O}_Z(v-d) \otimes_{\mathcal{O}_Z} \omega_Z)$ et $\bigoplus_{p=1}^{v-d-1} \mathrm{H}^0(S, \mathcal{L}^{\otimes p} \otimes_{\mathcal{O}_S} \omega_S)$.

L'application ξ' admet donc une section naturelle τ.

Pour montrer que l'application ξ est surjective, il suffit de montrer que pour toute section globale $\bar{\varphi}$ appartenant à $\mathrm{H}^0(S, \mathcal{I}_F^{(d)} \otimes_{\mathcal{O}_S} \mathcal{L}^{\otimes v-d-1} \otimes_{\mathcal{O}_S} \omega_S)$, la section φ, image de $\bar{\varphi}$ dans $\mathrm{H}^0(Z, \mathcal{O}_Z(v-d) \otimes_{\mathcal{O}_Z} \omega_Z)$ par la section τ, appartient en fait au sous-espace $\mathrm{H}^0(Z, \mathcal{A}^{(d)}(v-d) \otimes_{\mathcal{O}_Z} \omega_Z)$. Ceci est évident par l'égalité (2.17) et car le faisceau d'idéaux $\mathcal{A}^{(d)}$ est égal au faisceau $\mathcal{O}_Z$ en dehors du diviseur H.

Remarque II.5 *Pour montrer comment les propositions II.4 et II.5 sont une généralisation des résultats de Zariski ([Za.3]), nous allons les formuler différemment.*

Pour cela, nous allons introduire les notions suivantes.
(i) Un diviseur E sur S est de degré v s'il est défini par une section globale du faisceau $\mathcal{L}^{\otimes v}$, et de même un diviseur F sur $Z = \mathbf{P}(\mathcal{O}_S \oplus \mathcal{L})$ est de degré v s'il est défini par une section globale du faisceau $\mathcal{O}_Z(v)$.
(ii) Soit C un diviseur sur S, ayant uniquement des singularités isolées, alors pour tout nombre rationnel α, $-1 \leq \alpha \leq 0$, nous définissons la notion d'adjonction avec exposant α de la manière suivante :
un diviseur E sur S est adjoint au diviseur C avec exposant α si pour tout point o appartenant au lieu singulier $\mathrm{Sing}(C)$ de C, le diviseur E est défini au voisinage du point o par un élément $\bar{\varphi}$ de l'anneau local $\mathcal{O}_{S,o}$ vérifiant $\beta_{C,o}(\bar{\varphi}) > \alpha$.

En particulier, comme le faisceau $\mathcal{A}_{-1}$ est égal à $\mathcal{O}_S$, tout diviseur E sur S est adjoint à C avec exposant -1, et comme le faisceau $\mathcal{A}_0$ est égal à l'idéal d'adjonction $\mathcal{A}_C$ de C dans S, un diviseur E est adjoint à C si et seulement si il est adjoint à C avec exposant 0.

Dans le cas où C est une courbe ayant pour seuls points singuliers des points doubles ordinaires et des points cuspidaux, un diviseur E sur la surface S est adjoint à C avec exposant $-1/6$ si et seulement si E passe par tous les points cuspidaux de C, c'est à dire avec la terminologie de Zariski, si et seulement si le diviseur E est pré-adjoint à C.

Nous pouvons donner la nouvelle formulation des propositions précédentes.

Propositions II.4 et II.5 *La trace sur le diviseur H de Z du système linéaire complet formé des diviseurs F de degré $v - d$ dans Z tels que $F + dH$ soit adjoint à X, est le système linéaire complet formés des diviseurs E sur $H = S$, de degré $v - d - e(d)$ et adjoints à C avec exposant α, où α est égal à $-\dfrac{d+1}{n}$.*

Nous allons maintenant suivre les idées de Zariski, pour calculer dans le cas où la variété S est le plan projectif $\mathbf{P}^2$, c'est à dire le cas qu'il considère dans [Za.3], l'irrégularité q du revêtement cyclique X de S sans utiliser la variété $\tilde{X}$ et le faisceau dualisant $\omega_{\tilde{X}}$.

Nous montrerons ensuite comment il est possible de généraliser cette démonstration au cas d'une variété lisse projective S quelconque.

Pour tout entier d, $0 \leq d \leq n$, où n est le degré du revêtement cyclique X de S, nous appellons r_d la caractéristique d'Euler-Poincaré du faisceau $\mathcal{A}^{(d)}(n-d) \otimes_{\mathcal{O}_Z} \omega_Z$:

$$r_d = \chi(Z, \mathcal{A}^{(d)}(n-d) \otimes_{\mathcal{O}_Z} \omega_Z).$$

Alors, grâce à la suite exacte (2.16) et à l'isomorphisme (2.18), nous trouvons pour tout d la nouvelle suite exacte:

$$0 \longrightarrow \mathcal{A}^{(d+1)}(n-d-1) \otimes_{\mathcal{O}_Z} \omega_Z \longrightarrow \mathcal{A}^{(d)}(n-d) \otimes_{\mathcal{O}_Z} \omega_Z \longrightarrow \mathcal{I}_F^{(d)} \otimes_{\mathcal{O}_S} \mathcal{L}^{\otimes n-d-1} \otimes_{\mathcal{O}_S} \omega_S \longrightarrow 0.$$

Nous en déduisons l'égalité $r_d - r_{d+1} = \chi(S, \mathcal{I}_F^{(d)} \otimes_{\mathcal{O}_S} \mathcal{L}^{\otimes n-d-1} \otimes_{\mathcal{O}_S} \omega_S)$, d'où pour tout entier $m \geq 1$:

$$(2.19) \qquad r_0 - r_m = \sum_{d=0}^{m-1} \chi(S, \mathcal{I}_F^{(d)} \otimes_{\mathcal{O}_S} \mathcal{L}^{\otimes n-d-1} \otimes_{\mathcal{O}_S} \omega_S).$$

Comme le faisceau $\mathcal{A}^{(p)}$ est égal à $\mathcal{O}_Z$ pour $p \geq n-1$ (cf. Remarque II.2 (ii)), le faisceau $\mathcal{A}^{(n)}(0) \otimes_{\mathcal{O}_Z} \omega_Z$ est isomorphe au faisceau dualisant ω_Z et nous avons l'égalité:

$$r_n = \chi(Z, \mathcal{A}^{(n)}(0) \otimes_{\mathcal{O}_Z} \omega_Z) = \chi(Z, \omega_Z).$$

Par définition le faisceau $\mathcal{A}^{(0)}$ est égal à l'idéal d'adjonction $\mathcal{A}_X$ de X dans Z, et comme le diviseur X est de "degré" n, le faisceau $\mathcal{A}^{(0)}(n) \otimes_{\mathcal{O}_Z} \omega_Z$ est isomorphe à $\mathcal{A}_X \otimes_{\mathcal{O}_Z} \omega_Z(X)$.

Nous déduisons de la suite exacte définissant l'idéal d'adjonction (cf. (2.4)) que la caractéristique d'Euler-Poincaré du faisceau $\mathcal{A}_X \otimes_{\mathcal{O}_Z} \omega_Z(X)$ vérifie:

$$r_0 = \chi(Z, \mathcal{A}_X \otimes_{\mathcal{O}_Z} \omega_Z(X)) = \chi(Z, \omega_Z) + \chi(X, \pi_*(\omega_{\tilde{X}})).$$

D'après le théorème de Grauert-Riemenschneider, les faisceaux $\mathrm{R}^j\pi_*(\omega_{\tilde{X}})$ sont nuls pour $j \geq 1$, et les caractéristiques d'Euler-Poincaré des faisceaux $\pi_*(\omega_{\tilde{X}})$ sur X et $\omega_{\tilde{X}}$ sur $\tilde{X}$ sont égales.

Nous déduisons alors de ce qui précède l'égalité: $r_0 - r_n = \chi(\tilde{X}, \omega_{\tilde{X}})$.

Grâce à l'égalité (2.19) pour $m = n$, et grâce à la proposition II.4 nous trouvons:

$$\chi(\tilde{X}, \omega_{\tilde{X}}) = \sum_{d=0}^{n-1} \chi\big(S, \mathcal{A}_\alpha((n-d-1-e(d))L) \otimes_{\mathcal{O}_S} \omega_S\big), \quad \text{avec} \quad \alpha = -\frac{d+1}{n}.$$

En posant $p = n-1-d$ et en utilisant la définition de l'entier $e(d)$, nous avons alors:

$$(2.20) \qquad \chi(\tilde{X}, \omega_{\tilde{X}}) = \sum_{p=0}^{n-1} \chi\big(S, \mathcal{A}_\alpha(\lceil \frac{mp}{n} \rceil L) \otimes_{\mathcal{O}_S} \omega_S\big), \quad \text{avec} \quad \alpha = \frac{p}{n} - 1.$$

De la même manière, pour tout entier d, $0 \leq d \leq n$, nous appelons s_d la dimension de l'espace des sections globales du faisceau $\mathcal{A}^{(d)}(n-d) \otimes_{\mathcal{O}_Z} \omega_Z$, c'est à dire:

$$s_d = \mathrm{h}^0(Z, \mathcal{A}^{(d)}(n-d) \otimes_{\mathcal{O}_Z} \omega_Z).$$

Grâce à la suite exacte précédente et à la surjectivité de l'application ξ, nous avons l'égalité $s_d - s_{d+1} = \mathrm{h}^0(S, \mathcal{I}_F^{(d)} \otimes_{\mathcal{O}_S} \mathcal{L}^{\otimes n-d-1} \otimes_{\mathcal{O}_S} \omega_S)$ pour $0 \leq d \leq n-2$, d'où pour tout entier m, $1 \leq m \leq n-1$:

$$(2.21) \qquad s_0 - s_m = \sum_{d=0}^{m-1} \mathrm{h}^0(S, \mathcal{I}_F^{(d)} \otimes_{\mathcal{O}_S} \mathcal{L}^{\otimes n-d-1} \otimes_{\mathcal{O}_S} \omega_S).$$

Nous déduisons de l'isomorphisme entre les faisceaux $\rho^*(\omega_S \otimes_{\mathcal{O}_S} \mathcal{L}) \otimes_{\mathcal{O}_Z} \mathcal{O}_Z(-1)$ et $\mathcal{A}^{(n-1)}(1) \otimes_{\mathcal{O}_Z} \omega_Z$, et des égalités (2.1) et (2.2) que pour tout $j \geq 0$, les faisceaux images directes $\mathrm{R}^j \rho_*(\mathcal{A}^{(n-1)}(1) \otimes_{\mathcal{O}_Z} \omega_Z)$ sont nuls, d'où:

$$(2.22) \qquad \mathrm{H}^j(Z, \mathcal{A}^{(n-1)}(1) \otimes_{\mathcal{O}_Z} \omega_Z) = 0 \quad \forall j \geq 0.$$

Les entiers $s_{n-1} = \mathrm{h}^1(Z, \mathcal{A}^{(n-1)}(1) \otimes_{\mathcal{O}_Z} \omega_Z)$ et $r_{n-1} = \chi(Z, \mathcal{A}^{(n-1)}(1) \otimes_{\mathcal{O}_Z} \omega_Z)$ sont donc nuls.

De même nous déduisons des relations $\rho_*(\omega_Z) = 0$ et $\mathrm{R}^1\rho_*(\omega_Z) \simeq \omega_S$ l'égalité $\mathrm{H}^0(Z, \omega_Z) = 0$ et les isomorphismes $\mathrm{H}^j(Z, \omega_Z) \simeq \mathrm{H}^{j-1}(S, \omega_S)$ pour $1 \leq j \leq d+1$.

Si nous supposons X irréductible, l'application canonique de $\mathrm{H}^d(\tilde{X}, \omega_{\tilde{X}})$ dans $\mathrm{H}^{d+1}(Z, \omega_Z)$ est un isomorphisme et le groupe $\mathrm{H}^{d+1}(Z, \mathcal{A}_X \otimes_{\mathcal{O}_Z} \omega_Z(X))$ est nul.

Nous en déduisons aussi la nouvelle suite exacte:

$$\begin{aligned} \mathrm{H}^{d-1}(Z, \omega_Z) \longrightarrow & \mathrm{H}^{d-1}(Z, \mathcal{A}_X \otimes_{\mathcal{O}_Z} \omega_Z(X)) \longrightarrow \mathrm{H}^{d-1}(\tilde{X}, \omega_{\tilde{X}}) \\ & \longrightarrow \mathrm{H}^d(Z, \omega_Z) \longrightarrow \mathrm{H}^d(Z, \mathcal{A}_X \otimes_{\mathcal{O}_Z} \omega_Z(X)) \longrightarrow 0. \end{aligned}$$

Nous pouvons maintenant énoncer le résultat qui généralise le théorème de Zariski ([Za.3] fin du paragraphe II, page 499), c'est à dire le corollaire I.1 pour $S = \mathbf{P}^2$ et $r = 1$.

Proposition II.6 *L'irrégularité q du revêtement cyclique X de degré n du plan projectif $\mathbf{P}^2$, ramifié au dessus de la courbe C et de la droite à l'infini L est égale à:*

$$q := \dim \mathrm{H}^1(\tilde{X}, \omega_{\tilde{X}}) = \sum_{p=0}^{n-1} \mathrm{h}^1\Big(\mathbf{P}^2, \mathcal{A}_\alpha(\Big\lceil \frac{pm}{n} \Big\rceil - 3)\Big) \quad \textit{avec} \quad \alpha = \frac{p}{n} - 1.$$

Démonstration : Nous allons montrer que pour toute surface projective lisse S et pour tout d, $0 \leq d \leq n-2$, le groupe

$$\mathrm{H}^2\Big(S, \mathcal{I}_F^{(d)} \otimes_{\mathcal{O}_S} \mathcal{L}^{\otimes n-d-1} \otimes_{\mathcal{O}_S} \omega_S\Big) = \mathrm{H}^2\Big(S, \mathcal{A}_\alpha\Big(\Big\lceil \frac{pm}{n} \Big\rceil L\Big) \otimes_{\mathcal{O}_S} \omega_S\Big),$$

avec $p = n - d - 1$ et $\alpha = \dfrac{p}{n} - 1$, est nul.

Comme le fermé F_α de S défini par $\mathcal{A}_\alpha$ est inclus dans le lieu singulier de la courbe C, nous déduisons de la suite exacte suivante:

$$0 \longrightarrow \mathcal{A}_\alpha \otimes_{\mathcal{O}_S} \omega_S\left(\left\lceil \frac{pm}{n} \right\rceil L\right) \longrightarrow \omega_S\left(\left\lceil \frac{pm}{n} \right\rceil L\right) \longrightarrow \mathcal{O}_{F_\alpha} \otimes_{\mathcal{O}_S} \omega_S\left(\left\lceil \frac{pm}{n} \right\rceil L\right) \longrightarrow 0$$

un isomorphisme entre les groupes

$$\mathrm{H}^2\left(S, \mathcal{A}_\alpha \otimes_{\mathcal{O}_S} \omega_S\left(\left\lceil \frac{pm}{n} \right\rceil L\right)\right) \text{ et } \mathrm{H}^2\left(S, \omega_S\left(\left\lceil \frac{pm}{n} \right\rceil L\right)\right).$$

Ce groupe est nul car le diviseur L est très ample sur la surface S et car $\left\lceil \frac{pm}{n} \right\rceil$ est strictement positif. Nous avons alors pour $0 \le d \le n-2$:

$$\begin{aligned} \mathrm{h}^0(S, \mathcal{I}_F^{(d)} \otimes_{\mathcal{O}_S} \mathcal{L}^{\otimes n-d-1} \otimes_{\mathcal{O}_S} \omega_S) - \chi(S, \mathcal{I}_F^{(d)} \otimes_{\mathcal{O}_S} \mathcal{L}^{\otimes n-d-1} \otimes_{\mathcal{O}_S} \omega_S) \\ = \mathrm{h}^1(S, \mathcal{I}_F^{(d)} \otimes_{\mathcal{O}_S} \mathcal{L}^{\otimes n-d-1} \otimes_{\mathcal{O}_S} \omega_S). \end{aligned}$$

Nous déduisons de la nullité des groupes $\mathrm{H}^j(Z, \mathcal{A}^{(n-1)}(1) \otimes_{\mathcal{O}_Z} \omega_Z)$ (cf (2.22)) que la différence $s_0 - r_0$ est égale à $(s_0 - s_{n-1}) - (r_0 - r_{n-1})$, c'est à dire d'après les relations (2.19) et (2.21) et d'après l'égalité précédente:

$$s_0 - r_0 = \sum_{d=0}^{n-2} \mathrm{h}^1(S, \mathcal{I}_F^{(d)} \otimes_{\mathcal{O}_S} \mathcal{L}^{\otimes n-d-1} \otimes_{\mathcal{O}_S} \omega_S).$$

Nous pouvons encore écrire cette égalité sous la forme:

$$\begin{aligned} \mathrm{h}^0(Z, \mathcal{A}_X \otimes_{\mathcal{O}_Z} \omega_Z(X)) - \chi(Z, \mathcal{A}_X \otimes_{\mathcal{O}_Z} \omega_Z(X)) + \mathrm{h}^1(S, \omega_S) \\ = \sum_{d=0}^{n-1} \mathrm{h}^1(S, \mathcal{I}_F^{(d)} \otimes_{\mathcal{O}_S} \mathcal{L}^{\otimes n-d-1} \otimes_{\mathcal{O}_S} \omega_S). \end{aligned}$$

Si l'application de $\mathrm{H}^1(Z, \omega_Z)$ dans $\mathrm{H}^1(Z, \mathcal{A}_X \otimes_{\mathcal{O}_Z} \omega_Z(X))$ déduite du diagramme (2.4) est nulle, le nombre $\mathrm{h}^0(Z, \mathcal{A}_X \otimes_{\mathcal{O}_Z} \omega_Z(X)) - \chi(Z, \mathcal{A}_X \otimes_{\mathcal{O}_Z} \omega_Z(X))$ est égal à la différence $\mathrm{h}^1(\tilde{X}, \omega_{\tilde{X}}) - \mathrm{h}^2(Z, \omega_Z)$, c'est à dire à $\mathrm{h}^1(\tilde{X}, \omega_{\tilde{X}}) - \mathrm{h}^1(S, \omega_S)$.

Nous déduisons alors de l'égalité précédente la relation:
(2.23)

$$q = \sum_{d=0}^{n-1} \mathrm{h}^1(S, \mathcal{I}_F^{(d)} \otimes_{\mathcal{O}_S} \mathcal{L}^{\otimes n-d-1} \otimes_{\mathcal{O}_S} \omega_S) = \sum_{p=0}^{n-1} \mathrm{h}^1(S, \mathcal{A}_\alpha \otimes_{\mathcal{O}_S} \omega_S \otimes_{\mathcal{O}_S} \mathcal{L}^{\otimes \lceil \frac{pm}{n} \rceil}),$$

avec $\alpha = \frac{p}{n} - 1$.

C'est le cas en particulier si la variété S est le plan projectif $\mathbf{P}^2$ car nous avons alors $\mathrm{H}^1(Z, \omega_Z) \simeq \mathrm{H}^0(\mathbf{P}^2, \omega_{\mathbf{P}^2}) = 0$. Comme le faisceau ample $\mathcal{L}$ est le faisceau $\mathcal{O}_{\mathbf{P}^2}(1)$ et comme le faisceau dualisant ω_S est égal à $\mathcal{O}_{\mathbf{P}^2}(-3)$, nous trouvons le résultat cherché.

Pour retrouver le corollaire I.1 dans le cas où la variété S est une variété lisse projective quelconque, de dimension $d \geq 2$, nous devons utiliser l'isomorphisme entre les faisceaux $\pi_*(\omega_{\tilde{X}})$ et $\bar{\sigma}_*(\omega_{\tilde{X}})$, et plus précisément l'isomorphisme:

$$\rho_*\pi_*(\omega_{\tilde{X}}) \simeq \sigma_*\bar{\psi}_*(\omega_{\tilde{X}}) \simeq \bigoplus_{p=0}^{n-1} \sigma_*\left(\omega_{\tilde{S}} \otimes_{\mathcal{O}_{\tilde{S}}} \mathcal{L}^{[p]}\right).$$

Lemme II.3 *Pour tout entier $j \geq 0$ nous avons l'égalité:*

$$\mathrm{h}^j(\tilde{X}, \omega_{\tilde{X}}) = \mathrm{h}^j(S, \omega_S) + \mathrm{h}^j(Z, \mathcal{A}_X \otimes_{\mathcal{O}_Z} \omega_Z(X)).$$

Démonstration : Nous déduisons du diagramme (2.4) la suite exacte:

$$0 \longrightarrow \rho_*(\mathcal{A}_X \otimes_{\mathcal{O}_Z} \omega_Z(X)) \longrightarrow \rho_*\pi_*(\omega_{\tilde{X}}) \longrightarrow \mathrm{R}^1\rho_*(\omega_Z) \longrightarrow \mathrm{R}^1\rho_*(\mathcal{A}_X \otimes_{\mathcal{O}_Z} \omega_Z(X)) \longrightarrow 0.$$

L'application $\rho_*\pi_*(\omega_{\tilde{X}}) \longrightarrow \mathrm{R}^1\rho_*(\omega_Z)$ correspond grâce à l'isomorphisme précédent à la projection du faisceau $\bigoplus_{p=0}^{n-1} \sigma_*(\omega_{\tilde{S}} \otimes_{\mathcal{O}_{\tilde{S}}} \mathcal{L}^{[p]})$ sur le facteur direct $\sigma_*(\omega_{\tilde{S}} \otimes_{\mathcal{O}_{\tilde{S}}} \mathcal{L}^{[0]}) = \omega_S \simeq \mathrm{R}^1\rho_*(\omega_Z)$.

En effet il suffit de remarquer que l'application $\rho_*(\omega_X) \longrightarrow \mathrm{R}^1\rho_*(\omega_Z)$ déduite de la suite exacte d'adjonction correspond à la projection de

$$\rho_*(\omega_X) = \bigoplus_{p=0}^{n-1}(\omega_S \otimes_{\mathcal{O}_S} \mathcal{L}^{\otimes p})$$

sur le facteur direct $\omega_S \simeq \mathrm{R}^1\rho_*(\omega_Z)$, et que l'application de $\pi_*(\omega_{\tilde{X}})$ dans ω_X induit une bijection du facteur direct $\sigma_*(\omega_{\tilde{S}} \otimes_{\mathcal{O}_{\tilde{S}}} \mathcal{L}^{[0]})$ de $\rho_*\pi_*(\omega_{\tilde{X}})$ dans le facteur direct ω_S de $\rho_*(\omega_X)$.

Nous déduisons de ce qui précède que l'image directe $\rho_*(\mathcal{A}_X \otimes_{\mathcal{O}_Z} \omega_Z(X))$ est isomorphe au faisceau $\bigoplus_{p=1}^{n-1} \sigma_*(\omega_{\tilde{S}} \otimes_{\mathcal{O}_{\tilde{S}}} \mathcal{L}^{[p]})$ et que l'image directe supérieure $\mathrm{R}^1\rho_*(\mathcal{A}_X \otimes_{\mathcal{O}_Z} \omega_Z(X))$ est nulle.

Nous trouvons ainsi une suite exacte scindée de faisceaux sur S:

$$0 \longrightarrow \rho_*(\mathcal{A}_X \otimes_{\mathcal{O}_Z} \omega_Z(X) \longrightarrow \rho_*\pi_*(\omega_{\tilde{X}}) \longrightarrow \mathrm{R}^1(\omega_Z) \longrightarrow 0,$$

dont nous déduisons l'égalité $\mathrm{h}^j(\tilde{X}, \omega_{\tilde{X}}) = \mathrm{h}^{j+1}(Z, \omega_Z) + \mathrm{h}^j(Z, \mathcal{A}_X \otimes_{\mathcal{O}_Z} \omega_Z(X))$.

En utilisant l'isomorphisme entre les groupes $\mathrm{H}^j(S, \omega_S)$ et $\mathrm{H}^{j+1}(Z, \omega_Z)$ nous trouvons pour tout entier $j \geq 0$ l'égalité cherchée.

Nous avons aussi besoin d'un résultat qui généralise la proposition II.5 à tous les groupes de cohomologie.

Lemme II.4 *L'application $\xi: \mathrm{H}^j(Z, \mathcal{A}^{(d)}(n-d) \otimes_{\mathcal{O}_Z} \omega_Z) \longrightarrow \mathrm{H}^j(S, \mathcal{I}_F^{(d)} \otimes_{\mathcal{O}_S} \mathcal{L}^{\otimes n-d-1} \otimes_{\mathcal{O}_S} \omega_S)$ est surjective pour tout $j \geq 0$ et pour tout d, $0 \leq d \leq n-2$.*

Démonstration : L'application ξ est définie grâce à la suite exacte de faisceaux sur Z:

$$0 \longrightarrow \mathcal{A}^{(d+1)}(n-d-1) \otimes_{\mathcal{O}_Z} \omega_Z \longrightarrow \mathcal{A}^{(d)}(n-d) \otimes_{\mathcal{O}_Z} \omega_Z \longrightarrow \mathcal{I}_F^{(d)} \otimes_{\mathcal{O}_S} \mathcal{L}^{\otimes n-d-1} \otimes_{\mathcal{O}_S} \omega_S \longrightarrow 0.$$

Pour calculer l'image directe par ρ des faisceaux $\mathcal{A}^{(d)}(v-d) \otimes_{\mathcal{O}_Z} \omega_Z \simeq \mathcal{A}^{(d)}(v-d-2) \otimes_{\mathcal{O}_Z} \rho^*(\omega_S \otimes_{\mathcal{O}_S} \mathcal{L})$, nous allons utiliser la description de Z comme réunion des deux ouverts Z^+ et Z^- (2.5).

Comme l'idéal d'adjonction $\mathcal{A}_X$ est égal à $\mathcal{O}_Z$ en dehors du diviseur H la restriction de $\mathcal{A}^{(d)}(u)$ à l'ouvert Z^+ est isomorphe à $\bigoplus_{p\geq 0} \mathcal{L}^{\otimes p}$, et d'après l'égalité (2.17) sa restriction à l'ouvert Z^- est isomorphe à $\bigoplus_{q\leq 0} \mathcal{I}_F^{(-q+d)} \otimes_{\mathcal{O}_S} \mathcal{L}^{\otimes q+u}$.

Nous en déduisons l'image directe du faisceau $\mathcal{A}^{(d)}(u)$ par le morphisme ρ:

$$\rho_*(\mathcal{A}^{(d)}(u)) = \bigoplus_{p=0}^{u} \mathcal{I}_F^{(p+d)} \otimes_{\mathcal{O}_S} \mathcal{L}^{\otimes u-p} \quad \text{pour} \quad u \geq 0,$$

$$\rho_*(\mathcal{A}^{(d)}(u)) = 0 \quad \text{pour} \quad u < 0.$$

La section h de $\mathcal{O}_Z(1)$ définit une application de $\mathcal{A}^{(d+1)}(u-1)$ dans $\mathcal{A}^{(d)}(u)$, qui est une bijection sur l'ouvert Z^+ et qui correspond sur l'ouvert Z^- à l'inclusion naturelle de $\bigoplus_{p\geq 0} \mathcal{I}_F^{(p+1+d)} \otimes_{\mathcal{O}_S} \mathcal{L}^{\otimes u-p-1}$ dans $\bigoplus_{p\geq 0} \mathcal{I}_F^{(p+d)} \otimes_{\mathcal{O}_S} \mathcal{L}^{\otimes u-p}$.

Par image directe nous en déduisons pour tout entier $u \geq 0$ une suite exacte scindée:

$$0 \longrightarrow \rho_*(\mathcal{A}^{(d+1)}(u-1)) \longrightarrow \rho_*(\mathcal{A}^{(d)}(u)) \longrightarrow \mathcal{I}_F^{(d)} \otimes_{\mathcal{O}_S} \mathcal{L}^{\otimes u} \longrightarrow 0,$$

et un isomorphisme entre les faisceaux $\mathrm{R}^1\rho_*(\mathcal{A}^{(d+1)}(u-1))$ et $\mathrm{R}^1\rho_*(\mathcal{A}^{(d)}(u))$.

Nous trouvons alors que pour tout entier $d \leq n-2$ le faisceau $\mathrm{R}^1\rho_*(\mathcal{A}^{(d)}(n-d-2))$ est isomorphe au faisceau image directe $\mathrm{R}^1\rho_*(\mathcal{A}^{(n-1)}(-1))$ qui est nul car $\mathcal{A}^{(n-1)}$ est égal à $\mathcal{O}_Z$, d'où:

$$\mathrm{R}^1\rho_*(\mathcal{A}^{(d)}(n-d-2)) = 0 \quad \forall d,\ 0 \leq d \leq n-1.$$

Nous déduisons de ce qui précède la suite exacte:

$$0 \longrightarrow \mathrm{H}^j(Z, \mathcal{A}^{(d+1)}(n-d-1) \otimes_{\mathcal{O}_Z} \omega_Z) \longrightarrow \mathrm{H}^j(Z, \mathcal{A}^{(d)}(n-d) \otimes_{\mathcal{O}_Z} \omega_Z) \longrightarrow \mathrm{H}^j(S, \mathcal{I}_F^{(d)} \otimes_{\mathcal{O}_S} \mathcal{L}^{\otimes n-d-1} \otimes_{\mathcal{O}_S} \omega_S) \longrightarrow 0,$$

pour tout entier $j \geq 0$ et pour tout d vérifiant $0 \leq d \leq n-2$.

Nous appelons s_d^j la dimension du groupe de cohomologie $\mathrm{H}^j(\mathcal{A}^{(d)}(n-d) \otimes_{\mathcal{O}_Z} \omega_Z)$, c'est à dire:

$$s_d^j = \mathrm{h}^j(Z, \mathcal{A}^{(d)}(n-d) \otimes_{\mathcal{O}_Z} \omega_Z).$$

Avec les notations précédentes nous avons $r_d = \sum_{j=0}^{\dim(Z)} (-1)^j s_d^j$ et $s_d = s_d^0$.

Grâce au lemme II.4 nous avons pour tout j et pour tout d, $0 \leq d \leq n-2$, l'égalité $s_d^j - s_{d+1}^j = \mathrm{h}^j(S, \mathcal{I}_F^{(d)} \otimes_{\mathcal{O}_S} \mathcal{L}^{\otimes n-d-1} \otimes_{\mathcal{O}_S} \omega_S)$, d'où pour tout entier m, $1 \leq m \leq n-1$:

$$s_0^j - s_m^j = \sum_{d=0}^{m-1} \mathrm{h}^j(S, \mathcal{I}_F^{(d)} \otimes_{\mathcal{O}_S} \mathcal{L}^{\otimes n-d-1} \otimes_{\mathcal{O}_S} \omega_S).$$

Par définition le nombre s_0^j est égal à $\mathrm{h}^j(Z, \mathcal{A}_X \otimes_{\mathcal{O}_Z} \omega_Z(X))$, et le nombre s_{n-1}^j est nul (2.22). Nous déduisons alors de l'égalité précédente pour $m = n-1$:

$$\mathrm{h}^j(Z, \mathcal{A}_X \otimes_{\mathcal{O}_Z} \omega_Z(X)) = \sum_{d=0}^{n-2} \mathrm{h}^j(S, \mathcal{I}_F^{(d)} \otimes_{\mathcal{O}_S} \mathcal{L}^{\otimes n-d-1} \otimes_{\mathcal{O}_S} \omega_S).$$

Grâce au lemme II.3, en ajoutant $\mathrm{h}^j(S, \omega_S)$ au deux termes de l'égalité précédente nous trouvons:

$$\mathrm{h}^j(\tilde{X}, \omega_{\tilde{X}}) = \sum_{d=0}^{n-1} \mathrm{h}^j(S, \mathcal{I}_F^{(d)} \otimes_{\mathcal{O}_S} \mathcal{L}^{\otimes n-d-1} \otimes_{\mathcal{O}_S} \omega_S),$$

d'où la proposition suivante.

Proposition II.7

$$\dim \mathrm{H}^j(\tilde{X}, \omega_{\tilde{X}}) = \sum_{p=0}^{n-1} \mathrm{h}^j(S, \mathcal{A}_\alpha \otimes_{\mathcal{O}_S} \mathcal{L}^{\otimes \lceil \frac{pm}{n} \rceil} \otimes_{\mathcal{O}_S} \omega_S), \quad \textit{avec} \quad \alpha = \frac{p}{n} - 1.$$

Nous voulons maintenant obtenir une généralisation du théorème d'annulation de Zariski (fin du paragraphe III.7, page 503 de [Za.3]), c'est à dire démontrer les propositions I.2 et I.3 dans le cas où la variété S est le plan projectif complexe $\mathbf{P}^2_{\mathbf{C}}$ en utilisant la méthode de Zariski.

Dans la suite nous supposons que S est le plan projectif $\mathbf{P}^2_{\mathbf{C}}$ et que C est une courbe irréductible de degré m sur $\mathbf{P}^2_{\mathbf{C}}$.

Nous pouvons alors utiliser le résultat d'annulation suivant:

Théorème ([Za.2]) *Soit C une courbe irréductible du plan projectif complexe $\mathbf{P}^2_{\mathbf{C}}$ et soit X le revêtement cyclique de $\mathbf{P}^2_{\mathbf{C}}$ ramifié au dessus de la courbe C et de la droite à l'infini L.*
Si le degré n du revêtement est de la forme $n = r^\alpha$, avec r un nombre premier, l'irrégularité q du revêtement est nulle.

Remarque II.6 *Pour démontrer les propositions I.2 et I.3, nous avons utilisé la théorie de Hodge et un théorème d'annulation pour les groupes de cohomologie singulière $\mathrm{H}^j(U, \mathbf{C})$ si U est un ouvert affine complexe de dimension*

$d < j$.
Dans ce paragraphe, pour avoir des résultats d'annulation analogues, nous utilisons le théorème précédent qui est de nature topologique.

Nous reprenons les notations précédentes, pour tout nombre α, $-1 \leq \alpha \leq 0$, nous appelons $\mathcal{A}_\alpha$ le faisceau d'idéaux de $\mathcal{O}_S$ défini par:

$$\mathcal{A}_\alpha = \{\varphi \,/\, \forall x \in \mathrm{Sing}(C) \quad \beta_{C,x}(\varphi) > \alpha\}.$$

Les faisceaux $\mathcal{A}_\alpha$ vérifient les propriétés suivantes:
i) $\alpha \geq \beta \Longrightarrow \mathcal{A}_\alpha \subset \mathcal{A}_\beta$, en particulier pour tout nombre α, $-1 \leq \alpha \leq 0$, nous avons $\mathcal{A}_0 \subset \mathcal{A}_\alpha \subset \mathcal{A}_{-1}$, où le faisceau $\mathcal{A}_0$ est l'idéal d'adjonction $\mathcal{A}_C$ de la courbe C dans S et où le faisceau $\mathcal{A}_{-1}$ est égal à $\mathcal{O}_S$.
ii) pour $\alpha > \beta$, les faisceaux $\mathcal{A}_\alpha$ et $\mathcal{A}_\beta$ sont différents si et seulement si il existe un point x de $\mathrm{Sing}(C)$ et un germe φ de $\mathcal{O}_{S,x}$ vérifiant $\alpha \geq \beta_{C,x}(\varphi) > \beta$, c'est à dire si et seulement si $Sp(C) \bigcap [\alpha, \beta[\neq \emptyset$, où $Sp(C)$ le spectre singulier de la courbe C.

Dans la suite nous appelons (α_j) les éléments de $Sp(C)$ compris entre -1 et 0, plus précisément nous posons $-1 = \alpha_0 < \alpha_1 < \ldots < \alpha_{r-1} < \alpha_r = 0$, où $\alpha_r = 0$ appartient à $Sp(C)$ si et seulement si au moins une des singularités de C n'est pas irréductible.

Les exposants α_j sont des nombres rationnels et pour tout j, $0 \leq j \leq r$, nous posons $\alpha_j = \dfrac{p_j}{q_j} - 1$, avec $p_j \geq 0$, $q_j > 0$ et p_j et q_j premiers entre eux.

Nous déduisons de la propriété (ii) des faisceaux $\mathcal{A}_\alpha$ la relation suivante:

$$\alpha_{j-1} \leq \alpha < \alpha_j \quad \Longrightarrow \quad \mathcal{A}_\alpha = \mathcal{A}_{\alpha_{j-1}}.$$

Alors, d'après la proposition II.6, l'irrégularité q du revêtement cyclique X de degré n de $\mathbf{P}^2_{\mathbf{C}}$ ramifié au dessus de C, est égale à:

$$(2.24) \quad q = \sum_{j=1}^{r} \Bigg(\sum_{p_{j-1} \leq p < p_j} \mathrm{h}^1\Big(\mathbf{P}^2_{\mathbf{C}}, \mathcal{A}_{\alpha_{j-1}}\big(\Big\lceil \frac{pm}{n} \Big\rceil - 3\big)\Big)\Bigg) \quad \text{avec} \quad p_j = n(\alpha_j + 1).$$

Lemme II.5 *Soit C une courbe plane projective complexe irréductible de degré m, et soit α un nombre compris entre -1 et 0, alors il existe un entier v vérifiant la propriété suivante:*

$$\mathrm{H}^1(\mathbf{P}^2_{\mathbf{C}}, \mathcal{A}_\alpha(m-3-k)) \neq 0 \quad \Longleftrightarrow \quad k \geq v.$$

De plus l'entier v est strictement positif.

Démonstration : Pour tout entier u, nous avons une inclusion du faisceau $\mathcal{A}_\alpha(u)$ dans $\mathcal{A}_\alpha(u+1)$ telle que le support du conoyau soit de dimension nulle, d'où la relation:

$$\mathrm{H}^1(\mathbf{P}^2_{\mathbf{C}}, \mathcal{A}_\alpha(u)) = 0 \quad \Longrightarrow \quad \mathrm{H}^1(\mathbf{P}^2_{\mathbf{C}}, \mathcal{A}_\alpha(u+1)) = 0,$$

ce qui donne l'existence de l'entier v.

Comme l'idéal d'adjonction $\mathcal{A}_C$ C est inclus dans le faisceau $\mathcal{A}_\alpha$, avec un conoyau inclus dans $\mathrm{Sing}(C)$, nous déduisons du théorème de régularité de l'adjointe, c'est à dire de la nullité de $\mathrm{H}^1(\mathbf{P}^2_{\mathbf{C}}, \mathcal{A}_C(m-3))$, la nullité du groupe $\mathrm{H}^1(\mathbf{P}^2_{\mathbf{C}}, \mathcal{A}_\alpha(m-3))$.

Nous en déduisons que l'entier v est strictement positif.

Pour tout exposant α_j appartenant à $Sp(C)$, avec $-1 \leq \alpha_j \leq 0$, nous appelons v_j l'entier défini grâce au lemme II.5, c'est à dire l'entier vérifiant:

$$\mathrm{H}^1(\mathbf{P}^2_{\mathbf{C}}, \mathcal{A}_{\alpha_j}(m-3-k)) \neq 0 \quad \Longleftrightarrow \quad k \geq v_j. \tag{2.25}$$

Comme $\mathcal{A}_{\alpha_j}$ est inclus $\mathcal{A}_{\alpha_{j-1}}$, nous avons pour tout u l'implication suivante:

$$\mathrm{H}^1(\mathbf{P}^2_{\mathbf{C}}, \mathcal{A}_{\alpha_j}(u)) = 0 \quad \Longrightarrow \quad \mathrm{H}^1(\mathbf{P}^2_{\mathbf{C}}, \mathcal{A}_{\alpha_{j-1}}(u)) = 0,$$

d'où l'inégalité: $v_{j-1} \geq v_j$.

Nous déduisons de l'égalité (2.24) que l'irrégularité q du revêtement cyclique est non nulle si et seulement si il existe un exposant α_j et un entier p, $0 \leq p \leq n-1$, vérifiant:

$$\alpha_j \leq \frac{p}{n} - 1 < \alpha_{j+1} \quad \text{et} \quad \left\lceil \frac{pm}{n} \right\rceil \leq m - v_j.$$

Proposition II.8 *Pour tout exposant α_j appartenant au spectre singulier de la courbe C, l'entier v_j défini précédemment vérifie l'inégalité:*

$$-m\alpha_j \leq v_j. \tag{2.26}$$

Démonstration : Supposons qu'il existe un exposant α_j tel que $-m\alpha_j > v_j$. Nous pouvons alors trouver un nombre premier n et un entier p, $0 \leq p \leq n-1$, qui vérifient les inégalités:

$$\alpha_j \leq \frac{p}{n} - 1 < \alpha_{j+1} \quad \text{et} \quad v_j < -m\Big(\frac{p}{n} - 1\Big).$$

De la première relation nous déduisons que les faisceaux $\mathcal{A}_\alpha$ et $\mathcal{A}_{\alpha_j}$ sont égaux pour $\alpha = \dfrac{p}{n} - 1$. De la deuxième relation nous déduisons l'inégalité $\left\lceil \dfrac{pm}{n} \right\rceil \leq m - v_j$, donc d'après (2.25) que le groupe $\mathrm{H}^1\Big(\mathbf{P}^2_{\mathbf{C}}, \mathcal{A}_\alpha(\left\lceil \dfrac{pm}{n} \right\rceil)\Big)$ est non nul.

L'irrégularité q du revêtement cyclique de degré n de $\mathbf{P}^2_{\mathbf{C}}$ est alors non nulle, ce qui est impossible pour n premier d'après le théorème d'annulation de Zariski.

Nous allons maintenant montrer comment retrouver les propositions I.2 et I.3 dans le cas d'une courbe complexe irréductible C de $\mathbf{P}^2_{\mathbf{C}}$ et pour $r = 1$.

Proposition II.9 *L'irrégularité q du revêtement cyclique X de $\mathbf{P}^2_{\mathbf{C}}$ ramifié au dessus de la courbe C et de la droite à l'infini L est non nulle si et seulement si il existe un exposant α_j, $\alpha_j = \dfrac{p_j}{q_j} - 1$ avec p_j et q_j premiers entre eux, appartenant au spectre singulier $Sp(C)$ qui vérifie les deux propriétés suivantes:*
1) $-m\alpha_j = v_j$, en particulier le dénominateur q_j doit diviser le degré m de C;
2) le dénominateur q_j divise le degré n du revêtement cyclique X.

Démonstration : L'irrégularité q du revêtement est non nulle si et seulement si il existe un exposant α_j et un entier p, $0 \le p \le n-1$, tels que le groupe $\mathrm{H}^1\Big(\mathbf{P}^2_{\mathbf{C}}, \mathcal{A}_{\alpha_j}(\lceil \dfrac{pm}{n} \rceil - 3)\Big)$ soit non nul.

Par définition de l'entier v_j, ce groupe est non nul si et seulement si nous avons l'inégalité $\lceil \dfrac{pm}{n} \rceil \le m - v_j$. Nous avons alors les relations:

$$(\alpha_j + 1)m \le \frac{pm}{n} \le \left\lceil \frac{pm}{n} \right\rceil \le m - v_j.$$

Nous déduisons de la propositions II.8 que toutes ces inégalités sont en fait des égalités, d'où $v_j = -m\alpha_j$ et $\alpha_j = \dfrac{p}{n} - 1$.

III POLYNÔME D'ALEXANDER

Pour déduire des résultats précédents sur l'irrégularité q des revêtements cycliques de $\mathbf{P}^2_{\mathbf{C}}$, des informations sur le groupe fondamental $\pi_1(\bar{Y})$ du complémentaire de la courbe C dans $\mathbf{P}^2_{\mathbf{C}}$, nous devons introduire le polynôme d'Alexander $\Delta_C(t)$.

Nous renvoyons aux articles de Libgober ([Li.1] et [Li.2]) et de Loeser et l'auteur ([Lo.-Va.]) pour la définition du polynôme d'Alexander d'une courbe plane C.

Rappelons le théorème 4.1 de [Lo.-Va.] qui donne le polynôme d'Alexander $\Delta_C(t)$ d'une courbe C en fonction de la dimension des groupes de cohomologie des faisceaux $\mathcal{A}_\alpha(\mu)$.

Soit C une courbe de degré m du plan projectif $\mathbf{P}^2_{\mathbf{C}}$, ayant r composantes irréductibles, et soit A_C la partie du spectre singulier de la courbe C comprise entre -1 et 0 et incluse dans $(1/m)\mathbf{Z}$, c'est à dire $A_C = \{\alpha \in Sp(C) / -1 < \alpha < 0 \text{ et } m\alpha \in \mathbf{Z}\}$, nous avons alors:

$$\Delta_C(t) = (t-1)^{r-1} \prod_{\alpha \in A_C} (\Delta_\alpha(t))^{l_\alpha},$$

où les polynômes Δ_α et les entiers l_α sont définis par les formules suivantes:

$$\Delta_\alpha(t) = (t - \exp(2\pi i\alpha))(t - \exp(-2\pi i\alpha))$$
$$l_\alpha = \dim \mathrm{H}^1\left(\mathbf{P}^2_{\mathbf{C}}, \mathcal{A}_\alpha(m(\alpha+1)-3)\right).$$

Le polynôme d'Alexander Δ_C détermine l'irrégularité de tout revêtement cyclique X de $\mathbf{P}^2_{\mathbf{C}}$ ramifié au dessus de $C \bigcup L$.

Proposition III.1 *L'irrégularité q du revêtement cyclique X de degré n du plan projectif $\mathbf{P}^2$, ramifié au dessus de la courbe C et de la droite à l'infini L est égale à:*

$$q = \sum_{\alpha \in A_C(n)} l_\alpha,$$

avec $A_C(n) = \{\alpha \in Sp(C) / -1 < \alpha < 0 \text{ , } m\alpha \in \mathbf{Z} \textit{ et } n\alpha \in \mathbf{Z}\} = A_C \bigcap \frac{1}{n}\mathbf{Z}.$

Démonstration : D'après les propositions I.2 et I.3 (ou la proposition II.8 si la courbe C est irréductible), nous savons que le groupe $\mathrm{H}^1(\mathbf{P}^2_{\mathbf{C}}, \mathcal{A}_\alpha(\lceil m(\alpha + 1)\rceil - 3))$ est nul si le nombre α n'appartient pas au spectre singulier $Sp(C)$ de C ou si α ne vérifie pas $m\alpha \in \mathbf{Z}$ et $n\alpha \in \mathbf{Z}$.

Par conséquent les groupes de cohomologie $\mathrm{H}^1\left(\mathbf{P}^2_{\mathbf{C}}, \mathcal{A}_\alpha(\lceil m(\alpha+1)\rceil - 3)\right)$ sont nuls pour $\alpha \notin A_C(n)$, et nous déduisons la proposition III.1 de la proposition II.6.

Nous pouvons maintenant montrer le résultat annoncé à la fin de l'introduction, et qui généralise au cas d'une courbe non irréductible un théorème de Zariski (fin du paragraphe IV.9, page 509 de [Za.3]).

Proposition III.2 *Si le groupe fondamental du complémentaire de la courbe C dans $\mathbf{P}^2_{\mathbf{C}}$ est commutatif, l'irrégularité de tout revêtement cyclique de $\mathbf{P}^2_{\mathbf{C}}$ ramifié au dessus de C et d'une droite à l'infini L transverse à C, est nulle.*

Démonstration : Si le groupe fondamental $\pi_1(\mathbf{P}^2_{\mathbf{C}} \setminus C)$ est commutatif alors le polynôme d'Alexander $\Delta_C(t)$ de la courbe C est égal à $(t-1)^{r-1}$ (Remarque (3), page 165 de [Lo.-Va.]).

D'après le théorème 4.1 de [Lo.-Va.] nous en déduisons que pour tout α appartenant à A_C nous avons: $\mathrm{H}^1(\mathbf{P}^2_{\mathbf{C}}, \mathcal{A}_\alpha(m(\alpha+1)-3)) = 0$, d'où le résultat.

Bibliographie

[De.]: P. Deligne, Théorie de Hodge, II et III, Publ. Math. I.H.E.S. 40 (1971), 5-58 et 44 (1975), 5-77.

[Dm.]: M. Demazure, Anneaux gradués normaux, Introduction à la théorie des singularités I, Méthodes algébriques et géométriques, Travaux en cours 37 (1988), 35-69.

[En.]: F. Enriques, Sulla costruzione delle funzioni algebriche di due variabili posse- denti una data curva di diramazione, Ann. di mat. pura ed appl. 4 (1923), 185-198.

[Es.]: H. Esnault, Fibre de Milnor d'un cône sur une courbe plane singulière, Invent. Math. 68 (1982), 477-496.

[Es.-Vi.1]: H. Esnault et E. Viehweg, Revêtements cycliques, Algebraic threefolds, Proceedings Varenna 1981, Lect. Notes Math. 947 (1982), 241-250.

[Es.-Vi.2]: H. Esnault et E. Viehweg, Logarithmic De Rham complexes and vanishing theorems, Invent. Math. 86 (1986), 161-194.

[Li.1]: A. Libgober, Alexander polynomial of plane algebraic curves and cyclic multiple planes, Duke Math. J. 49 (1982), 833-851.

[Li.2]: A. Libgober, Alexander invariants of plane algebraic curves, Proc. Symp. Pure Math. 40 (1983), Part 2, 135-143.

[Lo.-Va.]: F. Loeser et M. Vaquié, Le polynôme d'Alexander d'une courbe plane projective, Topology 29 (1990), 163-173.

[Me.Te.]: M. Merle et B. Teissier,Condition d'adjonction, d'après Duval, Séminaire sur les Singularités des Surfaces, Lect. Notes Math. 777 (1980), 229-245.

[Va.]: M. Vaquié, Irrégularité des revêtements cycliques des surfaces projectives non singulières, Am. J. Math. 114 (1992), 1187-1199.

[Vr.1]: A. Varchenko, The asymptotics of holomorphic forms determine a mixed Hodge structure, Soviet Math. Dokl. 22 (1980), 772-775.

[Vr.2]: A. Varchenko, Asymptotic Hodge structure in the vanishing cohomology, Math. U.S.S.R. Izv. 18 (1982), 469-512.

[Vi.1]: E. Viehweg, Rational singularities of higher dimensional schemes,

Proc. Am. Math. Soc. 63 (1977), 6-8.

[Vi.2]: E. Viehweg, Vanishing theorems, J. Reine Angew. Math. 335 (1982), 1-8.

[Za.1]: O. Zariski, On the problem of existence of algebraic functions of two variables possessing a given branch curve, Am. J. Math. 51 (1929), 305-328.

[Za.2]: O. Zariski, On the linear connection index of the algebraic surfaces $z^n = f(x, y)$, Proc. Nat. Sc. 15 (1929), 494-501.

[Za.3]: O. Zariski, On the irregularity of cyclic multiple planes, Ann. Math. 32 (1931), 485-511.

Laboratoire de Mathématiques, UA 762 du CNRS
Ecole Normale Supérieure, Paris.

Printed in the United States
By Bookmasters